physical chemistry of nucleic acids

physical chemistry of nucleic acids

VICTOR A. BLOOMFIELD
University of Minnesota
DONALD M. CROTHERS
Yale University
IGNACIO TINOCO, Jr.
University of California, Berkeley

Harper & Row, Publishers
New York, Evanston,
San Francisco, London

Sponsoring Editor: John A. Woods
Project Editor: Lois Wernick
Designer: Michel Craig
Production Supervisor: Valerie Klima

PHYSICAL CHEMISTRY OF NUCLEIC ACIDS

Library of Congress Cataloging in Publication Data
Bloomfield, Victor A
Physical chemistry of nucleic acids.
Includes bibliographies.
1. Nucleic acids. I. Crothers, Donald M., joint author. II. Tinoco, Ignacio, joint author. III. Title. [DNLM: 1. Chemistry, Physical. 2. Nucleic acids. QU55 B655p 1974]
QD433.B44 547'.596 73-8373
ISBN 0-06-040779-4

Dedicated to our wives
Clara, Leena, Joan

contents

preface

The nucleic acids and their constituents are key molecules in the storage and transmission of genetic information, in the readout of this information during protein synthesis, and as substrates, cofactors, or allosteric effectors in a host of other metabolic processes. Because of this central importance, the nucleic acids have been subject to extraordinarily intensive scrutiny by investigators using a wide range of physical and chemical techniques.

We have aimed in this book to present the methods and results of the physical chemical study of nucleic acids, in a fundamental and comprehensive fashion. We have attempted to lay particular stress on general principles, to cite particular results only when they seem firmly established (not an easy matter to assess in a rapidly changing field), and to provide, by fairly detailed discussion of the theoretical basis of the major experimental techniques, a firm foundation for future work.

The book is directed at graduate students and research scientists who are directly concerned with nucleic acids or who, in neighboring fields of physical chemistry, polymer chemistry, biochemistry, biophysics, and molecular biology, may be interested in how their own studies may illuminate, or be illuminated by, the physical chemical properties of nucleic acids. We have assumed that the reader has a basic knowledge of physical chemistry and of the biochemistry of the nucleic acids.

We have attempted to reconcile the often conflicting demands, in a book of finite size, of an advanced research-level monograph and of a textbook that will be useful (when properly supplemented) in graduate courses on biophysical chemistry and on the chemistry of nucleic acids. To this end, we have provided a substantial amount of pedagogical and mathematical detail on topics which have not been extensively covered in standard textbooks. With other more standard topics, particularly spectroscopy and some aspects of polymer physical chemistry, we have contented ourselves with a brief discussion of basic principles before proceeding to applications and have directed the reader to appropriate references for a more detailed treatment of fundamentals.

It is inevitable, and not undesirable, that the authors of a book such as this will emphasize topics close to their own research interests. It is fortunate, therefore, that the research interests of the present authors complement each other to a fair extent—hydrodynamics, light scattering, and dimensional statistics of polynucleotides (Bloomfield); helix-coil transitions and ligand binding reactions (Crothers); and spectroscopic properties and thermodynamics (Tinoco)—so that the book as a whole is, we hope, reasonably comprehensive and balanced. Although each author took primary responsibility for the material indicated above, all chapters were scrutinized and revised by each author to provide coordination and avoid overlap.

No book like this can be written without the active assistance of many colleagues. Professor Charles Cantor and Dr. Gary Felsenfeld read the entire book in manuscript, and Professor James Wang read most of it. Each made many valuable suggestions. Peter and Susan Berget prepared the index. We are also grateful to the many people who sent us preprints and original photographs and who graciously granted permission for use of published figures and tables. We finally acknowledge our students and co-workers, whose comments often improved our understanding of the material in this book.

V. A. B.
D. M. C.
I. T.

chapter 1
introduction

The key insight into the biological function of nucleic acids at the molecular level is claimed by Watson[1] when in 1952 he taped up a paper sheet above his desk saying:

$$\text{DNA} \longrightarrow \text{RNA} \longrightarrow \text{protein}$$

This slogan represented the transfer of genetic information from the sequences of nucleotides in DNA molecules to the sequences of amino acids in proteins. The details of the processes symbolized by the arrows have since been extensively studied. A slightly more explicit description of the chemical reaction involved is:[2]

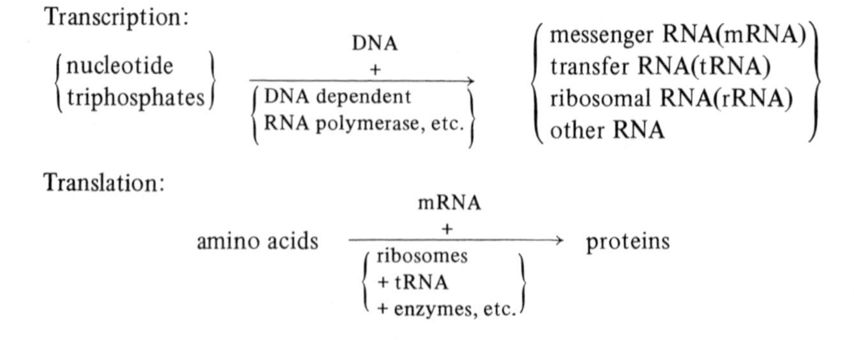

TABLE 1-1 THE GENETIC DICTIONARY[a]

First base	Second base				Third base
	U	C	A	G	
U	Phe	Ser	Tyr	Cys	U
	Phe	Ser	Tyr	Cys	C
	Leu	Ser	Stop	Stop	A
	Leu	Ser	Stop	Trp	G
C	Leu	Pro	His	Arg	U
	Leu	Pro	His	Arg	C
	Leu	Pro	Gln	Arg	A
	Leu	Pro	Gln	Arg	G
A	Ile	Thr	Asn	Ser	U
	Ile	Thr	Asn	Ser	C
	Ile	Thr	Lys	Arg	A
	Met or start	Thr	Lys	Arg	G
G	Val	Ala	Asp	Gly	U
	Val	Ala	Asp	Gly	C
	Val	Ala	Glu	Gly	A
	Val or start	Ala	Glu	Gly	G

[a] The structures and full names of the amino acids are given in Table 1-2.

The four ribonucleotide triphosphates are polymerized on the DNA templates by RNA polymerase to form many different RNA molecules. Only one strand of the DNA molecule is copied. A single DNA molecule contains the templates for many different molecules. The DNA base sequences which specify the beginning and end of the RNA molecules are not known at this time. Apparently protein molecules are bound to those sites and control the action of RNA polymerase[3,4]. The code for transcribing DNA into RNA is simple

DNA ⟶	RNA
adenine (A)	uracil (U)
cytosine (C)	guanine (G)
guanine (G)	cytosine (C)
thymine (T)	adenine (A)

The translation step is more complicated. Up to 20 amino acids are polymerized to form proteins. The sequence of amino acids is governed by the sequence of base triplets on the messenger RNA. All but 3 of the 64 possible base triplets are read by transfer RNA molecules, each of which carries an amino acid. The three special triplets specify termination of the protein chain; they are recognized by specific proteins.[5] The genetic dictionary for translating from RNA base sequences to protein amino acid sequences is given in Table 1-1. Table 1-2 gives the dictionary for translating

TABLE 1-2 THE GENETIC DICTIONARY[a]

Amino Acid	Triplet	Amino Acid	Triplet
$H-$ glycine	GGX	$^{-}O(O=)CCH_2-$ glutamic acid	GAPy
CH_3- alanine	GCX	$^{-}O(O=)CCH_2CH_2-$ aspartic acid	GAPu
CH_3 $CH-$ CH_3 valine	GUX	$O=C(NH_2)CH_2-$ glutamine	AAPy
CH_3 $CHCH_2-$ CH_3 leucine	CUX UUPu	$O=C(NH_2)CH_2CH_2-$ asparagine	CAPu
CH_3CH_2 $CH-$ CH_3 isoleucine	AUPy AUA	$H_2NCH_2CH_2CH_2CH_2-$ lysine	AAPu
$C_6H_5-CH_2-$ phenylalanine	UUPy	(imidazolyl, N, N–H)$-CH_2-$ histidine	CAPy
$HO-C_6H_4-CH_2-$ tyrosine	UAPy	$H_2NC(=NH_2)NHCH_2CH_2CH_2-$ arginine	CGX AGPu
(indolyl, N)$-CH_2-$ tryptophan	UGG	$HOCH_2-$ serine	AGPy UCX

TABLE 1-2 – *continued*

Amino Acid	Triplet	Amino Acid	Triplet
$HSCH_2-$ cysteine	UGPy	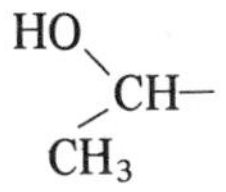 $HO\backslash CH-/CH_3$ threonine	ACX
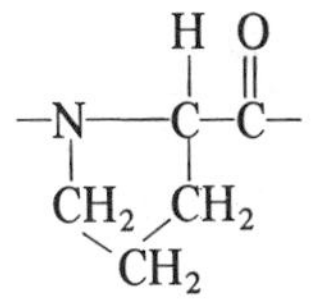 $-N-C(H)-C(=O)-$, CH_2, CH_2, CH_2 proline	CCX	$CH_3SCH_2CH_2-$ methionine	AUG
Start ≡ formyl methionine	AUG GUG	Stop	UAPu UGA

[a]The structure of the side chain of each amino acid is given. X = any base, Pu = purine, Py = pyrimidine.

from amino acids to base triplets. The structures of the side groups of the amino acids are also shown in Table 1-2. Protein synthesis begins at an AUG or GUG which is recognized by a tRNA carrying *N*-formyl methionine.[6] As these triplets also code for amino acids in the interior of the polypeptide, it is clear that additional starting signals are present in messenger RNA.[7] The formyl methionine is subsequently removed from the completed polypeptide chain.

The translation of the genetic information takes place on ribosomes. The messenger RNA, the transfer RNAs and all the enzymes and protein factors interact with the ribosomes and each other to produce a polypeptide. More than one region of the messenger RNA is being translated at the same time so more than one ribosome is attached to the messenger RNA.[8] An electron microscope picture of these polyribosomes is shown in Fig. 1-1.

Both DNA and RNA must replicate in order to store and propagate the genetic information. Furthermore, RNA is sometimes transcribed into DNA.[9] Therefore, a more complete mnemonic for the molecular biology of nucleic acids is:

$$\circlearrowright\text{DNA} \rightleftharpoons \circlearrowright\text{RNA} \longrightarrow \text{protein}$$

The physical chemistry of nucleic acids includes knowledge of the structures of all the polynucleotides involved. The conformations, or secondary and tertiary structures, at each step in the various reactions are of

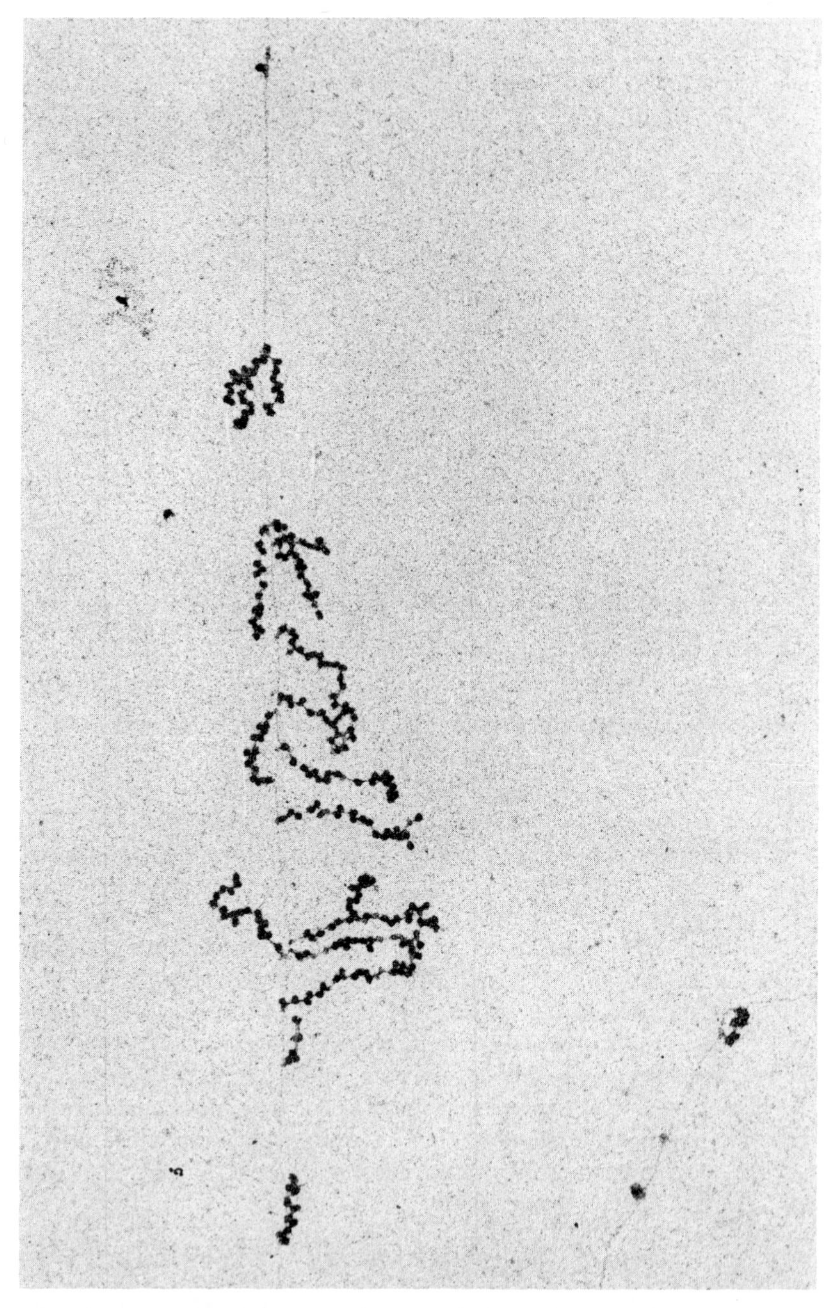

FIGURE 1-1 The transcription of DNA into RNA and the translation of the RNA into proteins. [From O. L. Miller, Jr., B. A. Hamkalo, and C. A. Thomas, Jr., *Science*, **169**, 392 (1970). Reprinted with permission.]

particular importance. These structures determine what reactions occur and how fast they happen. The interactions of the nucleic acids with each other, with proteins, and with small molecules control the process.

In the succeeding four chapters we have presented the fundamentals of the determination of structure in nucleic acids. These include X-ray and light scattering, spectroscopic techniques, and hydrodynamic methods. There follow two chapters on the thermodynamics, statistical mechanics, and kinetics of reactions of nucleic acids. The last chapter is a detailed description of physical chemical studies on one type of ribonucleic acid: transfer RNA.

We have attempted to stress general principles and to cite particular results only when they seem firmly established.

The plan of this book has been to proceed from the simple to the complex; from monomers to oligomers to polymers; from nucleic acids in their native, unperturbed states to nucleic acids whose conformations have been perturbed by some chemical or physical agent; from "pure" nucleic acids to nucleic acids interacting with other molecules.

It is worthwhile to list some of the important topics we have covered only briefly, or not at all. These include the organic chemical, as opposed to the physical chemical, behavior of the nucleic acids, and the biological functioning and metabolism of the nucleic acids. To learn about these subjects and to obtain a more balanced view, the reader is referred to earlier books on nucleic acids. These include the three-volume set entitled *The Nucleic Acids* edited by E. Chargaff and J. N. Davidson (Vols. I, II, 1955; Vol. III, 1960, Academic Press); *The Chemistry of Nucleic Acids* by D. O. Jordan (1960, Butterworth); *Polynucleotides* by R. F. Steiner and R. F. Beers, Jr. (1960, Elsevier) and *The Chemistry of Nucleosides and Nucleotides* by A. M. Michelson (1963, Academic Press). A short and very useful summary of the biological function of nucleic acids at the molecular level is *The Biochemistry of Nucleic Acids* by J. N. Davidson (6th ed., 1970, John Wiley). To obtain a good historical perspective, *Nucleic Acids* by P. A. Levene and L. W. Bass (1931, Chemical Catalog Co.) is recommended.

The physical chemistry of nucleic acids combined with proteins and other macromolecules in functional complexes such as chromosomes, ribosomes, viruses, and mitochondria and membrane-bound replication complexes are not discussed in this book. These topics are very important and interesting. However, their study is just beginning; we leave them for the future. To keep abreast of these developments one must avidly read the daily newspapers and the weekly magazines such as *Science, Nature*, and *Time*. The more lasting scientific contributions are reviewed in such annual publications as *Progress in Nucleic Acid Research and Molecular Biology* edited by J. N. Davidson and W. E. Cohn (Academic Press). The annual *Cold Spring Harbor Symposia on Quantitative Biology* have often had very timely collections of articles on a particular aspect of nucleic acids.

REFERENCES

1 J. D. Watson, *The Double Helix*, Atheneum, New York (1968), p. 153.
2 J. D. Watson, *The Molecular Biology of the Gene*, W. A. Benjamin, New York (1965).
3 A. A. Travers and R. R. Burgess, *Nature,* **222**, 537 (1969).
4 J. W. Roberts, *Nature,* **224**, 1168 (1969).
5 E. M. Scolnick and C. T. Caskey, *Proc. Nat. Acad. Sci., U.S.,* **64**, 1235 (1969).
6 H. P. Ghosh, D. Söll, and H. G. Khorana, *J. Mol. Biol.,* **25**, 275 (1967).
7 J. A. Steitz, *Nature,* **224**, 957 (1969).
8 J. R. Warner, A. Rich, and C. E. Hall, *Science,* **138**, 1399 (1962).
9 H. M. Temin and S. Mizutani, *Nature*, **226**, 1211 (1970); D. Baltimore, *Nature*, **226**, 1209 (1970).

chapter 2
monomers

I CONFIGURATION AND CONFORMATION

A. Introduction

The distinction between configuration and conformation made by the organic chemist is the one we will use here. Configuration refers to the covalent bonding in a molecule; conformation refers to the three-dimensional structure of a molecule. The configuration is constant for a particular molecule; the conformation depends on the environment (temperature, solvent, etc.) of the molecule. As a final distinction we remark that organic chemists and biochemists are usually more interested in configuration, while physical chemists are interested in conformation.

The large amount of chemical work necessary to establish configuration of the nucleic acid components is described in Volume I of *The Nucleic Acids.*[1] The conformations of the nucleosides and nucleotides are still not very well known. We will discuss what is known at present and describe the methods used to study the problem.

B. Bases

We will consider the four bases in DNA and RNA which are responsible for coding genetic information. These are adenine, cytosine, guanine, and

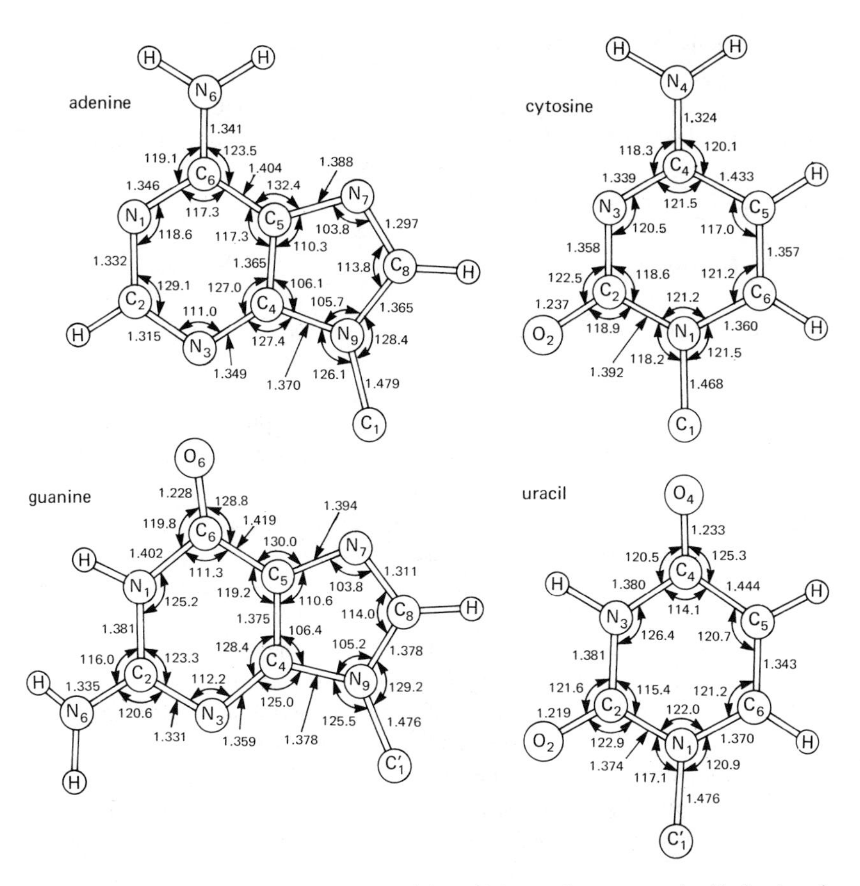

FIGURE 2-1 The structures of the nucleic acid bases drawn to scale. Each atom is numbered and average bond lengths and bond angles are given. [Published with permission from D. Voet and A. Rich, *Progr. Nucleic Acid Res. Mol. Biol.,* **10**, 183 (1970).]

thymine for DNA and adenine, cytosine, guanine, and uracil for RNA. They are drawn to scale in Fig. 2-1 from average bond angles and bond distances deduced from several crystal structures. The numbering system for the base atoms is the standard *Chemical Abstracts* one; Cl′ refers to the sugar carbon of the nucleoside. One should note that, as drawn, the numbering system is clockwise for the pyrimidines (cytosine and uracil), but counterclockwise for the purines (guanine and adenine). An older convention used the same numbering system for the pyrimidines as used presently for the six-membered ring in purines. Some papers on X-ray diffraction of nucleic acids still use the old system, so one should check the nomenclature carefully before using published coordinates for bases.

Other bases sometimes occur in certain DNA or RNA molecules. The largest number of uncommon bases has been found in transfer RNA molecules and new ones are still being found. They include *N*-alkyl-substituted bases, deaminated adenine and guanine (hypoxanthine and xanthine) 5,6-dihydrouracil, and thiouracil (see Fig. 10-7). DNA sometimes contains 5-methylcytosine or 5-hydroxymethylcytosine.

C. Nucleosides

The structure of the nucleosides found in RNA is shown in Fig. 2-2. A ribonucleoside is a derivative of ribose as characterized by the configurations at the asymmetric carbons C2′ and C3′. The base is attached in the β configuration which means it is on the same side of the sugar ring as the 5′ carbon. The naturally occurring nucleosides in RNA contain D-ribose. The 4 asymmetric carbons in a ribonucleoside can produce 16 different stereoisomers. The four possibilities at the 2′ and 3′ carbons relative to C5′ produce ribose, arabinose, xylose, and lyxose (see Fig. 2-2). The base can be attached α or β. Finally, each sugar can be either the D or L form. In RNA, however, only the β-D-ribosyl nucleosides are found. For the four coding bases their names are adenosine, cytidine, guanosine, and uridine; if thymine occurs as a ribonucleoside it is called ribosylthymine or 5-methyluridine.

DNA contains 2′ deoxynucleosides with the same configuration as the ribonucleosides. That is, they are β-D-2′-deoxyribonucleosides. Figure 2-2 can represent the deoxyribonucleosides found in DNA if the 2′ oxygen is deleted. As there is one less asymmetric carbon atom, the naturally occurring deoxyribonucleosides in DNA are only one out of eight possible stereoisomers. Their names are deoxyadenosine, deoxycytidine, deoxyguanosine, and thymidine. Other nucleosides occur in DNA and RNA. These include the nucleosides formed from the odd bases mentioned earlier. In addition, pseudouridine occurs in transfer RNA. Pseudouridine is the ribonucleoside in which uracil is connected to the C1′ of the ribose by the C5 carbon instead of the usual N1 nitrogen. Ribonucleosides in which the O2′ oxygen of the ribose is methylated are also found.

Specifying the structure of the planar nucleic acid bases was easy; it is very complicated for the nucleosides. The ring atoms of the sugar are not planar and the out-of-plane atoms depend on the compound. There are also five single bonds not in the sugar ring which can rotate: C1′–base, C2′–O2′, C3′–O3′, C4′–C5′, and C5′–O5′. One would like to know which conformations are probable, but the most useful information is knowing which conformations are not possible. This greatly simplifies building model of polynucleotide structures and makes it easy to dismiss large classes of structures.

Analysis[2–5] of X-ray crystal data has shown that four of the sugar ring

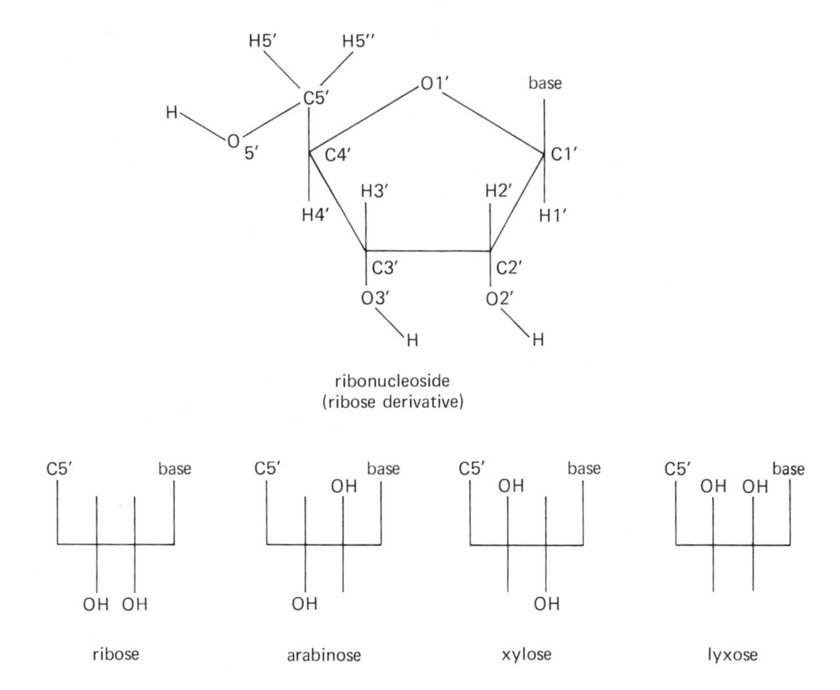

FIGURE 2-2 The configuration of the ribonucleosides found in RNA. The deoxyribonucleosides found in DNA are identical except that they lack O2′. The configuration of other possible pentoses is also shown.

atoms form a plane to within ±0.04 Å and the fifth ring atom is significantly displaced. The four nearly planar atoms are C1′, C4′, O1′ and either C2′ or C3′. If the out-of-plane ring atom is on the same side of the plane as the 5′ carbon, the conformation is designated *endo*. The conformation is *exo* if the out-of-plane ring atom is on the opposite side from the 5′ carbon. This is illustrated in Fig. 2-3. The out-of-plane C2′ or C3′ atoms are about 0.5 Å above or below the plane; C2′ *endo* or C3′ *endo* are the most commonly found, while C2′ or C3′ *exo* are rare.[4] Evidence for the puckering of the ribose and deoxyribose rings of nucleoside in solution had been found earlier by Jardetsky.[6,7] Analysis of nuclear magnetic resonance data on ribonucleosides and deoxyribonucleosides led to the proposal of conformations with C2′ or C3′ *endo* or *exo*. The spin-spin coupling constant between the H1′ and H2′ can be related to the angle between the H2′–C2′–C1′ plane and the H1′–C1′–C2′ plane (the H1′–H2′ dihedral angle). This angle depends on the conformation.

The general conclusion to be drawn from the X-ray data for crystals and the NMR data for solutions is that either C2′ or C3′ will be out of the plane determined by C1′, C4′ and O1′. The energy differences between the

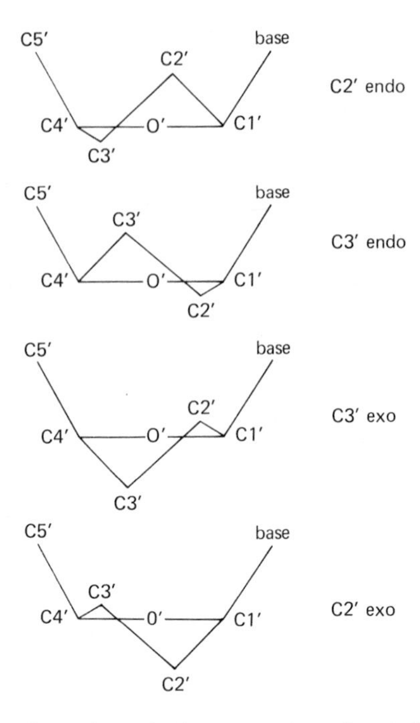

FIGURE 2-3 The conformation of the sugar rings found in crystals of various nucleosides. Four atoms of the sugar ring are nearly in a plane: C1′, C4′, O1′ and either C2′ or C3′. The out-of-plane atom is designated *endo* or *exo* as illustrated.

various *endo* and *exo* forms and the barriers to interconversion in solution are not known. It would be useful to know the fraction of each conformation present and the rate of exchange between conformations in solution.

The orientation of the base relative to the sugar in a nucleoside has been studied extensively. Donohue and Trueblood[8] originally labeled the angle specifying this orientation as ϕ_{CN} (see Fig. 2-4). It is now labeled[2,3] χ. There is no general agreement about the quantitative definition of this angle, therefore one must be very careful in using published values. Following Sundaralingam[3] we choose χ as the angle between the O1′–C1′–N plane and the C1′–N–C6 plane for pyrimidines (or the C1′–N–C8 for purines). It is zero when bond O1′–C1′ is *cis* planar to bond N–C6 in pyrimidines (or N–C8 in purines). It is positive for clockwise rotation of the N–C6 bond (or N–C8 bond) as one looks from C1′ to N. For ϕ_{CN} (of Donohue and Trueblood[8]) negative, $\chi = -\phi_{CN}$; for ϕ_{CN} positive, $\chi = 360 - \phi_{CN}$. Analysis of X-ray data[3] shows that the pyrimidine nucleosides have values of χ between 0° and 72°. This means the carbonyl oxygen (O2) of uracil and cytosine is away from the ribose. This conformation is called *anti*; it is shown in Fig. 2-4. Purine nucleosides mainly are *anti* ($3° < \chi < 55°$), with the

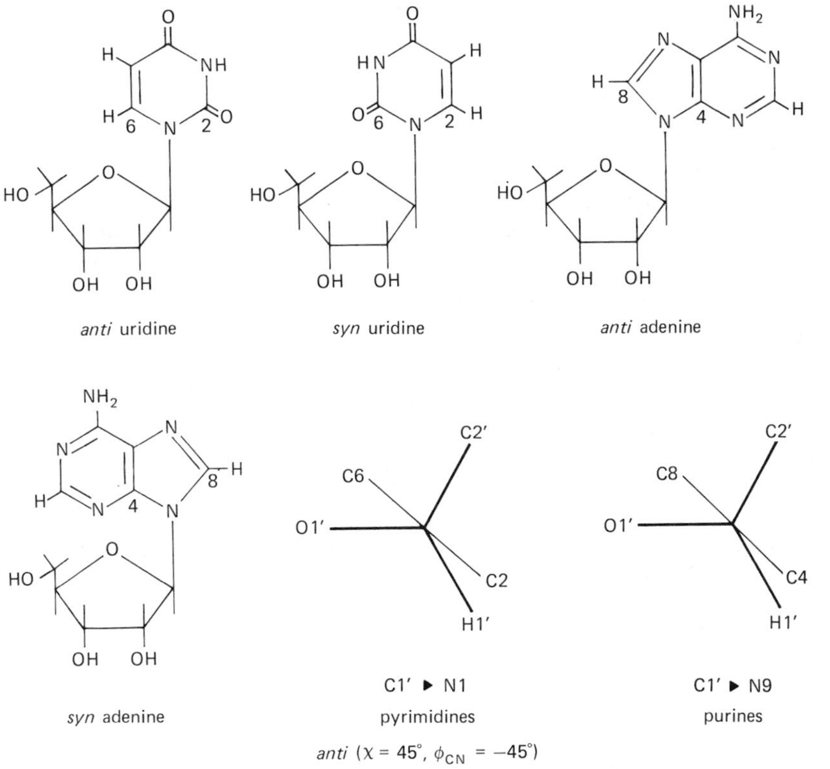

FIGURE 2-4 Illustrations of conformations around the glycosidic bond. The most probable conformation is shown in projection; one looks along the glycosidic bond from C1′ to N.

six-membered ring of the purine away from the ribose, but some are *syn* ($\chi \approx 120°$; $210° < \chi < 260°$). These conformations are also shown in Fig. 2-4. The *anti* conformation is also illustrated in a projection in which one looks along the C1′ to N bond. Of course, *anti* refers to a range of conformations, but the projection in Fig. 2-4 represents the center of the range.

Haschemeyer and Rich[9] have calculated possible conformations on the basis of van der Waals contacts between atoms. One chooses a set of allowed distances between atoms from analysis of crystal structures. Calculated conformations which have distances shorter than this are ruled impossible. For purine nucleosides almost any value of χ is allowed. For pyrimidine nucleosides an *anti* range of $\chi = 25–105°$ is allowed; a *syn* range is much more sterically hindered. It is only expected if some other stabilization favors it. These steric calculations thus agree with the published X-ray data for crystals: Essentially all the pyrimidine nucleosides are *anti*; but some purine nucleosides are *syn*. It is evident that pyrimidines will be more hindered in the *syn* position because the O2 carbonyl oxygen crowds the H2′ and H3′ of the sugar, whereas the N3 of the purines is farther away.

Attempts have been made to calculate more quantitative values of the relative stability of *syn* and *anti* conformations.[10,11] The relative potential energy of each nucleoside was calculated as a function of χ. Potential energies of the form $ar^{-1} + br^{-4} + cr^{-6} + dr^{-12}$ (corresponding to charge-charge, charge-induced dipole, London, and repulsive interactions, respectively) were calculated for each pair of nonbonded atoms. For purine nucleosides the calculated difference between *syn* and *anti* was only 1–2 kcal/mole. The actual magnitude and sign of the difference depended on whether the base was adenine or guanine and whether the sugar was 2′ *endo* or 3′ *endo*. For pyrimidine nucleosides the *anti* conformation was calculated to be more stable than the *syn* by 5–7 kcal/mole. The calculated hindrance to rotation of χ was much higher for pyrimidine nucleosides than for purine nucleosides. In these calculations the main factor favoring the *anti* conformation for pyrimidines was the coulombic repulsion between the negatively charged O2 of the base and the negatively charged O1′ of the sugar. Relative stabilities for *syn* and *anti* have also been calculated by Jordan and Pullman[12] using extended Hückel molecular orbitals. They conclude that adenosine, cytidine, and uridine are *anti*, but guanosine is *syn*.

All the calculations must be taken with strong reservations. The calculations are very crude even for the molecule in the gas phase, and no attempt is made to consider the solvent.

Studies of optical rotatory dispersion have provided some information about the *syn* and *anti* conformations in solution. Emerson et al.[13] found that the pyrimidine nucleosides have a positive Cotton effect near 260 nm. Analogous compounds in which the pyrimidine is covalently linked in an *anti* conformation also have positive Cotton effects, but if the pyrimidine is *syn*, the Cotton effect is negative. For example, if the carbonyl oxygen O2 of uracil is linked to the C3′ or C5′ carbon of the ribose (*syn*), a negative Cotton effect results. If the C6 of uracil is linked to the O5′ oxygen of the ribose (*anti*), a positive Cotton effect is seen. Emerson et al.[14] have also studied purine nucleosides. These compounds have a negative Cotton effect, as does 8,5′-cycloadenosine in which the C8 of adenine is linked to C5′ (*anti*). The conclusion is that all the nucleosides are *anti* in solution. In contrast, Klee and Mudd[15] think that adenine nucleosides may be *syn*. Their conclusions are based on optical rotatory dispersion of compounds in which the 5′ oxygen has been replaced by a sulfur. However, the large polarizability of the sulfur makes these compounds poor models for comparisons based on optical rotatory dispersion. One assumes that the model compounds have similar intrinsic optical properties and only changes in conformation are important.

As optical rotation is so sensitive to polarizability, model compounds in which oxygen is replaced by sulfur are suspect.

Very extensive studies of the circular dichroism (CD) of nucleosides have been made by Miles et al.[16,17] From both experimental and theoretical evidence they conclude that the sign of the CD band near 260 nm (B_{2u} band) is a good measure of the conformation around the glycosidic band, χ. For uracil and cytosine β-D-ribo- or -deoxyribonucleosides, there is a positive CD band for the *anti* conformations ($\chi = 0 \pm 80^\circ$) and a negative CD band for *syn.* For purines the sign of the CD is reversed. The authors make more quantitative predictions, but the reader must look at the original papers for a complete discussion of their methods and conclusions.

Additional evidence for *syn* or *anti* conformations in solution has come from nuclear magnetic resonance studies of nucleotides. These will be discussed in a later section.

Syn and *anti* refer to one of the five bonds which are capable of rotating 360° in nucleosides. Discussion of the other bonds is more pertinent to the conformation of nucleotides and polynucleotides, and is discussed in the following section.

D. Nucleotides

From each possible ribonucleoside three mononucleotides can be made; the phosphate ester can be made of the 2′, 3′, or 5′ hydroxyl. From a deoxyribonucleoside only the 3′ or 5′ phosphate ester can be made. The names of the common nucleotides are adenylic acid, cytidylic acid, guanylic acid, and uridylic acid for the ribonucleotides and deoxyadenylic acid, deoxycytidylic acid, deoxyguanylic acid, and thymidylic acid for the deoxynucleotides. The position of the phosphate is specified by a 2′, 3′, or 5′ before the name of the compound. Alternatively, the compounds can be named as derivatives of nucleosides, such as adenosine-3′-phosphate.

Nucleoside cyclic monophosphates are also possible; the phosphate is esterified to two hydroxyl groups. An example is adenosine-3′,5′-cyclic phosphate.

At pH 7 the phosphate is completely ionized so that replacing a hydrogen by the phosphate does not introduce any new rotatable bonds. The same type of conformation questions exist in the nucleotides as in the nucleosides. Furthermore, the presence of the phosphate does not seem to produce any special changes in conformation. There does not seem to be any correlation in crystals between *endo, exo, syn*, or *anti* and the presence of the phosphate. The optical rotatory dispersion of the nucleosides and corresponding nucleotides are similar. Evidence about conformations of nucleotides therefore should apply to the nucleosides.

The possible orientations about each of the bonds in nucleotide

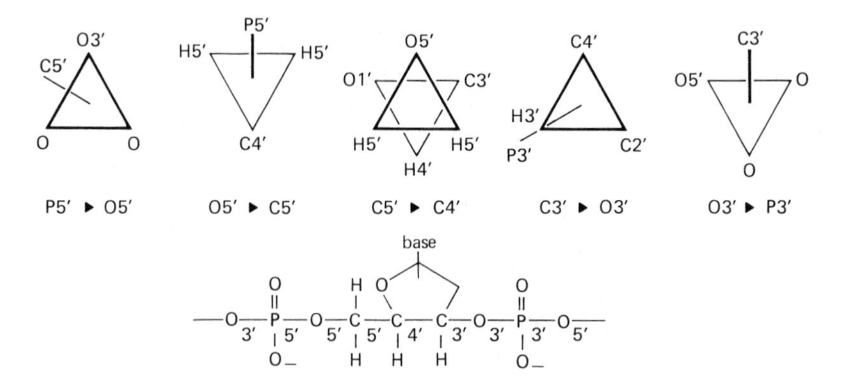

FIGURE 2-5 Probable conformations about bonds in nucleotides. One looks along a bond from A to B (A ▶ B) and notes the projection of the other three substituents on tetrahedral carbon or phosphorus, or the one other substituent on oxygen.

derivatives have been deduced from X-ray crystal data[2-5] on nucleosides, mononucleotides, and polynucleotides. Just as for the glycosidic bond (C1'—N) a range of angles is possible. The data can be summarized, however, by showing the most likely orientation about each bond; this is done in Fig. 2-5. In the figure one looks along a bond from A to B (A ▶ B) and notes the position of the other three substituents on tetrahedral carbon or phosphorus, or the one other substituent on oxygen. Figures 2-3 to 2-5 thus illustrate the most likely conformation for a nucleotide either as a monomer or in a

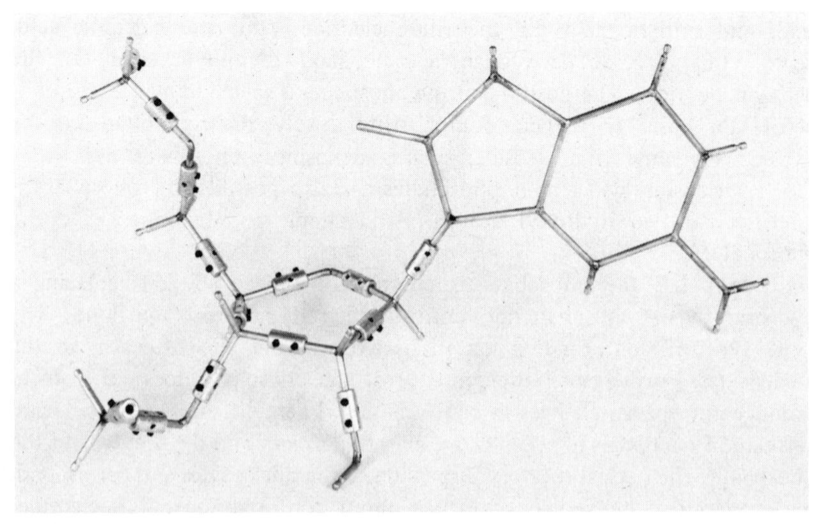

FIGURE 2-6 A picture of 5',3'-guanosine diphosphate with the bonds in their most likely positions as deduced from X-ray studies.

polymer. Figure 2-6 shows a picture of 5′,3′-guanosine diphosphate with the base *anti*, the sugar 2′ *endo*, and the other bonds in their likely positions as given in Fig. 2-5. It should be reemphasized that these conformations are not definite. The X-ray data on many crystals have simply been compiled to show which conformations appear most often.[2–5]

There is correlation between successive bonds in a nucleotide; if one bond is fixed, other bonds are limited in orientation. The limitations can be deduced from X-ray studies[3] or one can attempt to calculate them.[10,18]

Evidence for the conformations present in solution has come mainly from nuclear magnetic resonance, and the optical studies discussed earlier. The NMR data will be discussed in detail later. It is clear that the conformation in solution can be affected by pH[19] and salt concentration.[20] Probably, the conformation of a nucleotide in solution is not rigid, but includes a range of structures. Support for this conclusion comes from nuclear magnetic resonance experiments.[21]

II ELECTRONIC PROPERTIES

A. Introduction

In this section we will discuss the many spectroscopic methods which have been applied to nucleic acids. These include absorption (both ultraviolet and infrared), circular dichroism and optical rotatory dispersion, light emission and scattering, and magnetic resonance. The experimental findings for the individual nucleic acid components are presented both because of their intrinsic interest, and because they are necessary for the understanding of the results on the polynucleotides. The theory of each method is briefly sketched. For a thorough discussion the interested reader is referred to the books and review articles listed for each method. Useful books which cover the methods described here occur in the series, *Technique of Organic Chemistry*. The pertinent ones are Volume I, *Physical Methods of Organic Chemistry*,[22] and Volume IX, *Chemical Applications of Spectroscopy*.[23]

B. Charge Distributions

Not many experimental data are available on the charge distributions in nucleic acid bases. Table 2-1 gives measured values of permanent dipole moments of pyrimidine, 9-butylpurine and three substituted bases. With this scarcity of experiment we must turn to calculations for the needed data. Alberte and Bernard Pullman and co-workers[24] have made thorough calculations on charge distributions. They have calculated the net charge on

TABLE 2-1 MEASURED DIPOLE MOMENTS OF NUCLEIC ACID BASES

Compound	Solvent	Dipole moment (Debye)	Ref.
Pyrimidine	Dioxane	2.4	*a*
9-Butylpurine	Dioxane	4.3	*b*
9-Butyladenine	CCl_4	3.0	*b*
1,3-Dimethyluracil	Dioxane	3.9	*b*
1-Methyl, 5-bromouracil	Dioxane	4.5	*b*

[a] C. P. Smyth, *Dielectric Behavior and Structure*, McGraw-Hill, New York (1955).
[b] H. DeVoe and I. Tinoco, Jr., *J. Mol. Biol.*, **4**, 500 (1962).

each atom by various methods. One method obtains the net charge as the sum of the net charge of the σ electrons and the net charge of the π electrons. The σ electrons are treated by the method of Del Re[25]; the π electrons are treated by the Pariser, Parr, Pople self-consistent field method.[26] However, both of these methods were optimized by choosing parameters which gave calculated values of dipole moments and ionization potentials which agreed with measured reference compounds. The reference compounds (such as pyridine, pyrimidine, pyrrole, aniline, formaldehyde, formamide, 2-pyridone) were similar to the bases. The semiempirical approach gives one some confidence in the results. Table 2-2a shows values of Berthod and Pullman.[27] An improved treatment of the π electrons[28] gives slightly different values for the dipole moments except for adenine whose value is significantly lowered (and made less like experiment). The calculations give magnitudes of the dipole moments which are in reasonable agreement with experiment, and they give directions of dipole moments which cannot be tested yet. The calculations actually lead to values for the net charge on each atom.[24] The data are illustrated in Fig. 2-7 where the directions and magnitudes of the calculated dipole moments are compared with the signs of the charges. The calculated charges range from −0.7 electronic charge to +0.5 electronic charge. Atoms with net charges between +0.1 and −0.1 are labeled with a zero. The most obvious conclusion is that characterizing the complicated charge distribution by a dipole is a gross oversimplification.

Other less empirical methods of calculating charge distributions have been used in which either all valence electrons[24] or all electrons[29] are considered. The various methods agree qualitatively, but it is too early to tell which method is best. Table 2-2b gives the results of the least empirical calculation[29]; the numbers are in remarkable agreement with those in Table 2-2a. The calculated charges on each atom, however, are not in such good agreement.

Besides knowing the static distribution of charges, it is useful to know the mobility of the charge distribution, that is, the polarizability. Discussion of this is left to the next section on ultraviolet optical properties.

TABLE 2-2a CALCULATED[a] DIPOLE MOMENTS OF NUCLEIC ACID BASES

Compound	π Dipole moment (Debye)	Direction of μ_π[b]	σ Dipole moment (Debye)	Direction of μ_σ[b]	Total dipole moment (Debye)	Direction of μ[b]	Alternate dipole moment (Debye)[c]
9-Methyladenine	2.7	74°	0.4	83°	3.1	75°	2.0
9-Methylguanine	5.7	−33°	1.2	−46°	6.9	−35°	7.2
1-Methylcytosine	5.3	96°	1.8	99°	7.1	97°	7.1
1,3-Dimethyluracil	3.3	27°	0.9	73°	3.9	36°	4.0
1-Methylthymine	2.8	29°	0.9	73°	3.5	39°	4.0

[a] H. Berthod and A. Pullman, *J. Chim. Phys.*, **55**, 942 (1965).

[b] The direction of the dipole moment is specified by the angle the negative end of the dipole makes with the C4 to C5 axis in purines or the N1 to C4 axis in pyrimidines. Positive angles are measured counterclockwise for the bases as drawn in Fig. 2-1.

[c] H. Berthod, C. Giessner-Prettre, and A. Pullman, *Theoret. Chim. Acta*, **5**, 53 (1966). These values are calculated for the unsubstituted bases. The value for thymine is from A. Denis and A. Pullman, *Theoret. Chim. Acta*, **7**, 110 (1967).

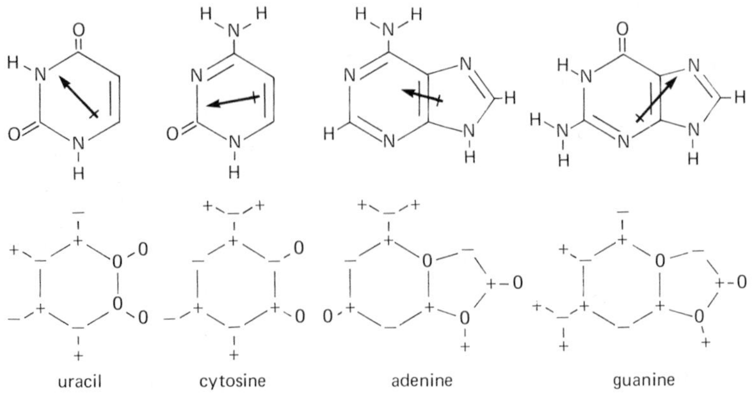

uracil cytosine adenine guanine

FIGURE 2-7 The calculated permanent dipole moments and net charges on each nucleus for the nucleic acid bases. [From B. Pullman and A. Pullman, *Prog. Nucleic Acid Res. Mol. Biol.*, **9**, 328 (1969).]

The charge distribution in a nucleoside and a nucleotide can be calculated by semiempirical methods similar to those applied to the bases.[10] The charge distribution in a base does not change much on forming the nucleoside. The calculated charge distribution in the sugar has the hydroxyl oxygens negative, the ether oxygen less negative, the carbons and their hydrogens neutral, and the hydroxyl hydrogens positive. The range of charges is from −0.5 to +0.3 of an electron.

C. Ultraviolet Optical Properties

1. ABSORPTION, REFRACTION, AND REFLECTION

Numbers and types of bands / The ultraviolet optical properties help characterize the electronic structure of each molecule. These properties can be related to the electronic wave functions and energies. In addition, of course, they provide very useful analytical methods for identifying and studying the nucleic acid bases, the nucleosides, and the nucleotides. It is impractical to present the large amount of data available on these compounds as a function of pH, solvent, and temperature. We have chosen to present figures for the properties of the ribonucleosides of adenine, guanine, cytosine, and uracil, and the deoxyribonucleoside of thymine. Mainly spectra obtained in neutral, aqueous solution at room temperature will be given.

Figure 2-8 gives the absorption spectra in water (pH near 7 where the bases are uncharged) for adenosine, guanosine, cytidine, uridine, and thymidine.[30] The spectra are similar, with intense absorption bands for wavelengths less than 300 nm. The positions and magnitudes of the peaks are

TABLE 2-2b CALCULATED[a] DIPOLE MOMENTS OF NUCLEIC ACID BASES

Compound	Total dipole moment (Debye)	Direction of μ^b
Adenine	2.6	84°
Guanine	6.9	−34°
Cytosine	6.4	99°
Thymine	3.3	32°

[a] E. Clementi, J. M. Andre, M. C. Andre, D. Klint, and D. Hahn, *Acta Phys.*, **27**, 493 (1969).
[b] Same convention as Table 2-2a.

listed in Table 2-3. The longest wavelength maxima of these spectra generally occur near 260 nm, which explains the use of optical absorption measurements at this wavelength for quantitative analytical determination of nucleic acid content in unknown samples. The spectrum is essentially that of the base. The base, its nucleoside or deoxynucleoside and its phosphate derivatives all have very similar spectra. The sugar and phosphate increase the absorbance and cause a slight shift to longer wavelength in general, but the shape of the curve is unchanged.[30] Ionizing the bases by changing the pH causes a marked change in spectrum.

Absorption of the bases has been measured in the vacuum ultraviolet region.[31] The spectra show intense bands for all bases to the lowest wavelengths (120 nm) measured.

In order to understand a spectrum, the following questions should be answered. How many electronic transitions are present? What electrons are involved in the transitions: $\pi \rightarrow \pi^*$, $n \rightarrow \pi^*$, etc? A more quantitative question which is very important for the interpretation of the optical properties of polynucleotides is what the change is in electron distribution for each transition. An experimental answer to this question is obtained by the determination of the direction and magnitude of the transition electric dipole moment.

The minimum number of electronic transitions present is determined simply by counting the number of maxima in the spectrum (see Table 2-3). However, there may be many more transitions which are not seen either because of low intensity or because of overlapping of two bands. To estimate the number of bands we might expect in these compounds above 180 nm, we can compare them with compounds whose spectra have been assigned. For example, benzene has three $\pi \rightarrow \pi^*$ transitions above 180 nm and naphthalene has four.[32] This is roughly consistent with the number seen so we do not expect that there is more than one hidden $\pi \rightarrow \pi^*$ transition in any base. The number of $n \rightarrow \pi^*$ transitions which are not seen may be large. There are two

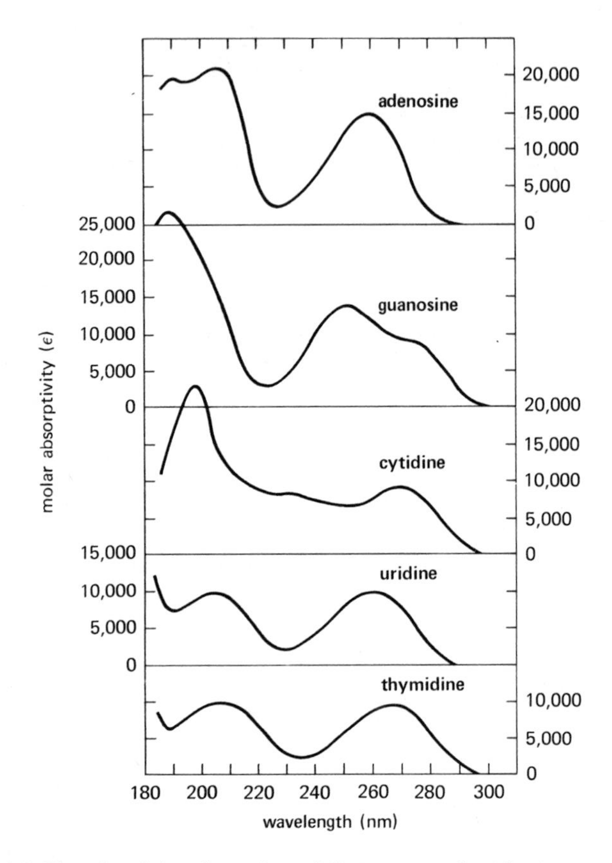

FIGURE 2-8 The ultraviolet absorption of five mononucleosides in neutral, aqueous solution. [Data from D. Voet, W. B. Gratzer, R. A. Cox, and P. Doty, *Biopolymers*, 1, 193 (1963).]

types of $n \rightarrow \pi^*$ transitions likely in these molecules: the nonbonding electrons may come from the ring nitrogens or the carbonyl oxygens. The transition involving the ring nitrogen is approximately an $sp^2(n)$ to $p(\pi^*)$ transition and is allowed because of its $s \rightarrow p$ character. Adenine can have three allowed $n \rightarrow \pi^*$ transitions (from N1, N3, and N7); guanine can have two (N3, N7); cytosine can have one (N3), and uracil or thymine none. The transition of the carbonyl oxygen is like a $p_x(n)$ to $p_z(\pi^*)$; it is forbidden and should have a molar absorptivity of less than 100. The small amount of absorption can come either from vibrations of the molecule or from the influence of the rest of the molecule which destroys the pure p_x to p_z character of the transition. Guanine can have one forbidden $n \rightarrow \pi^*$ transition (from O6); cytosine can have one (O2); and uracil or thymine can have two (O2, O6). A nitrogen $n \rightarrow \pi^*$ absorption has been seen as a long wavelength tail in both purine and pyrimidine.[33] In carbonyl compounds and simple amides an oxygen $n \rightarrow \pi^*$

TABLE 2-3 THE SPECTRA OF NUCLEOSIDES IN NEUTRAL AQUEOUS SOLUTION[a,b]

	λ max[b]	ε max	λ max	ε max	λ max	ε max
Adenosine	259.5	14,900	206.3	21,200	190	19,800
Guanosine	276	9,000	252.5	13,700	188.3	26,800
Cytidine	271	9,100	230	8,200	198	23,200
Uridine	261	10,100	205	9,800		
Thymidine	267	9,700	206.5	9,800		

[a] D. Voet, W. B. Gratzer, R. A. Cox, and P. Doty, *Biopolymers*, **1**, 193 (1963).
[b] The maximum wavelength (λ max) is given in nanometers and the maximum absorptivity (ε max) is given in liters per mole centimeter.

absorption has been seen as the longest wavelength absorption.[34] However, these have been seen in the gas phase or in nonpolar solvents; in polar solvents the absorption disappears. Presumably hydrogen bonding to the lone-pair electrons stabilizes them, lowers the ground state of the transition, and thus shifts the absorption to lower wavelengths. In aqueous solutions of bases, nucleosides, or nucleotides, no convincing evidence has been reported for an $n \to \pi^*$ absorption. Miles et al.[35] suggested that the 230 nm peak in cytidine was an $n \to \pi^*$ because it disappeared in acid solution. The proton bound at N3 would be expected to shift an $n \to \pi^*$ involving this nitrogen to low wavelengths. However, they later questioned their suggestion.[16] Proof that this transition is not an $n \to \pi^*$ comes from the polarized reflection spectra of Callis.[36] He finds that the 230 nm cytosine absorption is polarized in the plane of the base and must therefore be a $\pi \to \pi^*$ transition. Even for substituted bases in nonpolar solvents (such as 9-methyladenine in methylcyclohexane[37] no evidence was found for $n \to \pi^*$ transitions. Apparently in the usual solvents the $n \to \pi^*$ transitions of the bases are either too weak to be seen in absorption or are buried under the $\pi \to \pi^*$ transitions. Under special conditions evidence for $n \to \pi^*$ absorption can be found, however.[38,39]

Other transitions involving nonbonded electrons and σ electrons are expected only at wavelengths less than 180 nm. For a more detailed comparison of the base bands to the benzene spectrum and for discussion of solvent shifts, pH effects, etc., the reader is referred to the original literature.[30,37,40]

Direction of polarization of each band / When polarized light is absorbed by an oriented molecule, the absorption depends on the direction of polarization of the light. For each band the direction of maximum absorption in the molecule is the direction of polarization of the band. This direction is helpful in assigning transitions; the $\pi \to \pi^*$ transitions must be polarized in the plane of the bases, and the allowed $n \to \pi^*$ transitions are polarized perpendicular to the plane.

The usual method of determining directions of polarization is to study a single crystal of the substance. The absorption or reflection of light polarized along the crystal axes can be measured. If the crystal geometry is favorable, the directions of polarization along the molecules can be determined. Unfortunately these directions apply to the molecule in the crystal and in general they need not be the same directions for the isolated molecule or the molecule in solution. There are large shifts in absorption maxima and absorption intensity, but it is not known how large a shift in absorption direction occurs. Measurements and calculations on purine crystals lead to changes of up to 15° between the oriented gas model and a model which includes intermolecular interaction in the crystal.[41] At the present we will have to accept as correct the experimental directions obtained from single crystals treated as oriented gases. Dilute single crystals in which the chromophores are separated would be a good experimental method to obtain useful directions of polarization. Other methods of orientation can also be used. Oriented fibers can give useful information.[42] Electric dichroism or birefringence can give the direction of polarization of a band relative to the permanent electric dipole moment. This method has been recently applied to nucleic acid bases.[43] Fluorescence studies can give relative directions of polarization; this will be discussed in a later section. Analysis of rotational fine structure of an electronic band for a molecule in the gas phase can give the direction of polarization relative to the moments of inertia of the molecule. This technique has recently been applied to molecules nearly as large as the nucleic acid bases.[44]

Stewart et al.[45] have studied the spectra and structures of single crystals of 9-methyladenine, 1-methylthymine, and a 1:1 complex of these molecules. For 9-methyladenine two $\pi \rightarrow \pi^*$ bands were found above 230 nm: The longest wavelength band is intense and polarized along the short (C4–C5) axis; the other band is weak and polarized approximately perpendicular to the first. Below 230 nm there is a strong long axis polarized band. The one long wavelength $\pi \rightarrow \pi^*$ band in 1-methylthymine is polarized close to the Nl–C4 axis; the next $\pi \rightarrow \pi^*$ band (below 230 nm) is approximately perpendicular to the first. Measurement of the absorption perpendicular to the plane of the bases in the adenine–thymine complex gives information about $n \rightarrow \pi^*$ absorption. Absorption occurs beginning at about 300 nm and rises as far as it was possible to make measurements (~230 nm). The molar absorptivity is low ($\epsilon < 1000$) and is assigned to the three allowed nitrogen $n \rightarrow \pi^*$ transitions in adenine. In the crystal N7 is hydrogen bonded, but N1 and N3 are not. There was no evidence of $n \rightarrow \pi^*$ absorption above 310 nm: The value of the molar absorptivity was less than 2 between 310 nm and 400 nm.

Eaton and Lewis[46] studied 1-methyluracil crystals. Their results for the uracil derivative were consistent with those of Stewart and Davidson[45] for the thymine derivative. However, they found a weak (molar absorptivity ≈ 200) band at 264 nm in 1-methyluracil which was polarized perpendicular to the plane of the ring. This band was assigned to a carbonyl $n \to \pi^*$ transition.

Absorption and reflection spectra of single crystals of cytosine monohydrate and 1-methylcytosine have been measured.[36] The anomalous dispersion of the reflection (which corresponds to the maximum of the absorption) was polarized nearly along the N1—C4 axis of cytosine for both the 270 nm and 230 nm band. That is, unlike the other bases, the first two $\pi \to \pi^*$ transitions in cytosine are nearly parallel to each other.

Recent work on guanine hydrochloride and 9-ethylguanine indicates that the longest wavelength band is polarized along the short axis, and the second band is perpendicular to the first.[47]

A quantitative discussion of the directions of polarization will be given in the section on theory after other optical methods are discussed. The new qualitative information obtained from crystals is that the adenine band at 260 nm is actually two $\pi \to \pi^*$ bands. The presence of $n \to \pi^*$ absorption polarized perpendicular to the base plane was detected below 300 nm in adenine and uracil. The absorption is less than 10% of the $\pi \to \pi^*$ absorption and it will be shifted to lower wavelengths in aqueous solution.

2. CIRCULAR DICHROISM AND OPTICAL ROTATORY DISPERSION

The difference in molar absorptivity for right and left circularly polarized light (circular dichroism) is shown in Fig. 2-9 for four ribonucleosides.[48] This difference is about one part in 10^4 of the average absorptivity. Each circular dichroism band must correspond to an electronic transition in the molecule. Comparing the circular dichroism spectrum with the absorption spectrum, one notices some new bands to the blue of the maximum in the absorption. These bands provide some evidence for the existence of $n \to \pi^*$ transitions in this region.

It is clear from Fig. 2-9 that the magnitude of the circular dichroism is not proportional to the absorption. In fact, the pyrimidine nucleosides have the smaller absorption and the larger circular dichroism. The probable explanation for this is the greater hindrance to rotation of the pyrimidine base around the bond to the sugar. That is, the base in the pyrimidine nucleoside is relatively fixed, while the purine base can have a wide range of angular orientation, thereby decreasing the effective asymmetry which contributes to optical activity. The circular dichroism of the deoxyribonucleosides is similar, but smaller than the corresponding ribonucleosides. This is expected because of the one less asymmetric center present. The addition of phosphates to make mononucleotides does not have a large effect on the circular dichroism. Measurement of the rotation of plane polarized light, the optical rotatory dispersion, gives the same information in principle as the circular dichroism. In fact the rotatory dispersion was measured and

TABLE 2-4 OPTICAL ROTATORY DISPERSION IN THE VISIBLE FOR NUCLEOSIDES: PARAMETERS[a] IN THE DRUDE EQUATION $[\phi] = k/(\lambda^2 - \lambda_c^2)$

	k	λ_c
Adenosine	−48.6	226
Guanosine	−41.0	214
Cytidine	20.0	329
Uridine	4.4	346
Thymidine	12.0	276

[a] J. T. Yang, T. Samejima, and P. K. Sarkar, *Biopolymers,* **4**, 623 (1966). The solutions were neutral (pH 7.5–7.9) in 0.15 M KF or 0.1 M phosphate buffer.

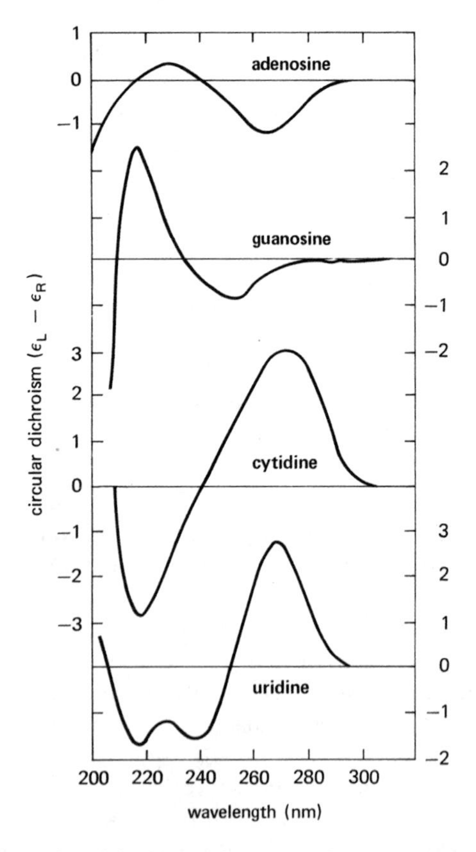

FIGURE 2-9 The circular dichroism of four monoribonucleosides in neutral aqueous solution. [Data from M. M. Warshaw and C. R. Cantor, *Biopolymers,* 9, 1079 (1970).]

analyzed before the circular dichroism.[49] The rotatory dispersion can sometimes be measured and found useful when it is impractical or impossible to measure the circular dichroism. Examples are solutions in which the absorbance is too high to measure the circular dichroism. The rotatory dispersion in the visible for the nucleosides can be fit by a one-term Drude equation.[49]

$$[\phi] = \frac{k}{(\lambda^2 - \lambda_c^2)}$$

where $[\phi]$ is the molar rotation in units of 100 deg liter/mole centimeter. Values of k and λ_c for nucleosides are given in Table 2-4.

3. MAGNETIC CIRCULAR DICHROISM AND MAGNETIC OPTICAL ROTATORY DISPERSION

In a magnetic field all matter shows circular dichroism and optical rotatory dispersion. This phenomenon is known as the Faraday effect. The external magnetic field causes the electrons of the sample to tend to move in helical paths when they are excited by the incident light. The effect depends on interactions involving all the transitions in the sample. The magnitude of the effect is proportional to the component of the magnetic field parallel to the direction of propagation of the light. Figure 2-10 gives the magnetic circular dichroism for five nucleosides.[50] The magnetic circular dichroism and the natural circular dichroism add; therefore, the natural circular dichroism had to be subtracted from the observed values to obtain the spectra shown.[50] No new qualitative information on the number of electronic transitions is seen, but the composite nature of the 260 nm adenine peak is very obvious. Unlike the natural circular dichroism, the magnetic circular dichroism is not very sensitive to conformation. The spectra are essentially the same for the bases and the nucleosides.

4. FLUORESCENCE AND PHOSPHORESCENCE

The emission of light by the bases, nucleosides, and nucleotides[51] can give further information about the nature of the electronic transition in the bases. Fluorescence spectra (emission from an excited state with paired electrons – a singlet state) and phosphorescence spectra (emission from an excited state with two unpaired electrons – a triplet state) are given in Fig. 2-11 for the 5′-nucleotides.[52] The solvent is a 1:1 mixture of ethylene glycol–water at pH 7 and 80°K. At room temperature in neutral aqueous solution the fluorescence is extremely weak and the phosphorescence is nonexistent. Even at liquid nitrogen temperatures the emission intensities are very low. The curves are very similar for the nucleosides. The fact that the fluorescence is measurable immediately gives information about $n \rightarrow \pi^*$ states. Fluorescence occurs from the lowest energy excited singlet state of the molecule; if this state is an $n \rightarrow \pi^*$ state no fluorescence is expected,[53]

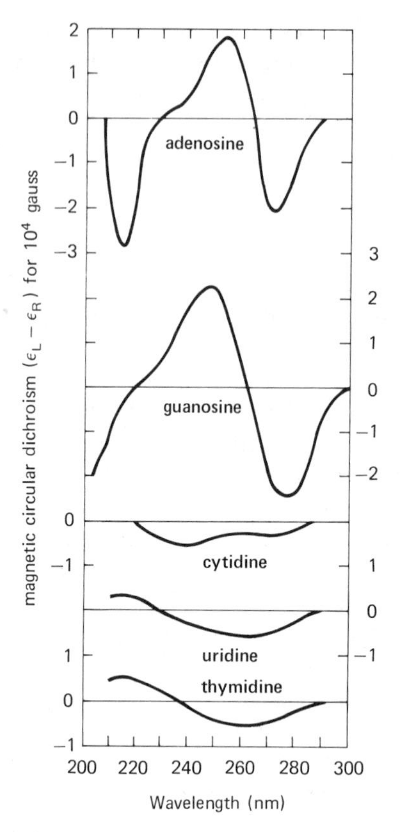

FIGURE 2-10 The magnetic circular dichroism of five nucleosides in neutral aqueous solution. The natural circular dichroism has been subtracted. [Data from W. Voelter et al., *J. Amer. Chem. Soc.*, **26**, 6163 (1968).]

although $n \to \pi^*$ fluorescence does occur rarely. The reason for this is essentially the same reason which makes the $n \to \pi^*$ absorption weak: The probability of transition is small. Therefore the presence of fluorescence implies that there are no $n \to \pi^*$ absorption bands for the bases in aqueous solution at wavelengths longer than the $\pi \to \pi^*$ absorption bands. Quantitative information about the emission is given in Table 2-5. Thymidylic acid is interesting in that it does not phosphoresce when it is excited in dilute solution. Apparently the transfer of excitation from its singlet to its triplet state is very improbable. However, the triplet can be excited by transfer of energy from other molecules.[54] The phosphorescence properties quoted in the table for thymidylic acid were determined after excitation by acetone. Furthermore, the phosphorescence of DNA is essentially the same as that from thymidylic acid.[54] As thymidylic acid has the lowest energy triplet state of the four

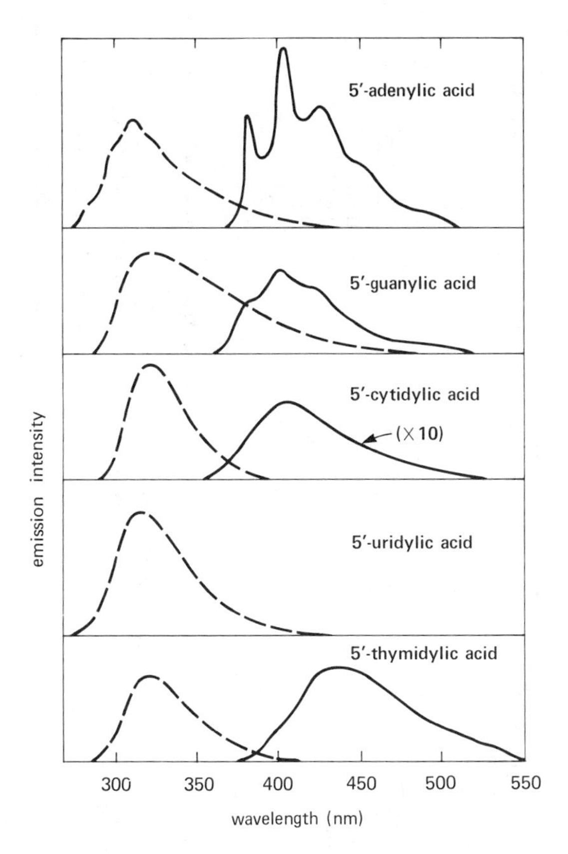

FIGURE 2-11 The fluorescence (dashed lines) and phosphorescence (solid lines) of five mononucleotides in an ethylene glycol–water glass at 80°K. [Data from J. Eisinger, *Photochem. Photobiol.*, 7, 597 (1968).]

nucleotides (see Table 2-5), this is not unreasonable. Apparently the triplet states of the other nucleotides are either quenched or they transfer excitation to thymidylic acid.

Further evidence that the DNA triplet state is essentially due to thymine comes from electron spin resonance studies of the unpaired electrons.[54] The spin signals from the excited DNA and thymidylic acid are very similar. Analysis of the electron spin resonance signal indicates that the triplet state is a $\pi \rightarrow \pi^*$ state, not an $n \rightarrow \pi^*$ state.[54]

The excitation spectrum of each molecule can also be measured. One monitors the emission at a constant wavelength and varies the wavelength of the exciting light. The excitation spectrum generally reproduces the absorption spectrum of the emitting molecule. If the polarization of the emitted light is measured relative to that of the exciting light, the relative polarization

TABLE 2-5 EMISSION PROPERTIES OF MONONUCLEOTIDES AT 80°K IN AN ETHYLENE GLYCOL TO WATER (1:1) SOLVENT AT pH 7[a]

	Fluorescence			Phosphorescence		
	Energy[b] (cm^{-1})	Quantum yield[c]	Lifetime[d] (nsec)	Energy[b] (cm^{-1})	Quantum yield[c]	Lifetime[d] (sec)
5'-Adenylic acid	35,200	0.0043	2.8	26,700	0.006	2.4
5'-Guanylic acid	34,000	0.066	~5	27,200	0.042	1.3
5'-Cytidylic acid	33,700	0.042	–	27,900	0.005	0.34
5'-Uridylic acid	34,900	0.007	–	–	~0	–
5'-Thymidylic acid	34,100	0.097	3.2	26,300[e]	~0	0.33[e]

[a] J. Eisinger, *Photochem. Photobiol.*, **7**, 597 (1968).

[b] This energy corresponds to the transition between the ground vibrational levels of the two electronic states; it is essentially the short wavelength limit of the emission.

[c] The quantum yield is the ratio of the total number of photons emitted to the number of photons absorbed.

[d] The emitted light intensity usually decays exponentially with time after the exciting light is turned off. The lifetime is the time it takes to reach $1/e$ of its initial value.

[e] The phosphorescence properties of thymidylic acid were studied by using acetone as a sensitizer to excite the triplet state [A. A. Lamola, M. Guéron, T. Yamane, J. Eisinger, and R. G. Shulman, *J. Chem. Phys.*, **47**, 2210 (1967)].

of the absorbing and emitting bands can be determined. The molecule must be fixed rigidly in a glass or crystal so that the state of polarization of the emitted light depends on the bands, not the rate of rotatory diffusion of the molecule. From polarization fluorescence measurements in glasses Callis et al.[36,55] have concluded that the two longest wavelength bands in guanine are polarized perpendicular to each other, but that the two longest wavelength bands in 5-methylcytosine are polarized approximately parallel to each other. The long wavelength peak in adenine was found to consist of two electronic transitions, the lower wavelength one having only 10% of the intensity of the higher wavelength one. These results are in agreement with the polarized absorption and reflection spectra mentioned earlier.

5. THEORY

We have been discussing optical properties of molecules, but we have not quantitatively related them to molecular structure. Here we will sketch the quantitative relations which link the optical properties to electronic energy levels and transition moments. The properties discussed are absorption, magnetic circular dichroism, and natural circular dichroism. They are related (in order) to refractive index, Faraday rotation, and natural rotation by the Kronig-Kramers transform. Emission properties (fluorescence, phosphorescence) are also discussed.

The interaction of light with matter can be characterized by the interaction of the electromagnetic field with the charged particles of the sample. We need only consider the interaction of the electrons with the vector potential of the electromagnetic field. As the electric vector **E** of the field is in the same direction as the vector potential, we can characterize the field by its electric vector **E**. All the optical properties can be related to the probability of a transition from state O to state A being induced in the sample by the light of frequency ν. This probability of transition is proportional to[56]

$$\mathrm{Re}\langle \psi_{\mathrm{A}}{}^{*} \exp\,[i2\pi\nu \mathbf{k}\cdot\mathbf{R}/c]\;\mathbf{P}\cdot\mathbf{E}\psi_{\mathrm{O}}d\tau\rangle^{2}\;\delta(\nu-\nu_{\mathrm{OA}}) \tag{2-1}$$

$\psi_{\mathrm{O}},\psi_{\mathrm{A}}$ are the wave functions for the sample in its ground and excited states. They will in general depend on the coordinates of all the electrons in the sample. The wave function may be imaginary, so the asterisk tells us to take the complex conjugate.

R is the sum of all position vectors for the electrons in the sample

P is the sum of all linear momentum operators for the electrons in the sample

k, **E** are vectors which characterize the light. **k** is a unit vector along the direction of propagation and **E** is the electric vector.

ν, c are the frequency and the velocity of the light in vacuum
$\int d\tau$ indicates an integral taken over the volume of the system
$\langle\ \rangle$ are brackets which remind us to average the integral over the accessible states of the sample
Re means take the real part of the expression
$\delta(\nu - \nu_{OA})$ is the delta function. It is necessary because a transition can only occur if the frequency of the light corresponds to the energy difference between two states of the system ($h\nu_{OA} = E_A - E_O$).

For most samples (including those containing macromolecules) the distance between significantly interacting electrons is small compared to the wavelength of light, ($\lambda = c/\nu$), therefore the exponential is expanded.

$$\exp[i2\pi\nu\mathbf{k} \cdot \mathbf{R}/c] = 1 + i2\pi\nu\mathbf{k} \cdot \mathbf{R}/c \qquad (2\text{-}2)$$

The first term is about 100 times larger than the second. For absorption and emission the effect of the second term is nearly always negligible; its effect might be 1 part in 10^4. Neglect of the second and higher terms in the expansion leads to the electric dipole approximation for the transition probability.

$$\text{Re}\,\langle \int \psi_A{}^* \boldsymbol{\mu} \cdot \mathbf{E}\psi_O d\tau \rangle^2\, \delta(\nu - \nu_{OA}) \qquad (2\text{-}3)$$

The transition electric dipole is

$$\boldsymbol{\mu}_{AO} = \boldsymbol{\mu}_{OA} = \int \psi_A{}^* \boldsymbol{\mu} \psi_O d\tau \qquad (2\text{-}4)$$

$\boldsymbol{\mu} = e\mathbf{R}$ is the sum of the electric dipole operators for the electrons in the sample. The electronic charge is e.

To obtain Eq. (2-3), $\mathbf{P}_{AO} = \int \psi_A{}^* \mathbf{P} \psi_O d\tau$ has been replaced by its equivalent $(2\pi i m/e)\ \nu_{OA}\boldsymbol{\mu}_{OA}$ for exact wave functions.[56] This replacement follows from the definition of momentum as mass (m) times velocity ($d\mathbf{R}/dt$). For exact wave functions $d\boldsymbol{\mu}_{OA}/dt = i2\pi\nu_{OA}\ {}_{OA}$. It is traditional to use the transition dipole $\boldsymbol{\mu}_{OA}$ even though for inexact wave functions $\mathbf{P}_{AO}$ is usually more accurate.[57,58]

Absorption / The molar absorptivity or molar extinction coefficient $\epsilon(\nu)$ is defined as the absorbance per unit molar concentration and unit length.

$$\epsilon(\nu) = \frac{A}{lm} = \frac{\log I_0/I}{lm} \qquad (2\text{-}5)$$

A is absorbance
$\log I_0/I$ is the base 10 logarithm of the incident intensity divided by the transmitted intensity of the light of frequency ν
l is path length in centimeters
m is molar concentration in moles/liter

It is related to molecular properties through the transition probability.[59]

$$\epsilon(\nu) = \frac{8\pi^3 \nu N_0}{6909\, ch} \operatorname{Re} \sum_{A \neq O} \mu_{OA}{}^2\, \delta(\nu_{OA} - \nu) \tag{2-6}$$

We implicitly remember to average over all molecules in the sample. If our sample is a hydrogen atom we can accurately predict the positions and intensities of absorption lines from the known wave functions. We need only calculate the corresponding $_{OA}$ and ν_{OA}. For a molecule in a solvent the states A can be considered to be vibronic (vibrational-electronic). The rotational-translational states and the interactions with the solvent can be treated as giving rise to a particular width to each vibronic transition.[60] The absorption spectrum is then made up of a sum of vibronic transitions calculable from a vibronic ground state O and vibronic excited states A. The shape of each vibronic absorption band is not calculated, but is assumed, or taken from experiment. Similarly, we can assume electronic and vibrational states are separable. This is the Born-Oppenheimer approximation; it rests on the fact that the electronic motion is much faster than the nuclear motion. One then considers the spectrum to be a sum of electronic transitions. The states O and A refer now to electronic wave functions and the shape of each electronic absorption band is assumed or taken from experiment. In each case, we simply replace the delta functions by appropriate shape functions whose integrated area is unity. For example, we could use a gaussian shape and write Eq. (2-6) as

$$\epsilon(\nu) = \frac{8\pi^3 \nu N_0}{6909\, ch} \operatorname{Re} \sum_{A \neq O} \frac{\mu_{OA}{}^2}{\sqrt{\pi}\,\theta_A} \exp[-(\nu - \nu_{OA})^2/\theta_A{}^2] \tag{2-7}$$

θ_A is the width of gaussian band A, which is assumed small compared to ν_{OA}

The wave functions ψ_O and ψ_A now refer to electronic states of solute molecules in a particular solvent.

If we do make this assumption, we need to know only electronic wave functions for the molecule to predict the spectrum. Of course we do not have very good wave functions, but we can still predict changes in spectra caused by changes in molecular structure. The main use we will make of these equations is to relate the spectrum of a polynucleotide to its mononucleotides, or to correlate the spectra of different conformations of polynucleotides (see Chapter 4). These changes in molecular structure are slight and can be treated as perturbations. One useful method is to measure the transition electric dipole moments and transition frequencies for the mononucleotides and then to calculate how these molecular properties are changed in the polynucleotide.

The transition frequency for an electronic transition can be chosen as the frequency at the maximum absorption. If the absorption peaks are overlapping, the spectrum must first be resolved, for example, into a sum of gaussian-shaped bands. The magnitude of the transition electric dipole moment is obtained by integrating the absorption band. The dipole strength is defined as the square of the transition electric dipole moment. From Eq. (2-6) we see that it is equal to:

$$D_{OA} = \boldsymbol{\mu}_{OA}^2 = \frac{6909\, ch}{8\pi^3 N_0} \int \frac{\epsilon d\nu}{\nu}$$
$$= 9.180 \times 10^{-39} \int \frac{\epsilon d\nu}{\nu} \tag{2-8}$$

In Eq. (2-8) we give the numerical value corresponding to centimeter-gram-second–electrostatic units. However, the transition dipole is often given in units of debyes squared (10^{-18} cgs-esu). To obtain the absolute direction of μ_{OA} one needs to use polarized light and oriented molecules, as the transition probability depends on $\boldsymbol{\mu}_{OA} \cdot \mathbf{E}$ [see Eq. (2-3)]. Another measure of the total absorption in an electronic band is the oscillator strength, f. It is defined to be unitless.[32]

$$f = 2303 \frac{mc^2}{\pi e^2 N_0} \int \epsilon d\nu \qquad (\nu \text{ in cm}^{-1})$$
$$= 4.318 \times 10^{-9} \int \epsilon d\nu \tag{2-9}$$

m is the mass of the electron

Roughly, f equals the fraction of an electron taking part in the transition; it has a magnitude of about 0.5 for the 260 nm absorption band in nucleic acids.

Magnetic circular dichroism / All absorbing samples become circularly dichroic in a magnetic field parallel to the direction of propagation of the light. To calculate the circular dichroism we need to know the difference in extinction coefficients for right and left circularly polarized light. The electric vector for circularly polarized light propagating along the **k** direction can be written as[61]:

$$\mathbf{E} = \text{constant}\ (\mathbf{i} \pm i\mathbf{j})$$

i, **j** are unit vectors perpendicular to **k**
i is the imaginary ($\sqrt{-1}$)

The plus sign corresponds to right circularly polarized light and the minus sign to left. Using Eq. (2-6) we write the circular dichroism as:

$$\epsilon_L(\nu) - \epsilon_R(\nu)$$
$$= \frac{8\pi^3 \nu N_0}{6909\, ch} \text{Re} \sum_{A \neq O} \{[\boldsymbol{\mu}_{OA} \cdot (\mathbf{i} - i\mathbf{j})]^2 - [\boldsymbol{\mu}_{OA} \cdot (\mathbf{i} + i\mathbf{j})]^2\}\, \delta(\nu_{OA} - \nu)$$
$$= \frac{8\pi^3 \nu N_0}{6909\, ch} \text{Re} \sum_{A \neq O} [-2i\boldsymbol{\mu}_{OA} \cdot \mathbf{i})\, (\boldsymbol{\mu}_{OA} \cdot \mathbf{j})]\, \delta(\nu_{OA} - \nu) \tag{2-10}$$

This equation and similar succeeding ones could equally well be written with a gaussian shape replacing the delta function as in Eq. (2-7). If $\boldsymbol{\mu}_{OA}$ is real, then the circular dichroism is zero, because of the factor of i in Eq. (2-5). In the absence of a magnetic field, $\boldsymbol{\mu}_{OA}$ is real and to this approximation there is no circular dichroism. In the presence of a magnetic field $\boldsymbol{\mu}_{OA}$ becomes complex and (magnetic) circular dichroism can occur. A complex $\boldsymbol{\mu}_{OA}$ corresponds to a helical motion of the electrons excited by light in the magnetic field. The usual form for the magnetic circular dichroism is obtained if the magnetic field is treated as a perturbation.

$$\frac{\epsilon_L(\nu) - \epsilon_R(\nu)}{H_0} = \frac{8\pi^3 \nu N_0}{6909\, ch} \operatorname{Im} \sum_{A \neq O} \boldsymbol{\mu}_{OA} \cdot \left[\sum_{B \neq A} \frac{\boldsymbol{\mu}_{OB} \times \mathbf{m}_{AB}}{h(\nu_{OB} - \nu_{OA})} - \sum_{B \neq O} \frac{\boldsymbol{\mu}_{AB} \times \mathbf{m}_{OB}}{h\nu_{OB}} \right] \delta(\nu_{OA} - \nu) \qquad (2\text{-}11)$$

H_0 is the magnitude of the static magnetic field along the direction of propagation of the light

m is the transition magnetic dipole operator; it is equal to $(e/2mc) \sum_i \mathbf{r}_i \times \mathbf{p}_i$. $\mathbf{r}_i$ and $\mathbf{p}_i$ are the position and linear momentum vectors of electron i. $\mathbf{r}_i \times \mathbf{p}_i$ is therefore the angular momentum.

This term is the only nonzero term for molecules which have nondegenerate energy levels; molecules with no axis of symmetry. Other terms[63] arise from effects of the magnetic fields on the degenerate energy levels.

The equation has introduced another molecular property: the transition magnetic dipole moment $\mathbf{m}_{OA}$. Actually, in Eq. (2-11) transition moments from the ground state are important ($\boldsymbol{\mu}_{OA}$, $\boldsymbol{\mu}_{OB}$, and $\mathbf{m}_{OB}$), but also transition moments between excited states occur ($\mathbf{m}_{AB}$ and $\boldsymbol{\mu}_{AB}$). The magnetic transition moment is a vector like $\boldsymbol{\mu}_{OA}$, but it is a pure imaginary for real wave functions because, in the operator formulation of quantum mechanics, the momentum

$$\mathbf{p} = \frac{h}{2\pi i} \nabla = \frac{h}{2\pi i} \left(\frac{\partial}{\partial x} \mathbf{i} + \frac{\partial}{\partial y} \mathbf{j} + \frac{\partial}{\partial z} \mathbf{k} \right) \qquad (2\text{-}12)$$

is a pure imaginary, and therefore so is $\mathbf{r} \times \mathbf{p}$. This means

$$\mathbf{m}_{OA} = -\mathbf{m}_{AO} = \int \psi_0{}^* \mathbf{m}\, \psi_A \, d\tau \qquad (2\text{-}13)$$

The magnetic transition dipole is a molecular property that is very difficult to measure directly. No convenient optical property depends solely on $\mathbf{m}_{OA}$. It also has another difference from the electric dipole transition moment. If the coordinates of each particle in a molecule change sign ($x \to -x$, $y \to -y$, $z \to -z$), the electric dipole also changes sign, but the magnetic dipole does not. This has important consequences on the discussion of natural and magnetic circular dichroism.

We can define a magnetic rotational strength analogous to the dipole strength.

$$F_{OA} = \text{Im}\, \boldsymbol{\mu}_{OA} \cdot \left[\sum_{B \neq A} \frac{\boldsymbol{\mu}_{OB} \times \mathbf{m}_{AB}}{h(\nu_{OB} - \nu_{OA})} - \sum_{B \neq O} \frac{\boldsymbol{\mu}_{AB} \times \mathbf{m}_{OB}}{h\nu_{OB}} \right]$$
$$= \frac{6909\, ch}{8\pi^3 N_0} \left(\frac{1}{H_0}\right) \int \left(\frac{\epsilon_L - \epsilon_R}{\nu}\right) d\nu \tag{2-14}$$

It is clear that the magnetic circular dichroism is a much more complicated function of molecular transitions than the absorption. In the perturbation notation of Eq. (2-14) the sums over B tell us that every possible transition in the molecule contributes to the magnetic rotational strength corresponding to a single absorption band. The frequency denominators weight the transitions near ν_{OA} heavily, but it is difficult to kow how many terms must be considered in the sums over B. The angular factors involved in the dot and cross vector products show that the magnetic circular dichroism may be more sensitive to conformation than absorption which depends only on the magnitude of $\boldsymbol{\mu}_{OA}$. However, enantiomers will have identical magnetic rotational strengths, because on inversion of a molecule neither the product of two $\boldsymbol{\mu}$'s nor $\mathbf{m}$ changes sign, and therefore $\boldsymbol{\mu}_{OA} \cdot \boldsymbol{\mu}_{OB} \times \mathbf{m}_{AB}$ and $\boldsymbol{\mu}_{OA} \cdot \boldsymbol{\mu}_{AB} \times \mathbf{m}_{OB}$ remain invariant.

Natural circular dichroism / Equation (2-10) states that the circular dichroism is zero for molecules in the absence of magnetic fields. As we know that some molecules (those which are different from their mirror images) are circularly dichroic, we must see if Eq. (2-1) leads to this conclusion. To obtain Eq. (2-10) we expanded the exponential [Eq. (2-2)] and dropped the second term. This is satisfactory for discussion of absorption, emission, and magnetic circular dichroism. But as the first term alone does not contribute to natural circular dichroism, the second term must now be considered. The absorption $\epsilon(\nu)$ becomes proportional to

$$\text{Re} < \int \psi_A^* \left[\mathbf{P} \cdot \mathbf{E} + \frac{i2\pi\nu}{c} \mathbf{k} \cdot \mathbf{R P} \cdot \mathbf{E} \right] \psi_0 d\tau >^2 \delta(\nu - \nu_{AO}) \tag{2-15}$$

In the absence of magnetic fields ($\boldsymbol{\mu}_{OA}$ real), the terms in Eq. (2-10) are zero and the first nonzero terms are

$$\epsilon_L - \epsilon_R(\nu) = \frac{32\pi^3 \nu N_0}{6909\, ch} \text{Im} \sum_{A \neq O} \boldsymbol{\mu}_{OA} \cdot \mathbf{m}_{AO}\, \delta(\nu_{AO} - \nu) \tag{2-16}$$

The $\mathbf{m}_{AO}$ vectors appear, but they do not come from the perturbations of a static magnetic field. Instead they arise from components of the $(\mathbf{k} \cdot \mathbf{R}\ \mathbf{P} \cdot \mathbf{E})_{AO}$ term, that is, the effect of the electromagnetic field of the light.

Natural circular dichroism is easier to interpret than magnetic circular dichroism, but not as easy to interpret as absorption. Conversely, circular dichroism is more sensitive to conformation than absorption is. There are two different molecular vectors to be affected by the structure, instead of just the magnitude of one vector as in absorption. However, the main difference is that when the molecule is replaced by its mirror image the circular dichroism changes sign ($\boldsymbol{\mu}_{OA} \rightarrow -\boldsymbol{\mu}_{OA}$, $\mathbf{m}_{OA} \rightarrow \mathbf{m}_{OA}$). This means that any molecule which is identical to its mirror image has zero circular dichroism. Furthermore, enantiomers will have equal and opposite circular dichroism. This is exactly consistent with observations on optically active molecules. A rotational strength corresponding to each transition is defined analogous to dipole and magnetic rotational strengths.

$$R_{OA} = \mathrm{Im}\, \boldsymbol{\mu}_{OA} \cdot \mathbf{m}_{AO} = \frac{6909\, ch}{32\pi^3 N_0} \int \frac{\epsilon_L - \epsilon_R}{\nu} d\nu$$
$$= 2.295 \times 10^{-39} \int \frac{\epsilon_L - \epsilon_R}{\nu} d\nu \qquad (2\text{-}17)$$

The implication of the previous equations is that one can calculate the absorption, magnetic circular dichroism, or natural circular dichroism of a sample from the energies and wave functions of all its states. As these energies and wave functions are not usually available, the equations are used instead to obtain correlations between different types of spectra and between similar molecules.

Fluorescence and phosphorescence[64,65] / We have emphasized so far that the probability of a transition from state O → A in an electromagnetic field is the same as from A → O. Therefore, equations for absorption and emission are similar. However, in emission we are able to see processes very difficult to study in absorption. Phosphorescence is the transition from an excited triplet state to the ground singlet state. The process is improbable and the small signal-to-noise ratio plus impurity absorption make the corresponding singlet to triplet absorption difficult to measure. The time dependence of emission, another parameter useful in characterizing a sample, is also not available in absorption studies.

When a sample absorbs light the energy may be used in different ways. It can raise the temperature of the sample; it can be transformed into chemical energy; or finally it can be reradiated as light. The ratio of the amount of light reradiated to the amount absorbed is the quantum yield. If we ignore the photochemistry, the two competing processes are conversion to heat and emission as light. There is a slight indefiniteness about the exact infrared wavelength we call heat, but usually the distinction is obvious. When a molecule is excited it quickly ($\sim 10^{-13}$ seconds) returns to the lowest vibrational state of the excited electronic state. Whether the energy is now emitted or not depends greatly on the environment and temperature. The stronger the interaction with the environment the lower the probability of emission. This means polar molecules in polar solvents (nucleic acids in water) will have low quantum yields. High temperatures will also decrease the quantum yield. The number and type of vibrations possible in the molecule and solvent will affect the quantum yield. Increasing the symmetry of the molecule and solvent or replacing hydrogen by deuterium will tend to increase the quantum yield. This qualitative discussion can be formalized by writing the following kinetic equations for the deexcitation of molecule A*.

$$\begin{aligned} A^* &\xrightarrow{k_r} A + \text{photon} \\ A^* &\xrightarrow{k_T} A + \text{heat} \end{aligned} \tag{2-18}$$

The rate constants for radiation and thermal deexcitations are k_r and k_T. The quantum yield Q is then the ratio

$$Q = \frac{\text{quanta emitted}}{\text{quanta absorbed}} = \frac{k_r}{k_r + k_T} \tag{2-19}$$

The value of k_r is chosen to be the intrinsic fluorescence rate of the isolated molecule; it is directly related to the transition probability. Therefore, it will be small for an $n \rightarrow \pi^*$ or triplet excited state. The value of k_T is very dependent on the environment.

The time dependence of the emission after the exciting light is removed is easily derived from the equations. The emission intensity is proportional to the number of photons produced per second.

$$I(\text{emitted}) \propto \frac{d(\text{photon})}{dt} = k_r A^* \tag{2-20}$$

The concentration of excited molecules will decrease exponentially with time after the exciting light is removed.

$$\begin{aligned} \frac{dA^*}{dt} &= -(k_r + k_T)A^* \\ A^* &= A_0^* \exp\left[-(k_r + k_T)t\right] \end{aligned} \tag{2-21}$$

Therefore the intensity emitted also decreases exponentially.

$$I(\text{emitted}) \propto \exp\left[-t/\tau\right] \tag{2-22}$$

The fluorescence lifetime is $\tau = (k_r + k_T)^{-1}$. It is related to the intrinsic fluorescence lifetime τ_0 of the isolated molecule ($\tau_0 = 1/k_r$) through the quantum yield [see Eq. (2-19)].

$$\tau = Q\tau_0 \tag{2-23}$$

The wavelength dependence of the emitted light can be characterized by the dipole strengths of the transitions from the ground vibrational state of

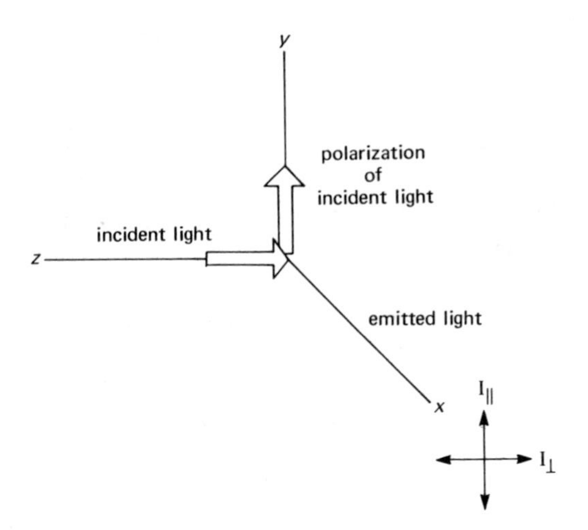

FIGURE 2-12 Definition of $I_{\|}$ and $I_{\perp}$ for the polarization of emitted light. The light is incident along z; it is polarized along y and observed along x.

the excited electronic state to excited vibrational states of the ground electronic state. The emission will therefore be at longer wavelengths than the absorption. However, when emission occurs from the same electronic band which is originally excited by absorption, the long wavelength limit of absorption will approximately equal the short wavelength limit of emission. If this does not occur, then the excitation must have been transferred to a different electronic band before emission. This method can be used to detect new electronic bands.

The polarization of the emitted light also gives information about hidden electronic bands. More importantly, it can tell us about relative orientations of electronic transition dipole moments. A measure of the polarization of the emission is the polarization ratio p. For light incident along z and polarized along y the emission can be detected along x (see Fig. 2-12). The value of p is defined in terms of emission intensities polarized along y and z ($I_y = I_{\|}, I_z = I_{\perp}$).

$$p = \frac{I_{\|} - I_{\perp}}{I_{\|} + I_{\perp}} \tag{2-24}$$

The emitted intensity polarized along any direction is proportional to the cosine squared of the angle between the emitting transition dipole and that direction [see Eq. (2-3)]. The distribution of excited molecules is proportional to the cosine squared of the angle between the absorbing transition dipole and the plane of polarization. For a rigid sample (such as a frozen glass) in which the emitting vector makes an angle of α with the absorbing vector, and in which the absorbing dipoles are oriented at random, the general equation is:

$$p = \frac{3\cos^2\alpha - 1}{\cos^2\alpha + 3} \tag{2-25}$$

The limiting values are $p = 1/2$ for $\alpha = 0$ and $p = -1/3$ for $\alpha = 90°$. Because of experimental difficulties and overlapping of absorption bands, these limits are never reached. However, empirical corrections can be made so that the angle α between two transition moments in a molecule can be obtained from the polarization ratio p.[66]

The time, polarization, and wavelength dependence of emitted light from nucleic acids can provide a great deal of useful information. Here we have only discussed what is pertinent to transition moment directions; we will mention other applications in later chapters. A potentially useful new parameter to study for asymmetric molecules, such as nucleotides, is the circular polarization of emitted light.[67]

Kronig-Kramers transforms / An exact relation can be applied to our previous equations to obtain results for the refraction of light and the rotation of plane polarized light. Refraction and absorption (or rotation and circular dichroism) can be represented by the real and imaginary parts of a complex parameter. The real and imaginary parts of this analytic function are not independent, since they ultimately depend on the same properties of the sample. In fact, they are integral transforms of each other.[60,68] The results are

refractive index (unitless)

$$n(\lambda) - 1 = \frac{\lambda^2}{2\pi^2}(2.303\text{ m})\int_0^\infty \frac{\epsilon(\lambda')\cdot d\lambda'}{\lambda^2 - \lambda'^2} \tag{2-26}$$

m is the concentration in moles/liter
λ is wavelength in centimeters
ϵ is molar absorptivity in liters/mole centimeter

rotation of polarized light

$$[\phi](\lambda) = \frac{2}{\pi}\int_0^\infty \frac{[\theta](\lambda')\cdot \lambda' d\lambda'}{\lambda^2 - \lambda'^2} \tag{2-27}$$

$[\theta]$ is molar ellipticity; it has the same units as molar rotation $[\phi]$ (degree liter/mole centimeter) x 100.
$[\theta]$ is 3298 $(\epsilon_L - \epsilon_R)$

The equations relate the value of an optical property at wavelength λ to measurements of other properties at wavelengths λ'. The refractive index, optical rotatory dispersion, and Faraday rotation can be calculated from the measured absorbance, natural circular dichroism, and magnetic circular dichroism, respectively. As the measurements are made over a limited wavelength range, the equations mean in practice that the contribution of a few bands to the refraction or rotation can be obtained. The results at wavelengths other than those measured can be predicted. The reverse transformations can be very useful for predicting absorption bands in the vacuum ultraviolet and for determining rotational strengths.

Absorbancy

$$\epsilon(\lambda) = \left(\frac{-8}{2.303\ \mathrm{m}}\right)\int_0^\infty \frac{[n(\lambda') - 1]d\lambda'}{\lambda^2 - \lambda'^2} \tag{2-28}$$

Circular dichroism

$$\epsilon_L - \epsilon_R(\lambda) = \left(\frac{-2}{3298\pi\lambda}\right)\int_0^\infty \frac{[\phi](\lambda')\lambda'^2 d\lambda'}{\lambda^2 - \lambda'^2} \tag{2-29}$$

The practical use of these equations in nucleic acid research has been discussed by Thiéry.[69] The shape of the rotatory dispersion curves corresponding to simple circular dichroism curves is shown in Fig. 2-13. The sign of each circular dichroism curve can be positive or negative. The corresponding rotatory dispersion curves (sometimes called Cotton effect curves) are also called positive or negative. To avoid confusion in describing the curves, the extrema in the absorption or circular dichroism curves are labeled maxima and minima, whereas the rotatory dispersion extrema are labeled peaks and troughs.

Reflection spectra, refractive index, and polarizability / It is very difficult to measure the absorption of highly absorbing crystals because of the very thin crystals required. An equivalent method of studying the positions, magnitudes, and polarization directions of absorption bands is to measure the reflection spectra. The reflection coefficient R (the ratio of the intensity of light reflected to that incident) depends on the refractive index and the absorbancy.[41,70]

$$R = \frac{(n-1)^2 + K^2}{(n+1)^2 + K^2} \tag{2-30}$$

K is directly proportional to the molar absorptivity ϵ.

The above equation holds for light incident at right angles to a sample in vacuum (or air, whose refractive index is also effectively 1). As n and ϵ have been related to transition moments in Eqs. (2-26) and (2-6), it is apparent

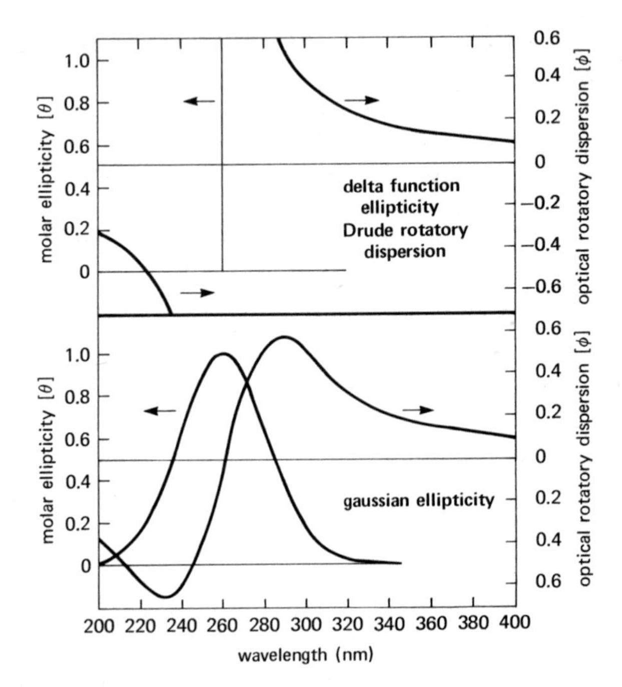

FIGURE 2-13 The circular dichroism (ellipticity) and optical rotatory dispersion of a representative band centered at 260 nm. The upper curve shows the rotatory dispersion calculated from the Kronig-Kramers transform for an ideal, infinitely sharp band. The lower curve shows the Kronig-Kramers transform of a gaussian band similar to the absorption band of a nucleic acid component. Note that the rotatory dispersion is still significant into the visible wavelength region and that the magnitudes of ellipticity and rotatory dispersion are similar.

that one can determine transition moment directions by measuring the reflection of polarized light from oriented samples.[36,41]

Although the refractive index is defined by Eq. (2-26), a more useful (though approximate) relation is the Lorentz-Lorenz equation.

$$\frac{n^2 - 1}{n^2 + 2} = \frac{4\pi}{3} N\alpha \tag{2-31}$$

N is number of molecules per cubic centimeter
α is polarizability in cubic centimeters

The polarizability depends on the sum of transition moments; for delta function absorption bands, as in Eq. (2-6), the average polarizability is

$$\alpha(\nu) = \frac{2}{3h} \sum_{\mathrm{A}} \frac{\nu_{\mathrm{OA}} \boldsymbol{\mu}_{\mathrm{OA}}{}^2}{\nu_{\mathrm{OA}}{}^2 - \nu^2} \tag{2-32}$$

The polarizability can also be defined as the dipole induced by a unit electric field. As the dipole and the field are vectors, the polarizability in general is a tensor. That is, the induced dipole is not necessarily along the direction of the field. The polarizability tensor α can be written as

$$\boldsymbol{\alpha}(\nu) = \frac{2}{h} \sum_{\mathrm{A}} \frac{\nu_{\mathrm{OA}}\, \boldsymbol{\mu}_{\mathrm{OA}}\, \boldsymbol{\mu}_{\mathrm{OA}}}{\nu_{\mathrm{OA}}{}^2 - \nu^2} \tag{2-33}$$

Therefore the direction dependence of the polarizability is related to the directions of the transition moments.

Transition moments / We have related a great many optical properties to the electric dipole transition moments of the system. For a transition from O to A the electric dipole transition moment is $\boldsymbol{\mu}_{\mathrm{OA}}$

$$\boldsymbol{\mu}_{\mathrm{OA}} = \int \psi_{\mathrm{O}}{}^{*}\, \boldsymbol{\mu}\, \psi_{\mathrm{A}}\, d\tau$$

$$\boldsymbol{\mu} = \sum_i e_i \mathbf{r}_i \tag{2-34}$$

The ψ_{O} and ψ_{A} are wave functions for the sample, and the dipole operator is a sum over all particles with charge e_i in the sample. Although the optical properties will depend on all the molecules in the sample, we usually try to simplify the theoretical problem by focusing on one particular type of solute molecule. The solvent is either treated empirically or ignored. The optical property of the sample is then interpreted as the property of the solute molecule multiplied by the number of solute molecules in the sample. For a solute such as a nucleotide, the further approximation is made that only the chromophore (the base) is important.

We will summarize some of the methods which have been used to obtain wave functions for molecules as large as the nucleic acid bases.[24] The object is to find solutions to the Schrödinger equation.

$$H\psi = E\psi$$

$$H = \frac{-h^2}{8\pi^2 m} \sum_i \nabla_i{}^2 - e^2 \sum_{i,j} \frac{Z_j}{R_{ij}} + e^2 \sum_{i>j} \frac{1}{r_{ij}} \tag{2-35}$$

$$\nabla^2 = \text{Laplacian} = \frac{\partial^2}{\partial x^2} + \frac{\partial^2}{\partial y^2} + \frac{\partial^2}{\partial z^2}$$

e, m is charge, mass of electron
Z_j is the charge on nucleus j
R_{ij} is the distance from electron i to nucleus j
r_{ij} is the distance from electron i to electron j

The nuclei are assumed to be fixed in the equilibrium configuration of the molecule, therefore the desired ψ's are the electronic wave functions. It is impractical to obtain a wave function which is a function of the $4N$ coordinates (3 space, 1 spin) for N electrons. One usually assumes instead that the wave function can be made up of molecular orbitals, each one describing only one electron. Each electron has two possible spins; therefore, each space molecular orbital describes two electrons. These orbitals can be determined by a series of successive approximations. An estimate is made of all molecular orbitals except one ($N - 1$ orbitals). This determines the average positions of all the electrons except the one of interest. The Schrödinger equation can now be solved to give its molecular orbital. This orbital plus $(N - 2)$ other orbitals are then used to find a better approximation for a second electron. The process is continued until we have second approximations for all electrons. These new, improved molecular orbitals are used to obtain a better approximation for the first electron. Eventually no further change in the molecular orbitals occurs. In this method each electron is assumed to move in the average field of all the other electrons. It is called the self-consistent field method.

The form of the functions used for the molecular orbitals is a linear combination of atomic orbitals. That is, each electron in the molecule is assumed to distribute itself around the nuclei as if it was partly a $1s$ electron of each hydrogen atom, and a $1s$, $2s$, and $2p$ electron of each carbon, nitrogen, and oxygen in the molecule. Higher energy atomic orbitals are usually not considered. For planar molecules each molecular orbital contains either $1s$, $2s$, $2p_x$ and $2p_y$ atomic orbitals (if the plane of the molecule is in the x,y plane), or only $2p_z$ atomic orbitals. The molecular orbitals containing only the p_z atomic orbitals are called π orbitals; the other molecular orbitals are called σ orbitals. Although the σ and π electrons interact and affect their respective orbitals, they can be treated separately. For each molecular orbital the coefficients of the atomic orbitals are varied to produce a minimum in the energy of that orbital. This variational procedure gives the best approximation using these orbitals to the correct value of the energy.

As an illustration of the method, let us consider in detail a typical calculation of the π orbitals for adenine. Each molecular orbital is a sum of the 10 p_z atomic orbitals of the 10 nonhydrogen nuclei in adenine. There are 12 π electrons in these orbitals. Each carbon contributes one π electron; N1, N3, and N7 contribute one π electron; but N9 and the amino nitrogen each contribute two π electrons. The atoms N1, N3, and N7 each have two nonbonding electrons, two σ electrons, and one π electron. The atoms N7 and the amino nitrogen each have three σ electrons and two π electrons.

Berthod et al.[28] have applied the linear-combination-of-atomic-orbitals–self-consistent-field–molecular-orbitals, LCAO–SCF–MO, method to the nucleic acid bases. The many parameters needed were adjusted by calculating known properties of analogous compounds. The energies for the ten molecular orbitals are plotted in Fig. 2-14; the two electrons in each space

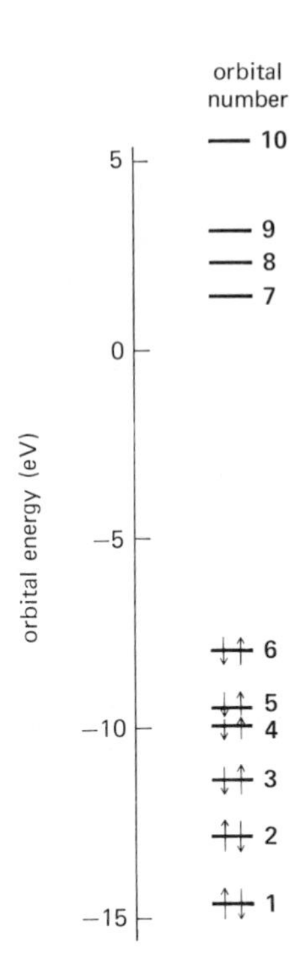

FIGURE 2-14 Calculated π molecular orbital energies for the ground state of adenine. The data were furnished by Dr. A. Pullman. See H. Berthod, C. Giessner-Prettre, and A. Pullman, *Theoret. Chim. Acta*, **5**, 53 (1966).

orbital in the ground state of the molecule are indicated. Each molecular orbital is specified by the coefficients of the atomic orbitals which comprise it. The excited states of the molecule can be approximated by promoting an electron from an occupied molecular orbital to an empty orbital. The energy of the molecule in each state is the sum of occupied orbital energies plus interaction terms. The interaction terms depend on the spin of the electrons; therefore, the excited state energies will depend on whether all electrons are paired (singlet states), or whether two electrons are unpaired (triplet states). The LCAO–SCF–MO method gives the best approximation for the ground state energy, but no variational procedure is carried out to optimize the excited states. Better approximations are obtained for the excited states if they are chosen as linear combinations of the calculated molecular orbitals. The coefficients of the molecular orbitals are varied to give the lowest (best) energies for the excited states. This improvement is called configuration interaction (CI) or configuration mixing (CM).

Berthod et al.[71] have applied this LCAO–SCF–MO–CI method to the nucleic acid bases. The calculated excited state energies are shown in Fig.

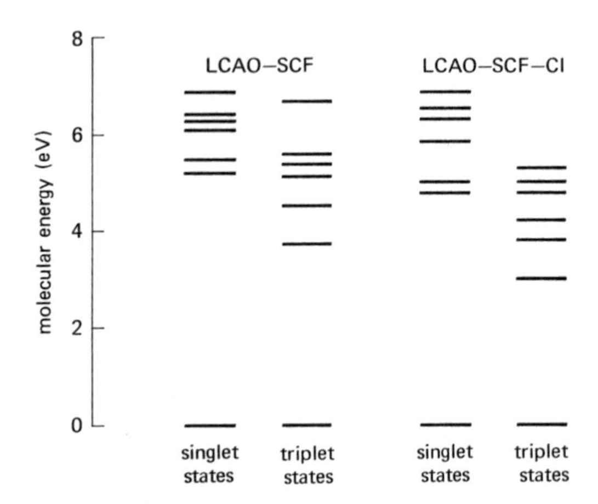

FIGURE 2-15 Calculated energies for π electron excited states of adenine. Linear combination of atomic orbitals self-consistent field calculations (LCAO-SCF) from H. Berthod, C. Giessner-Prettre, and A. Pullman, *Theoret. Chim. Acta,* **5**, 53 (1966); LCAO-SCF-configuration interaction calculations from the same authors, *Inter. J. Quantum Chem.*, **1**, 123 (1967).

2-15 and compared with those obtained before configuration interaction. There is a significant decrease in magnitude of the calculated energies of the lowest excited states; the variational principle tells us this is an improvement in these energies. The corresponding wave functions have been used to calculate electric dipole transition moment vectors. The magnitudes and transition energies for the singlet states are compared with experiment in Fig. 2-16. The agreement is encouraging in that the general trends are reproduced. Comparison of transition moment directions with experiment is more difficult. This is made clear in the discussion of the experimental transition moment directions a little later.

The calculated triplet energy levels have not been tested very thoroughly. Either the very weak absorption bands due to transitions from the ground singlet state to the excited triplet states would have to be measured, or else triplet–triplet absorption bands from the first excited triplet state would need to be observed. The position of the first excited triplet can be compared with experiment. As seen in Table 2-6 the agreement is poor; however, both experiment and calculation indicate that thymine has the lowest lying triplet state.

Experimental transition moment directions have been obtained from optical studies of crystals. As discussed earlier, this gives transition moment directions for the whole crystal. Published transition moments for molecules have been obtained by assuming the crystal was an oriented, dilute gas. That is, the properties of the crystal were assumed to be the sum of the isolated

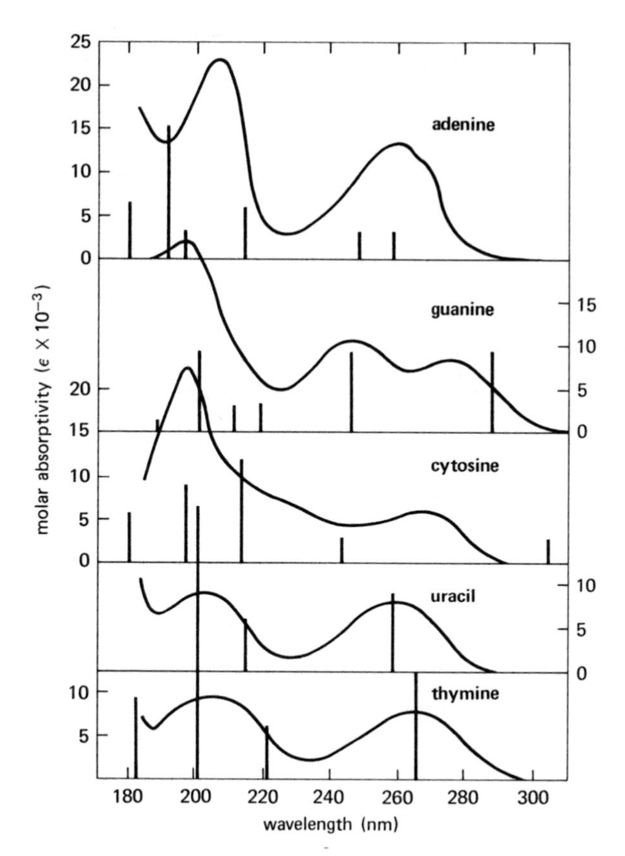

FIGURE 2-16 Comparison of experimental spectra with calculated transitions and magnitudes of their oscillator strengths. The length of each bar is proportional to the calculated oscillator strength. The experimental spectra are from D. Voet et al., *Biopolymers,* **1**, 193 (1963); the calculated values are from H. Berthod, C. Giessner-Prettre, and A. Pullman, *Inter. J. Quantum Chem.,* **1**, 123 (1967).

molecule properties. There is, therefore, some uncertainty in the experimental directions.[41]

Polarized fluorescence measurements on rigid, dilute solutions give relative transition directions. One essentially finds whether bands are mainly parallel or perpendicular to each other. This semiqualitative information can be relied upon. The main conclusion is that the first two bands for adenine, guanine, uracil, and thymine are perpendicular to each other, but for cytosine they are parallel to each other.[36] The results are summarized in Table 2-7. The directions are compared with calculated values in Fig. 2-17. The agreement is fair except for guanine. The conclusion is that more work is needed both experimentally and theoretically. Accurate data on transition

TABLE 2-6 CALCULATED AND EXPERIMENTAL LOWEST TRIPLET ENERGY LEVEL (RELATIVE TO GROUND STATE) FOR NUCLEIC ACID BASES

	Calculated frequency (cm^{-1})[a]	Experimental frequency (cm^{-1})[b]
Adenine	24,200	26,700
Guanine	20,970	27,200
Cytosine	18,550	27,900
Uracil	16,940	?
Thymine	16,130	26,300

[a] H. Berthod, C. Giessner-Prettre, and A. Pullman, *Inter. J. Quantum Chem.*, **1**, 123 (1967) and A. Denis and A. Pullman, *Theoret. Chim. Acta.*, 7, 110 (1967).
[b] See Table 2-5.

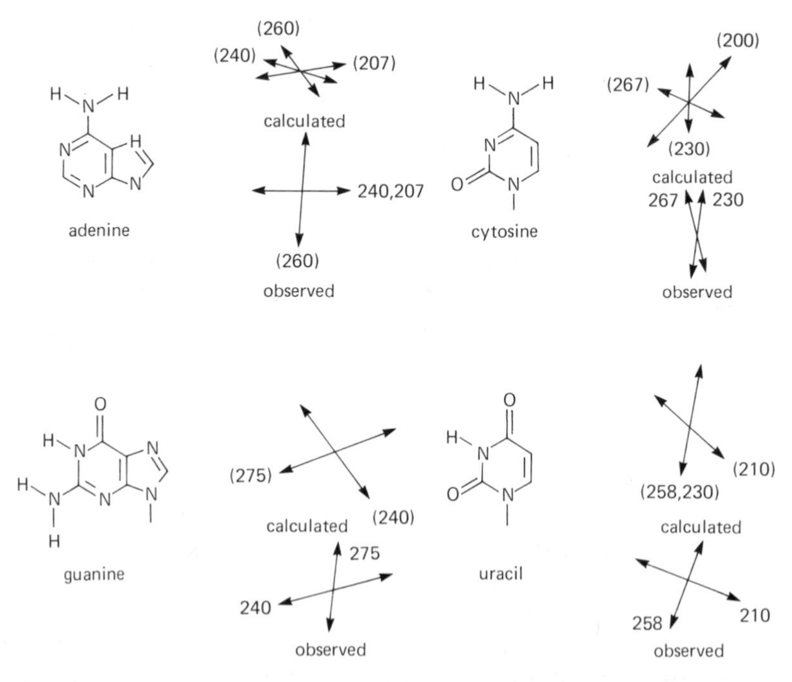

FIGURE 2-17 Comparison of experimental and calculated directions of transition electric dipole moments. References for the experimental data are given in Table 2-7; the calculated directions are from H. Berthod, C. Giessner-Prettre, and A. Pullman, *Inter. J. Quantum Chem.*, **1**, 123 (1967).

TABLE 2-7 EXPERIMENTAL DIRECTIONS OF NUCLEIC ACID BASE TRANSITION MOMENTS

Base	λ(nm)	Direction[a]	Comments
Adenine	260 240 207	$-3^\circ \pm 3^\circ$ $\sim 90^\circ$ $\sim 90^\circ$	The direction of the 260 nm band comes from studies of crystal 9-methyladenine (Refs. 1, 2).[b] The other directions are from polari fluorescence studies (Ref. 3).
Guanine	275 248	$-6^\circ \pm 4^\circ$ $-75^\circ \pm 4^\circ$	L. B. Clark, U. California, San Diego, Personal communication and Ref. 6.
Cytosine	267 230	$+12^\circ \pm 3^\circ$ $-5^\circ \pm 3^\circ$	The directions of the bands come from studies of crystalline cyto monohydrate and 1-methylcytosine (Ref. 4). Fluorescence studies 5-methylcytosine show the two bands are parallel to each other (Ref. 5).
Uracil and thymine	270 210	0° or $+7^\circ$ for uracil; -19° for thymine perpendicular to first band	The direction of the 270 nm band comes from studies of crystal 1-methylthymine (Ref. 1) and 1-methyluracil (Ref. 6). Polarized fluoresce and absorption shows the next band is perpendicular to the first (Refs. 4–6)

[a] The angle is measured relative to the C4–C5 axis in purines, or the N1–C4 axis in pyrimidines. Positive angles are measured counterclockwise w the bases are oriented as shown in Fig 2-1.

[b] References:

1 R. F. Stewart and N. Davidson, *J. Chem. Phys.*, **39**, 255 (1963).
2 R. F. Stewart and L. H. Jensen, *J. Chem. Phys.*, **40**, 2071 (1964).
3 P. R. Callis, E. J. Rosa, and W. T. Simpson, *J. Amer. Chem. Soc.*, **86**, 2292 (1964).
4 P. R. Callis and W. T. Simpson, *J. Amer. Chem. Soc.*, **92**, 3593 (1970); T. P. Lewis and W. A. Eaton, *J. Amer. Chem. Soc.*, **93**, 2054 (1971).
5 P. R. Callis, Ph.D. Thesis, University of Washington (1965).
6 W. A. Eaton and T. P. Lewis, *J. Chem. Phys.*, **53**, 2164 (1970).
7 P. R. Callis, B. Fanconi and W. T. Simpson, *J. Amer. Chem. Soc.*, **93**, 6679 (1971).

moments of the bases are essential to an understanding of the optical properties of the polynucleotides. Data on transitions in the sugar and phosphate groups would also be useful, but they are less critical.

Ionization energies / The energy required to remove completely the first electron from a base, the ionization energy, is of interest. The London interaction is dependent on this energy and in general the ionization energy can characterize the electron mobility of a molecule and therefore some of its chemistry. The ionization energy of some bases in the gas phase has been measured with a mass spectrometer.[72] The ion current of a base was measured as a function of ionizing voltage; the ionizing voltage was calibrated by use of a rare gas of known ionization energy. The results are shown in Table 2-8 and compared with the highest filled π molecular orbital calculated by A. Pullman et al.[71,73] The relative order of the bases is calculated correctly; guanine is calculated to be the most easily ionized, and uracil is calculated and measured to be the most difficult to ionize. The lowest empty π molecular orbital is related to the electron affinity of an aromatic molecule. The electron affinity is the energy needed to add an electron to a molecule to produce a negative ion. The calculated results predict that it is relatively easy to add an electron to uracil or thymine, but difficult to add an electron to adenine or guanine. There are no suitable experimental results for comparison.

It is well to emphasize the very approximate nature of all the wave functions which have been used to calculate permanent dipole moments, transition dipole moments, ionization energies, etc. Undoubtedly the wave functions will improve as computers get faster and bigger. However, the published calculations have been useful and so it is worthwhile to discuss them.

D. Infrared Spectroscopy and Raman Scattering

1. INFRARED ABSORPTION

The infrared absorption in the region of 1500 cm^{-1} to 1750 cm^{-1} (double bond stretching) of four mononucleotides is given in Fig. 2-18. The nucleotides are dissolved in D_2O at neutral pH; the strong absorption of H_2O in this region precludes its use as a solvent. It is also useful to note that the molar absorptivity is a factor of ten lower than in the ultraviolet region. Infrared spectra have been used to assign the tautomeric forms of the bases in aqueous solutions.[74,75] Comparison with model compounds in which mobile protons have been replaced with methyl groups shows that the bases are indeed in the keto and amino form (as shown in Fig. 2-1) in water. By comparing the effect of acid and base on these spectra with other model compounds, the sites of protonation and ionization can be determined. This

TABLE 2-8 IONIZATION ENERGIES AND ELECTRON AFFINITIES FOR THE NUCLEIC ACID BASES

Base	Measured[a] ionization energy (eV)	Calculated[b] ionization energy (eV)	Calculated[b] electron affinity (eV)
Adenine	8.9 ± 0.1	7.92	1.52
Guanine	?	7.59	1.47
Cytosine	8.9 ± 0.2	8.16	0.87
Uracil	9.8 ± 0.2	9.15	0.38
Thymine	9.4 ± 0.1	8.80	0.41

[a] C. Lifschitz, E. D. Bergmann, and B. Pullman, *Tetrahed. Lett.*, **46**, 4583 (1967).
[b] H. Berthod, C. Giessner-Prettre, and A. Pullman, *Theoret. Chim. Acta*, **5**, 56 (1966). A. Denis and A. Pullman, *Theoret. Chim. Acta*, **7**, 110 (1967).

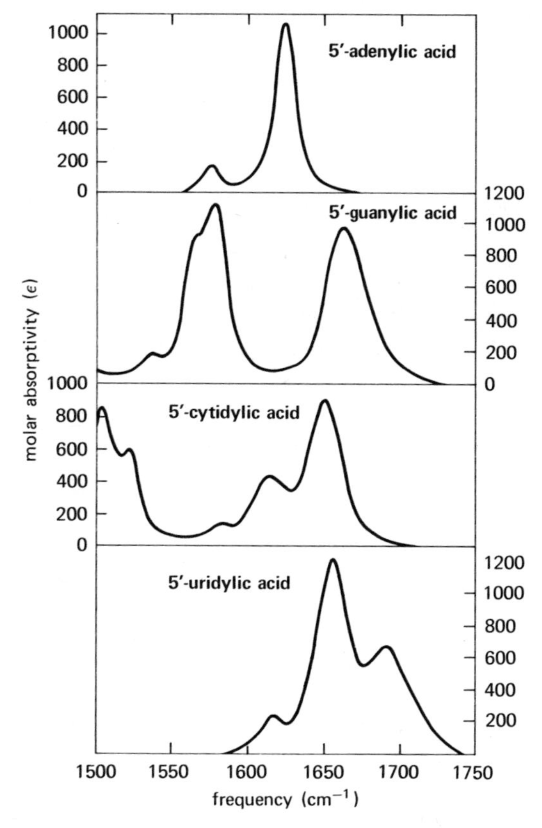

FIGURE 2-18 Infrared absorption spectra of four monoribonucleotides in neutral D_2O solution. [Data reprinted with permission from G. J. Thomas, *Biopolymers,* 7, 325 (1969). © 1969 by John Wiley & Sons, Inc.].

will be discussed in the section on p*K* values; here it is sufficient to state that protonation and ionization occur first on the ring nitrogens, not the amino groups.

The infrared data can obviously be used for analytical purposes. However, its main use in nucleic acid research follows from the fact that shifts occur in frequencies and molar absorptivities when two bases pair. This allows study of double strand content in polynucleotides.[74,76,77]

Infrared studies of base pairing in nonaqueous solvents such as chloroform have been made using the individual bases.[78,79] The absence of water allows significant hydrogen-bonded complex formation to occur. It was found that complex formation between each adenine and uracil derivative was always larger than the self-association of either molecule. It is very difficult to extrapolate the quantitative findings to aqueous solutions.

The infrared spectra of the sugar and phosphate groups of nucleotides are extremely difficult to study in either H_2O or D_2O. Infrared study of fibers of polynucleotides has been useful in defining the orientation of the phosphate groups in helical conformations, as will be discussed in Chapter 4.

It should be obvious that all the optical properties discussed for the ultraviolet region could in principle be measured in the infrared. In fact, they have not; the experimental difficulties will be very great.

2. RAMAN SCATTERING

The vibrational Raman scattering of ribonucleic acid derivatives as a function of pH and pD in H_2O and D_2O has been studied by Lord and Thomas.[80] An advantage this method has over infrared measurements is that H_2O can be used in the interesting frequency range of 3200–200 cm^{-1}. However, as in infrared, frequency and intensity changes can be seen when the bases ionize or base-pair. Conclusions about tautomeric forms of the bases and sites of ionization are consistent with the infrared data. Studies on polynucleotides have been hampered by the large amount of Rayleigh scattered light; laser light sources simplify the elimination of this light and now allow useful measurements.[81]

3. THEORY

The theory of infrared absorption due to vibrational transitions is of course exactly analogous to ultraviolet absorption. Vibrational energy levels correspond to stationary states of vibration of the molecule. For some molecules a vibration can be assigned to a single group in the molecule such as a particular double bond. For the nucleic acid bases the observed vibrations are not very localized and are assigned generally to conjugated carbonyls or to ring vibrations. However, empirically, hydrogen bonding at a known site can be correlated with an observed change in infrared spectrum. The intensity of absorption depends on the electric dipole transition moment as before. To get an explicit expression for the infrared transition moment one makes the Born-Oppenheimer approximation. The molecular wave function is written as a product of an electronic function dependent on electronic coordinates (r) and a vibrational function dependent on nuclear coordinates (q).

$$\psi(r, q) = \psi_0(r)\, \phi_{0v}(q)$$

$\psi_0(r)$ is the electronic wave function in the ground electronic state
$\phi_{0v}(q)$ is the vibrational wave function in the v vibrational state of the ground electronic state

The dipole moment operator is expanded around the equilibrium nuclear positions (q_0).

$$\boldsymbol{\mu}(r, q) = \boldsymbol{\mu}(r) + \frac{\partial\, \boldsymbol{\mu}(r)}{\partial\, q}(q - q_0) + \ldots \tag{2-37}$$

The transition moment for a pure vibrational transition from v to v' is

$$\begin{aligned} \boldsymbol{\mu}_{vv'} &= \int \psi_0 \frac{\partial\, \boldsymbol{\mu}}{\partial\, q} \psi_0 d\tau \int \phi_{0v}\, q\, \phi_{0v'}\, d\tau \\ &= \frac{\partial\, \boldsymbol{\mu}_{00}}{dq} \int \phi_{0v}\, q\, \phi_{0v'}\, d\tau \end{aligned} \tag{2-38}$$

This equation shows that for vibrational absorption there must be a change in permanent dipole moment of the molecule with the nuclear motion ($\partial\boldsymbol{\mu}_{00}/\partial q$).

Raman scattering is the inelastic scattering of light by molecules. The frequency difference between the incident light and the scattered light corresponds to energy levels within the molecule. These levels can be translational, rotational, vibrational, or electronic, but the vibrational Raman scattering is the most commonly and easily studied.

The intensity of inelastic scattering of light which results in a transition in the molecule from state O to A can be written as[82]

$$I(\text{scattered}) = \frac{128\pi^5}{9c^4}(\nu + \nu_{\text{OA}})^4\, I_0 \sum_{i,j=1}^{3} (\alpha_{ij})_{\text{OA}}^2 \tag{2-39}$$

I_0 is the intensity of the incident light of frequency ν
$\nu + \nu_{\text{OA}}$ is the frequency of scattered light

The scattering is proportional to the square of the transition polarizability tensor $(\alpha)_{\text{OA}}$, which is characterized by its components α_{ij}. This is exactly analogous to the absorption being proportional to the square of the transition dipole vector. Components of the transition polarizability tensor are

$$(\alpha_{ij})_{\text{OA}} = \frac{1}{h}\sum_{\text{B}}\left[\frac{(\mu_i)_{\text{AB}}(\mu_j)_{\text{OB}}}{\nu_{\text{OB}} - \nu} + \frac{(\mu_i)_{\text{OB}}(\mu_j)_{\text{AB}}}{\nu_{\text{AB}} + \nu}\right] \tag{2-40}$$

α_{ij} is one of the nine components of the polarizability tensor
μ_i and μ_j are one of the three components of the dipole vector
$_{AB}$, $_{OB}$ are transition dipoles defined in Eqs. (2-4) or (2-34)

If state O and state A are the same, then Eq. (2-39) gives the intensity of Rayleigh scattered light and Eq. (2-4) gives the components of the polarizability tensor defined in Eq. (2-33).

To get an explicit expression for the vibrational Raman effect, one can proceed as in the treatment of the infrared absorption. If the vibrational transition moment of Eq. (2-38) is used, one finds that the scattering depends on the change of the electronic polarizability with nuclear motion. For simple molecules one often finds that a particular vibration is allowed in infrared absorption, but forbidden in Raman scattering, and vice versa. However, for the nucleic acid bases the nuclear motions generally change both the dipole moment and the polarizability. This means each transition should be seen both in Raman and infrared absorption.

E. Magnetic Resonance

1. NUCLEAR MAGNETIC RESONANCE

Every nucleus with a magnetic moment can absorb radiation at characteristic frequencies in a static magnetic field. The characteristic frequency is directly proportional to the magnetic field at the nucleus; it is therefore sensitive to the chemical environment of each nucleus. As relative absorption frequencies can be measured to 1 part in 10^8 or better, NMR is a very powerful measure of structure.

Any isotope which occurs in nucleic acids and is magnetic is potentially useful. This includes ^1H, ^2H, ^{13}C, ^{14}N, ^{15}N, ^{17}O, and ^{31}P. (see Table 2-9). In addition, the cations present as counterions such as Li^+, Na^+, Tl^+, Mg^{2+}, Ca^{2+} all have magnetic isotopes whose NMR can be utilized. Most of the work so far has been with proton magnetic resonance; it has the highest sensitivity. Furthermore, because of its spin of 1/2, the spectrum obtained is the simplest possible. Other nuclei with spin 1/2 include ^{13}C, ^{15}N, ^{31}P, and ^{205}Tl. The thallium nucleus may turn out to be particularly useful, because it is similar to K^+ in its chemistry and has a high natural abundance and NMR sensitivity.[83]

Figure 2-19 shows the proton magnetic resonance (PMR) spectrum of four nucleosides in D_2O; only protons attached to carbon remain. A volume of about 0.5 ml and a concentration of at least 0.01 are needed for routine measurements; however, newer techniques can reduce the amount needed by a factor of 10. Each proton or group of identical protons (such as the CH_3 in thymine) is characterized by an absorption band. The absorption positions in the figure are given as chemical shifts relative to an external standard of

TABLE 2-9 MAGNETIC NUCLEI OCCURRING IN NUCLEIC ACIDS[a]

Nucleus	Natural abundance (%)	Relative sensitivity for equal no. of nuclei at const. field	Relative NMR frequency at const. field	Spin
^{1}H	100	1	100,000	1/2
^{2}H	0.02	1×10^{-2}	15,350	1
^{13}C	1	2×10^{-2}	25,140	1/2
^{14}N	100	1×10^{-3}	7,220	1
^{15}N	0.4	1×10^{-3}	10,130	1/2
^{17}O	0.07	3×10^{-2}	13,560	5/2
^{31}P	100	7×10^{-2}	40,480	1/2

[a] From J. A. Pople, W. G. Schneider, and H. J. Bernstein, *High-Resolution Nuclear Magnetic Resonance*, McGraw-Hill, New York (1959).

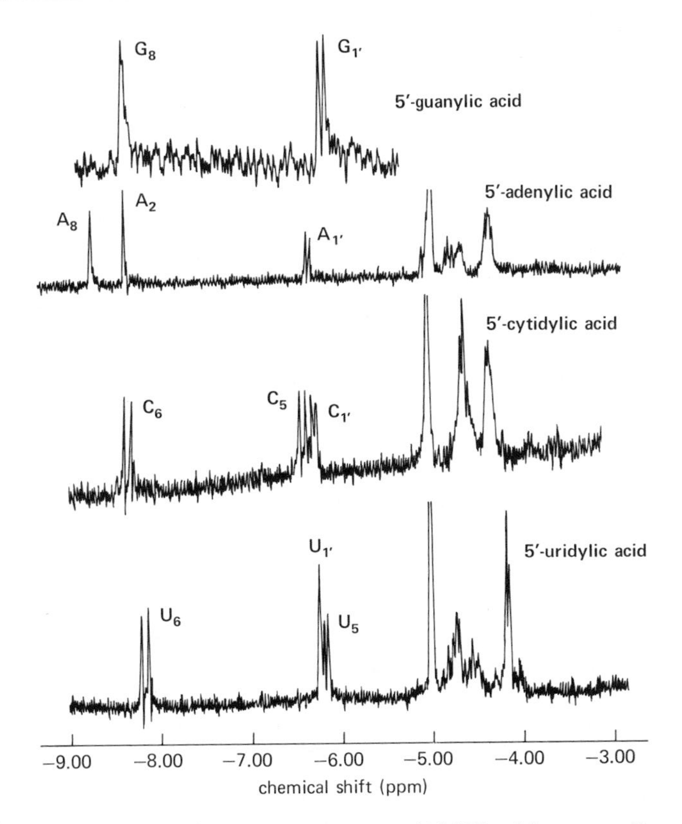

FIGURE 2-19 Proton magnetic resonance spectra at 100 MHz of four monoribonucleotides in neutral D_2O solution at 33°C. The chemical shift is expressed in parts per million (ppm) from hexamethyldisiloxane. [From Fig. 1 of I. C. P. Smith, T. Yamane, and R. G. Shulman. Reproduced by permission of the National Research Council of Canada from the *Canad. J. Biochem.*, **47**, 480 (1969).]

hexamethyldisiloxane. The chemical shift δ is defined in terms of the magnetic field H required for absorption at constant frequency or in terms of the absorption frequency ν at constant magnetic field.

$$\delta \equiv \frac{H(\text{reference}) - H(\text{sample})}{H(\text{reference})} \times 10^6$$

$$\delta \equiv \frac{\nu(\text{sample}) - \nu(\text{reference})}{\nu(\text{reference})} \times 10^6$$

The chemical shift is measured relative to some standard substance characterized by H(reference) or ν(reference). The magnitude of δ is given in parts per million (ppm) and shifts of 1–10 ppm are common. Figure 2-20 gives chemical shifts for four ribonucleosides and one ribonucleotide extrapolated to infinite dilution in neutral D_2O. The base protons and the H1′ of the ribose are well separated from the residual HDO protons in the water; they are given in the figure. The original assignments were made by Jardetsky and Jardetsky.[7] The H2′, H3′, and H4′ of the ribose occur close to HDO and to each other. They have been assigned in adenine nucleotides, with H2′ having the largest shift and H4′ the smallest.[84] The H5′ protons show the least shift of all.

Some of the proton bands are split into two or more peaks by interaction with neighboring protons. Obvious doublets in Fig. 2-19 are H5 split by H6 and H6 split by H5 in uridine and cytidine. The ribose H1′ split by H2′ and H5′ split by H4′ are also visible in all the spectra. The H2, H8 of adenine and H8 of guanine are not split, but the H2′, H3′, and H4′ are split into many components. The band of a proton (or group of protons) which interacts with n identical protons is split into $n + 1$ bands. The magnitude of the splitting is characterized by J, the spin–spin coupling constant in hertz. From the definition of δ it is clear that the separation in hertz between protons of different chemical shift will be directly proportional to the magnetic field. The splitting J, however, is independent of the field; it depends solely on the interaction between nuclei. This interaction is mainly transmitted through the bonds and falls off rapidly with distance. Usually only protons on adjacent nuclei give significant splittings. That is why H1′ is split by H2′ but not by H8, H6, or H3′.

Nuclear magnetic resonance has been used to study the conformation of nucleosides and nucleotides.[20,85–88] Addition of a phosphate to the 5′ hydroxyl of a nucleoside causes a shift in the H8 proton of adenine and guanine, but does not shift the H2 proton. This is expected for an *anti* conformation which places the H8 close to the 5′ phosphate and the H2 far away (see Fig. 2-21). Evidence that 5′ uridylic and cytidylic acid are *anti* is that the H6 of uridine and cytidine are significantly shifted by forming the 5′ phosphates. The shifts are in the expected direction for the negative

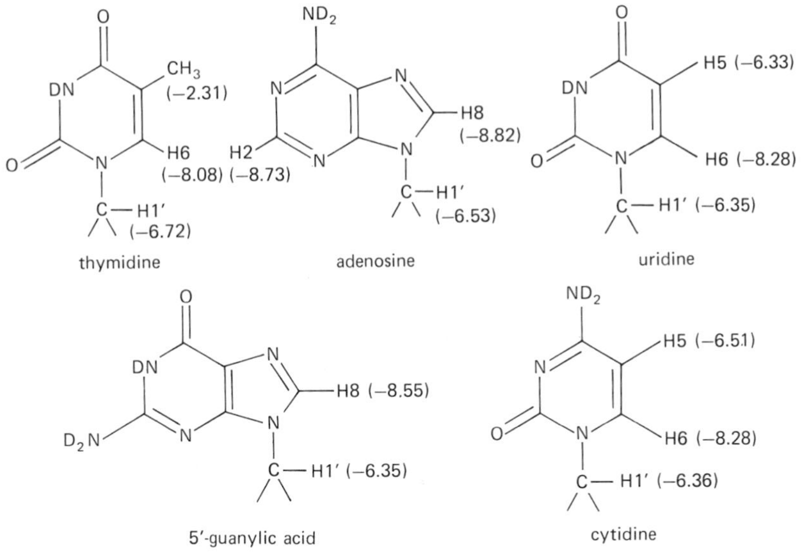

FIGURE 2-20 Proton chemical shifts (in parts per million from an external tetramethylsilane capillary) for four nucleosides and one nucleoside at infinite dilution in neutral D_2O at 30–35°C. Data for the nucleosides are from P. O. P. Ts'o, M. P. Schweizer, and D. P. Hollis, *Ann. New York Acad. Sci.*, 1969; data for the nucleotide are from P. O. P. Ts'o et al., *Biochemistry*, 8, 997 (1969).

phosphate and increase as the charge on the phosphate is changed from zero $(ROPO_3H_2)$ to one $(ROPO_3H)^-$ to two $(ROPO_3)^{2-}$. To ensure that the effect of the phosphate is mainly coulombic and a good measure of the distance between the phosphate and the effected proton, 2′ and 3′ phosphates were also studied. The shifts of H8 and H2 of the purines caused by 2′ or 3′ phosphate groups were small as predictable from Fig. 2-21. Furthermore, the effect of the 5′ phosphate was larger on H6 than on H5 of pyrimidines as predicted by a coulombic model. Similar results were obtained with both deoxyribo- and ribonucleosides.

The conclusion that the bases of nucleosides or nucleotides prefer the *anti* conformation in aqueous solution is strong. However, there is also good evidence that the base can exist in a range of conformations including *syn*. High concentrations of salts can change the average conformation of uridine and uridylic acid.[20] Double resonance experiments[21] can give information about distances between protons. One proton is irradiated while one looks for an enhancement in absorption of another. Using this technique Hart and Davis[21] conclude that the *anti* range is less favored by adenosine than by guanosine.

The splitting of the H1′ by the H2′ proton can be related to the

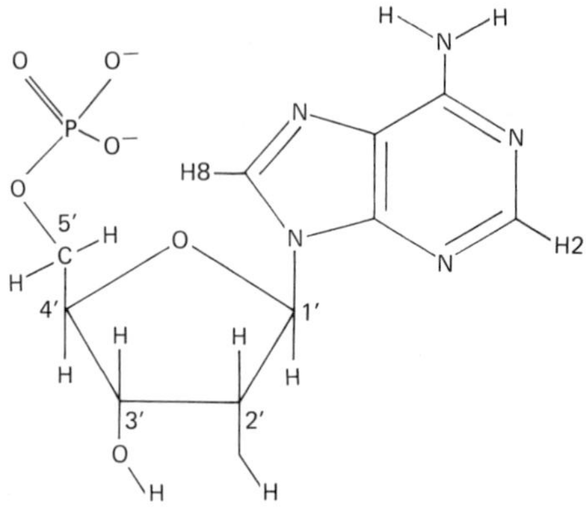

FIGURE 2-21 The *anti* conformation of 5′-adenylic acid at neutral pH. Note that H8 is near the 5′ phosphate, but that 2′ or 3′ phosphates cannot approach very near to H8. [See M. P. Schweizer et al., *J. Amer. Chem. Soc.*, **90**, 1042 (1968).]

conformation of the ribose ring.[87] The experimental coupling constants $J_{H1'-H2'}$ are: 5–6 Hz for purine ribonucleosides and ribonucleotides; 3–5 Hz for pyrimidine ribonucleosides and ribonucleotides; about 7 Hz for all deoxyribonucleosides and deoxyribonucleotides. Karplus[89] calculated the coupling constant expected for the H—C—C—H system as a function of the dihedral angle. Jardetzky and Jardetzky[7] used the results to predict a coupling constant of 6.9 Hz for 2′-*endo* and 1.7 Hz for 3′-*endo* (see Fig. 2-3). One might conclude that the purine ribosyl derivatives and the deoxyribosyl derivatives are in the 2′-*endo* conformation, but that the pyrimidine ribosyl derivatives tend to 3′-*endo*. More theoretical work is needed to calculate coupling constants and more experimental work on $J_{H2'-H3'}$, $J_{H3'-H4'}$ should be obtained before definite conclusions on ribose ring conformation are decided.

The splitting of H5′, H5″ by P in 5′-nucleotides has been interpreted as indicating that the phosphorus is mainly *trans* to C4′.[87,90] Compounds of known conformation were used to calibrate the coupling constant. This conclusion is in good agreement with the crystal structure data, as illustrated in Fig. 2-5.

2. ELECTRON SPIN RESONANCE

Nucleic acids have all electrons paired, as do most stable molecules; therefore, they do not have electron spin resonance signals. However, nucleic acid structure has been probed by introducing unpaired electrons. Metal ions which have incompletely filled *d* orbitals (transition metal ions) have unpaired electrons. These include the biologically interesting Fe^{2+}, Fe^{3+}, Mn^{2+}, Mn^{3+}, Co^{2+}, Co^{3+}, and Cu^{2+}. Organic molecules which *are* stable with unpaired electrons (free radicals) can be bound to nucleic acids. Finally, unpaired electrons can be induced in any molecule by irradiation with ultraviolet light or other high energy particles.

Shulman et al.[52,54] have studied the ESR of the nucleic acid bases, nucleosides, and nucleotides irradiated with ultraviolet light. For these molecules in an ethylene glycol glass at 77°K the light produces triplet states (two unpaired electrons) which last for about a second. At room temperature no triplets are seen. The electron spin signal is analogous to an NMR signal except the absorption bands are very broad and occur at a very different magnetic field and frequency range. Each base has a characteristic spectrum and lifetime. The study of triplet states by ESR complements their study by phosphorescence measurements (see Section IIC, 4). Two interesting facts have emerged from these studies. Thymine and thymine derivatives do not show an ESR signal when irradiated at low concentrations in neutral glasses. However, a thymine aggregate or thymine in the presence of sensitizers such as acetone does show an ESR signal. This means the triplet state of thymine can be formed by transfer of excitation from another triplet state, but not from its own singlet.[54] Furthermore, the ESR signal of irradiated DNA is identical with that of sensitized thymidylic acid. This is reasonable as thymine does have the lowest energy triplet and triplet energy transfer is efficient to thymine. However, before the sensitized thymine ESR was found, it was very puzzling to have the DNA triplet ESR signal unlike the ESR signal from the three bases which formed triplets.

The positive ion radical of chlorpromazine has been shown to intercalate between the bases in DNA from the changes in its ESR signal.[91] Nitroxide radicals of the form (R,R′)NO are quite stable in water and have been used to probe protein structure[92]; however, as yet little work has been done on nucleic acids. Chapter 10 discusses some of the work on transfer RNA.

3. NUCLEAR QUADRUPOLE RESONANCE

Just as nuclei with magnetic dipoles can orient in a magnetic field, so can nuclei with electric quadrupoles orient in an electric field gradient. The electric field gradient is produced by the electron distribution near the nucleus; therefore, NQR spectra can give information about electron distribution in molecules. All nuclei with spin greater than 1/2 have a quadrupole moment. The naturally abundant nitrogen isotope (^{14}N) has a quadrupole moment. The ^{14}N NQR from the nucleic acid bases would furnish useful tests of electronic wave functions for these molecules. However, the experimental difficulties are such that so far only simpler molecules such as pyridine have been studied.[93] With present methods of measurement, only pure materials can be studied. Aqueous solutions of bases, nucleosides, and nucleotides are as yet out of reach of the method.

4. THEORY[94]

The magnetic moment of a nucleus is a vector which is usually written as:

$$_{N} = g_N \beta_N \ \mathbf{I}$$

$$\beta_N = \text{nuclear magneton} = \frac{eh}{4\pi c M_{\text{proton}}}$$

Each nucleus is characterized by its spin **I** (an integral multiple of 1/2) and its nuclear g factor which is a unitless parameter defined by this equation. The magnetic moment of an electron is written as:

$$\boldsymbol{\mu}_e = -g\beta \ \mathbf{S}$$

$$\beta = \text{magneton} = \frac{eh}{4\pi c M_{\text{electron}}}$$

The spin **S** of an electron is 1/2 and the g value depends on the environment of the electron. In a magnetic field each magnetic dipole can orient and absorb energy from incident electromagnetic radiation. For the usual magnetic fields of 10–50 kG, nuclei absorb in the radiofrequency (FM) range and electrons absorb in the radar (microwave) range. For resonance absorption $h\nu = g\beta H$ for electrons and $h\nu = g_N \beta_N H$ for nuclei, where H is the magnetic field. The magnetic field interacting with the magnetic moment is not the external magnetic field applied. This fact makes NMR and ESR chemically useful.

For NMR measurements the magnetic field at a particular nucleus is characterized by the chemical shift δ, defined earlier. It is related to the theoretically calculable shielding factor σ

$$\delta = [\sigma(\text{reference}) - \sigma(\text{sample})] \times 10^6$$

where σ measures the difference between the applied magnetic field and the magnetic field at the nucleus.

$$\sigma = \frac{H(\text{applied}) - H(\text{nucleus})}{H(\text{applied})}$$

It is clear that a nucleus which is less shielded than the reference nucleus has a positive chemical shift.

Formulas exist for calculating the chemical shift at a nucleus from the wave function of the electrons near the nucleus. However, it is probably more useful to have a qualitative understanding of the factors involved. We will discuss only the hydrogen nucleus.

The most important factor for protons is simply the electron density in its 1s orbital; this is the local diamagnetic effect. A higher electron density leads to a larger shielding, an upfield shift for the magnetic field and thus a less positive δ. Any influence which increases the electron density at a proton will cause an upfield shift. Conversely, any influence which decreases the electron density will decrease the shielding, cause a downfield shift, and an increase in δ. These simple considerations are useful in assigning NMR peaks to protons and interpreting NMR spectra. A pertinent example is the downfield shift of the H8 in adenine caused by increasing the negative charge on the phosphate of 5′-adenylic acid from zero to one to two. Coulombic repulsion decreases the shielding at H8.

Another important factor is the anisotropy of the magnetic susceptibility of each group near the proton of interest. The magnetic susceptibility is the magnetic analog of the electrical polarizability. It characterizes the magnetic dipole induced by a magnetic field. In general the magnetic susceptibility is a tensor with three different principal values. To simplify the discussion we will consider a group which has cylindrical symmetry. Assuming point magnetic dipole interactions, one obtains the following equation for the magnetic anisotropy effect.

$$\Delta\sigma = \frac{1}{3R^3}(1 - 3\cos^2\theta)(\chi_{\parallel} - \chi_{\perp})$$

$\chi_{\parallel}, \chi_{\perp}$ are magnetic susceptibilities (which have negative signs) parallel and perpendicular to the symmetry axis of the group

R is the distance between the proton and the center of charge of group

θ is the angle between the symmetry axis of the group and the vector from the group to the proton

A special case of magnetic anisotropy which is usually considered separately is the "ring current" effect in benzene, adenine, and other aromatic molecules. Because of the relative freedom of π electrons in the plane of the molecule, aromatic molecules have a much larger magnetic susceptibility out of plane than in plane. For benzene $|\chi_{\parallel}| > |\chi_{\perp}|$; the symmetry axis is perpendicular to the plane of the ring. One sees that for protons in the plane of the ring θ is 90°, $(1-3\cos^2\theta)$ is one and $\Delta\sigma$ is negative. For protons above or below the center of the ring θ is 0°, $(1-3\cos^2\theta)$ is -2 and $\Delta\sigma$ is positive. The shifts can be large; the maximum values for benzene range from -2 ppm in plane to +4 ppm out of plane. Ring current shifts have been very useful in applying NMR to the conformation of dinucleoside phosphates involving adenine and guanine. These bases are sufficiently aromatic to produce large shifts for protons of bases stacked above them. Uracil and cytosine do not show these ring current effects.

Ts'o and co-workers[95] have made extensive calculations of chemical shifts for all the protons on the nucleic acid bases. They are in reasonable agreement with experiment and help provide tests of calculations of wave functions for bases.

Many peaks in an NMR spectrum occur as groups of lines called multiplets. These multiplets are caused by spin–spin splittings due to interactions of magnetic nuclei. For protons the experimental chemical shifts (δ) are of the order of 10 ppm. This means 100 Hz shifts for NMR at 100 MHz. Interaction of protons with other magnetic nuclei causes spin–spin splitting (J) of the peaks by about 1–10 Hz. Complete analysis of a complex NMR spectrum is in general difficult, but when differences in chemical shifts are large compared to spin–spin splittings ($\nu_A - \nu_B \geq 10\, J_{AB}$), the analysis is simple. First-order theory can be used and the effects of chemical shifts and spin–spin splittings are additive.

The spin–spin interaction leading to the splitting depends on the magnetic dipoles of the two nuclei and is transmitted through the electrons. That is, the nuclear magnetic moment polarizes the spins (magnetic moments) of its surrounding electrons. These couple to neighboring electrons which in turn polarize another nucleus. The value of the spin–spin coupling constant J depends on the number and type of bonds between the nuclei, the hybridization, the dihedral angles, and the electronegativity of substituents. Proton–proton coupling through more than three single bonds is usually not observable. For H_A–C–C–C–H_B, J_{AB} is usually less than 0.05 Hz. There are exceptions to this rule and when double or triple bonds are involved the coupling strength may be much higher.

A very useful calculation of coupling constants for the system H_A–C_A–C_B–H_B has been made by Karplus.[89] He used an approximate valence bond treatment for this ethane fragment to obtain the coupling constant as a function of the dihedral angle ϕ.

$$J = C_0 + C_1 \cos \phi + C_2 \cos 2\phi$$

ϕ is the angle between the H_A–C_A–C_B plane and the C_A–C_B–H_B plane

The calculated parameters were $C_0 = 4$, $C_1 = -0.5$, $C_2 = 4.5$ Hz. He pointed out that these parameters would depend on the other substituents on the carbons. This equation has been very useful in structure determination and the parameters have been determined empirically for certain classes of molecules. The conformations of the ribose ring in nucleosides have been investigated primarily through the H1′–H2′ coupling constants.

Sometimes it is sufficient to know the total number of protons contributing to each peak and the number of protons which cause the splitting into multiplets. The area of each peak, or sum of multiplet area, is proportional to the number of magnetically equivalent protons of this peak. Interactions among magnetically equivalent protons leads to no splitting. However, if one or more magnetically equivalent protons interact with another group of n equivalent protons, the peak will split into a multiplet of $n + 1$ lines. The $n + 1$ lines correspond to the $n + 1$ possible values for the total magnetic moment of n nuclei. The probability of each magnetic moment for the group is given by one of the coefficients of the binomial expansion. Thus interaction with one nucleus leads to a doublet with intensity ratio 1:1. Interaction with two magnetically equivalent nuclei leads to a triplet with intensity of 1:2:1. Three nuclei produce 1:3:3:1, etc. It should be reemphasized that all this discussion is based on spin–spin couplings being small compared to chemical shift differences between the groups of protons. As the chemical shift differences get smaller, the distribution of intensities in the multiplet becomes distorted and finally the first-order theory cannot be used. As the chemical shift frequencies are directly proportional to the magnetic field, one can in principle use higher fields to regain the simple first-order effects.

The intensity of absorption per proton depends on the transition probability at the appropriate frequency just as in optical spectroscopy. However, a new factor occurs in NMR which is not important in optical spectroscopy. The intensity of absorption also depends on the difference in the number of protons aligned with the field and against the field. In optical spectroscopy the number of molecules in the ground state is much larger than in the excited state and the number is essentially independent of light intensity. In NMR the number in the two states is nearly equal at equilibrium in the magnetic field and the ratio approaches 1 as the radiation intensity increases; saturation occurs. Saturation is opposed by processes which tend to restore equilibrium. The rate of approach to equilibrium can be determined by measuring the intensity needed to saturate the absorption. A more direct method is to measure the rate of attainment of equilibrium after a short pulse of radiation is applied. The kinetics are first order and lead to a characteristic time, T_1 for each proton. This time is called the spin-lattice relaxation time or longitudinal relaxation time. Another characteristic time for each peak is the transverse relaxation time, T_2. It is a measure of the width of each absorption peak. These times can be related to such factors as the presence of paramagnetic ions in the system and to the freedom of motion of the group to which the protons are attached. Both factors have been studied in nucleic acid systems; it is sufficient to mention here that the slower the motion of a proton due to translation, rotation, and vibration of its group, the wider the proton absorption band will be. For protons on rigid macromolecules like native DNA, the peaks are so broad no absorption is visible.

Nuclear magnetic resonance is also useful in studying rates of reactions. A chemical change of one species to another will cause some NMR lines to decrease in intensity as others increase. Of higher potential usefulness in nucleic acids is the ability to study rates of reversible reactions at equilibrium. For example, keto–enol tautomerisms and interconversion of different conformations of cyclic compounds have been studied.[95] *Syn* and *anti* or *exo* and *endo* conversions might be studied in mononucleotides. Stacking and unstacking might be monitored in dinucleotides. Double strand to single strand transitions could be followed in polynucleotides. Very little research of this type has been done yet.

Consider a proton which can exist in two sites either by direct exchange of the proton or by a change in conformation of the molecule. If the rate of exchange between sites is fast, only one NMR peak is seen; if the rate of exchange is slow, two peaks are seen. Fast exchange means that the frequency of exchange is larger than the frequency separation between the two proton peaks. This frequency is of order of magnitude of 100 Hz for a 10 ppm chemical shift; that is, times of 0.01 second. The position of the peak is the weighted average of the peaks for the two sites *A* and *B*.

$$\bar{\omega} = \omega_A f_A + \omega_B f_B$$

The angular frequencies at the sites are ω_A and ω_B; the fraction of time spent on each site is f_A and f_B. The proton peaks seen in nucleotides so far have all corresponded to the fast-exchange model. Apparently conformational changes at room temperature are rapid. For slow exchange two peaks would be seen with widths dependent on the rate of exchange.

For a quantitative discussion of the material in this section the reader is referred to the books mentioned earlier.[94]

III THERMODYNAMIC PROPERTIES

A. Introduction

Thermodynamics tells us about the properties and interactions of many molecules. We can study the interactions of the nucleic acid components with themselves, with the solvent, and with other molecules. We would like to understand the interactions of all nucleic acids in terms of the forces between the few components.

Good general textbooks include *Thermodynamics* by Lewis and Randall[96] and *An Introduction to Statistical Thermodynamics* by Hill.[97]

B. Interactions in Water

The thermodynamics of the bases, nucleosides, and nucleotides in aqueous solutions characterize their interactions with water, with themselves, and with other solutes. Consider the formation of a complex between two species

$$\mathrm{A} + \mathrm{B} \rightleftharpoons \mathrm{A} \cdot \mathrm{B}$$

The equilibrium constant is

$$K = \frac{(a_{AB}^{eq})}{(a_A^{eq})(a_B^{eq})}$$

and the standard free energy is

$$\Delta G^\circ = -RT \ln K = \Delta H^\circ - T\Delta S^\circ$$

To obtain ΔH° and ΔS° separately, the temperature dependence of the equilibrium constant can be measured at constant pressure

$$\frac{\partial \ln K}{\partial (1/T)} = -\frac{\Delta H^\circ}{R}$$

Alternatively the ΔH° can be obtained from calorimetric measurements.

For dilute solutions the standard state for solutes is usually chosen as a solution of unit molality, but with the extrapolated properties of the infinitely dilute solution. The activity of the solute approaches the molality as the molality approaches zero.

$$a = \gamma m$$

$$\lim_{m \to 0} \gamma = 1$$

The ratio of a/m at any concentration defines the activity coefficient γ. The equilibrium constant can now be written for solutions as

$$K_{\text{solution}} = \frac{(m_{AB}^{eq})}{(m_A^{eq})(m_B^{eq})} \frac{(\gamma_{AB})}{(\gamma_A)(\gamma_B)} = K_m K_\gamma$$

The activity coefficients can be used to calculate the dependence of equilibrium concentrations on total concentration. The activity coefficients depend on the interactions between the solute molecules in solution. Net attraction between solutes gives activity coefficients less than 1; net repulsion gives activity coefficients greater than 1. However, except for electrolytes, no very useful theories of activity coefficients are available.

Ts'o et al.[98] have measured the activity coefficients of uridine and cytidine in water at 25°C. They used a vapor phase osmometer to obtain values of γ from 0.94 at 0.05 m to about 0.5 at 0.7 m. Purine nucleosides could not be studied accurately because of limited solubility; however, they showed similar behavior.[99] It is useful to note that below about 0.01 m (or M) the solutions of nucleosides are nearly ideal, with γ approximately = 1.

One way to interpret a γ of less than 1 is to assume that association takes place. The data for the nucleosides are consistent with an association to form aggregates in which each step has the same equilibrium constant.

$$A + A \rightleftharpoons A_2 \quad K_1$$

$$A_2 + A \rightleftharpoons A_3 \quad K_2$$

$$A_3 + A \rightleftharpoons A_4 \quad K_3$$

$$\vdots$$

$$K_1 = K_2 = K_3 = \quad \cdots \quad K$$

Values of K and ΔG° for these aggregations in water at 25°C are given in Table 2-10.[100,101] Nuclear magnetic resonance studies of these solutions show that the aggregation occurs by the stacking of bases, not by hydrogen bonding. Equimolal mixtures of the Watson-Crick base pairs, deoxyadenosine and thymidine or deoxyguanosine and deoxycytidine, show less interaction than the purine nucleosides. This is consistent with the main interaction being stacking, but specific hydrogen bonding could also be involved.[101] Chemical shifts for the base proton peaks increase, because of the shielding by the bases above and below them. Table 2-10 then is a measure of the interactions leading to stacking. Uridine stacks the least; purines stack more than pyrimidines; and deoxyribonucleosides stack more than ribonucleosides. Other data[100-103] show that methylation of the bases increases stacking.

It would be useful to know the ΔH° and ΔS° for those stacking interactions. If ΔH° is approximately zero or positive (unfavorable for stacking) and ΔS° dominates the free energy, the interaction is called hydrophobic.[104] The aggregation occurs not because of attraction between bases (ΔH° negative), but because of ΔS° positive. For a hydrophobic interaction the aggregation would increase with increasing temperature. Bases,[102] mononucleosides,[103] and dinucleoside phosphates (see Chapter 3) all give ΔH° of stacking in the range of -5 to -10 kcal/mole. Thus the stacking interaction is mainly attractive; it is not primarily due to hydrophobic forces.

One way to learn more about the forces involved in an interaction is to consider the reaction to take place by a different path. It is often possible to pick a path which allows a better understanding of the original reaction. In the scheme shown in Fig. 2-22 an alternate path through the gas phase is shown. The thermodynamics of the gas phase reaction are clear. For a negative standard free energy the ΔH° must be negative, because ΔS° must be negative for the formation of the complex. The reaction of

$$A(\text{soln}) \longrightarrow A(\text{gas}) + \text{solvent}$$

has been discussed by Sinanoğlu.[105] He points out that in dilute solution its

TABLE 2-10 EQUILIBRIUM CONSTANTS AND STANDARD FREE ENERGY CHANGES FOR THE ASSOCIATION OF NUCLEOSIDES AT 25°C IN WATER

	K(molal)[a]	$\Delta G°$ (kcal)[a]
Uridine[b]	0.61	+0.29
Uridine[c]	0.70	+0.21
Cytidine[b]	0.87	+0.08
Deoxycytidine[c]	0.91	+0.06
Thymidine[c]	0.91	+0.06
Purine[b]	2.1	−0.44
Adenosine[b]	4.5	−0.90
2′-Deoxyadenosine[b]	4.7−7.5	−0.92−1.2
2′-Deoxyadenosine[c]	12	−1.5

[a] The reaction is $A + A \rightleftharpoons A_2$.
[b] P. O. P. Ts'o, in *Molecular Associations in Biology*, B. Pullman, ed., Academic Press, New York, (1968), p. 43.
[c] T. N. Solie and J. A. Schellman, *J. Mol. Biol.*, **33**, 61 (1968).

equilibrium constant is Henry's law constant $k_A{}^H$ (the proportionality between vapor pressure of a solute and its concentration).

$$p_A = k_A{}^H m_A$$

As the free energy change is independent of path, we can write

$$\Delta G°(\text{solution}) = \Delta G°(\text{gas}) + \Delta G°(\text{solvent effect})$$

$$K_{\text{soln}} = K_{\text{gas}} \frac{(k_A{}^H) \cdot (k_B{}^H)}{k_{AB}{}^H}$$

where $\Delta G°$ (solvent effect) is the standard free energy change of removing A

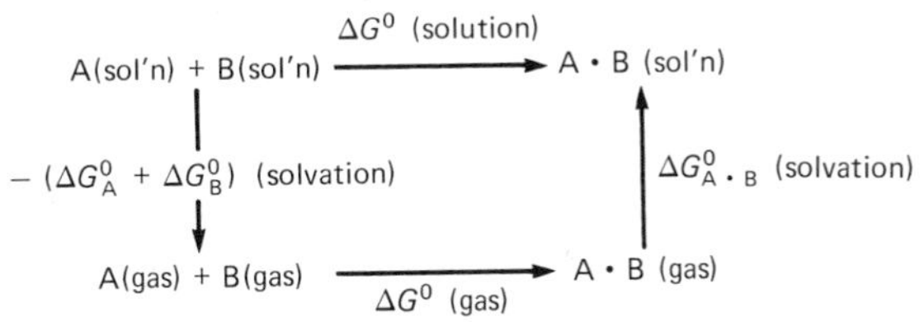

ΔG^0 (solution)
A(sol'n) + B(sol'n) → A · B (sol'n)
$-(\Delta G^0_A + \Delta G^0_B)$ (solvation)
$\Delta G^0_{A \cdot B}$ (solvation)
A(gas) + B(gas) → A · B (gas)
ΔG^0 (gas)

$$\begin{aligned}\Delta G^0 \text{ (solvent effect)} &= [G^0_A \text{ (gas)} - G^0_A \text{ (solution)}] + [G^0_B \text{ (gas)} - G^0_B \text{ (solution)}] \\ &\quad - [G^0_{A \cdot B} \text{ (gas)} - [G^0_{A \cdot B} \text{ (solution)}] \\ &= \Delta G^0_{A \cdot B} \text{ (solvation)} - \Delta G^0_A \text{ (solvation)} - \Delta G^0_B \text{ (solvation)}\end{aligned}$$

FIGURE 2-22 A cycle illustrating how the thermodynamics of a reaction in solution can be equated to the sum of solvation and gas phase reactions.

and B from the solvent into the gas and replacing A · B into the solvent from the gas.

1. THEORY

Just as the Schrödinger equation in principle tells us how to calculate any optical property of a system, knowledge of the energy levels of a system also leads to all the thermodynamic properties. For a reaction in the gas phase the contributing factors to the thermodynamics can be easily enumerated, if not calculated. Let us look at the familiar reaction of ideal gases A and B to form a complex A · B at temperature *T*.

$$\text{A(gas)} + \text{B(gas)} \rightleftharpoons \text{A}\cdot\text{B(gas)}$$

It is convenient to write the standard free energy change at temperature *T* as:[106]

$$\Delta G_T^\circ = \Delta H_0^\circ + (\Delta G_T^\circ - \Delta H_0^\circ)$$

ΔG_T° = free energy change at temperature *T* for reactants at 1 atm pressure going to products at 1 atm pressure

$\Delta H_0^\circ = \Delta E_0^\circ$ is the standard enthalpy or energy change at absolute zero

The ΔH_0° is independent of temperature and can be calculated from the electrical interactions between A and B in the geometry of A · B. The other term is written[107] in two parts:

$$(\Delta G_T^\circ - \Delta H_0^\circ)_{\text{translational}} = -RT\left[(3/2)\ln\frac{M_{AB}}{(M_A)(M_B)} - (5/2)\ln T\right] + 7.283\,T\ (\text{cal})$$

$$(\Delta G_T^\circ - \Delta H_0^\circ)_{\text{internal}} = -RT\ln\frac{Q_{AB}}{(Q_A)(Q_B)}$$

$Q = \sum_i g_i \exp[-\epsilon_i/kT]$ is the internal partition function for each molecule

g_i is the degeneracy of energy level ϵ_i

The first part is the contribution of the translational entropy; it depends on the molecular weights of the product and reactants. Although it is a large term (calculated to be about +12 kcal for dimerization of adenine at 25°C), it can be calculated accurately and it is essentially constant for all combinations of bases. The internal contribution to $(\Delta G_T^\circ - \Delta H_0^\circ)$ is smaller than the translational one and is made up of the rotational, vibrational, and electronic terms. The rotational contribution will dominate; it can be calculated from the moments of inertia of A, B, and A · B.[108] The internal $(\Delta G_T^\circ - \Delta H_0^\circ)$ may be negligible and will definitely be essentially constant.

As $\Delta H_0{}^\circ$ is the only part of $\Delta G_T{}^\circ$ which depends on the properties of the individual bases, it is the only part which has been calculated. $\Delta H_0{}^\circ$ is the energy of interaction of two molecules in a particular geometry. The molecules are usually assumed to be rigid, but the interaction should in principle be calculated with each nucleus undergoing its zero-point motion. Probably the correction would only be important for hydrogen atoms. The contributions to $\Delta H_0{}^\circ$ can be written as[109]:

$$\begin{aligned}\Delta H_0{}^\circ &= \text{electrostatic interactions } (E_{\rho\rho} + E_{\rho\mu} + E_{\mu\mu}) \\ &\quad + \text{polarization interactions } (E_{\rho\alpha} + E_{\mu\alpha}) \\ &\quad + \text{London interactions } (E_L) \\ &\quad + \text{short-range repulsion interactions } (E_S)\end{aligned}$$

$$E_{\rho\rho} = \sum_i \sum_{j>i} \frac{\rho_i \rho_j}{R_{ij}}$$

$$E_{\rho\mu} = \sum_i \sum_{j\neq i} \rho_i \mu_j C_{i,j}$$

$$E_{\rho\alpha} = -\frac{1}{2} \sum_i \sum_{j\neq j} \sum_{l=1}^{3} \rho_i{}^2 \alpha_{jl} C_i{}^2{}_{,jl}$$

$$E_{\mu\mu} = \sum_i \sum_{j>i} \mu_i \mu_j G_{i,j}$$

$$E_{\mu\alpha} = -\frac{1}{2} \sum_i \sum_{j\neq i} \sum_{l=1}^{3} \mu_i{}^2 \alpha_{jl} G_i{}^2{}_{,jl}$$

where

$$C_{i,j} = \frac{1}{R_{ij}{}^3}\, \mathbf{e}_j \cdot \mathbf{R}_{ij}$$

$$G_{i,j} = \frac{1}{R_{ij}{}^3}\left[\mathbf{e}_i \cdot \mathbf{e}_j - \frac{3}{R_{ij}{}^2}(\mathbf{e}_i \cdot \mathbf{R}_{ij})(\mathbf{e}_j \cdot \mathbf{R}_{ij})\right]$$

Here ρ_i and $\boldsymbol{\mu}_i$ are the charge and permanent dipole of group i; R_{ij} is the distance between groups i and j; α_{jl} is the component of the polarizability of group j along the principal polarization axis l; $C_{i,j}$ and $G_{i,j}$ are geometric factors involving unit vectors $\mathbf{e}_j$ and $\mathbf{e}_j$ which lie along the group dipoles $\boldsymbol{\mu}_i$ and $\boldsymbol{\mu}_j$ ($G_{i,jl}$ involves unit vectors along $\boldsymbol{\mu}_j$ and $\boldsymbol{\alpha}_{jl}$, etc.).

The London energy can be written

$$E_L = -\frac{h}{4} \sum_i \sum_{j>i} \sum_{l=1}^{3} \sum_{m=1}^{3} \frac{\nu_i \nu_j}{\nu_i + \nu_j} \alpha_{il} \alpha_{jm} G_{il}{}^2{}_{,jm}$$

where ν_i is the characteristic frequency determined from the dispersion of the refractive index of group i. Other slightly different expressions for the London energy have been used.[110]

The short-range repulsion energy can be written with either a theoretically correct exponential distance dependence or a more practical inverse power of a distance.

$$E_S = \sum_i \sum_{j>i} A_{ij} \exp\,[-a_{ij}R_{ij}]$$

or

$$E_S = \sum_i \sum_{j>i} \frac{B_{ij}}{R_{ij}^{12}}$$

The parameters A_{ij} and a_{ij} in the exponential term are obtained by fitting known crystal structures[111] and by requiring that the sum of the London attraction and the short-range repulsion energies is a minimum at the sum of the van der Waals radii of i and j. For the R^{-12} term the second condition fixes the value of B_{ij} to give

$$E_S = -\frac{1}{2} \sum_i \sum_{j>i} \frac{(R_i^0 + R_j^0)^6}{R_{ij}^6} \quad E_L$$

R_i^0 is the van der Waals radius for group i

The main practical advantage of an R^{-12} term for E_S is that this ensures a dominant repulsion for very small internuclear distances. For the exponential term an absurd attraction from the London R^{-6} term will dominate for small enough R.

There have been many calculations of ΔH_0° for nucleic acid bases in various arrangements; discussion and pertinent references can be found in *Molecular Associations in Biology*.[112] Coplanar arrangements of bases[113] and stacked arrangements[114] have been calculated. The ten possible nearest-neighbor combinations of Watson-Crick base pairs have been calculated.[109,112] The earlier calculations[109] were made using a point dipole approximation for the charge distribution on each base (one calculated $E_{\mu\mu} + E_{\mu\alpha} + E_L + E_S$). This is a poor approximation for the interactions between molecules whose dimensions are larger than the shortest distance between them. The newer calculations[112,113] are done by assigning a point monopole to each nucleus in the molecule (one calculates $E_{\rho\rho} + E_{\rho\alpha} + E_L + E_S$). Table 2-11 gives the calculated energies ΔH_0° for the ten nearest-neighbor Watson-Crick bases in B form DNA geometry.[112] These values do not include the short-range repulsion energies. Guanine-cytosine base pairs are calculated to be stronger than adenine-thymine base pairs. A rough value of the contribution of the G · C to the total interaction energy for each possible

TABLE 2-11 CALCULATED INTERACTION ENERGY $\Delta H_0{}^\circ$ (kcal/mole of base pair) FOR THE TEN NEAREST-NEIGHBOR BASE PAIRS IN B-FORM DNA[a]

	$\overrightarrow{\text{GpC}}$ | | $\overleftarrow{\text{CpG}}$	$\overrightarrow{\text{CpG}}$ | | $\overleftarrow{\text{GpC}}$	$\overrightarrow{\text{GpG}}$ | | $\overleftarrow{\text{CpC}}$	$\overrightarrow{\text{GpA}}$ | | $\overleftarrow{\text{CpT}}$	$\overrightarrow{\text{CpA}}$ | | $\overleftarrow{\text{GpT}}$	$\overrightarrow{\text{CpT}}$ | | $\overleftarrow{\text{GpA}}$	$\overrightarrow{\text{GpT}}$ | | $\overleftarrow{\text{CpA}}$	$\overrightarrow{\text{ApA}}$ | | $\overleftarrow{\text{TpT}}$	$\overrightarrow{\text{TpA}}$ | | $\overleftarrow{\text{ApT}}$	$\overrightarrow{\text{ApT}}$ | | $\overleftarrow{\text{TpA}}$
Stacking interactions	−11.3	− 8.5	− 7.7	− 9.9	− 7.2	− 7.0	− 7.0	− 7.4	− 6.1	− 5.0
Pairing interactions	−19.2	−19.2	−19.2	−12.2	−12.2	−12.2	−12.2	− 5.5	− 5.5	− 5.5
Total interactions	−30.5	−27.7	−26.9	−22.1	−19.4	−19.2	−19.2	−12.9	−11.6	−10.5

[a]From B. Pullman, in *Molecular Associations in Biology*, B. Pullman, ed., Academic Press, New York (1968), p. 17.

nearest-neighbor is -14 kcal. The corresponding contribution of A · T is about -6 kcal. As one sees from Table 2-11, two G · C base pairs have a $\Delta H_0{}^\circ$ of ~ -28 kcal, one G · C and one A · T have ~ -20 kcal and two A · T have ~ -12 kcal. Significant (up to -3 kcal) sequence effects are also calculated. Stacking energies of two bases amount to about -5 kcal/mole of base.

Many improvements can be made in the calculation of $\Delta H_0{}^\circ$ and $\Delta G_T{}^\circ$. However, the types of interactions are known and the methods are available. (For a review of intermolecular forces see *Intermolecular Forces*, edited by Hirschfelder.)[115] Calculation of the effect of solvent on the reaction is a much more difficult problem. DeVoe[116] reviews the problem in general. Various detailed methods have been proposed,[117-120] but no very convincing calculations have been carried out.

One may well ask why anyone would want to calculate free energies when it is possible to measure them. The dominant reason is that we want to know the free energy changes for reactions that cannot be studied directly. If a base is part of a large molecule, it may not be possible to measure its contribution to the overall reaction. Even if a free energy change could be measured, it might be easier to make the calculation. Furthermore, one may not know what the structure of the A · B product is. Finally, when calculations are made that agree with the experiments which have been done (see Table 2-10), we will feel confident that we understand the results and can predict new results.

One possible approach to the problem is to combine calculations of the gas phase reaction with experiments on the solvation reaction. We have most confidence in the gas phase calculations and least confidence in the solvation calculations. It might be possible (though very difficult) to measure the Henry's law constants for the solutes A and B. The vapor pressure of solute in equilibrium with a solution of known concentration would need to be measured. An indirect method which shows promise is to measure the vapor pressure of the pure solid and the solubility of the pure solid. The thermodynamics are illustrated by the following equation.

$$A(\text{gas}) \xrightarrow{\Delta G^\circ\,(\text{condensation})} A(\text{solid}) \xrightarrow{\Delta G^\circ\,(\text{solution})} A(\text{solution})$$

$$\begin{aligned}
\Delta G^\circ\,(\text{solvation}) &= \Delta G^\circ\,(\text{condensation}) + \Delta G^\circ\,(\text{solution}) \\
\Delta G^\circ\,(\text{condensation}) &= -\Delta G^\circ\,(\text{vaporization}) \\
&= RT \ln\,(\text{vapor pressure of solid}) \\
\Delta G^\circ\,(\text{solution}) &= -RT \ln\,(\text{activity of solute in the saturated solution}) \\
&\cong -RT \ln\,(\text{molality of saturated solution})
\end{aligned}$$

As the vapor pressure may have to be measured at a higher temperature than the solubility, one can measure

$$\Delta H^\circ \text{ (vaporization)} = -R \left[\frac{d \ln \text{(vapor pressure)}}{\partial(1/T)}\right]$$

and

$$\Delta S^\circ = \frac{\Delta H^\circ - \Delta G^\circ}{T}$$

at the higher temperature. The ΔG° (condensation) can then be calculated at any temperature if one assumes that ΔH° and ΔS° are independent of temperature. The advantage of this method is that once the thermodynamics of vaporization of the solid are measured, the solvation effects for any solvent can be measured simply from the solubility in that solvent. Table 2-12 gives the small amount of data available for free energies of condensation and solution of nucleic acid components. More data and more accurate data are greatly needed. One sees that the standard free energies of solvation for adenine or 9-methyladenine in water are about −12 to −13 kcal/mole; for uracil it is about −8 kcal/mole. These numbers can be used to estimate free energies of solvation for A · B complexes from the surface and volume dependence of the solvation free energies.[117] It is clear that much more experimental and theoretical work is needed in the thermodynamics of association.

C. Ionization

The equilibrium binding of hydrogen ions, metal ions, and other species to nucleic acid components has a marked effect on the structure of the nucleic acids. We will only consider hydrogen ion binding here; metal ion binding is discussed in Chapter 9.

The dissociation of a molecule can be characterized by its pK values.

$$\text{AH} \rightleftharpoons \text{A}^- + \text{H}^+$$

$$\text{p}K = -\log K = -\log \frac{(a_{\text{H}^+})(a_{\text{A}^-})}{(a_{\text{AH}})}$$

Usually one replaces the activities by molarities and reports an apparent pK at a particular ionic strength. The ratio of $(m_{A^-})/(m_{\text{AH}})$ is determined either by direct volumetric titration or by spectrophotometric methods. The hydrogen ion activity is determined by a pH meter. Many pK values have been measured,[121] but agreement is found only to ±0.1. Of course the apparent pK depends on the ionic strength. Figure 2-23 illustrates the

TABLE 2-12 THE THERMODYNAMICS OF SOLVATION OF NUCLEIC ACID COMPONENTS IN WATER. DATA ARE GIVEN FOR CONDENSATION FROM GAS TO SOLID AND FOR SOLUTION FROM SOLID TO AQUEOUS SOLUTION

	ΔG°_{298} (condensation) (kcal/mole)	ΔG°_{298} (solution) (kcal/mole)	ΔH° (condensation) (kcal/mole)	ΔH° (solution) (kcal/mole)	ΔS° (condensation) (cal/mole deg)	ΔS° (solution) (cal/mole deg)
Adenine	−15 to −16[a]	+3.2[b]	−24 to −26[a]	+8[b]	−30 to −34[a]	+16[b]
9-Methyladenine	−14 to −15[a]	+2[d]	−26 to −29[a]	+7.0[c,d]	−41 to −48[a]	+16.7[d]
Uracil	~−10[a]	+2[d]	−20[a]		~−33[a]	
1-Methyluracil		+1.1[d]		+6.3[d]		+17.6[d]
Thymine		+2.1[d]				
1-Methylthymine		+1.9[d]		+5.3[c,d]		+11.3[d]
1-Methylcytosine		+0.6[d]	−36[a]	+5.3[d]		+15.9[d]
Adenosine		+2.3[d]		+8.0[d]		+19[d]
Deoxyguanosine		+2.7[d]		+10.3[d]		+26[d]
Uridine		+0.3[d]		+9.6[d]		+31[d]
Thymidine		+1.1[d]		+7.6		+22[d]
Cytidine		+0.6[d]		+7.8[d]		+24[d]

[a] L. B. Clark, G. G. Peschel, and I. Tinoco, Jr., *J. Phys. Chem.*, **69**, 3615 (1965).
[b] P. O. P. Ts'o, I. S. Melvin, and A. C. Olson, *J. Amer. Chem. Soc.*, **85**, 1289 (1963).
[c] S. J. Gill, D. B. Martin, and M. Downing, *J. Amer. Chem. Soc.*, **85**, 706 (1963).
[d] D. Carroll, University of California, Berkeley, personal communication.

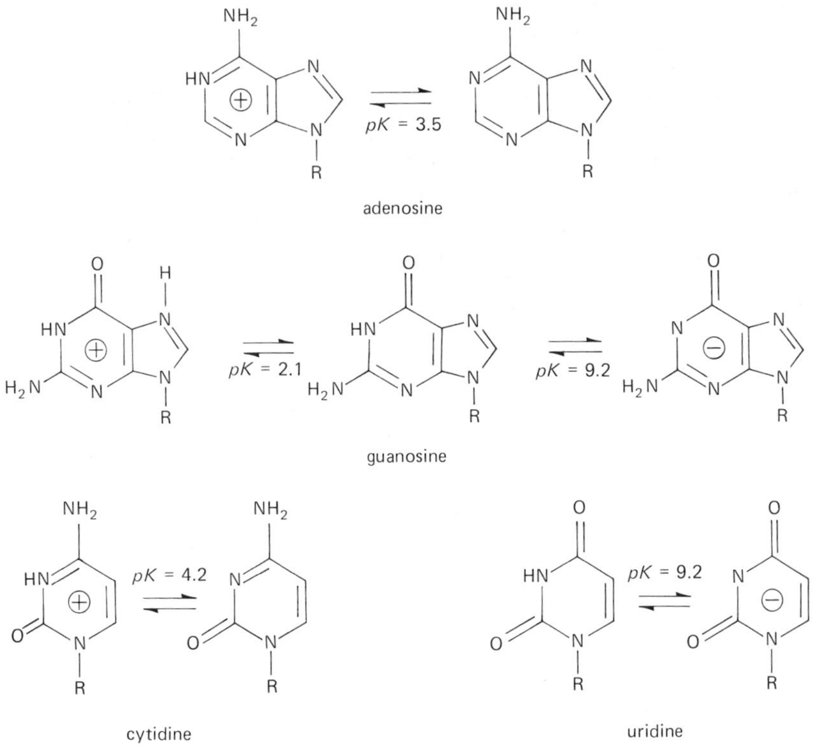

pK = 3.5

adenosine

pK = 2.1 pK = 9.2

guanosine

pK = 4.2

cytidine

pK = 9.2

uridine

FIGURE 2-23 Sites and *pK* values for protonation and ionization of four mononucleosides.

protonation and ionization of the bases in four mononucleosides. A great deal of work went into the determination of the sites of protonation. The evidence for protonation at the ring nitrogens as shown includes: (1) NMR of the compounds in dimethylsulfoxide and other aprotic solvents; (2) IR of the compounds; (3) UV of the compounds and their methylated derivatives; and (4) X-ray crystal studies. These data are reviewed by Tinoco and Holcomb[122]; Shapiro[123]; Voet and Rich[124]; and Ts'o et al.[87] Discussion of the tautomeric equilibria in the neutral bases is given by Wolfenden.[125] The ionization of the ring proton in guanosine and uridine is also well established. No attempt was made in Fig. 2-23 to depict the electron distribution in the monoanion forms of guanosine and uridine. It is clear that the negative charge will tend toward the oxygen atoms and will thus produce fully aromatic six-membered rings. Each of the nucleosides also has a pK = 12.3–12.5 for an ionizable sugar proton.[121]

There is a slight increase in pK for deoxyadenosine and deoxyguanosine over adenosine and guanosine.[126] This was interpreted by the authors as evidence of hydrogen bonding between N3 of the purine and the 2′ OH of ribose, but other convincing evidence was not given.

There is a slight increase in pK in going from a nucleoside to a nucleotide. This is larger for a 5′ phosphate than a 2′ or 3′ phosphate. The explanation of this effect is simple. Electrostatic attraction of the negatively charged phosphate reduces the ionizing ability of the proton.[127] The 5′ phosphate is nearer than the 2′ or 3′ phosphate.

In a mononucleotide two more ionizing protons are introduced. One occurs about pK = 1 and the other about pK = 6.[121]

$$\mathrm{RO\overset{\overset{\displaystyle O}{\|}}{P}OH}\ (\text{with } \mathrm{OH}) \xrightleftharpoons{\mathrm{p}K\,=\,0.7-1.0} \mathrm{RO\overset{\overset{\displaystyle O}{\|}}{P}O^-}\ (\text{with } \mathrm{OH}) \xrightleftharpoons{\mathrm{p}K\,=\,6.0-6.3} \mathrm{RO\overset{\overset{\displaystyle O}{\|}}{P}O^-}\ (\text{with } \mathrm{O^-})$$

Polynucleotides have only the first pK (~1.0) for nonterminal phosphates.

Excited state pK values (pK*) can be measured by using fluorescence or phosphorescence measurements to determine $(m_{A^-}{}^*)/(m_{AH}{}^*)$. Fluorescence studies lead to the pK* of the first excited singlet state and phosphorescence studies give the pK* of the first excited triplet state. Another method uses the measured excitation energies of AH and A^- to calculate pK*.[128] The free energy change between the excited state and the ground state can be obtained from the average of the absorption frequency at the maximum and the emission (fluorescence or phosphorescence) frequency at the maximum. This frequency can be called the average excitation frequency. Using a simple thermodynamic cycle one can derive the following equation

$$\mathrm{p}K^* = \mathrm{p}K - \frac{h(\nu_{AH} - \nu_{A^-})}{2.303\ \mathrm{k}T}$$

where ν_{AH} and ν_{A^-} are the average excitation frequencies. With the frequencies measured in cm^{-1} the equation is written[128]

$$\mathrm{p}K^* = \mathrm{p}K - \frac{0.625\,(\nu_{AH} - \nu_{A^-})}{T}$$

For 5′-thymidylic acid pK* − pK = 2.0 for the first excited singlet state and 5.7 for the triplet. For 5′-adenylic acid pK* − pK = 3.3 for the singlet and −5.3 for the triplet. These studies were done at 77°K in ethylene glycol–water mixtures.[53] It may be possible to use these data to help test calculated excited state charge distributions.

REFERENCES

1 E. Chargaff and J. N. Davidson, eds., *The Nucleic Acids*, Vol. I, Academic Press, New York (1955).
2 S. Arnott, *Progress in Biophysics and Molecular Biology*, Vol. 21, Pergamon Press, New York (1970), p. 265.
3 M. Sundaralingam, *Biopolymers,* **7**, 821 (1969).
4 S. Arnott, S. D. Dover, and A. J. Wonacott, *Acta Cryst.,* **B25**, 2192 (1969).
5 S. Arnott and D. W. L. Hukins, *Nature,* **224**, 886 (1969).
6 C. D. Jardetsky, *J. Amer. Chem. Soc.,* **83**, 2919 (1961).
7 C. D. Jardetsky and O. Jardetsky, *J. Amer. Chem. Soc.,* **82**, 222, 229 (1960).
8 J. Donohue and K. N. Trueblood, *J. Mol. Biol.,* **2**, 363 (1960).
9 A. E. V. Haschemeyer and A. Rich, *J. Mol. Biol.,* **27**, 369 (1967).
10 R. C. Davis, Ph.D. Thesis, University of California, Berkeley (1967).
11 I. Tinoco, Jr., R. C. Davis, and S. R. Jaskunas, in *Molecular Associations in Biology*, B. Pullman, ed., Academic Press, New York (1968), p. 77.
12 F. Jordan and B. Pullman, *Theoretica Chim. Acta,* **9**, 242 (1968).
13 T. R. Emerson, R. J. Swan, and T. L. V. Ulbricht, *Biochemistry,* **6**, 843 (1967).
14 T. R. Emerson, R. J. Swan, and T. L. V. Ulbricht, *Biochem. Biophys. Res. Commun.,* **22**, 505 (1966).
15 W. A. Klee and S. H. Mudd, *Biochemistry,* **6**, 988 (1967).
16 D. W. Miles, M. J. Robbins, R. K. Robbins, M. W. Winkley, and H. Eyring, *J. Amer. Chem. Soc.,* **91**, 824, 831 (1969).
17 D. W. Miles, W. H. Inskeep, M. J. Robbins, M. W. Winkley, R. K. Robbins, and H. Eyring, *J. Amer. Chem. Soc.,* **92**, 3872 (1970).
18 V. Sasisekharan, A. V. Lakshminarayanan, and G. N. Ramachandran, in *Conformation of Biopolymers*, Vol. 2, G. N. Ramachandran, ed., Academic Press, New York, (1967), p. 641; A. V. Lakshminarayanan and V. Sasisekharan, *Biopolymers,* 8 475, 589 (1969).
19 W. Guschlbauer and Y. Courtois, *FEBS Letters,* **1**, 183 (1968).
20 J. H. Prestegard and S. I. Chan, *J. Amer. Chem. Soc.,* **91**, 2843 (1969).
21 P. A. Hart and J. P. Davis, *J. Amer. Chem. Soc.,* **91**, 512 (1969); F. E. Hruska, A. A. Grey, and I. C. P. Smith, *J. Amer. Chem. Soc.,* **92**, 214 (1970).
22 A. Weissberger, ed., *Technique of Organic Chemistry*, Vol. I, *Physical Methods of Organic Chemistry*, Interscience, New York (1960).
23 W. West, ed., *Technique of Organic Chemistry,* Vol. IX, *Chemical Applications of Spectroscopy*, Interscience, New York (1956).
24 B. Pullman and A. Pullman, *Progress in Nucleic Acid Research and Molecular Biology*, Vol. 9, Academic Press, New York (1969), p. 327.
25 G. Del Re in *Electronic Aspects of Biochemistry*, B. Pullman, ed., Academic Press, New York (1964), p. 221.
26 J. A. Pople, *Trans. Faraday Soc.,* **49**, 1375 (1953); R. Pariser and R. G. Parr, *J. Chem. Phys.,* **21**, 466, 767 (1953).
27 H. Berthod and A. Pullman, *J. Chimie Phys.,* **55**, 942 (1965).
28 H. Berthod, C. Giessner-Prettre, and A. Pullman, *Theoret. Chim. Acta,* **5**, 53 (1966).
29 E. Clementi, J. M. Andre, M. C. Andre, D. Klint, and D. Hahn, *Acta Physica,* **27**, 493 (1969).
30 D. Voet, W. B. Gratzer, R. A. Cox, and P. Doty, *Biopolymers,* **1**, 193 (1963).
31 T. Yamada and H. Fukutome, *Biopolymers,* **6**, 43 (1968).

32 H. H. Jaffé and M. Orchin, *Theory and Applications of Ultraviolet Spectroscopy*, Wiley, New York (1962).
33 Ibid., p. 361.
34 Ibid., p. 180.
35 D. W. Miles, R. K. Robbins, and H. Eyring, *J. Phys. Chem.*, **71**, 3931 (1967).
36 P. R. Callis, Ph.D. Thesis, University of Washington (1965); P. R. Callis and W. T. Simpson, *J. Amer. Chem. Soc.*, **92**, 3593 (1970); T. P. Lewis and W. A. Eaton, *J. Amer. Chem. Soc.*, **93**, 2054 (1971).
37 L. B. Clark, G. G. Peschel, and I. Tinoco, Jr., *J. Phys. Chem.*, **69**, 3615 (1965).
38 C. A. Bush and H. A. Scheraga, *Biopolymers*, **7**, 395 (1969).
39 B. L. Tomlinson, Ph.D. Thesis, University of California, Berkeley (1968).
40 L. B. Clark and I. Tinoco, Jr., *J. Amer. Chem. Soc.*, **87**, 11 (1965).
41 H. H. Chen and L. B. Clark, *J. Chem. Phys.*, **51**, 1862 (1969).
42 J. Brahms, J. Pilet, H. Damany, and V. Chandrasekharan, *Proc. Nat. Acad. Sci.*, **60**, 1130 (1968); D. M. Gray and I. Rubenstein, *Biopolymers*, **6**, 1605 (1968).
43 K. Seibold and H. Labhart, *Biopolymers*, **10**, 2063 (1971).
44 A. Hartford, Jr., and J. R. Lombardi, *J. Mol. Spectrosc.*, **34**, 257 (1970).
45 R. F. Stewart and N. Davidson, *J. Chem. Phys.*, **39**, 255 (1963); R. F. Stewart and L. H. Jensen, *J. Chem. Phys.*, **40**, 2071 (1964).
46 W. A. Eaton and T. P. Lewis, *J. Chem. Phys.*, **53**, 2164 (1970).
47 P. R. Callis, B. Fanconi, and W. T. Simpson, *J. Amer. Chem. Soc.*, **93**, 6679 (1971); L. B. Clark, personal communication.
48 M. M. Warshaw and C. R. Cantor, *Biopolymers*, **9**, 1079 (1970).
49 J. T. Yang, T. Samejima, and P. K. Sarkar, *Biopolymers*, **4**, 623 (1966).
50 W. Voelter, R. Records, E. Bunnenberg, and C. Djerassi, *J. Amer. Chem. Soc.*, **26**, 6163 (1968).
51 J. W. Longworth, R. O. Rahn, and R. G. Shulman, *J. Chem. Phys.*, **45**, 2930 (1966).
52 M. Guéron and R. G. Shulman, *Ann. Rev. Biochem.*, **37**, 571 (1968); J. Eisinger, *Photochem. Photobiol.*, **7**, 597 (1968).
53 M. Kasha, *Discuss. Faraday Soc.*, **9**, 14 (1950); *Radiation Research Suppl.*, **2**, 243 (1960).
54 A. A. Lamola, M. Guéron, T. Yamane, J. Eisinger, and R. G. Shulman, *J. Chem. Phys.*, **47**, 2210 (1967); R. G. Shulman and R. O. Rahn, *J. Chem. Phys.*, **45**, 2940 (1966).
55 P. R. Callis, E. J. Rosa, and W. T. Simpson, *J. Amer. Chem. Soc.*, **86**, 2292 (1964).
56 L. I. Schiff, *Quantum Mechanics*, McGraw-Hill, New York (1968), p. 401.
57 A. Moscowitz, Ph.D. Thesis, Harvard University (1957).
58 W. Lamb, R. Young, and S. R. La Paglia, *J. Chem. Phys.*, **49**, 2868 (1968).
59 W. T. Simpson, *Theories of Electrons in Molecules*, Prentice-Hall, Englewood Cliffs, N.J. (1962), p. 171.
60 W. Moffitt and A. Moscowitz, *J. Chem. Phys.*, **30**, 648 (1959).
61 W. A. Shurcliff, *Polarized Light*, Harvard University Press, Cambridge (1962).
62 I. Tinoco, Jr. and C. A. Bush, *Biopolymers Symposia*, **1**, 235 (1964).
63 P. J. Stephens, *J. Chem. Phys.*, **52**, 3489 (1970).
64 L. Brand and B. Witholt, *Methods in Enzymology*, **11**, 766 (1967).
65 N. J. Turro, *Molecular Photochemistry*, W. A. Benjamin, New York (1965).
66 A. C. Albrecht, *J. Mol. Spectroscopy*, **6**, 84 (1961).
67 C. A. Emeis and L. J. Oosterhoff, *Chem. Phys. Lett.*, **1**, 129 (1967).
68 L. D. Landau and E. M. Lifshitz, *Electrodynamics of Continuous Media*, Addison-Wesley, Reading, Mass. (1960), p. 256.

69 J. Thiéry, *J. Chim. Phys.*, **65**, 98 (1965).
70 L. D. Landau and E. M. Lifshitz, op. cit., p. 274.
71 H. Berthod, C. Giessner-Prettre, and A. Pullman, *Int. J. Quantum Chem.*, **1**, 123 (1967).
72 C. Lifschitz, E. D. Bergmann, and B. Pullman, *Tetrahedron Lett.*, **46**, 4583 (1967).
73 A. Denis and A. Pullman, *Theoret. Chim. Acta*, **7**, 110 (1967).
74 H. T. Miles, *Proc. Nat. Acad. Sci., U.S.*, **47**, 791 (1961).
75 M. Tsuboi, K. Matsuo, T. Shimanouchi, and Y. Kyogoku, *Biochim. Biophys. Acta*, **55**, 1 (1962).
76 H. T. Miles and J. Frazier, *Biochem. Biophys. Res. Commun.*, **14**, 21 129 (1964).
77 G. J. Thomas, Jr., *Biopolymers*, **7**, 325 (1969).
78 Y. Kyogoku, R. C. Lord, and A. Rich, *Proc. Nat. Acad. Sci.*, **57**, 250 (1967).
79 Y. Kyogoku, R. C. Lord, and A. Rich, *J. Amer. Chem. Soc.*, **89**, 496 (1967).
80 R. C. Lord, and G. J. Thomas, Jr., *Spectrochimica Acta*, **23A**, 2551 (1967); *Biochim. Biophys. Acta*, **142**, 1 (1967).
81 B. Fanconi, B. Tomlinson, L. A. Nafie, E. W. Small, and W. L. Peticolas, *J. Chem. Phys.*, **51**, 3993 (1969); E. W. Small and W. L. Peticolas, *Biopolymers*, **10**, 69 (1971).
82 A. C. Albrecht, *J. Chem. Phys.*, **34**, 1476 (1961).
83 F. J. Kayne and J. Reuben, *J. Amer. Chem. Soc.*, **92**, 220 (1970).
84 I. Feldman and R. P. Agarwal, *J. Amer. Chem. Soc.*, **90**, 7329 (1968).
85 S. S. Danyluk and F. E. Hruska, *Biochemistry*, **7**, 1038 (1967).
86 M. P. Schweizer, A. D. Broom, P. O. P. Ts'o, and D. P. Hollis, *J. Amer. Chem. Soc.*, **90**, 1042 (1968).
87 P. O. P. Ts'o, M. P. Schweizer, and D. P. Hollis, *Ann. N.Y. Acad. Sci.*, **158**, 256 (1969).
88 P. O. P. Ts'o, N. Kondo, M. P. Schweizer, and D. P. Hollis, *Biochemistry*, **8**, 997 (1969).
89 M. Karplus, *J. Chem. Phys.*, **30**, 11 (1959); *J. Amer. Chem. Soc.*, **85**, 2870 (1963).
90 M. Tsuboi, F. Kuriyagawa, K. Matsuo, and Y. Kyogoku, *Bull. Chem. Soc. Japan*, **40**, 1813 (1967).
91 S. Ohnishi and H. M. McConnell, *J. Amer. Chem. Soc.*, **87**, 2293 (1965).
92 C. L. Hamilton and H. M. McConnell in *Structural Chemistry and Molecular Biology*, A. Rich and N. Davidson, eds., W. H. Freeman, San Francisco (1968), p. 115.
93 E. Scrocco in *Molecular Biophysics*, B. Pullman and M. Weissbluth, eds., Academic Press, New York (1965).
94 J. A. Pople, W. G. Schneider, and H. J. Bernstein, *High Resolution Nuclear Magnetic Resonance*, McGraw-Hill, New York (1959); A. Carrington and A. D. McLachlan, *Introduction to Magnetic Resonance*, Harper & Row, New York (1967); E. D. Becker, *High Resolution NMR*, Academic Press, New York (1969); F. A. Bovey, *Nuclear Magnetic Resonance Spectroscopy*, Academic Press, New York (1969).
95 P. O. P. Ts'o in *Fine Structure of Protons and Nucleic Acids*, G. Fasman and S. N. Timasheff, eds., Marcel Dekker, New York (1970).
96 G. N. Lewis and M. Randall, *Thermodynamics*, 2nd ed., rev. by K. S. Pitzer and L. Brewer, McGraw-Hill, New York (1961).
97 T. L. Hill, *An Introduction to Statistical Thermodynamics*, Addison-Wesley, Reading, Mass. (1960).
98 P. O. P. Ts'o, I. S. Melvin, and A. C. Olson, *J. Amer. Chem. Soc.*, **85**, 1289 (1963).
99 A. D. Broom, M. P. Schweizer, and P. O. P. Ts'o, *J. Amer. Chem. Soc.*, **89**, 3612 (1967).

100 P. O. P. Ts'o in *Molecular Associations in Biology*, B. Pullman, ed., Academic Press, New York (1968).
101 T. N. Solie and J. A. Schellman, *J. Mol. Biol.*, **33**, 61 (1968).
102 S. J. Gill, D. B. Martin, and M. Downing, *J. Amer. Chem. Soc.*, **85**, 706 (1963).
103 S. J. Gill, M. Downing, and G. F. Sheats, *Biochemistry*, **6**, 272 (1967).
104 W. Kauzmann, in *Advances in Protein Chemistry*, Vol. 14, Academic Press, New York (1959), p. 34.
105 O. Sinanoğlu in *Molecular Associations in Biology*, B. Pullman, ed., Academic Press, New York (1968).
106 G. N. Lewis and M. Randall, op. cit., p. 167.
107 Ibid., p. 421.
108 Ibid., p. 428.
109 H. DeVoe and I. Tinoco, Jr., *J. Mol. Biol.*, **4**, 500 (1962).
110 K. S. Pitzer, in *Advances in Chemical Physics*, Vol. 2, Interscience, New York (1959), p. 59.
111 P. Claverie in *Molecular Associations in Biology*, B. Pullman, ed., Academic Press, New York (1968).
112 B. Pullman in *Molecular Associations in Biology*, B. Pullman, ed., Academic Press, New York (1968).
113 H. A. Nash and D. F. Bradley, *J. Chem. Phys.*, **45**, 1380 (1966).
114 P. Claverie, B. Pullman, and J. Caillet, *J. Theoret. Biol.*, **12**, 419 (1966); P. Claverie, *J. Mol. Biol.*, **56**, 75 (1971).
115 J. O. Hirschfelder, ed., *Intermolecular Forces* (*Advances in Chemical Physics*, Vol. 12), Interscience, New York (1967).
116 H. DeVoe in *Structure and Stability of Biological Macromolecules*, S. N. Timasheff and G. D. Fasman, eds., Marcel Dekker, New York (1969).
117 O. Sinanoğlu in *Molecular Associations in Biology*, B. Pullman, ed., Academic Press, New York (1968).
118 O. Sinanoğlu in *Intermolecular Forces* (*Advances in Chemical Physics*, Vol. 12), J. Hirschfelder, ed., Interscience, New York (1967).
119 O. Sinanoğlu and S. Abdulnur, *Fed. Proc.*, **24**, S-12 (1965).
120 R. Lumry and S. Rajender, *Biopolymers*, **9**, 1125 (1970).
121 D. O. Jordan, *The Chemistry of Nucleic Acids*, Butterworth, Washington (1960).
122 I. Tinoco, Jr., and D. N. Holcomb, *Ann. Rev. Phys. Chem.*, **15**, 379 (1964).
123 R. Shapiro, in *Progress in Nucleic Acid Research and Molecular Biology*, Vol. 8, Academic Press, New York (1968), p. 73.
124 D. Voet and A. Rich, *Progress in Nucleic Acid Research and Molecular Biology*, Vol. 10, Academic Press, New York (1970), p. 183.
125 R. V. Wolfenden, *J. Mol. Biol.*, **40**, 307 (1969).
126 P. O. P. Ts'o, S. A. Rapaport, and F. J. Bollum, *Biochemistry*, **5**, 4153 (1966).
127 J. G. Kirkwood and F. H. Westheimer, *J. Chem. Phys.*, **6**, 506, 513 (1938).
128 A. Weller, *Prog. Reaction Kinetics*, **1**, 189 (1961).

chapter 3
oligonucleotides and single-stranded polynucleotides

I INTRODUCTION

We will discuss in this chapter the structure of oligonucleotides and single-stranded polynucleotides. This includes the conformation of the molecules, in particular the local structure determined by short-range interactions. The questions are: What conformations are present? What are the observable properties of the conformations? Experiments and theories about methods that give information about local structure in solution will be stressed.

II DINUCLEOTIDES

The simplest polynucleotide to consider is the dimer; it has most of the local interactions which characterize the polymer. The dinucleotides characteristic of natural nucleic acids will have a phosphate linking the 3′ and 5′ oxygens of the two sugars. A standard nomenclature identifies the base of the 3′-phosphate nucleotide followed by the base of the 5′-phosphate nucleotide. The standard prefixes are adenylyl, cytidylyl, guanylyl, and uridylyl for RNA and deoxyadenylyl, deoxycytidylyl, deoxyguanylyl, and thymidylyl for DNA. A dinucleotide must also have a phosphate group not linking the nucleosides; it can be on the free 5′ hydroxyl or the free 3′ hydroxyl. If neither phosphate is present, the dimer is called a dinucleoside phosphate. The standard abbreviation for dimers (and polymers) uses a p for phosphate and A, C, G, U, dA, dC, dG, T for the corresponding nucleosides. The p on the left of the nucleoside symbol means a 5′-phosphate; the p on the right means a 3′-phosphate. Examples are given in Fig. 3-1.

Dinucleoside phosphates have been studied extensively by means of absorption, CD, and ORD,[1-4] proton magnetic resonance (PMR)[5,6] and light emission.[7,8] The main conclusion is that the two bases interact strongly in dinucleoside phosphates in aqueous solution. The evidence for this is that the properties of the bases in a dinucleoside phosphate are different from those properties in the mononucleotides. Raising the temperature, adding alcohol to the solution, or changing the pH from neutrality changes the property reversibly to that approaching the constituent mononucleotides. This shows that the differences in dinucleotide properties are not due to the covalent bond formed in the 3′–5′ phosphate ester linkage. Figure 3-2 compares the

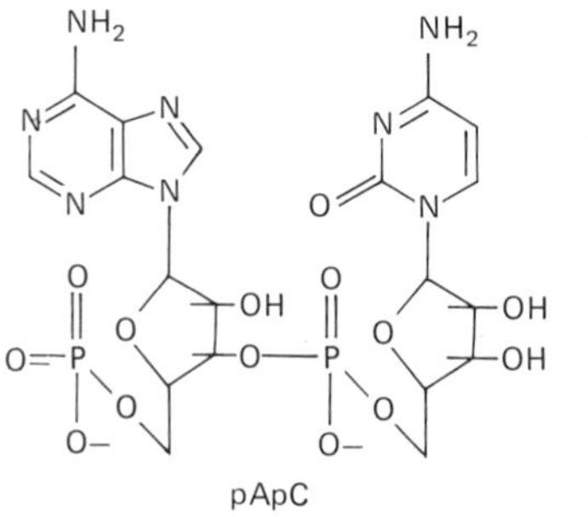

pApC
5′–phosphate adenylyl–(3′ → 5′)–cytidine

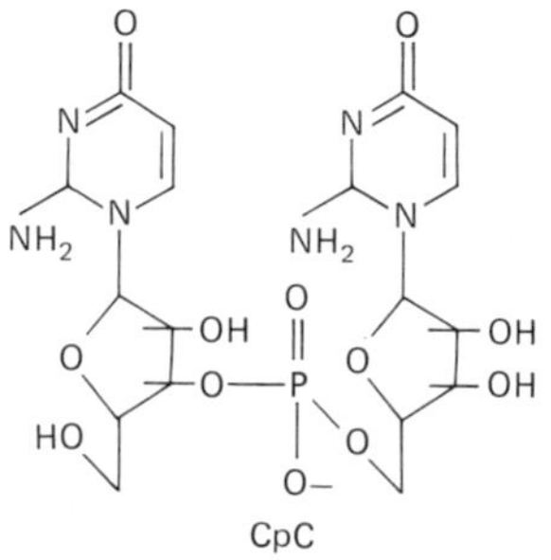

CpC
cytidylyl–(3′ → 5′)–cytidine

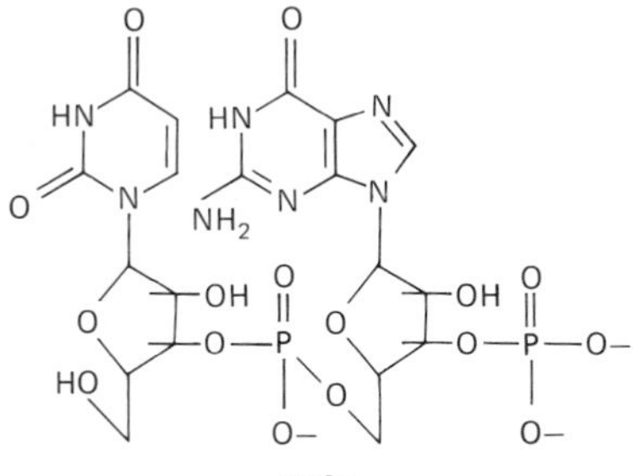

UpGp
uridylyl–(3′ → 5′)–guanosine–3′–phosphate

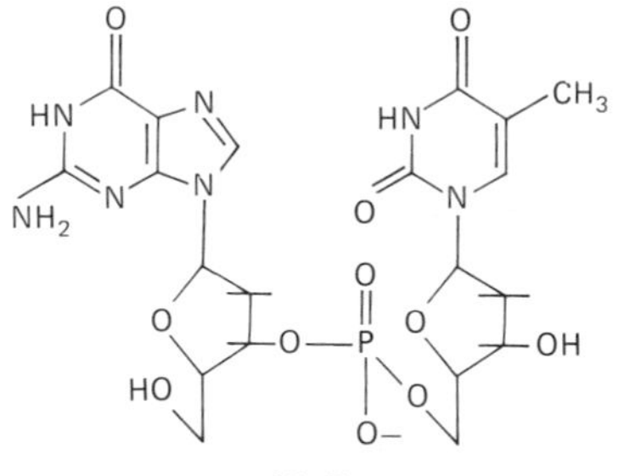

dGpT
deoxyguanylyl–(3′ → 5′)–thymidine

FIGURE 3-1 Abbreviations and nomenclature for dinucleotides and dinucleoside phosphates. pApC and UpGp are dinucleotides; CpC and dGpT are dinucleoside phosphates.

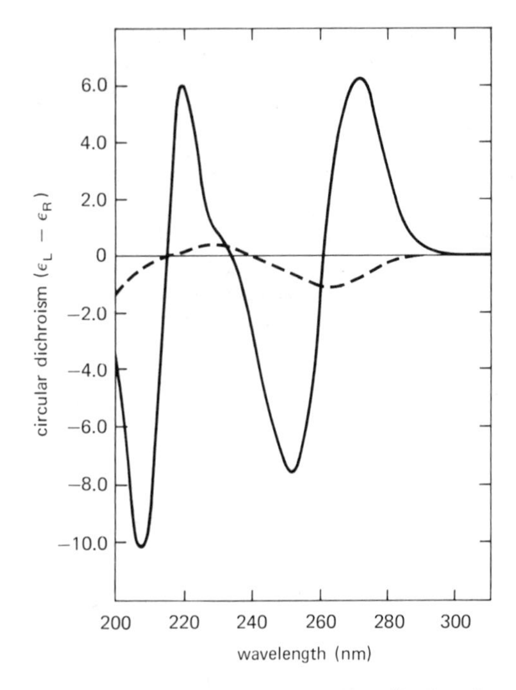

FIGURE 3-2 The circular dichroism of adenylyl-(3′ → 5′)-adenosine. ApA (solid line) compared with adenosine (dashed line) at neutral pH in aqueous solution at room temperature. The units of CD are the usual liters/mole centimeter with moles referring to moles of base; the dimer and the monomer are thus compared per monomer unit. [Data from M. M. Warshaw and C. R. Cantor, *Biopolymers*, 9, 1079 (1970).]

circular dichroism of adenylyl-(3′→5′)-adenosine, ApA, with adenosine; the adenosine CD is essentially the same as either 3′ or 5′ adenylic acid.[9] The marked change in shape of the CD from a single CD band at 260 nm for adenosine to the split band for ApA is characteristic of base–base interaction. The upfield shift in the proton magnetic resonance lines of H2 and H8 of adenine in ApA is further compelling evidence for interaction. The fact that the chemical shift occurs to a higher field requires that the bases in ApA be stacked. When the bases are stacked, the ring-current magnetic anisotropy of one base provides an induced magnetic field at the protons of the adjacent base which is counter to the external field; therefore, a higher external field is necessary to achieve resonance. Further evidence for stacking is the decrease of about 8% in the 260 nm absorption of the dimer relative to its monomers.[10]

The detailed structure of the stacked ApA is more difficult to determine, but some facts are clear. The sign of the circular dichroism indicates that the two bases begin to form a *right*-handed helix.[11] That is, as one moves forward along the stack, the transition moments (the bases) rotate clockwise. The proton magnetic resonance allows one to determine that both bases are preferentially *anti* in the same way that the base in adenylic acid was assigned *anti.*[5,12] An obvious hypothesis is that the bases in ApA have a conformation similar to that in a single strand of double-stranded DNA. It is difficult to be much more precise. We will discuss in some detail the application of two methods which have been used to try to deduce more precise conformations for dinucleoside phosphates.

A nuclear magnetic resonance spectrum of an aqueous dinucleoside phosphate contains a great deal of useful information about its conformation. In order to extract the information, the peaks must be assigned to protons. In ApA there are eight nonexchangeable protons in each nucleoside: H2, H8 of adenine, H1′, H2′, H3′, H4′, and two H5′ of ribose. The NMR peaks from the four base protons of the two adenines are easily distinguishable. The two H8 protons are assigned on the basis of their exchange with D_2O at high temperature. To assign each peak to the base on the 3′ linked or 5′ linked nucleoside is more difficult. In general, synthesis of molecules with specific deuterium substitution at each position would be needed. However, for ApA the peaks can be assigned on the basis of more indirect evidence.[12] The dominant factors controlling the chemical shifts of the base protons are the electric field of the phosphate and the ring current of the neighboring adenine. Studies on nucleosides show that the phosphate charge has a large effect on H8 for *anti* bases (see Chapter 2). Comparison of the chemical shifts with 5′ and 3′ adenylic acid leads to the assignment and the conclusion that both bases are mainly *anti.* Perturbation of the spectrum by binding Mn^{2+} to the phosphate confirms the assignment. The H2 protons are not affected much by the phosphate; their chemical shifts depend mainly on the neighboring base and therefore characterize the stacking. As the stacking can be decreased by raising the temperature, an assignment and conclusions about stacking can be made from the temperature dependence of the peaks.[12] The conclusion is again that both bases must be *anti*. Calculations of the magnitude of the shielding for each proton can in principle lead to reasonable models for the time average conformation at each temperature. Similar analysis of the spectrum of the 12 ribose protons for ApA can be done, but the assignment is much more difficult because of the overlap of most of the peaks with each other and with residual HDO.

Calculations of the circular dichroism of ApA can also give more information about possible conformations. The CD is calculated for various conformations and compared with experiment. The 260 nm band of adenine is essentially one electrically allowed transition which is split in ApA. The CD depends on the magnitude of the splitting and the sign and magnitude of the rotational strength for each peak. The splitting is calculated from Coulomb's law and the transition monopoles for the 260 nm transition of adenine

(Chapter 2). The rotational strengths depend on these transition monopoles, the relative orientation of the 260 nm transition dipoles, and interactions with absorption bands farther into the ultraviolet. Calculations of this type for ApA are only consistent with experiment for right-handed, stacked conformations very similar to those in a single strand of DNA.[13]

All 16 diribonucleoside phosphates have been studied in CD; they show behavior qualitatively similar to ApA. The conclusion is that they too are stacked in a DNA-like single strand. Optical properties of the 16 dideoxyribonucleoside phosphates have also been studied; they also have similar conformations in solution.[9] There are quantitative differences, however, as measured by the difference in measured property for the dinucleoside phosphate (both ribo and deoxyribo) compared with the monomers. The order of stacking seems to be G > A > C > U. This is consistent with the aggregation of the nucleosides in water (see Table 2-10), but in dinucleoside phosphates the stacking of cytosine is more nearly like the purine bases. A more precise statement may be that stacking in dinucleoside phosphates has the following characteristics: $G \approx A \approx C > U$.

The base sequence of the dinucleoside phosphate also affects the optical properties. A striking example is the CD of ApG and GpA shown in Fig. 3-3. The extent of stacking also appears to depend on sequence. The fact that the sequence isomers are different is not surprising if one looks at molecular models. Figure 3-4 shows the sequence isomers adenylyl-(3′→5′)-cytosine, ApC, and cytidylyl-(3′→5′)-adenosine, CpA, in structures corre-

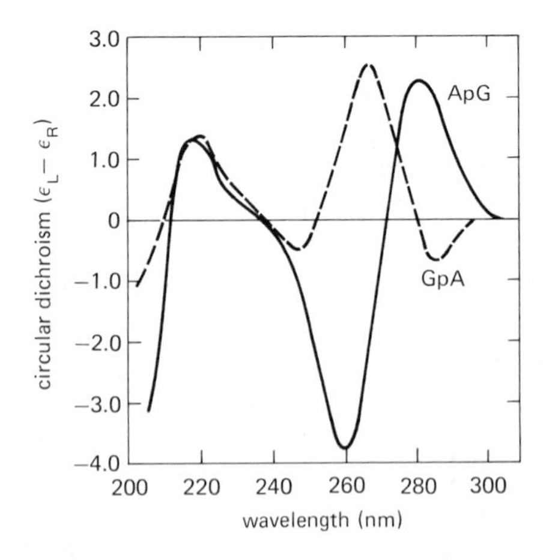

FIGURE 3-3 The circular dichroism of adenylyl-(3′ → 5′)-guanosine, ApG (solid line) and guanylyl-(3′ → 5′)-adenosine, GpA (dashed line) at neutral pH in aqueous solution at room temperature. [Data from M. M. Warshaw and C. R. Cantor, *Biopolymers*, **9**, 1079, (1970).]

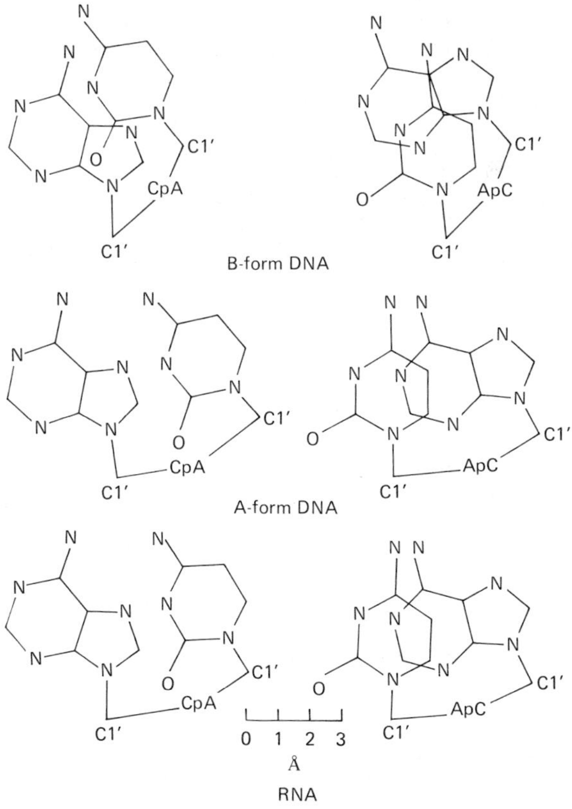

FIGURE 3-4 Sequence isomer dinucleoside phosphates with the conformations corresponding to a single strand of double-stranded A-form DNA, B-form DNA, and RNA. [Coordinates are from S. Arnott et al., *Acta. Cryst.*, **B25**, 2192 (1969).]

sponding to a single strand of three different double-stranded structures. The coordinates of the structures are taken from Arnott et al.[14] Although we do not know that the dinucleoside phosphates have any of these structures, it is clear that the relative orientation of the bases in the sequence isomers may be very different. In the structures for the dinucleoside phosphates proposed for the PMR work,[5,6] more base stacking than in Fig. 3-4 is present, but all the structures do have both bases *anti* and a right-handed helix.

The temperature dependence of the optical properties and the PMR have been studied. The interaction decreases with increasing temperature and the hope was to determine thermodynamic values for the stacking interaction. If a simple equilibrium exists between two forms, stacked and unstacked, any property of the system is the weighted sum of the properties of each form. The equilibrium constant, $K(T)$, for the reaction can then be easily related to any measured property as a function of temperature.

$$\text{unstacked} \underset{}{\overset{K}{\rightleftharpoons}} \text{stacked}$$

$$K(T) = \frac{\text{conc. stacked}}{\text{conc. unstacked}} = \frac{P(T) - P_U}{P_S - P(T)}$$

where $P(T)$ is the measured property of the equilibrium system at temperature T, P_U is the property of the completely unstacked molecule, and P_S is the property of the completely stacked molecule, both at temperature T. We have assumed activity coefficients are unity. There are many difficulties in applying this method to dinucleoside phosphates. Experimentally it is difficult to obtain P_U and P_S; theoretically one is worried by the assumption of only two states. It seems more reasonable that the dinucleoside phosphate in solution is a dynamic structure with the bases relatively free even in the "stacked" conformation. As the temperature increases the motion increases.

Nevertheless, it is worthwhile to consider the analysis of the optical properties with the two-state model; the validity of this approach has been discussed.[15] Most of the measurements were made in dilute buffer between 0 and 100°C[4]; some were made in 4.5 M KF down to −20°C[2]; and some were made in 25.2% LiCl down to −70°C[4]. Figure 3-5 shows the molar rotation of ApA at 262 nm over the temperature range −70°C to +90°C. To obtain the equilibrium constant, K, one also needs values for the molar rotation of the completely stacked and unstacked forms. The unstacked rotation was chosen as that of the component nucleosides or nucleotides at each temperature. The stacked rotation was assumed independent of temperature and chosen as the low temperature limit of the measured data. The standard thermodynamic values for the reaction were obtained from plots of $\log K$ vs. $1/T$.

$$\Delta H^\circ = -4.576 \frac{d(\log K)}{d(1/T)} \qquad \text{(in cal/mole)}$$

$$\Delta G^\circ = -4.576 T \log K \qquad \text{(in cal/mole)}$$

$$\Delta S^\circ = \frac{\Delta H^\circ - \Delta G^\circ}{T} \qquad \text{(in cal/deg mole)}$$

The rotation data for the 15 dinucleoside phosphates (GpG was not measured) fit reasonable straight lines for $\log K$ vs. $1/T$ plots.[4] The standard enthalpy for stacking ranged from −4.8 kcal/mole to −8.4 kcal/mole; the standard entropy of stacking varied from −17 cal/deg mole to −32 cal/deg mole; the standard free energy of stacking at 25°C varied from +0.9 kcal/mole to −1.7 kcal/mole. Using circular dichroism data for seven

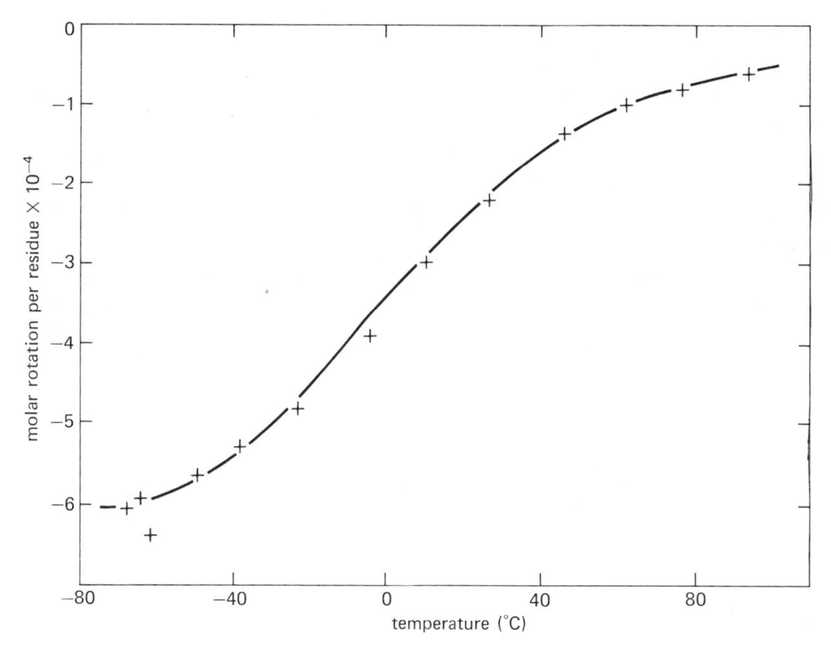

FIGURE 3-5 Molar rotation of ApA at 262 nm in 25.2% LiCl, pH 7, as a function of temperature. [Data are from R. C. Davis and I. Tinoco, Jr., *Biopolymers*, **6**, 223 (1968).]

dinucleoside phosphates, Brahms et al.[2] have found ΔH° from −6.1 to −8.0, ΔS° from −21 to −28, and ΔG° at 0°C from −0.2 to −0.7. Interpretation of the data with the two-state model thus requires about 7 kcal to unstack the bases in a dinucleoside phosphate. There is a gain of about 25 entropy units on unstacking. The standard free energy change near room temperature is about zero. The many uncertainties in the interpretation make a detailed discussion unprofitable. The absorption of a dinucleoside phosphate is less than the absorption of its constituent monomers. This hypochromicity is a measure of stacking whose temperature dependence leads to values of ΔH° for ApA from −8.5 to −10 kcal/mole.[4,16,17]

The PMR temperature dependence shows from chemical shifts that even at 95°C the bases in ApA are still perceptibly stacked.[12] As there is evidence even at low temperatures that more than one conformer is present in solution, the PMR temperature dependence is not interpreted as a simple, two-state process. If it were, the value obtained for ΔH° would be −11 kcal/mole, which is significantly higher than values obtained from optical rotation measurements. If the reaction were indeed a two-state process, all properties would give the same thermodynamic values; therefore, this provides further evidence against the two-state model.

It is interesting that the optical rotation data can be fit equally well by a model which is the opposite of the two-state model.[18] In this model the bases are assumed to be connected by a torsional spring. Temperature simply increases the torsional oscillation of the bases; they never unstack.

One can completely unstack dinucleoside phosphates by changing the pH so as to charge the bases[1,12,19] or by adding solvents other than water.[20] Both the optical and the PMR spectra become essentially identical to the appropriate monomer spectra.

If ionizing one of the bases in a dinucleotide causes complete unstacking, this provides a method for measuring the stacking equilibrium constant.[19] Consider the following equilibria.

$$\begin{array}{ccc} \text{unstacked} & \overset{K_s}{\rightleftharpoons} & \text{stacked} \\ K_i \searrow\!\!\nwarrow & & \swarrow\!\!\nearrow K_i K_s \\ & \text{unstacked ionized} & \end{array}$$

$$K_s = \frac{\text{(conc. stacked)}}{\text{(conc. unstacked)}} = \text{stacking equilibrium constant}$$

$$K_i = \frac{\text{(conc. unstacked) (conc. hydrogen ion)}}{\text{(conc. unstacked ionized)}}$$

$$= \text{ionization constant for unstacked base}$$

$$K_i K_s = \frac{\text{(conc. stacked) (conc. hydrogen ion)}}{\text{(conc. unstacked ionized)}}$$

$$= \text{ionization constant for stacked base}$$

By titrating the dinucleotide one can measure its degree of protonation, α, which depends on K_i and K_s.

$$\alpha = \frac{\text{(conc. of unstacked ionized)}}{\text{(total conc.)}}$$

If one assumes that K_i is equal to the ionization constant of the corresponding mononucleotide, one can determine K_s. The equation is

$$\text{pH} = \text{p}K_i + \log\left(\frac{1-\alpha}{\alpha}\right) - \log(1 + K_s)$$

More complicated equations can be derived if both bases of the dinucleotide ionize and if one considers stacking of the ionized forms.[19] Using this method Simpkins and Richards[19] have obtained a value of $K_s = 5.38$ for ApA at 20°C ($\Delta G^\circ = -0.99 \pm 0.005$ kcal/mole). For UpU they[21] find no stacking within experimental error. There are various approximations made in the application of this method, but it seems more direct than the analysis of the temperature-dependent optical properties.

The results of all the previous studies can be summarized easily. Dinucleoside phosphates in neutral aqueous solution tend to have the bases stacked. Although the structure is not rigid, the bases are usually in van der Waals contact. The conformation is analogous to that found in a single strand of double-stranded DNA or RNA. There are definite differences in the extent of base–base interaction which depend on the bases and the sequence; uracil stacks least, if at all. Water is important in stabilizing the structure, because alcohol (for example) will destroy the stacking.

III TRINUCLEOTIDES

It is useful to compare the various trinucleotides or trinucleoside diphosphates with the dinucleotides. There are 64 trimers possible from four different bases. Trinucleoside diphosphates and trinucleotides obtained from pancreatic ribonuclease[22] and T1 ribonuclease[23] hydrolyses of RNA have been studied. The bases in the trimers are stacked as in the dimers. In fact it is possible to calculate with a fair degree of accuracy the optical property of the trimers from the measured properties of the dimers.[22,24] For a trinucleoside diphosphate an optical property, P, per mole of nucleotide at any wavelength can be written as the identity.

$$3P(X\text{p}Y\text{p}Z) \equiv P(X) + P(Y) + P(Z) + I_{XY} + I_{YZ} + I_{XZ}$$

This equation defines the three interaction terms I_{XY}, I_{YZ}, and I_{XZ}; their values depend on the values of the measured monomer properties $P(X)$, $P(Y)$, and $P(Z)$. A similar equation can be written for a dinucleoside phosphate.

$$2P(X\text{p}Y) \equiv P(X) + P(Y) + I_{XY}'$$

If the orientation of the bases in the dimer and trimer are the same relative to each other and relative to the sugars, then I_{XY} should approximately equal I_{XY}'. Assuming this is so and ignoring I_{XZ} (assuming only nearest-neighbor interactions), we can write

$$3P(X\text{p}Y\text{p}Z) = 2P(X\text{p}Y) + 2P(Y\text{p}Z) - P(Y)$$

A test of this equation for ORD is shown in Fig. 3-6. It is clear that the equation (dotted lines) is a useful approximation. The best choice of molecule to represent the monomer contribution to the property is not clear. Neither the nucleoside nor nucleotide is ideal, because the nucleoside has no phosphate while the nucleotide has a doubly charged phosphate at neutral

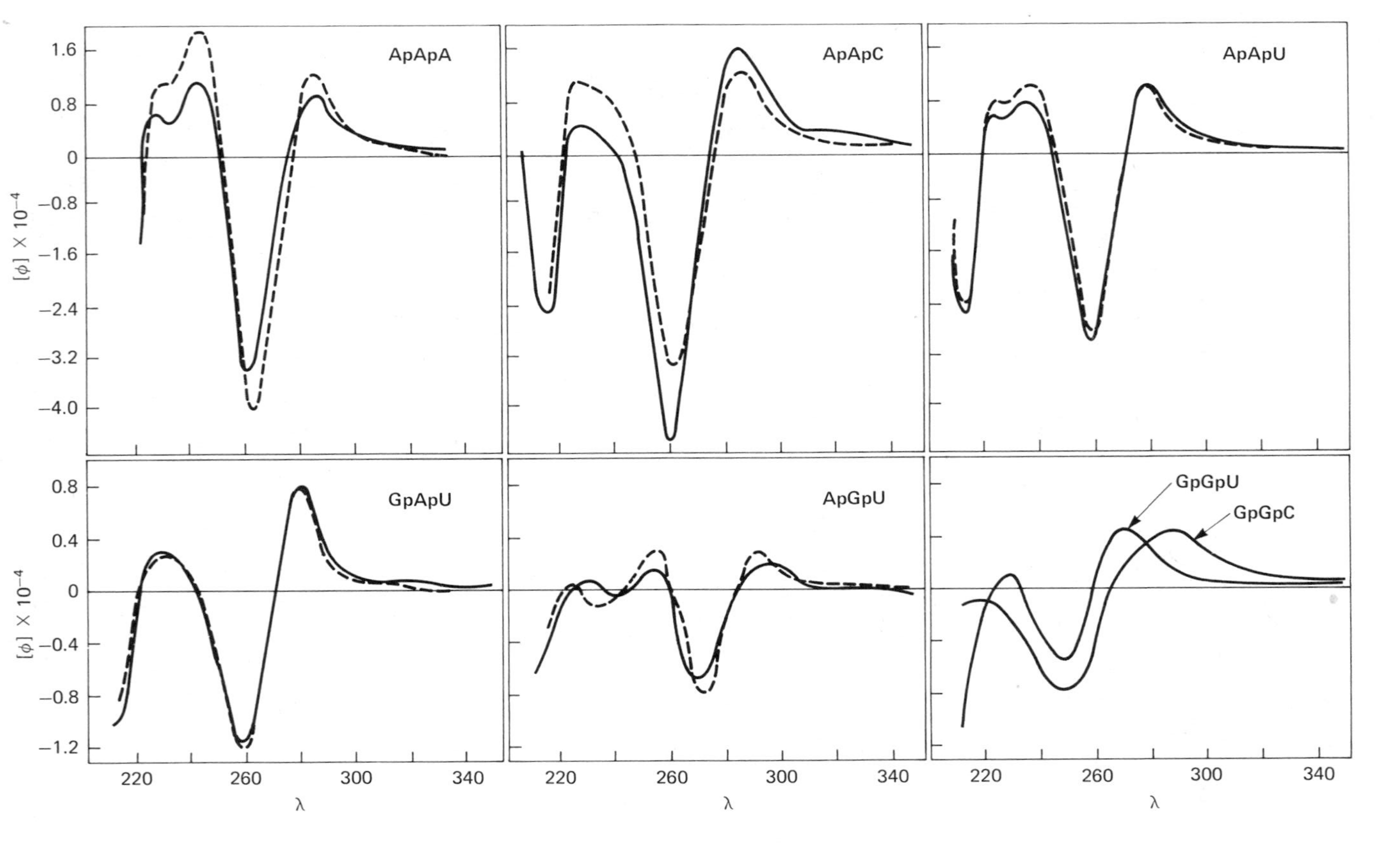

FIGURE 3-6 Molar rotation of trinucleoside diphosphates compared with that calculated from the appropriate sum of their constituent dinucleoside phosphates and mononucleosides. Solid line (————) is measured; dashed line (– – – – – –) is calculated. [Reprinted with permission from C. R. Cantor and I. Tinoco, Jr., *J. Mol. Biol.*, **13**, 65 (1965).]

pH. The dinucleoside phosphate has a single charge on the phosphate. As the critical requirement is that $I_{XY} = I_{XY}'$, it probably does not matter much which monomer is used. The other requirement for the equation to hold is that next nearest-neighbor interaction I_{XZ} be negligible. This may not be true for trimers of the type PupUpPu (Pu = purine). The ORD of the trimer ApUpG is very different from that calculated from the nearest-neighbor equation[25]; its sequence isomers ApGpU and GpApU are predicted correctly (see Fig. 3-6). The simplest explanation for this is that uracil stacks so slightly it swings out and allows adenine and guanine to interact. With the exceptions noted, the spectral properties of the trimers can be calculated from the dimers and monomers.

The important conclusion from the utility of the equation is that the local conformation of the trimers is very similar to that of the dimers. Although neither oligomer is rigid, both apparently have nearly the same time average orientation of the nearest-neighbor bases. This means that detailed knowledge of the structure of the dimers will be immediately applicable to oligomers and polymers.

IV POLYNUCLEOTIDES

A. Single-Stranded Structures

Many properties and characteristics of natural and synthetic polynucleotides have been studied. We will limit ourselves in this section to local, physical properties. This means mainly optical properties and other kinds of spectra. We can emphasize only the fundamental principles; for details and applications the reader is referred to the continuing review articles being published.[26-30]

1. RIBOPOLYMERS

No X-ray data have been interpreted for polynucleotides in terms of single strands. Apparently these structures are too disordered to give appropriate diffraction patterns. However, analysis of the optical properties of the homopolynucleotides poly A and poly C leads to a structure in which the bases are stacked similar to the conformation in a dinucleoside phosphate. The evidence for this includes the similarity in ORD between the polynucleotide and the dinucleoside phosphate. With the same assumptions used for deriving the equation relating dimers to trimers, one can obtain the following equation for the relation between a polymer property and its constituent dimer and monomer properties. *P*(   ) can represent the circular dichroism, absorption, or rotation at any wavelength for monomer, $P(\text{N})$, or dimer, $P(\text{NpN})$.

$$P(\text{poly N}) = 2P(\text{NpN}) - P(\text{N})$$

Figure 3-7 shows this relation for polyadenylic acid (poly A), polycytidylic acid (poly C), and polyuridylic acid (poly U). The agreement is reasonable; the discrepancies are probably due to next-nearest-neighbor interactions rather than different geometry of the nearest-neighbors. Further evidence for stacking in polynucleotides is summarized by Felsenfeld and Miles.[26]

The temperature dependence of the optical properties of poly A have been studied by many workers.[26] If one assumes no cooperativity in the stacking interaction (the stacking of each base is independent of the others), the data can be treated identically to that of ApA (see Section II). An equilibrium constant K is calculated from the data as before and log K is plotted vs. $1/T$. Using absorbance measurements, Leng and Felsenfeld[17] obtained $\Delta H^\circ = -13$ kcal for the stacking interaction in poly A. This is similar to the value of $\Delta H^\circ = -10$ kcal these authors found for ApA and is consistent with a small amount of cooperativity. Applequist and Damle[16] did consider cooperativity. They found $\Delta H^\circ = -9.4$ kcal/mole, $\Delta S^\circ = -29.3$ cal/deg mole and a cooperativity parameter, $\sigma = 0.6$. A noncooperative reaction has $\sigma = 1$; polypeptides have $\sigma \cong 10^{-4}$ for their highly cooperative helix-coil transition.

The structure that is consistent with all the data is that poly A and poly C tend to have stacked bases. The structure is not rigid, but is similar to the dinucleoside phosphates. Maybe a reasonable picture to visualize is one in which the bases are connected by springs which undergo torsional oscillation plus extention and compression. These hypothetical springs have very special characteristics which incorporate the constraints of the polynucleotide bond lengths and bond angles.

Polyuridylic acid (poly U) is special. It seems to have only very slight local structure in addition to that imposed by the sugar-phosphate backbone. Its optical properties are more nearly like the monomer properties than either poly A or poly C.[31] Figure 3-8 compares the ORD of poly U and uridylic acid with that of poly A and adenylic acid (all at neutral pH, room temperature). It is clear that the ORD of poly A is markedly different in shape and magnitude from adenylic acid, while poly U is similar to its monomer. At low temperature poly U does form an ordered double-stranded structure.[32]

Polyguanylic acid (poly G) has not been studied extensively; however its CD[33] is quite different from the CD of guanosine. Interpretation is difficult because poly G aggregates so strongly (as does oligo G)[34]; it is not clear that anyone has observed single-stranded regions of repeating guanine residues. We would expect poly G to stack similarly to poly A and poly C, but there is no direct evidence for this.

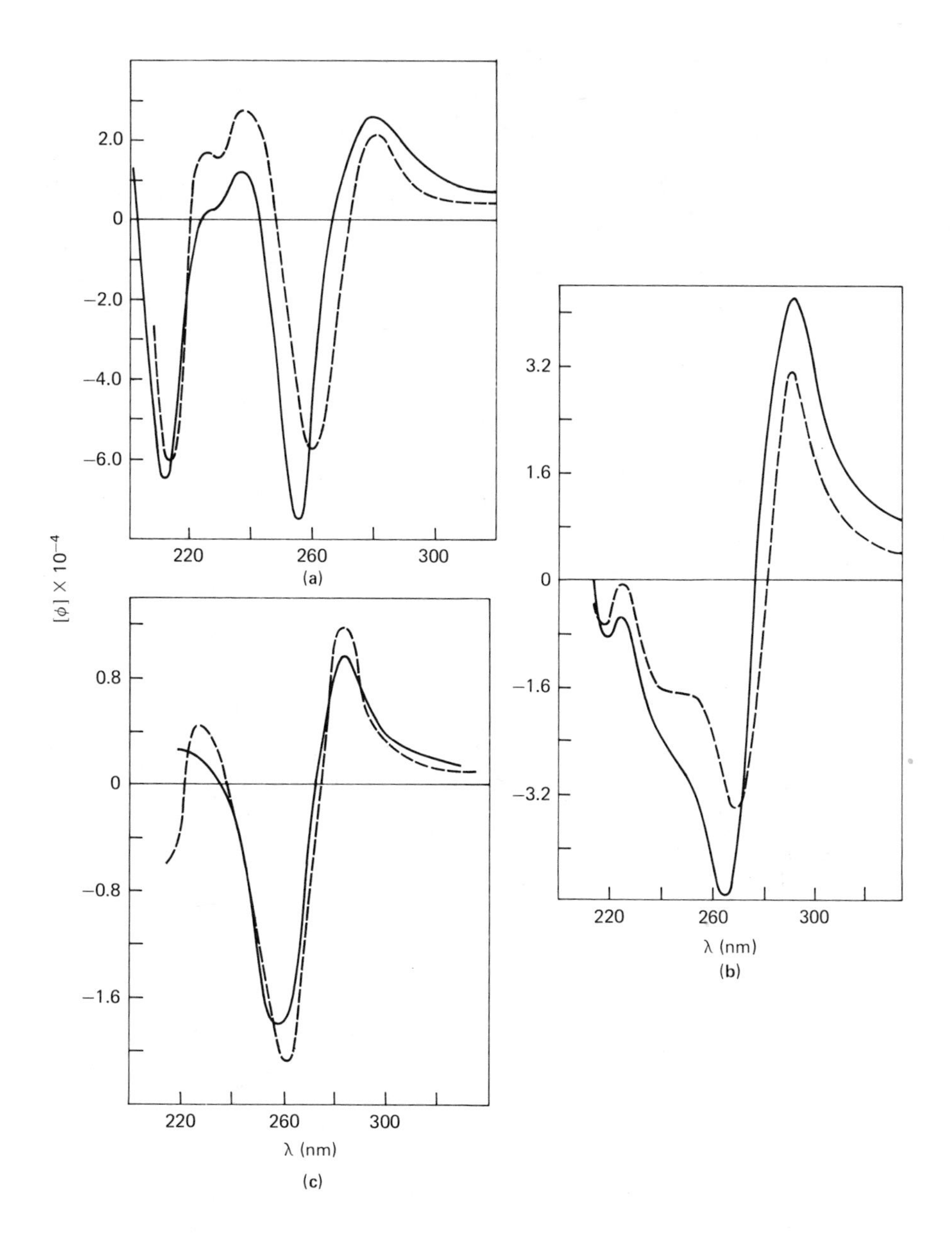

FIGURE 3-7 Molar rotation of polynucleotides at neutral pH compared with that calculated from the appropriate sum of their constituent dinucleoside phosphates and mononucleosides. Solid line (————) is measured; dashed line (– – – – – –) is calculated. (a) Poly A; data from D. N. Holcomb and I. Tinoco, Jr., *Biopolymers,* **3**, 121 (1965). (b) Poly C and (c) Poly U; data from P. K. Sarkar and J. T. Yang, *J. Biol. Chem.*, **240**, 2088 (1965). [Reprinted with permission from *J. Mol. Biol.*, **20**, 39 (1966).]

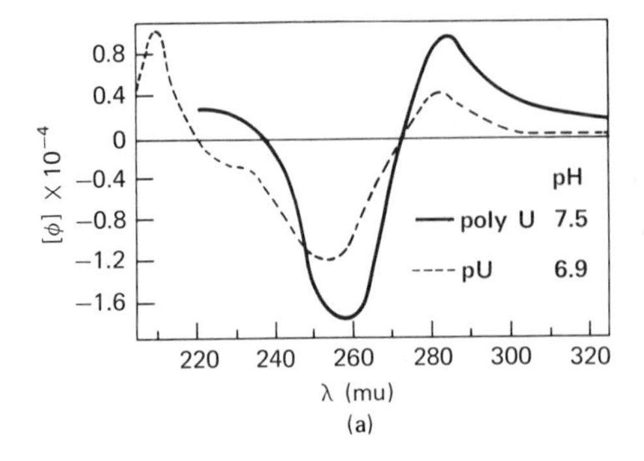

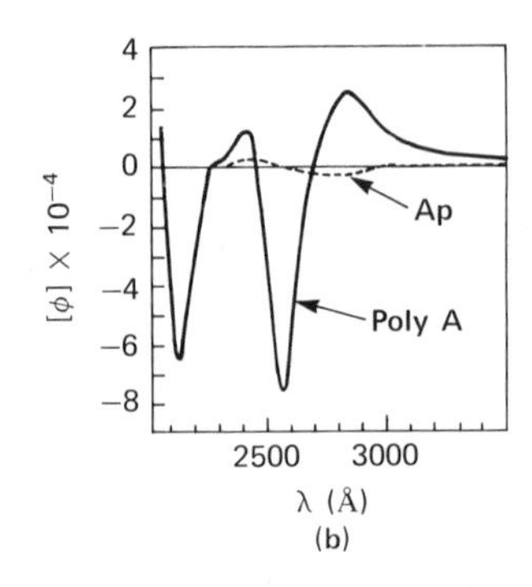

FIGURE 3-8 The optical rotatory dispersion (ORD) of polyadenylic acid and 2′,3′-adenylic acid on the left compared with the ORD of polyuridylic acid and 5′-uridylic acid on the right. All solutions were at neutral pH and room temperature. [Data from D. N. Holcomb and I. Tinoco, Jr., *Biopolymers*, 3, 121 (1965) and *J. Mol. Biol.*, **20**, 39 (1966). Reprinted with permission.]

Polymers containing more than one base also have properties predictable from the measured properties of oligomers. The generalization of the previous equation is

$$\mathrm{P(polymer)} = 2 \sum_{i=1}^{4} \sum_{j=1}^{4} f_{ij}\,\mathrm{P}(I\mathrm{p}J) - \sum_{i=1}^{4} f_i\,\mathrm{P}(I)$$

The mole fraction of each monomer in the polymer is f_i; the mole fraction of each dimer is f_{ij}. The total number of moles of dimers is one less than the number of moles of monomers; for a long polymer this distinction is negligible. If the nearest-neighbor frequencies are not known, they can be approximated by their most probable values; that is, f_{ij} is replaced by $f_i f_j$. This equation has been tested with synthetic polymers of alternating base sequence and with single-strand regions of natural polymers. An alternating copolymer of adenylic acid and cytidylic acid, poly AC, fits well, but poly GU does not.[25] It may be that the uracil bases swing out to allow the guanines to stack. There is evidence for similar behavior in trinucleotides[35] (see Section III). Nevertheless, it is possible to predict approximately the optical properties of natural RNA single strands from the equation. Fig. 3-9 shows the calculated and measured ORD of tobacco mosaic virus RNA at low ionic strength and neutral pH. The low salt prevents the formation of double-strand regions. Good agreement is found. This implies that the bases in the RNA have relative conformations like those in dinucleoside phosphates. This is true in general, although bases next to uracil may be different.[35] As the strongly stacked bases have the largest circular dichroism, they dominate the CD of the polymer.

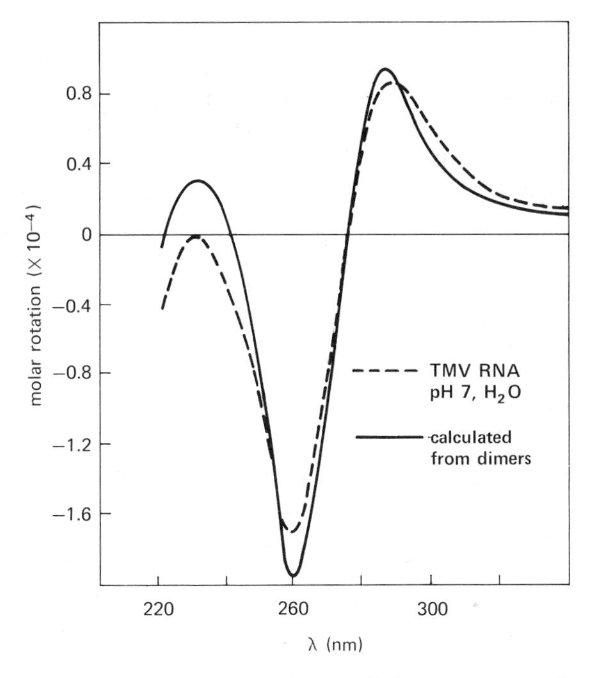

FIGURE 3-9 The optical rotatory dispersion of salt-free tobacco mosaic virus RNA at neutral pH and room temperature. The solid line is measured; the dashed line is calculated. [Data from *J. Mol. Biol.*, **20**, 39 (1966). Reprinted with permission.]

2. DEOXYRIBOPOLYMERS

Polydeoxyribonucleotides behave similarly, but not identical to, polyribonucleotides. Polydeoxyriboadenylic acid (poly dA) has a locally ordered structure as shown by its hypochromicity, its ORD and its CD.[36,37] The geometry of the ordered form is different from that of polyriboadenylic acid (poly A). This is clear from the fact that its CD is quite different and that it is more stable to unstacking by dioxane than is poly A.[37] How it differs in conformation from poly A is not known; it could just be that there is more overlap between adjacent bases in poly dA. Poly-2′-*O*-methyladenylic acid (poly-2′-*O*-mA) has a CD spectrum nearly identical to that of poly A and therefore different from poly dA.[38] Poly dA, poly rA, and poly-2′-*O*-mA all form double-stranded structures in the presence of acid.[37,38]

Polydeoxyribocytidylic acid (poly dC) is less ordered than polyribocytidylic acid (poly C). Its ORD is similar in shape to poly C, but has a much smaller magnitude. Its hypochromicity is smaller than poly C.[36] It is less stable to heat and to ethylene glycol unstacking.[37] Poly dC has stacking properties more consistent with the studies on nucleosides and bases. It is poly C which has an unexpectedly high amount of stacking. Studies of cytosine containing dimers with deoxyribose (dC) or arabinose (aC) in addition to ribose (C)[39] indicate that the presence of ribose in the 3′-linked unit of the dimer is needed for stacking. That is, CpC, CpA, and ApaC are stacked, but dCpdC, aCpaC, and aCpA are not.

Polythymidylic acid (poly dT) shows very little local order and is thus similar to polyuridylic acid (poly U). It does differ from poly U and poly rT in that at low temperatures in the presence of Mg^{2+} ion it does not form a double-stranded structure.[36]

3. CONCLUSIONS

An approximate knowledge of the single-stranded conformations has been obtained. The single strands are definitely not disordered polymers. Their local structures seem to be well defined and dependent on the base composition, the base sequence, and the nature of the sugar. The bases are stacked to form a helix similar to that which a single strand has in double-stranded structures.

The forces which determine the conformations of polynucleotides have been discussed earlier (Chapter 2) in relation to monomer properties. In polymers long-range interactions such as electrostatic effects from the phosphate groups become particularly important. To study solely short-range, local interactions one can use the "theta" conditions of Flory.[40] For a particular temperature and solvent the long-range interactions in the polymer become negligible; polymer-polymer effects are just balanced by polymer–solvent ones. For these conditions the polymer solution acts like an ideal thermodynamic solution. Felsenfeld and co-workers[41] have used theta conditions to study the unperturbed dimensions of polyadenylic and polyuridylic acid. They find strong stacking attraction between the bases in poly A at room temperature and below; poly U, however, does not show this base stacking. As the stacking in poly A diminishes with increasing temperature, its mean chain dimensions approach those of poly U. The chain dimensions for both polymers are considerably larger than those calculated if one assumes free rotation around the bonds (see Chapter 5). The conclusion is that bond orientations are highly restricted and that they are consistent with the orientations found for the monomers in crystals.[42]

We would recommend that an excellent way to learn about detailed conformations and the forces responsible for them in single-strand polynucleotides is to study conformations of oligonucleotides.

Another way of experimentally learning what local structures are present in a single-stranded polynucleotide is to study polymers of simple, known sequence. Measured properties of the model polymers can be used to calculate the property of a polymer of known sequence. Alternatively, measurements on a polynucleotide of unknown sequence can be used to obtain information about its near-neighbor sequences.

If the local structure is determined by the nearest-neighbor sequences only, then the 16 (4 squared) possible nearest-neighbor sequences are pertinent. However, for an infinite, or reentrant, polymer, only 13 nearest-neighbor frequencies are independent.[43] The three constraints which reduce the number of possible frequencies are (1) the sum of all the nearest-neighbor frequencies must be one and (2) two equations of the type:

$$f_{\mathrm{AC}} + f_{\mathrm{AG}} + f_{\mathrm{AU}} = f_{\mathrm{CA}} + f_{\mathrm{GA}} + f_{\mathrm{UA}}$$

where f_{ij} = mole fraction of dinucleotide in the polymer. The reader can easily convince himself that this must be true by counting nearest-neighbor frequencies in any circular oligonucleotide. This means that all properties which depend on this local structure can be calculated once the properties of 13 model polynucleotides have been measured. For example, the extinction coefficient at 260 nm could be calculated from its nearest-neighbor frequencies, f_{ij}. If all nearest-neighbor frequencies are not known, one can approximate the f_{ij} from the known base composition; this reduces the model compounds needed as in this approximation $f_{ij} = f_{ji}$.

It is clear that in general it will not be possible to use the measured property of the single-stranded polymer to determine the nearest-neighbor frequencies. If circular dichroism were to be used, it would be necessary to measure the CD for at least 13 wavelengths to get 13 equations. However, these equations could only be solved uniquely if each model polymer had a CD spectrum linearly independent of the others. This is not true for the present accuracy of measurement.

If longer range interactions are present so that more than nearest-neighbors are important, then more model polymers must be studied. For second nearest-neighbor properties (three bases interacting), 49 polymers are necessary to specify the properties of all possible polynucleotides.

REFERENCES

1 M. M. Warshaw and I. Tinoco, Jr., *J. Mol. Biol.*, **20**, 29 (1966).
2 J. Brahms, J. C. Maurizot, and A. M. Michelson, *J. Mol. Biol.*, **25**, 481 (1967).
3 A. M. Michelson, *Molecular Associations in Biology*, B. Pullman, ed., Academic Press, New York, p. 93 (1968).
4 R. C. Davis and I. Tinoco, Jr., *Biopolymers*, **6**, 223 (1968).
5 P. O. P. Ts'o, N. S. Kondo, M. P. Schweizer, and D. P. Hollis, *Biochemistry*, **8**, 997 (1969).

6 B. W. Bangerter and S. I. Chan, *J. Amer. Chem. Soc.*, **91**, 3910 (1969).
7 M. Guéron, R. G. Shulman, and J. Eisinger, *Proc. Nat. Acad. Sci., U.S.*, **55**, 1387 (1966).
8 C. Helene and A. M. Michelson, *Biochim. Biophys. Acta,* **142**, 12 (1967).
9 M. M. Warshaw and C. R. Cantor, *Biopolymers,* **9**, 1079 (1970); C. R. Cantor, M. M. Warshaw, and H. Shapiro, *Biopolymers,* **9**, 1059 (1970).
10 M. M. Warshaw and I. Tinoco, Jr., *J. Mol. Biol.,* **13**, 54 (1965).
11 C. A. Bush and I. Tinoco, Jr., *J. Mol. Biol.,* **23**, 601 (1967).
12 S. I. Chan and J. H. Nelson, *J. Amer. Chem. Soc.,* **91**, 168 (1969).
13 W. C. Johnson, Jr., M. S. Itzkowitz, and I. Tinoco, Jr., *Biopolymers,* **11**, 225 (1972).
14 S. Arnott, S. D. Dover, and A. J. Wonocott, *Acta Cryst.,* **B25**, 2192 (1969).
15 J. F. Brandts, *Structure and Stability of Biological Macromolecules*, S. N. Timasheff and G. D. Fasman, eds., Marcel Dekker, New York, p. 213 (1969); J. T. Powell, E. G. Richards, and W. B. Gratzer, *Biopolymers,* **11**, 235 (1972).
16 J. Applequist and V. Damle, *J. Amer. Chem. Soc.,* **88**, 3895 (1966).
17 M. Leng and G. Felsenfeld, *J. Mol. Biol.,* **15**, 455 (1966).
18 D. Glaubiger, D. A. Lloyd, and I. Tinoco, Jr., *Biopolymers,* **6**, 409 (1968).
19 H. Simpkins and E. G. Richards, *Biochemistry,* **6**, 2513 (1967).
20 S. L. Davis, Ph.D. Thesis, University of California, Berkeley (1965).
21 H. Simpkins and E. G. Richards, *J. Mol. Biol.,* **29**, 349 (1967).
22 C. R. Cantor and I. Tinoco, Jr., *J. Mol. Biol.,* **13**, 65 (1965).
23 Y. Inoue, S. Aoyagi, and K. Nakanishi, *J. Amer. Chem. Soc.,* **89**, 5701 (1967).
24 C. R. Cantor and I. Tinoco, Jr., *Biopolymers,* **5**, 821 (1967).
25 D. M. Gray, I. Tinoco, Jr., and M. J. Chamberlin, *Biopolymers,* **11**, 1235 (1972).
26 G. Felsenfeld and H. T. Miles, *Ann. Rev. Biochem.,* **36**, 407 (1967).
27 A. M. Michelson, J. Massoulie, and W. Guschlbauer, *Progress in Nucleic Acid Research and Molecular Biology*, Vol. 6, Academic Press, New York (1967), p. 83.
28 M. Guéron and R. G. Shulman, *Ann. Rev. Biochem.,* **37**, 571 (1968).
29 S. Beychok, *Ann. Rev. Biochem.,* **37**, 437 (1968).
30 J. T. Yang and T. Samejima, *Progress in Nucleic Acid Research and Molecular Biology,* Vol. 9, Academic Press, New York (1969), p. 224.
31 E. G. Richards, C. P. Flessel, and J. R. Fresco, *Biopolymers,* **1**, 431 (1963).
32 J. C. Thrierr and M. Leng, *Biochim. Biophys. Acta,* **182**, 575 (1969).
33 F. H. Wolfe, K. Oikawa, and C. M. Kay, *Canad. J. Biochem.,* **47**, 637 (1969).
34 F. Pochon and A. M. Michelson, *Proc. Nat. Acad. Sci., U.S.,* **53**, 1425 (1965).
35 J. Brahms, A. M. Aubertin, G. Dirheimer, and M. Grunberg-Manago, *Biochemistry,* **8**, 3269 (1969).
36 P. O. P. Ts'o, S. A. Rapoport, and F. J. Bollum, *Biochemistry,* **5**, 4153 (1966).
37 A. Adler, L. Grossman, and G. D. Fasman, *Proc. Nat. Acad. Sci., U.S.,* **57**, 423 (1967).
38 A. M. Bobst, P. A. Cerutti, and F. Rottman, *J. Amer. Chem. Soc.,* **91**, 1246 (1969).
39 J. Brahms, J. C. Maurizot, and J. Pilet, *Biochim. Biophys. Acta,* **186**, 110 (1969).
40 P. J. Flory, *Principles of Polymer Chemistry*, Cornell University Press, Ithaca, N.Y. (1953).
41 L. D. Inners and G. Felsenfeld, *J. Mol. Biol.,* **50**, 373 (1970); H. Eisenberg and G. Felsenfeld, *J. Mol. Biol.,* **30**, 17 (1967).
42 M. Sundaralingam, *Biopolymers,* **7**, 821 (1969).
43 D. M. Gray and I. Tinoco, Jr., *Biopolymers,* **9**, 223 (1970).

chapter 4
double-stranded polynucleotides

I INTRODUCTION

In this chapter we discuss the structure of DNA and other double-stranded polynucleotides. This includes base composition, which leads to the generalization of base complementarity, and the detailed elaboration of the basic proposal by Watson and Crick of a double helical structure for DNA. Among the questions to be answered are the parallel or antiparallel disposition of the two strands, the geometry of base pairing, structural parameters of the helixes, and differences between DNA and RNA. Since X-ray diffraction has played the predominant role in structure determination of double-stranded polynucleotides, we discuss the theory of fiber diffraction in some detail.

A brief historical survey may be interesting at this point. Several workers[1–5] intimately concerned with the early establishment of the basic ideas of DNA structure have summarized the events leading to the recognition by Watson and Crick that DNA, isolated native from nearly all biological sources, has a double helical structure in which adenine pairs with thymine and guanine with cytosine. Early measurements of the hydrodynamic properties of DNA solutions showed that such solutions had high viscosity and low birefringence. This indicated that DNA was a long, thin polymer.[6] The negative birefringence and ultraviolet dichroism of DNA solutions showed that the bases, which are responsible for the absorption of ultraviolet light, were aligned perpendicular to the long axis of the nucleic acid molecule.[7] Early studies by Astbury[8] of the X-ray diffraction from DNA fibers showed a 3.4 Å periodicity. Therefore, Astbury suggested that the bases were stacked 3.4 Å apart like a "pile of pennies." At the same time, electrometric titration studies indicated that the bases were joined by amino-carboxyl hydrogen bonds.[9] Thus, a picture arose of DNA as composed of polynucleotide chains joined by hydrogen bonds between the bases.[10] However, the number of such chains, and their geometrical disposition relative to one another, was still completely unclear.

Early X-ray studies of moist DNA fibers suggested that DNA in fact adopted a helical configuration,[11] although later studies unfortunately temporarily cast doubt on this.[12] The number of chains in the helix, if indeed it were a helix, remained uncertain, as stated above; and Pauling and Corey,[13] fresh from their success with the α-helix model of polypeptide chains, constructed three-chain helixes. These, however, did not agree with the X-ray data that were available.

Chargaff,[14] meanwhile, was investigating the chemical structure and composition of DNA and demonstrated the consistent relation %A = %T, %G = %C. This, in turn, suggested some type of pairing between the bases.[15] Finally, Watson and Crick,[16] having an existential faith both in the helicity of DNA, although the X-ray evidence for helixes was questionable at that time, and in the idea that DNA was two-stranded rather than three, built models in which A paired with T and G and with C on the inside of the helix with the phosphate groups on the outside, minimizing electrostatic repulsion. This had the virtue of agreeing with the chemical evidence and also seemed to be in satisfactory accord with most of the X-ray experimental data. This A · T, G · C base pairing was extremely important, because it had the consequence that the base pairs were of equal size and, thus, regardless of irregularities in base sequence along the chain, permitted construction of a helix of regular diameter. These regular helixes could then form the crystalline fiber preparations which were in fact found. Subsequent X-ray studies confirmed and refined the structure proposed by Watson and Crick.

II CHEMICAL STRUCTURE

A. Base Composition

As far as is known, all DNAs, with a few minor exceptions, have the same biological function. This is to encode genetic information which is passed on as a genotype to future generations, and which is deciphered to give the phenotype of enzymic and structural molecules which constitute the virus, cell, or organism. As a consequence of this functional similarity, nearly all DNAs have the same basic principles of structural organization. This is reflected in their base composition. A useful tabulation of base compositions of DNA from a wide variety of organisms is given by Mahler and Cordes[17]. Some generalizations are[14]:

1. The base composition of the DNA is characteristic of the organism from which it has been extracted. Different cells or tissues of the same organism have identical or closely similar base composition, indicating that their genetic inheritance is identical. The base composition, however, can vary from organism to organism within rather wide limits. The GC content generally ranges from 40 to 45% in mammals and from 25 to 75% in bacteria.

2. It is observed that closely related organisms exhibit similar base composition in their DNA. Much recent work has been devoted to extending this insight by demonstrating that base sequences, as well as base compositions, are similar among related organisms. These findings provide a basis for a chemical understanding of evolution.

3. Chargaff[14] pointed out a regularity which has had great implications for the elucidation of the three-dimensional structure of DNA. He noted that in nearly all DNAs examined the percentage of A equaled the percentage of T and the percentage of G equaled the percentage of C. This complementary relation between A and T and between G and C is a key piece of evidence supporting the structural proposal by Watson and Crick of the double-stranded, complementary base-paired structure of DNA. The relations $A = T$, $G = C$ yield by addition two interesting chemical consequences. The first is that $A + G = T + C$; or chemically, that the number of purines in the DNA duplex equals the number of pyrimidines. The second is that $A + C = G + T$; or that the number of amino groups in the DNA equals the number of keto groups.

4. It is interesting to note, however, that A is not equal to T, nor G equal to C, for the DNA isolated from the bacteriophage ϕX 174. Nor do these equalities hold for the A and B chains of phage alpha. This is an indication that these DNAs are single-stranded.

5. A very curious DNA is that found in crabs. Here we note that $(A + T)/(G + C) = 17.5$, or that the $G + C$ content is only 5.4%, so that this naturally occurring DNA is nearly entirely an AT copolymer which, it turns out, is strictly alternating. The biological utility of this chemical species is not evident.

B. Strand Orientation and Nearest-Neighbor Frequencies

Having established that native DNA is double-stranded, it then becomes important to determine whether the two strands are oriented parallel or antiparallel to one another. That is, if one strand is laid out so that it progresses from the 5′ terminus on the left to the 3′ terminus on the right, a parallel orientation is one in which the complementary strand also progresses from the 5′ on the left to 3′ on the right. An antiparallel configuration is one in which the complementary strand progresses from the 3′ terminus on the left to the 5′ terminus on the right. Watson and Crick[16] suggested the antiparallel orientation because it led to a regular helix backbone and because it provided nucleotides on both strands with equivalent environments.

However, more direct evidence for the antiparallelism of the two strands in DNA has come from chemical studies.[18] The method used is that of nearest-neighbor frequency determination developed by Kornberg and co-workers.[19] In this method a strand of DNA is used as a template on which to synthesize a new strand with the enzyme DNA polymerase. All four deoxyribonucleotide triphosphates are present as substrates, one of which, say X, is radioactively labeled in the innermost (α) phosphate:ppp*X. After synthesis, the DNA is degraded with micrococcal deoxyribonuclease and spleen phosphodiesterase to give the 3′-phosphorylated nucleotides as products. Thus, the radioactive phosphorus is transferred to the nearest neighbor of X. By measuring the fraction of P* which ends up attached to A, T, G, or C, one determines what fraction of the time an A, T, G, or C, respectively, was the nearest neighbor to X.

Using this method one can determine the nearest-neighbor frequencies in DNA. It is evident from inspection of Fig. 4-1(a) that if the strands run parallel, the following equalities for the nearest-neighbor frequencies must be satisfied:

ApT = TpA   GpA = CpT
ApG = TpC   GpT = CpA
ApC = TpG   GpC = CpG

On the other hand, if the strands run antiparallel, we have, as is evident from inspection of Fig. 4-1(b),

ApC = GpT   GpA = TpC
ApG = CpT   CpA = TpG

The results of nearest-neighbor frequency determination in DNA molecules from numerous organisms support the antiparallel polarity hypothesis. As will be shown later in this chapter, this has important consequences

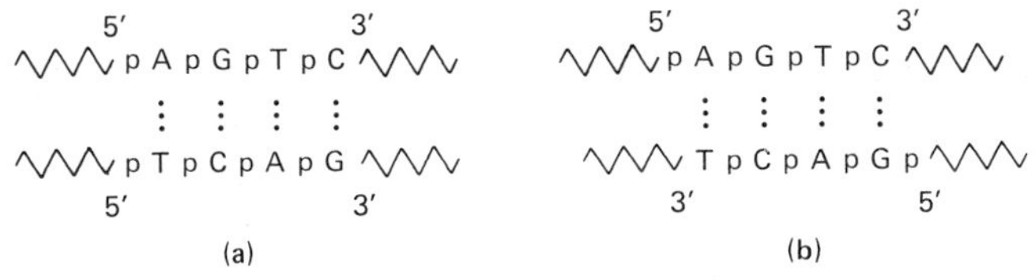

5′ 3′
pApGpTpC
pTpCpApG
5′ 3′
(a)

5′ 3′
pApGpTpC
TpCpApGp
3′ 5′
(b)

FIGURE 4-1 Illustration of nearest-neighbor relations for (a) parallel and (b) antiparallel strand orientation.

for the mechanism of DNA duplication, in which each strand serves as a template for a complementary copy of itself.

In summary, determinations of DNA base composition lead to the complementarity rules A = T, G = C, suggestive of base pairing and a two-stranded structure. Nearest-neighbor base frequency measurements indicate that the two strands are antiparallel. This point has been directly verified by X-ray diffraction, as is discussed later in the chapter.

III PHYSICAL STRUCTURE DETERMINATION

A. X-Ray Diffraction

An X-ray diffraction photograph of a moist fiber of the Li salt of DNA is shown in Fig. 4-2.[1] It is immediately evident that this photo, which is an unusually good one, contains many fewer spots than are commonly seen in photos obtained from crystals of small molecules or proteins. One can therefore determine much less atomic detail, and a necessary consequence is that in determination of molecular structures from fiber diffraction, certain stereochemical assumptions must be made. These are, basically, that monomeric structures, bond lengths, bond angles, and van der Waals distances are nearly the same in the polymer.

Figure 4-2, it will be shown shortly, is characteristic of diffraction from a helix. Thus the major information immediately available from X-ray diffraction studies of nucleic acids is the parameters of the helix. These are the pitch, or distance along the axis per turn of the helix; the step height, or vertical distance between adjacent residues; the number of residues per turn; the pitch angle; and the radius.

Some of these parameters are illustrated in Fig. 4-3(a). With good quality microcrystalline preparations, one can obtain information on packing of the fibers into the unit cell of the crystal, and thus on intermolecular interactions. Finally, with detailed consideration of the intensities of the diffraction maxima, and successive refinement of proposed structures, it is possible to get fairly accurate information on the disposition of the base, sugar, and phosphate groups in the helix, and on the helix sense or handedness.

Model building is essential in this work. The procedure followed is, roughly, to guess at a structure compatible with the measured helical parameters. A model of the structure is built and is examined for acceptable stereochemistry. The diffraction pattern of the model is then calculated, and checked against the observed positions and intensities of the diffraction photograph. The model is adjusted in an attempt to improve agreement with

FIGURE 4-2 X-ray diffraction pattern of LiDNA at 66% relative humidity. [Courtesy of Prof. M. H. F. Wilkins, Medical Research Council, King's College, London. Reprinted with permission.]

the experimental data, and this process is continued until satisfactory agreement is reached or until it is clear that the assumed starting structure cannot be modified, except fundamentally, to produce the observed pattern. It is difficult by such a procedure to show that a proposed structure is unique. Luckily, there are three forms of DNA – known as the A, B, and C forms – which are readily interconvertible with changes in relative humidity and counterion, and which are modifications of the same basic structure. This provides much greater confidence in the correctness of the basic structure than would be forthcoming were there only one form.

Before embarking on a discussion of the results obtained by X-ray diffraction studies on nucleic acids, it will be useful to discuss briefly the basic results of X-ray diffraction theory and their application to diffraction from helixes.[20,21]

The most familiar result is Bragg's law

$$n\lambda = 2d \sin \theta_b$$

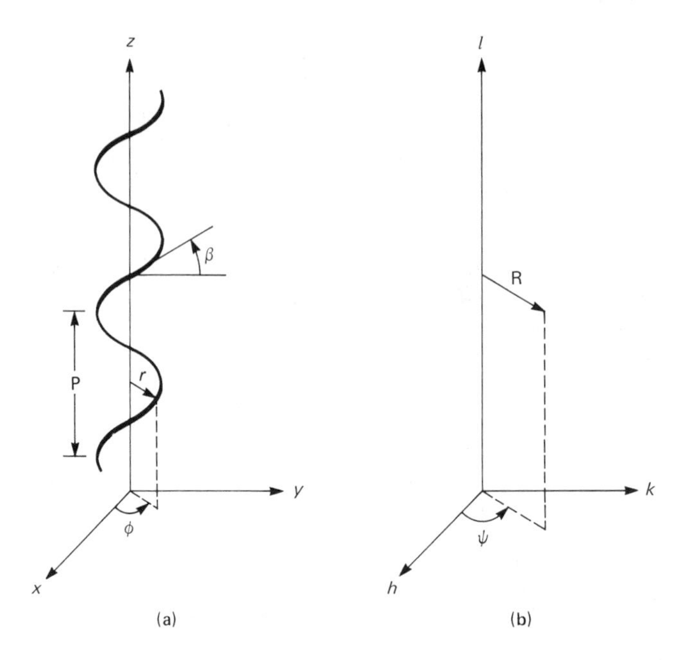

FIGURE 4-3 (a) Continuous helix of pitch P, radius r, and pitch angle β. (b) Cylindrical coordinate system in reciprocal space.

where d is the distance between scattering planes, θ_b is half the scattering angle, λ is the X-ray wavelength, and n is an integer. $\sin \theta_b$ is proportional to d^{-1}, so that small distances correspond to large angles and vice versa. Astbury's[8] early studies of X-ray diffraction from DNA fibers showed a strong reflection at an angle corresponding to 3.4 Å, which he correctly interpreted as being the distance between the planar bases stacked one upon the other in the fiber. This 3.4 Å reflection is evident as the very intense horizontal streak at the top and bottom of the photograph in Fig. 4-2.

A more detailed treatment of X-ray diffraction leads to the result that the scattering is described by the structure factor F_{hkl}, which is a three-dimensional Fourier transform of the electron density $\rho(x, y, z)$ in the unit cell:

$$F_{hkl} = \iiint \rho(x, y, z) \exp[2\pi i(hx + ky + lz)]\ dxdydz$$

h, k, l are the Miller indices, integers which index the points in reciprocal space.

The scattered intensity I_{hkl} is proportional to the square of the absolute value of the scattering amplitude:

$$I_{hkl} = |F_{hkl}|^2$$

It is these expressions which determine the intensities of the spots on the diffraction photograph, and relate them to the structure of the scattering object. The problem in structure determination by X-ray diffraction is to invert the Fourier analysis, by the inverse process called Fourier synthesis, to obtain $\rho(x, y, z)$.

Let us now specialize these general results to consider diffraction from helixes.[20,22-25] We consider first a continuous helix in which ρ is a constant along the helix backbone, whose value may arbitrarily be set equal to unity. Because of the helical geometry, it is convenient to transform to cylindrical coordinates, as shown in Fig. 4-3(a):

$$x = r\cos(2\pi z/P)$$

$$y = r\sin(2\pi z/P)$$

$$z = z$$

$$dxdydz = \text{constant}\cdot dz$$

r is the radius of the helix, and P is its pitch.

Then, neglecting constants which will appear only in calibration of intensities,

$$F_{hkl} = \int_0^P \exp\left[2\pi i\left(rh\cos\frac{2\pi z}{P} + rk\sin\frac{2\pi z}{P} + zl\right)\right]dz$$

We also transform to cylindrical coordinates in reciprocal space [Fig. 4-3(b)], a geometry appropriate to an X-ray film cylinder with its axis coincident with the fiber axis:

$$R^2 = h^2 + k^2$$

$$\tan\psi = k/h$$

Then the structure factor becomes

$$F(R, \psi, l) = \int_0^P \exp\left\{2\pi i\left[Rr\cos\left(\frac{2\pi z}{P} - \psi\right) + zl\right]\right\}dz \tag{4-1}$$

It can be shown that this integral equals zero unless

$$l = n/P \quad (n = 0, 1, 2, \ldots) \tag{4-2}$$

n is therefore an integer which indexes the layer lines that occur at heights n/P in reciprocal space. The smaller the pitch P, the larger the spacing of the layer lines. One then finds that the structure factor becomes

$$F(R, \psi, n/P) = \int_0^P \exp\left\{2\pi i\left[Rr\cos\left(\frac{2\pi z}{P} - \psi\right) + \frac{nz}{P}\right]\right\}dz$$
$$= J_n(2\pi Rr)\exp[in(\psi + \pi/2)] \tag{4-3}$$

and the intensity becomes

$$I(R, \psi, n/P) = [J_n(2\pi Rr)]^2$$

Thus the intensity of scattering from a continuous helix is independent of ψ. The functions $J_n(x)$ are nth order Bessel functions. Their behavior for small values of n is shown in Fig. 4-4. We see that they behave like damped oscillations, in which the first and highest maximum moves successively away from the meridian axis with increasing order. For a continuous helix, J_n contributes only on the nth layer line. Thus a prominent feature of the diffraction pattern from a continuous helix is a cross formed by the highest intensity reflections, as shown in Fig. 4-5(a).

The half-angle δ of this cross is approximately equal to the pitch angle β of the helix. These angles are defined by

$$\tan \beta = P/2\pi r$$

$$\tan \delta = R_{max}(n)/l(n)$$

where $R_{max}(n)$ is the value of R for which $J_n(2\pi Rr)$ reaches its highest maximum; $l(n)$ is given by Eq. (4-2). Examination of the numerical behavior of the Bessel functions indicates that after the first few orders the distance to the first maximum of $J_n(x)$, $x_{max}(n)$, is

$$x_{max}(n) = 2\pi r R_{max}(n) \approx 1.1n + 0.9$$

Thus

$$\tan \delta = \frac{x_{max}(n)/2\pi r}{n/P} = \frac{x_{max}(n)}{n} \tan \beta \approx 1.1 \tan \beta$$

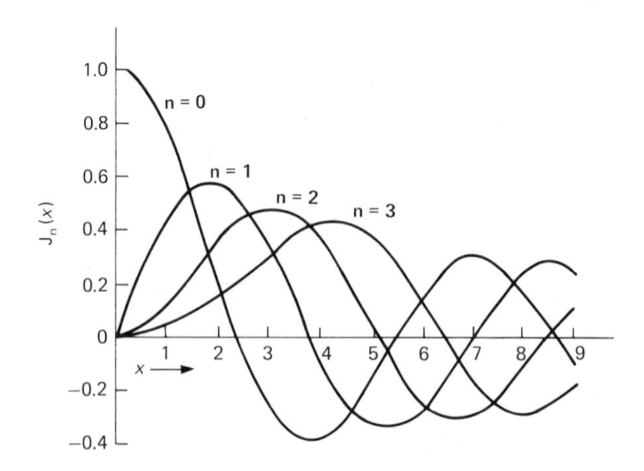

FIGURE 4-4 Zero- through third-order Bessel functions $J_n(x)$.

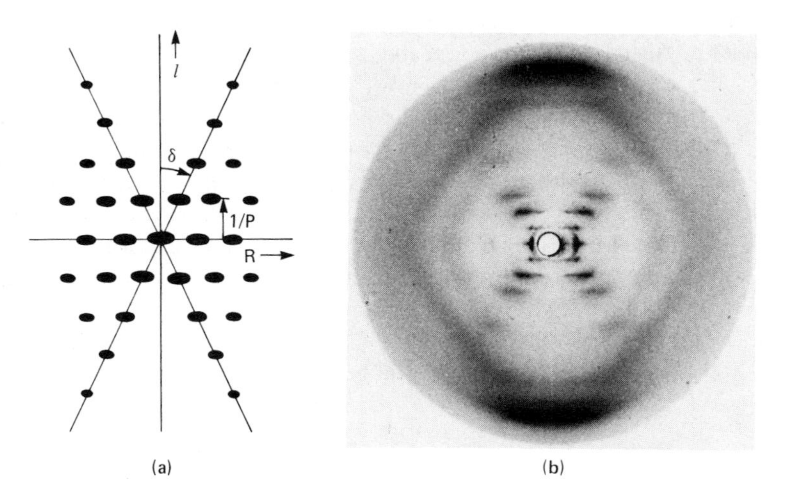

FIGURE 4-5 (a) Diffraction pattern calculated for a continuous helix. (b) Diffraction pattern from moist fiber of NaDNA. [From R. E. Dickerson, in *The Proteins*, 2nd ed., Vol. II, H. Neurath, ed., Academic Press, New York (1964). Reprinted with permission.]

Figure 4-5(b) shows the diffraction pattern from moist fibers of Na DNA. This pattern is strikingly similar to that calculated in Fig. 4-5(a), indicating that NaDNA behaves much like a continuous helix. Some additional features, which appear in Fig. 4-5(b) and more plainly in Fig. 4-2, are due to the 3.4 Å base stacking repeat distance and other structural details.

At the next level of complexity, we consider scattering from a discontinuous helix. That is, we imagine an infinitely thin helical backbone, which contributes nothing to the scattering, on which are arrayed scattering centers with a vertical spacing p. This distribution of electron density may be considered to be the product of two density functions: K, which is unity on the helix and zero otherwise; and H, which is unity on an infinite set of horizontal planes of spacing p, and zero otherwise. Thus KH represents a discontinuous helix, and the scattering amplitude is the Fourier transform of KH. That is, from Eq. (4-1),

$$F(R, \psi, l) = \int_0^P KH \exp\left\{2\pi i\left[Rr\cos\left(\frac{2\pi z}{P} - \psi\right) + zl\right]\right\} dz$$

We now use, without proof, the very important convolution theorem of Fourier analysis,[20,26] which states that the Fourier transform of the product of two functions equals the convolution of the transforms of the individual functions. The convolution operation on two functions f_1 and f_2 is denoted $f_1 * f_2$ and defined as

$$f_1 * f_2 = \int f_1(y - z) f_2(z) dz$$

Physically, this means to place the origin of f_1 at a point of f_2, multiply the values of f_1 and f_2 at that point, and continue over all points of f_2, summing (or integrating) the products.

We have already calculated the Fourier transform of the helix function K; it is Eq. (4-3). The Fourier transform of an infinite set of planes arrayed perpendicular to the z axis in fiber space, which is equivalent to the diffraction pattern of such a set of planes, can easily be shown to be zero save on an infinite set of points along the l axis of reciprocal space with spacing $1/p$. Thus the scattering from a discontinuous helix, which is the convolution of the transforms of K and H, is obtained by setting down Eq. (4-3) with its origin in reciprocal space successively at the points (0, 0, 0), (0, 0, $\pm 1/p$), (0, 0, $\pm 2/p$) . . . . Thus reflections will be observed only at heights

$$l = \frac{n}{P} + \frac{m}{p} \qquad (n, m = 0, \pm 1, \pm 2, \ldots) \tag{4-4}$$

This produces diamond-shaped regions above and below the center of the pattern, by repetition of the cross of Fig. 4-5(a), as is shown in Fig. 4-6. In contrast to the situation with the continuous helix, Bessel functions of several orders now contribute on each layer line.

It is perhaps useful at this point to work out several examples of the theory. The number of scattering units per turn of the helix is P/p. Let us first suppose this is an integer, M. Then Eq. (4-4) becomes

$$lP = n + nM$$

which is an equation to be solved in integers for n. In Fig. 4-6(a), M is 5. When $lP = 0$, the solutions are

$$(n, m) = (0, 0), (\pm 5, \mp 1), (\pm 10, \mp 2), \ldots$$

so that contributions from J_0, J_5, J_{10}, . . . will appear on the zeroth layer line. When $lP = 1$, the solutions are

$$(n, m) = (1, 0), (-4, 1), (6, -1), \ldots$$

so that contributions from J_1, J_4, J_6, . . . will appear on the first layer line. These features are shown in Fig. 4-6(a).

If the number of residues per turn, P/p, is not an integer but instead a rational fraction M/N, where M and N are integers, then the helix repeats in a vertical distance $C = NP$ in which there are N turns and M elements. In this case Eq. (4-4) becomes

$$Cl = Nn + Mm$$

In Fig. 4-6(b), the case is displayed with $M = 16$, $N = 3$. On the zeroth layer line, $Cl = 0$, the only solution to this equation is

$$(n, m) = (0, 0)$$

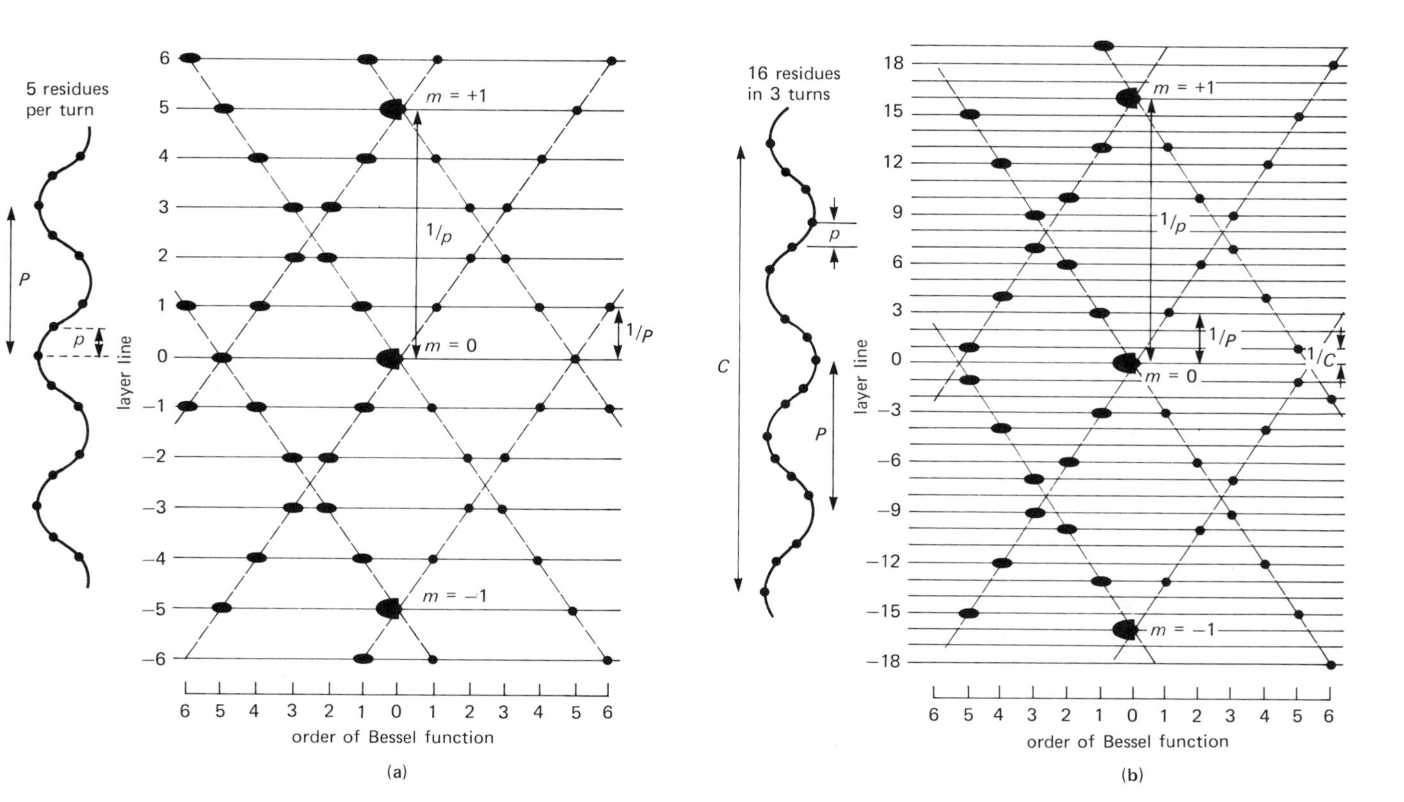

FIGURE 4-6 Calculated diffraction patterns for (a) a discontinuous helix with 5 units per turn, and (b) a discontinuous helix with $5^1/_3$ units per turn. The oval dark regions on the left halves of the diagrams correspond to the first maxima of the Bessel functions. [From H. R. Wilson, *Diffraction of X-Rays by Proteins, Nucleic Acids, and Viruses*, St. Martin's Press, New York (1966). Reprinted with permission from Edward Arnold, Ltd., London.]

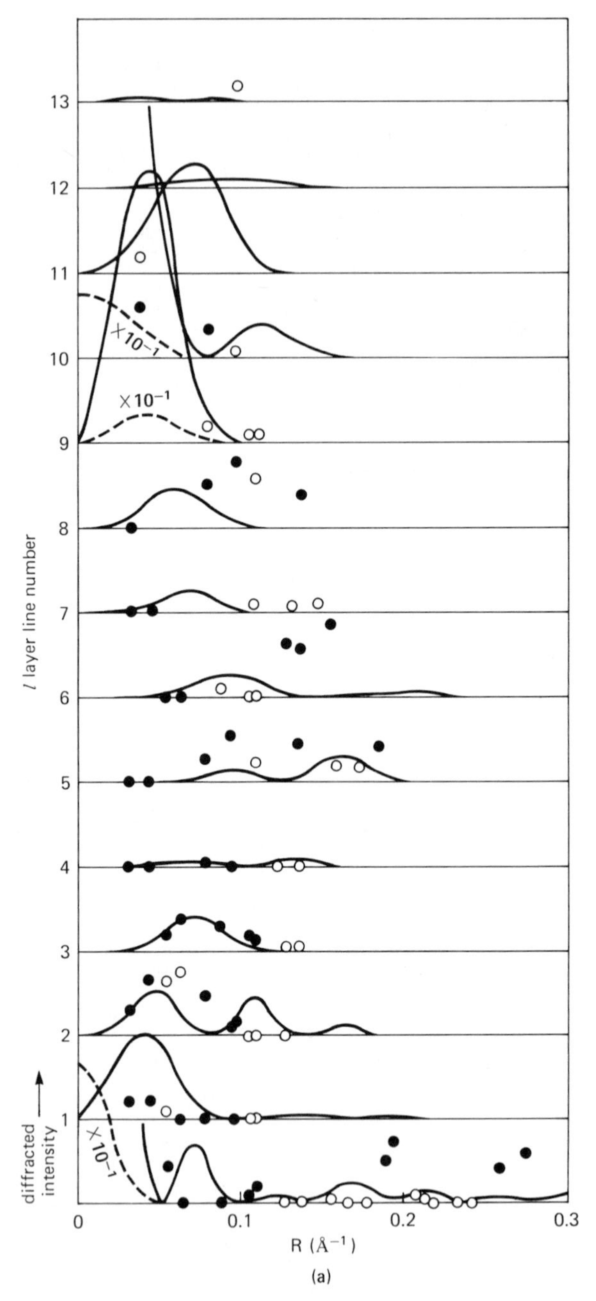

FIGURE 4-7 Comparison of observed and calculated intensities (averaged by rotation about the helix axis) for (a) original Watson-Crick model, and (b) Model 3 of Langridge

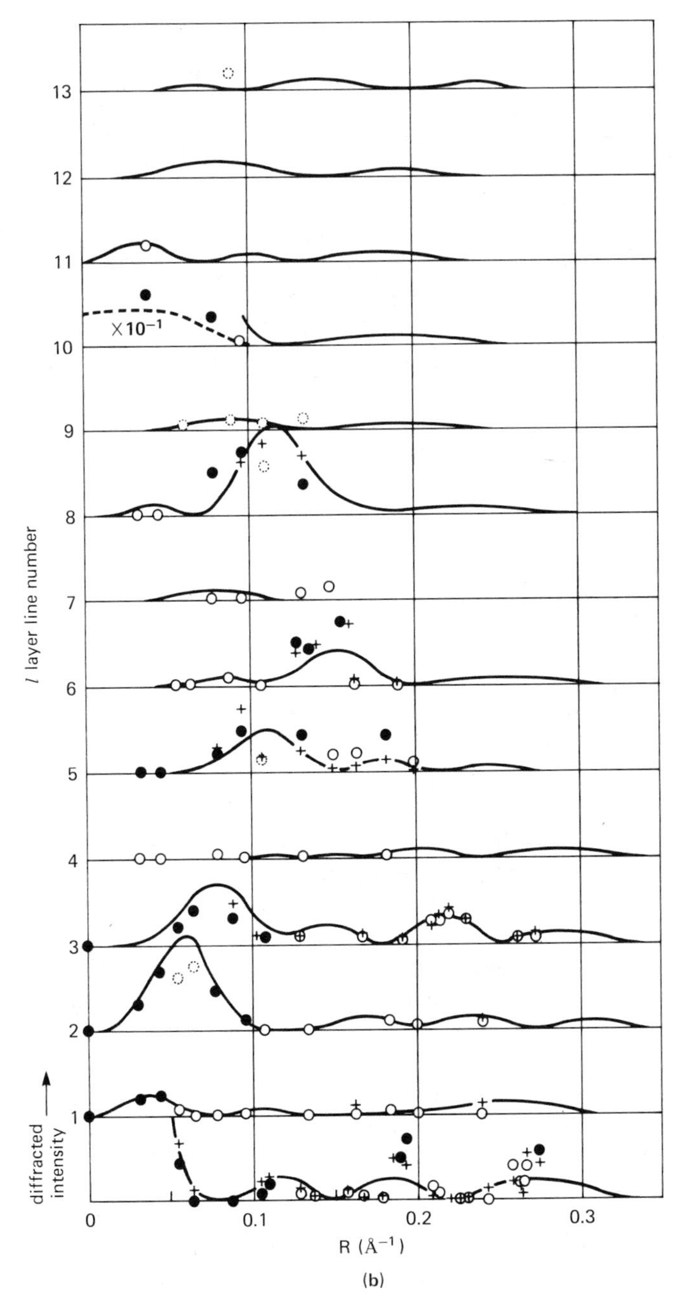

et al.[27]. Observed intensities are indicated •, O, ◌ in order of decreasing reliability. [From R. Langridge et al., *J. Mol. Biol.*, **2**, 38 (1960). Reprinted with permission.]

so only J_0 will contribute to scattering on the zeroth layer line. With $Cl = 1$, solutions are

$$(n, m) = (-5, 1), (11, -2), \ldots$$

so J_5, J_{11}, . . . contribute to scattering on the first layer line.

In the preceding discussion we have assumed that the scattering elements are simple, homogeneous objects of uniform density. If the elements have, instead, a complex internal structure, as have the nucleotides, then there will be additional variations in intensity of the various reflections, although their positions are determined by the helix geometry. It is here that model-building plays such a big role. There are simply not enough reflections at high angles, i.e., small distances, in a fiber diffraction photograph such as Fig. 4-2 to discern reliably all the atomic positions by the standard methods of X-ray diffraction. It is therefore necessary to build and adjust molecular models, keeping within acceptable stereochemical limits of bond lengths, bond angles, and nonbonded distances, until the structure which gives best agreement with the observed diffraction pattern is reached.

Figure 4-7 shows the improvement achieved in detailed fitting of the observed intensities in going from the original Watson-Crick model to the best model found by Langridge et al.[27] for the B form of LiDNA. Refinements were made by moving the base pairs closer to the helix axis and adjusting their tilt angle, rearranging the phosphate ester backbone somewhat, and adjusting the pucker of the sugar ring and its angle of rotation about the glycosidic bond. To obtain detailed agreement with the observed diffraction pattern, it was also necessary to take scattering by water in the moist fibers into account, and, with microcrystalline fibers, to consider packing of the molecules in the unit cell which leads to lattice effects on diffraction.

In well-oriented fibers, all of the molecules have their helical axes lined up in the same direction, although the relative vertical arrangement of the molecules, and their angles of rotation with respect to one another about their long axes, may be random. This axial alignment, along with some sort of fairly regular, often hexagonal, close packing of the fibers, leads to the equatorial reflections seen in Fig. 4-5. The spacing of these reflections indicates, according to Bragg's law, that the diameter of the DNA helix, measured to the phosphate groups, is about 20 Å. In microcrystalline fibers the packing is more regular in a vertical and angular sense, and the intermolecular arrangement may be studied in more detail. Figure 4-8 shows the packing of Li DNA, at 66% relative humidity, in the unit cell as deduced[1] from the diffraction pattern in Fig. 4-2.

After this brief discussion of the theoretical foundations of X-ray diffraction from helical fibers, we may proceed to survey the results which have been obtained by applying these ideas to nucleic acids and synthetic polynucleotides. There are several general references which cover this topic.[23,28-33,116]

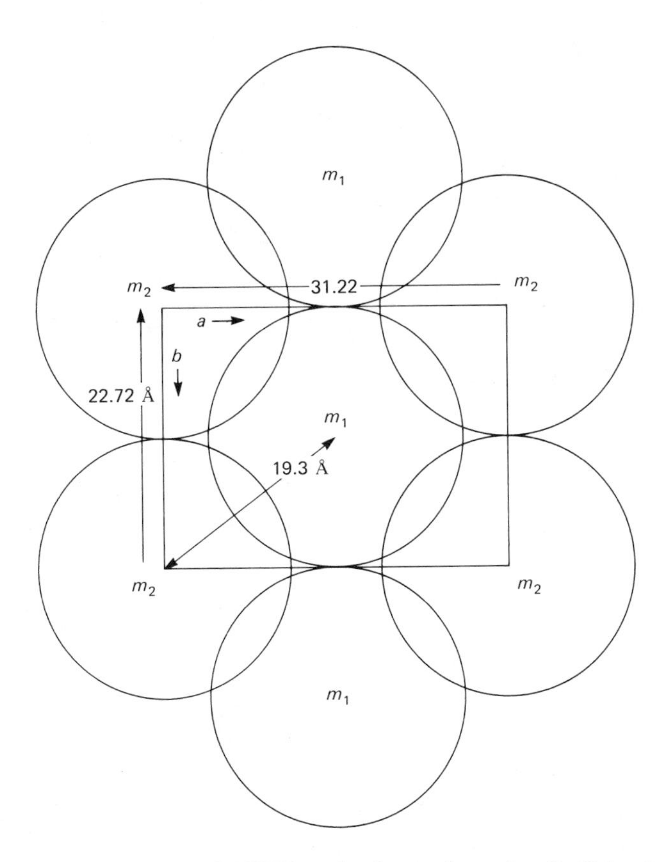

FIGURE 4-8 Arrangement of LiDNA molecules in the unit cell. Molecules m_1 are displaced in the c direction (out of the plane of the paper) by 11 Å relative to molecules m_2. [From R. Langridge et al., *J. Mol. Biol.*, **2**, 19 (1960). Reprinted with permission.]

Building on the experimental work of Wilkins, Franklin, and collaborators,[11,12] Watson and Crick[16] proposed that DNA is a right-handed double helix, with the phosphates on the outside, thereby minimizing electrostatic repulsion between the negatively charged groups, and with the bases paired in the helix interior. The base plane is roughly perpendicular to the helix axis, and also to the deoxyribose ring. The two chains of the helix run in opposite directions, so that the chains are equivalent if the molecule is turned end for end, and the nucleotides have the same environment regardless of which chain they are on. The backbone of the helix is regular, as required by the ability of DNA to crystallize, although the base sequence is random. This dilemma is resolved by noting that, although A, T, G, and C are all of different sizes and shapes, the hydrogen-bonded base pairs A · T and G · C are of the same size and shape (Fig. 4-9). With this base pairing scheme, models can be built in which both types of base pairs bear the same angular relationships to the

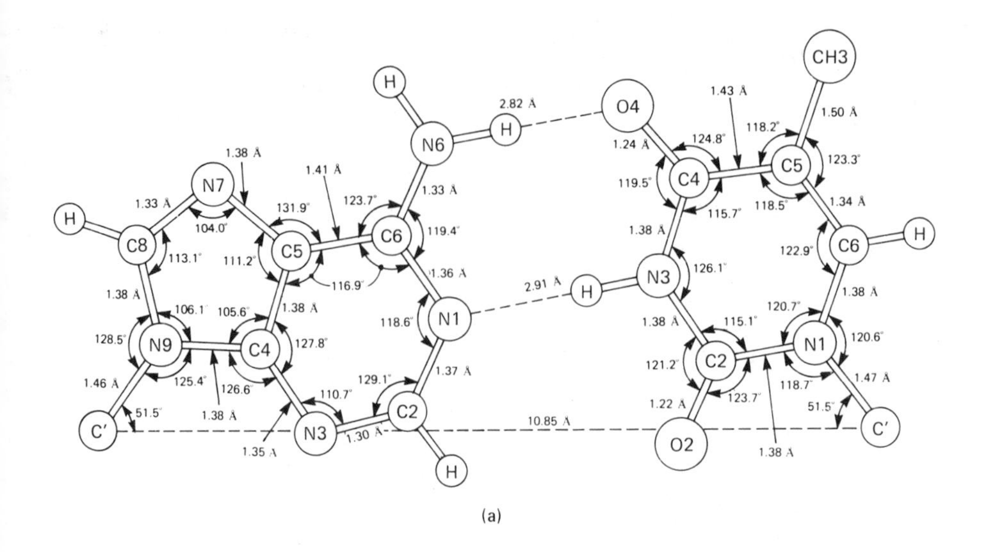

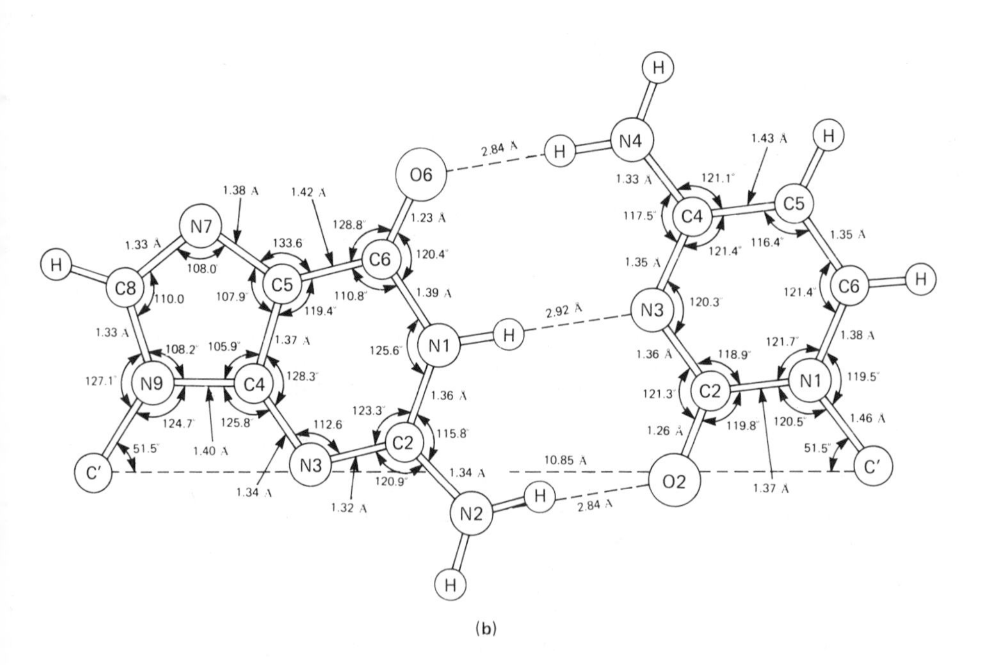

FIGURE 4-9 Watson-Crick base pairs (a) adenine-thymine and (b) guanine-cytosine. [Reprinted with permission from S. Arnott, S. D. Dover, and A. J. Wonacott, *Acta Cryst.*, **B25**, 2192 (1969).]

glycosidic bond from the base to C1′ of the sugar, the distance between the backbone chains remains constant, and the sugar-phosphate bonds are geometrically regular. The existence of three forms of DNA which conform to this basic plan with minor modifications, and which are interconvertible simply by changing the relative humidity or the cocrystallizing cation, constitutes strong evidence in its support.[34–37] This is particularly important considering the relative paucity of data obtainable from a single form.

Refinement of earlier work on the A, B, and C forms has led to the values for the helical parameters and atomic positions in these three forms which are given in Tables 4-1 and 4-2.

The B form,[1,27] on which most attention has been focused, exists at 92% relative humidity with the Na, K, and Rb salts of DNA, and at relative humidities of 60–92% with the Li salt. It has also been shown to occur in sperm cells[43] where it is complexed with basic proteins and polypeptides. The best crystalline diffraction patterns have been obtained with LiDNA at 66% relative humidity. Under these conditions the DNA crystallizes into an orthorhombic lattice with unit cell dimensions $a = 22.5$ Å, $b = 30.9$ Å, and $c = 33.7$ Å. This last value is measured along the helix axis, and is equal to the pitch P of the helix. For NaDNA, it has been found that $P = 34.6$ Å. In both cases there are exactly ten residues per turn of the helix, so the translation per residue p is 3.37 Å for LiDNA and 3.46 Å for NaDNA. As shown in Figs. 4-10(a) and 4-11, the heterocyclic bases are arrayed perpendicular to the helix axis in the B form. Table 4-1 also lists two synthetic double-stranded polydeoxynucleotides, poly dAT and poly dABrU, which also occur in the B configuration. Interestingly, synthetic polymers containing all purines on one strand and all pyrimidines on the other have X-ray patterns different from the B form.[44]

The A form[39] occurs at 75% relative humidity with the Na, K, and Rb salts of DNA, but never with the Li salt. It has also proved impossible to prepare the A form of DNA from T2 bacteriophage, which instead of cytosine has glycosylated hydroxymethylcytosine. The A form of NaDNA forms monoclinic crystals with unit cell dimensions $a = 22.24$ Å, $b = 40.62$ Å, $C(=P) = 28.15$ Å, and β, the angle between the a and c axes, $= 97.0°$. There are 11 residues per turn in the A form, with 2.55 Å of vertical translation between base pairs. The bases are tilted at an angle of 20° from perpendicular to the helix axis, as shown in Fig. 4-10(b) and 4-11. A DNA-RNA hybrid also adopts the A configuration.[40,41]

Study of the A form has been particularly valuable because of the information it has provided about the basic structure common to all three forms. In the first place, it was never possible to rule out on stereochemical grounds the possibility that the B form helix was left-handed. However, it has proved definitely impossible to devise a left-handed double helix that both fits the X-ray diffraction data on the A form and is stereochemically

TABLE 4-1 DIMENSIONS OF THE DIFFERENT FORMS OF DNA[a]

DNA	Pitch	Residues per turn	Translation per residue	Rotation per residue(°)	Angle between perpendicular to helix axis and bases (°)	Dihedral angle between base planes (°)	Furanose[b] out of plane atoms	ϕ_{CN}	Ref.[c]
A form, Na salt, 75% humidity	28.15 ± 0.16	11	2.55	32.7	20	16	C2′ = −0.13 C3′ = +0.53	−14.1°	39
B form, Na salt, 92% humidity	34.6	10	3.46	36					27
B form, Li salt, 66% humidity	33.7 ± 0.1	10	3.37	36	2	5	C2′ = +0.19[d] C3′ = 0.10	−86.7°	1
C form, Li salt, 66% humidity	31.0	9.3	3.32	39	6	10	C2′ = +0.41 C3′ = 0.05	−74.6°	42
DNA–RNA hybrid Na salt, 75% humidity	28.8 ± 0.5	11	2.62	32.7	~20				40
dAT,B form, Li salt, 66% humidity	33.4 ± 0.2	10	3.34	36					38
dABrU, B form, Li salt, 66% humidity	33.4 ± 0.2	10	3.34	36					38

[a] From D. R. Davies, *Ann. Rev. Biochem.*, **36**, 321 (1967). Reprinted with permission.
[b] Displacements from the plane containing C1′, O1′, and C4′. The plus sign indicates that the displacement is on the same side as the C5′ (endo).
[c] 39 Fuller et al., *J. Mol. Biol.*, **12**, 60 (1965).
27 Langridge et al., *J. Mol. Biol.*, **2** 38 (1960).
1 Langridge et al., *J. Mol. Biol.*, **2**, 19 (1960).
42 Marvin et al., *J. Mol. Biol.*, **3**, 547 (1961).
40 Milman et al., *Proc. Nat. Acad. Sci. U.S.*, **57**, 1804 (1967).
38 Davies and Baldwin, *J. Mol. Biol.*, **6**,251 (1963).
[d] Haschemeyer and Rich calculate C2′ to be 0.26 Å endo displaced from the least-squares plane through the remaining four atoms.

TABLE 4-2 COORDINATES OF THE NUCLEOTIDE ATOMS IN THE VARIOUS FORMS OF DNA[a]

Group	Atom	B form[26] r(Å)	B form[26] ϕ (Degrees)	B form[26] Z(Å)	A form[38] r(Å)	A form[38] ϕ (Degrees)	A form[38] Z(Å)	C form[41] r(Å)	C form[41] ϕ (Degrees)	C form[41] Z(Å)
Phosphate	P	9.05	58.8	−1.36	8.84	34.7	−6.49	9.05	68.9	−0.37
	O_1	8.88	62.9	0.04	9.60	36.5	−5.18	7.91	71.8	0.60
	O_2	9.14	65.3	−2.45	9.80	35.5	−7.64	9.20	75.5	−1.45
	O_3	10.33	54.2	−1.18	7.68	41.3	−6.55	10.32	67.3	0.40
	O_4	7.85	52.3	−1.43	8.60	24.9	−6.21	8.57	60.1	−1.02
Sugar	C_1	5.72	71.5	0.31	8.46	36.9	−1.67	5.48	88.0	0.88
	C_2	6.95	74.1	−0.56	9.16	45.0	−1.80	6.79	88.1	0.05
	C_3	8.13	71.1	0.35	9.75	44.9	−3.18	7.80	82.2	0.93
	C_4	7.58	70.4	1.80	9.81	53.0	−3.80	7.28	83.4	2.37
	C_5	8.08	79.1	2.63	8.96	38.7	−3.98	8.18	90.8	3.12
	O_5	6.13	70.0	1.70	8.87	31.5	−2.94	5.90	86.6	2.24
Purine	N_1	2.40	155.0	−0.14	2.91	14.5	−0.65	1.60	157.7	0.12
	C_2	1.70	124.0	−0.03	4.19	16.6	−0.64	1.12	103.3	0.28
	N_3	2.90	104.0	0.02	4.97	30.2	−1.07	2.35	88.5	0.44
	C_4	4.20	112.0	0.00	4.95	44.8	−1.55	3.34	106.1	0.45
	C_5	4.95	98.0	0.10	6.20	46.2	−1.79	3.58	128.6	0.29
	C_6	4.63	82.5	0.19	7.00	37.3	−1.52	2.90	150.2	0.12
	$O_6,(NH_2)_6$	3.30	79.0	0.16	6.33	27.8	−1.04	3.86	164.7	−0.03
	N_7	2.90	56.0	0.20	6.94	18.5	−0.62	4.93	127.8	0.33
	C_8	1.70	40.0	0.14	6.34	9.0	−0.23	5.38	114.6	0.51
	N_9	0.60	83.0	0.02	5.07	5.3	−0.22	4.67	101.8	0.56
Guanine	$(NH_2)_2$	2.25	0.80	0.20	7.31	2.0	0.20	0.83	17.2	0.28
Pyrimidine	N_1	3.20	140.0	−0.14	3.04	33.8	−1.04	2.98	126.9	0.28
	C_2	3.20	117.0	−0.03	4.31	35.7	−1.19	3.30	103.0	0.46
	O_2	2.40	96.0	0.05	5.20	24.2	−0.87	2.79	81.6	0.52
	N_3	3.34	76.0	0.16	6.50	27.2	−1.02	4.67	101.8	0.56
	C_4	3.40	55.0	0.23	7.40	20.7	−0.72	5.54	113.1	0.53
	C_5	4.63	82.5	0.19	7.00	37.3	−1.52	5.34	127.6	0.35
	C_6	5.17	97.0	0.10	6.39	47.0	−1.85	4.17	136.6	0.22
	$O_6,(NH_2)_6$	4.60	112.0	0.00	5.02	49.2	−1.69	4.51	152.9	0.05
Thymine	$(CH_3)_5$	5.33	121.0	−0.09	4.69	66.5	−2.17	6.74	133.9	0.29

[a] From Fuller et. al., *J. Mol. Biol.*, **12**, 69 (1965). Reprinted with permission.

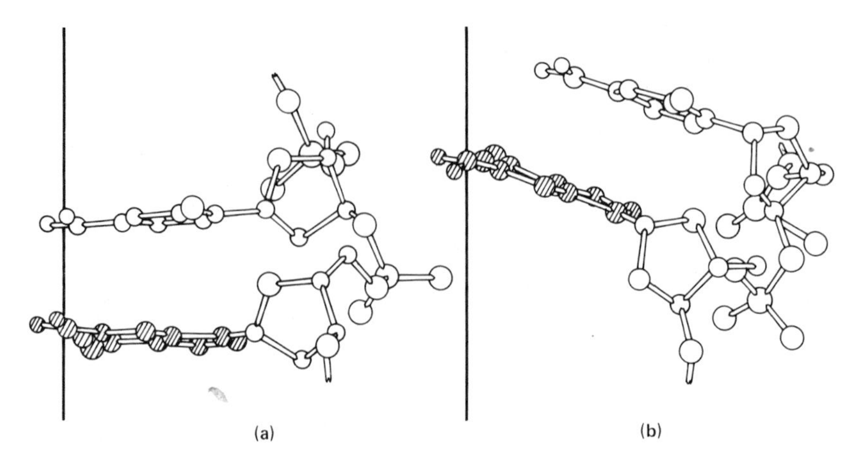

FIGURE 4-10 (a) Projection of two nucleotides in the B conformation of DNA showing bases horizontal and perpendicular to the helix axis. (b) Projection of two nucleotides in the A conformation of DNA showing bases inclined at 20°. [From L. D. Hamilton, *Nature*, **218**, 635 (1968). Reprinted with permission.]

plausible. Because of the ready interconvertibility among the forms, this proves that the B and C double helixes must also be right-handed. Second, the symmetry of the space group of the A form of DNA indicates the presence of a twofold rotation axis, or dyad axis, passing through and perpendicular to the helix axis. This is evidence for the antiparallel orientation of the two strands.

The C form of LiDNA[42] appears to be a drying artifact produced when the relative humidity drops below the value of 60%. The DNA fibers may pack in either a hexagonal or an orthorhombic lattice, depending on conditions. There is a nonintegral number of residues per turn, 9 1/3, in the C form, and the helix pitch is 31 Å. The configuration of the nucleotides is similar to that in the B form, but the base pairs are moved about 1.5 Å farther away from the helix axis and are tilted by about 6°, so that the narrow groove in the helix is made deeper.

The structures of double-stranded RNA, with complementary base composition, have been studied by X-ray diffraction for reovirus RNA,[45,46] wound tumor virus RNA,[47] the replicative form of MS2 phage RNA,[48] and rice dwarf virus RNA.[49] Further detailed studies of reovirus RNA[50,51] have revealed two different crystalline forms, denoted α and β. X-Ray diffraction from the double-stranded, complementary synthetic polyribonucleotide, poly(rA) · poly(rU), has also been examined.[52]

The helical parameters for these RNA molecules are summarized in Table 4-3. Duplex RNA appears to be in a conformation similar to the DNA A conformation as shown in Fig. 4-11, and does not change to another

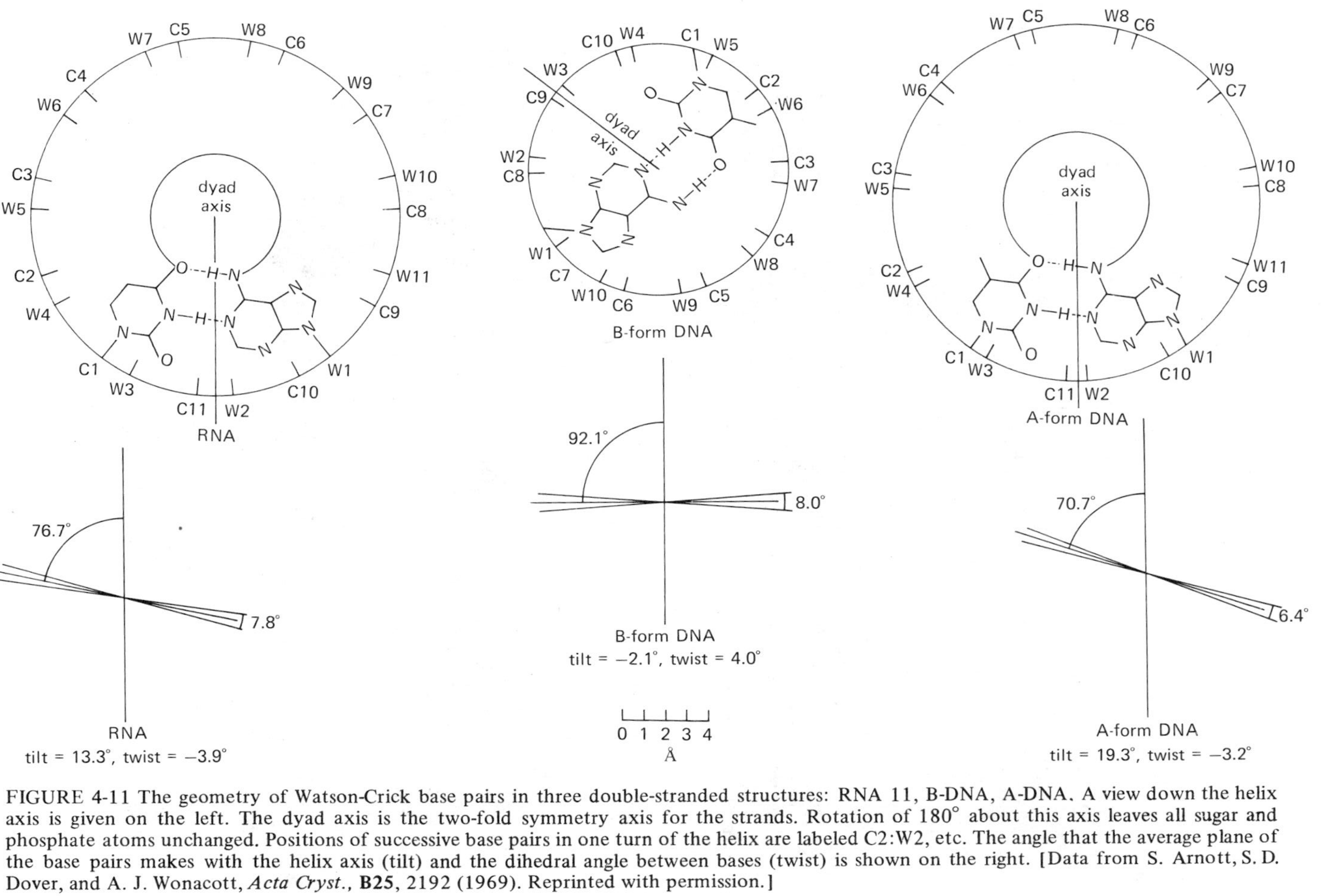

FIGURE 4-11 The geometry of Watson-Crick base pairs in three double-stranded structures: RNA 11, B-DNA, A-DNA. A view down the helix axis is given on the left. The dyad axis is the two-fold symmetry axis for the strands. Rotation of 180° about this axis leaves all sugar and phosphate atoms unchanged. Positions of successive base pairs in one turn of the helix are labeled C2:W2, etc. The angle that the average plane of the base pairs makes with the helix axis (tilt) and the dihedral angle between bases (twist) is shown on the right. [Data from S. Arnott, S. D. Dover, and A. J. Wonacott, *Acta Cryst.*, **B25**, 2192 (1969). Reprinted with permission.]

TABLE 4-3 DIMENSIONS OF RNA CRYSTAL FORMS[a]

RNA	Relative humidity (%)	"a" Equatorial spacing	"c" Fiber axis repeat	Base pairs per turn	Translation per residue (Å)	Ref.[b]
Yeast RNA fragments	75	40.7 ± 0.2	29.0 ± 0.2	10 or 11	2.9 or 2.64	46
Reovirus, α form	75	43.9 ± 0.2	30.0 ± 0.2	10 or 11	3.0 or 2.73	45,46
Reovirus, β form	75	39.9 ± 0.15	29.95 ± 0.15	10 or 11	3.0 or 2.73	46
Wound tumor virus	50–100	43.5	30.1 ± 0.3	10[c]	3.0[c]	47
Rice dwarf virus	75	40.0	30.5	10[c]	3.05[c]	49
Poly (rA) · Poly (rU)	92	a = 25.7 b = 31.5	30.9	10[c]	3.1[c]	52
Poly (rA) · Poly (rU)	92	a = 33Å	30.9	10[c]	3.1[c]	52

[a] From D. R. Davies, *Ann. Rev. Biochem.*, **36**, 321 (1967).
[b] [46] S. Arnott et al., *Nature*, **201**, 227 (1964).
[47] Tomita and Rich, *Nature*, **201**, 1160 (1964).
[49] Sato et al., *J. Mol. Biol.*, **16**, 180 (1966).
[52] Sasisekharan and Sigler, *J. Mol. Biol.*, **12**, 296 (1965).
[c] In view of the ambiguity concerning the number of residues per turn in the first three rows, and the general similarity of the intensities of all the diffraction patterns, it seems reasonable to suppose that there may be ambiguity in the assignments for these fibers also.

conformation with changing humidity or crystal form. It is uncertain at present whether there are 10 or 11 base pairs per turn of the helix. This uncertainty arises because a conformation with 10 residues per turn and the base pairs tilted 10° away from perpendicular to the helix axis is indistinguishable from a conformation with 11 residues per turn and a 14° tilt away from the helix axis, unless there is a very high degree of crystallinity. Differences in the rotation per residue (3.3°) and translation per residue (0.27 Å) are slight between the two models. However, details of the electron density distribution in Fourier synthesis, and considerations of intermolecular stereochemistry, tend to favor the 11-fold helix.[51] This again would be similar to the A form of DNA.

X-Ray diffraction has also been studied from samples of noncomplementary viral and cellular RNA, the latter mainly ribosomal RNA,[53,54] and from crystalline fragmented yeast ribosomal RNA which was originally thought to be transfer RNA.[55–58] These samples of RNA have diffraction patterns very similar in intensity distribution to the complementary double helical viral RNA discussed above, which leads one to believe that there are extensive Watson-Crick base-paired regions in ribosomal RNA. These may be preexistent or they may be artifacts of the procedure of preparation for crystallization.

An interesting feature of the X-ray diffraction studies on RNA is that there is no stereochemically feasible double-stranded configuration which agrees with the X-ray diffraction data that permits intramolecular hydrogen bonding involving the 2′-OH group of the ribose.[46,51] This hydroxyl group may, however, participate in intermolecular hydrogen bonding which influences crystallization.

Fairly extensive work has also been done on X-ray diffraction from synthetic polynucleotides. The helix dimensions of the synthetic polynucleotides which have been studied to date are summarized in Table 4-4.

We note first that double-stranded helixes formed between strands of complementary bases such as poly (rA) · poly (rU) and poly (rI) · poly (rC) have structures of the Watson-Crick type with 10–12 residues per turn and with antiparallel strands. Triple-stranded helixes such as poly (rA) · 2 poly (rU) and perhaps poly (rA) · 2 poly (rI) may also form, depending on the concentration of strands and the solution conditions. Structural models for these triple helixes are reviewed by Davies.[31]

Double-stranded helixes with pairing between identical bases, such as poly (rA), poly (rC), and poly (rI) form structures with *parallel* strands which are related by a twofold, or for poly (rI), a threefold rotation axis parallel to the helix axis. This is in distinct contrast to the other types of double helical polynucleotides discussed above, in which the strands run in antiparallel fashion.

Although we have emphasized Watson-Crick base pairing and the structures deduced from X-ray scattering of fibers, it is important to keep in mind other possible structures for double-stranded polynucleotides. Obviously, changes in conformation which still retain the double-stranded structure are to be expected in solution as a function of ionic strength and solvent. These changes are evident from the changes in CD with salt,[67] alcohol,[68] ethylene glycol,[69] etc. Forming covalently closed circles of super helical DNA,[70] or folding the DNA into a bacteriophage head[71] must also change the conformation. What about more basic changes?

Because of the importance of the Watson-Crick complementary base-pairing scheme, numerous efforts have been made to demonstrate its existence in mixed crystals of the bases and their derivatives. In single crystals of derivatives of G and C, the Watson-Crick pairing is indeed that which is found.[72,73] In crystals of derivatives of A and T, or A and U, on the other hand, an alternative type of pairing has been found, which has been called Hoogsteen pairing.[74,75] In this type of pairing, an NH . . . N hydrogen bond is formed between N1 and N7 rather than between N1 and N1 as in Watson-Crick pairing. These alternatives are compared in Fig. 4-12. Because of the observed regularity of the helix backbone, it is clear that in DNA itself, the pairing must either be all of the Watson-Crick type or all of the Hoogsteen type. Significantly, the Watson-Crick type of pairing has been found by X-ray analysis of a single crystal of the dinucleoside phosphate ApU.[117]

Arnott et al.[76] attempted to fit the electron density distribution in Li DNA, obtained using Fourier synthesis procedures, by assuming that base pairing was in fact of the Hoogsteen type. However, occurrence of a spurious

TABLE 4-4 DIMENSIONS OF THE POLYNUCLEOTIDE HELIXES[a]

Polynucleotide	Pitch	No. of residues per turn	Translation per residue	Rotation/residue	Ref.[b]
rA · rU	30.9	10[e]	3.1	36°	52
rA · 2rU	31.1	10	3.1	36°	59
rA · rI[c]	38.8	11.4	3.4	31°	60
rI · rC	36	12	3.0	30°	61,62
rC	37.3	12	3.11	30°	63
rA	30.4	8	3.8	45°	64
rA A form	31.8[d]	8.38	3.8	43°	65
rA B form	36.2	10.0	3.6	36°	65
rI	29.4	8.7	3.4	42°	65

[a] From D. R. Davies, *Ann, Rev. Biochem.*, **36**, 348 (1967). Reprinted with permission.
[b] 52 Sasisekharan and Sigler, *J. Mol. Biol.*, **12**, 296 (1965).
59 Sasisekharan et al., quoted in Davies.
60 Rich, *Nature,* **181**, 521 (1958).
61 Davies and Rich, *J. Amer. Chem. Soc.*, **80**, 1003 (1958).
62 Davies, Intern. Biophys. Congr.,.Stockholm, 1961.
63 Langridge and Rich, *Nature*, **198**, 725 (1963).
64 Rich et al., *J. Mol. Biol.,* **3**, 71 (1961).
65 Klug & Finch, personal commun.
[c] This helix may be rA · 2rI.
[d] These dimensions are calculated with the assumption of a twofold axis parallel to the helix axis.
[e] The number of residues per turn may be 11.

maximum in electron density in the center of the base-pair region, and severe stereochemical difficulties in constructing helical models based on Hoogsteen pairing, forced the conclusion that Watson-Crick type pairing was indeed correct.

The number of reasonable hydrogen-bonded structures between two bases which can be drawn among the four nucleosides is 29.[77,78] A reasonable hydrogen bond is one in which the three nuclei making up the bond are essentially linear and the bond distance between the end atoms is within 2.8–3.0 Å. Nonhydrogen-bonded contacts should be greater than about 3.1 Å. The stabilities of these various structures have been calculated and many of them have similar energies. For a review see the articles in Ref. 79. Why then are the Watson-Crick base pairs predominant in double-stranded nucleic acids? The Watson-Crick structures seem to be the only ones which allow a random sequence of bases along one strand. That is, other base pairs are stable, but they cannot be joined together arbitrarily to form a long double-stranded helix. For short regions in a long double strand, or for oligonucleotides, non-Watson-Crick base pairs should be considered. Short intermolecular double strands as found in messenger or transfer RNAs seem particularly likely locations for these base pairs. In the complex between the

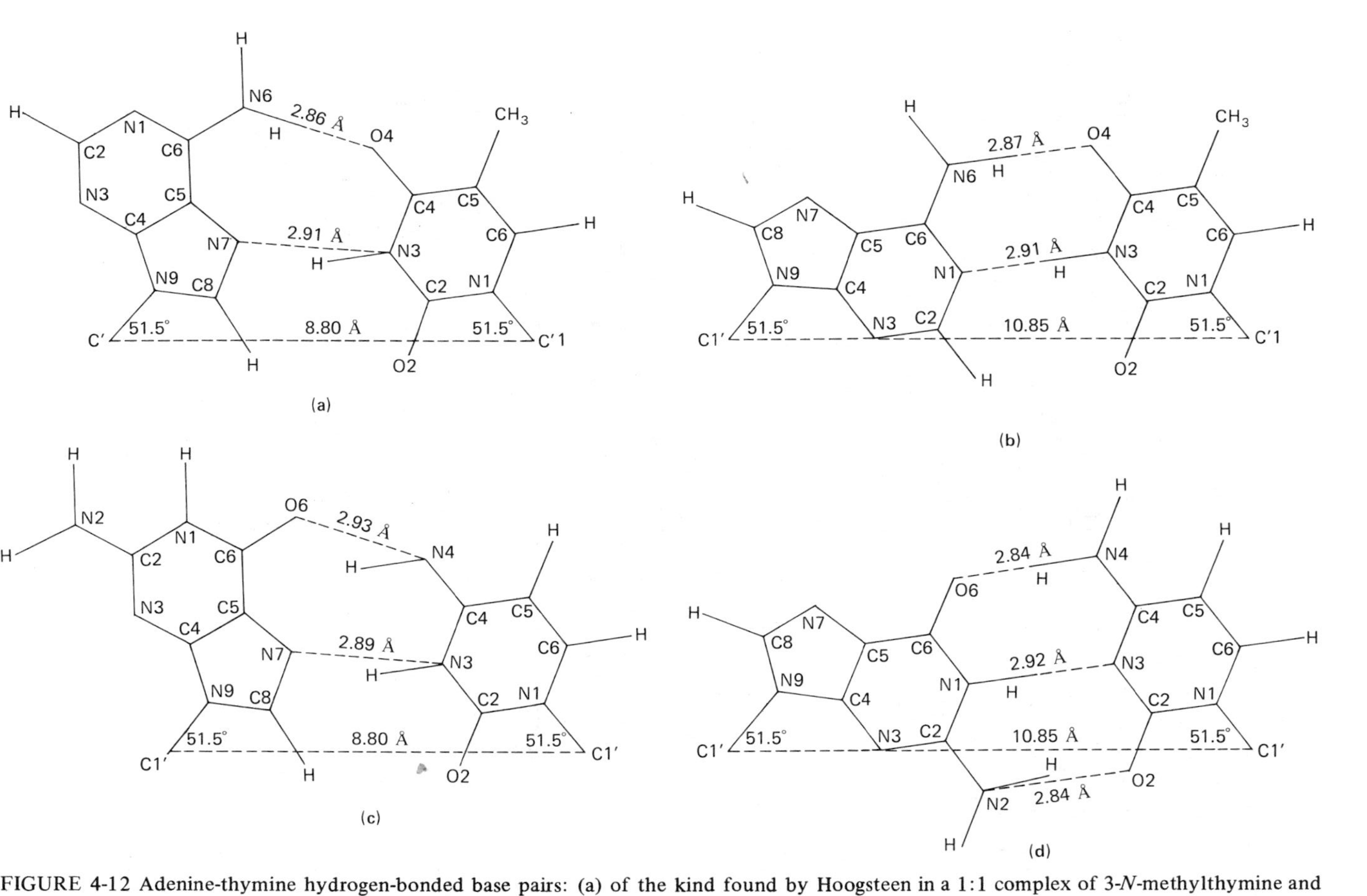

FIGURE 4-12 Adenine-thymine hydrogen-bonded base pairs: (a) of the kind found by Hoogsteen in a 1:1 complex of 3-*N*-methylthymine and 9-*N*-methyladenine; (b) postulated by Watson and Crick (dimensions refined by Arnott). Guanine-cytosine pairs: (c) Hoogsteen type; (d) Watson-Crick type. [From S. Arnott et al., *J. Mol. Biol.*, **11**, 391 (1965). Reprinted with permission.]

anticodon loop of transfer RNA and the messenger RNA codon involved in protein synthesis there are certain well-known non-Watson-Crick base pairs called wobble pairs (see Chapter 8).

It seems useful at this point to review the conclusions that have been reached about double-stranded nucleic acids from X-ray diffraction. Figure 4-11 shows the geometry of base pairs in double-stranded RNA (β-RNA 11) and the A and B forms of DNA.[44] For all naturally occurring nucleic acids the proposed structures have right-handed helixes of antiparallel strands. Also, only Watson-Crick base pairing (A · U or T, G · C) is present. The number of base pairs per turn of the helix, the angle the plane of the base pair makes with the helix axis, and the dimensions of the helix are variable, as shown in the figure. Only one structure (with 11 base pairs per turn) has been found for RNA double strands[33]; for DNA three forms dependent on humidity and salt concentration have been found. The structure found as the Na salt in low humidity has 11 base pairs per turn and is called the A form.[39] The B form[1,26] with ten base pairs per turn is present at high humidity. The C form has been found as the Li salt at low humidity; it has 9.3 base pairs per turn.[42] Hybrid double strands containing one strand of DNA and one of RNA seem to have an RNA or A form DNA-type structure.[40,41] Evidence other than X-ray data for these structures has been slight. It is important to obtain other evidence, particularly since X-ray data on fibers are usually not extensive enough to give as unequivocal results as obtained from single crystals.[80]

B. Infrared Dichroism

Early work on ultraviolet dichroism, as discussed earlier, showed that the heterocyclic bases were perpendicular to the long axis of the DNA molecule. By looking at infrared dichroism, it has been possible to get information on the phosphate group orientation in the helix backbone of polynucleotides and thus to obtain valuable confirmatory information supplementary to that obtainable from X-ray work.

Figure 4-13 shows the IR spectrum of oriented films of NaRNA from rice dwarf virus.[49] The assignment of the absorption bands is that the 1225 cm^{-1} absorption corresponds to the PO_2^- antisymmetric stretching mode,[81,82] the absorption at 1084 cm^{-1} corresponds to the PO_2^- symmetric stretching vibration[83,84]; the absorption at 1680 cm^{-1} in the deuterated film corresponds to an in-plane double bond stretching vibration of the nucleic acid bases.[49]

It is possible to estimate the angle θ between the helix axis and the transition moment of the particular transition from the dichroic ratio[85]

$$R = I(\parallel)/I(\perp) = \frac{2\cos^2\theta + g}{\sin^2\theta + g}$$

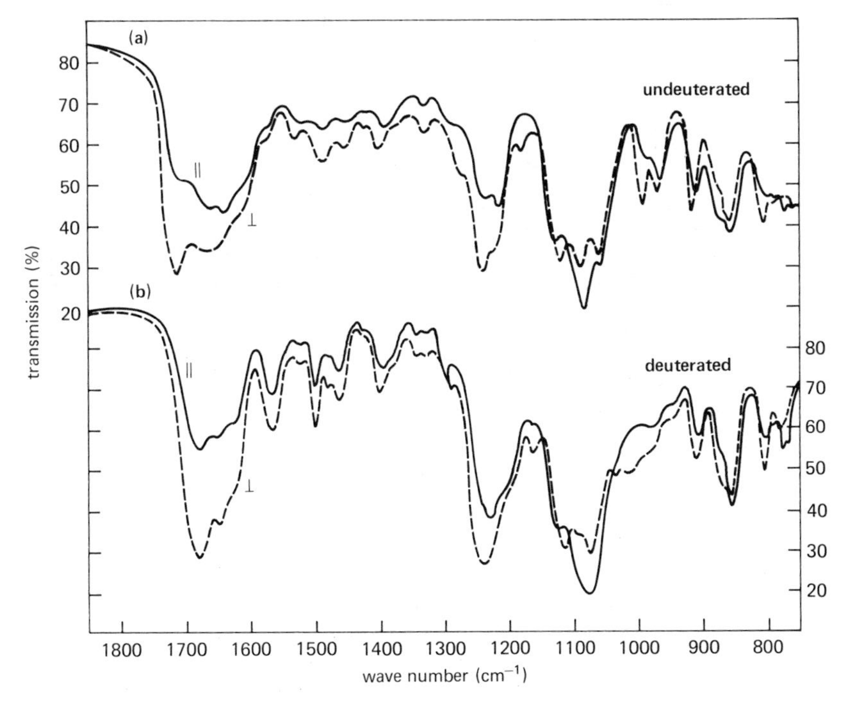

FIGURE 4-13 Infrared absorption spectra of oriented films of NaRNA from rice dwarf virus in the region 1850–750 cm^{-1}. (a) Undeuterated, 75% relative humidity; (b) deuterated, 75% relative humidity. Electric vector parallel (——) or perpendicular (– – –) to the fiber axis. [From T. Sato et al., *J. Mol. Biol.*, **16**, 180 (1966). Reprinted with permission.]

where g is a parameter that measures the deviation of the helixes in the film from perfect orientation. It is found that $R_{1225} = 0.56$, $R_{1084} = 1.8$, and $R_{1680} = 0.40$. Since the bases tilt from 5° to 15° away from perpendicular to the helix axis – that is, $\theta_{1680} = 85°$ to $75°$ – we can estimate that $g = 0.35$ to 0.68. We then calculate from the observed dichroic ratios for the phosphate group absorptions that the O . . . O line of the PO_2^- residue makes an angle of about 70° with the helix axis, while the bisector of the OPO angle makes an angle of about 40° with the helix axis.[49] For the structure of reovirus RNA determined by X-ray diffraction discussed above, the corresponding angles are 65° and 30° for the 10-fold helix, and 64° and 37° for the 11-fold helix.[51] Thus, agreement for both forms is within experimental error.

C. Ultraviolet Absorption and Circular Dichroism

1. INTRODUCTION

This section contains a quantitative discussion of the theory of the electronic optical properties.

The absorption of double-stranded DNA and RNA is shown in Fig. 4-14. For comparison, the spectra of the sum of their constituent mononucleosides are also shown. When the double strands are melted by high temperatures, their spectra approach that of their mononucleosides. The decrease in absorption relative to the sum of the components is called hypochromicity. The hypochromicity of double-stranded polynucleotides is a maximum near 260 nm; it decreases to zero near 290 nm and becomes hyperchromicity above 290 nm. The percent hypochromicity is defined as:

$$h(\lambda) = \left[1 - \frac{\epsilon_{\text{polymer}}(\lambda)}{\epsilon_{\text{monomers}}(\lambda)}\right] \cdot 100$$

$h(\lambda)$ is hypochromicity at wavelength λ

$\epsilon_{\text{polymer}}(\lambda)$ is the molar extinction coefficient (molar absorptivity) of the polymer per mole of base

$\epsilon_{\text{monomers}}(\lambda)$ is the molar extinction coefficient of the sum of the constituent monomers per mole of base

If the absorption of the polymer is greater than that of the monomers, $h(\lambda)$ becomes negative, and is called hyperchromicity. From Fig. 4-14 one can

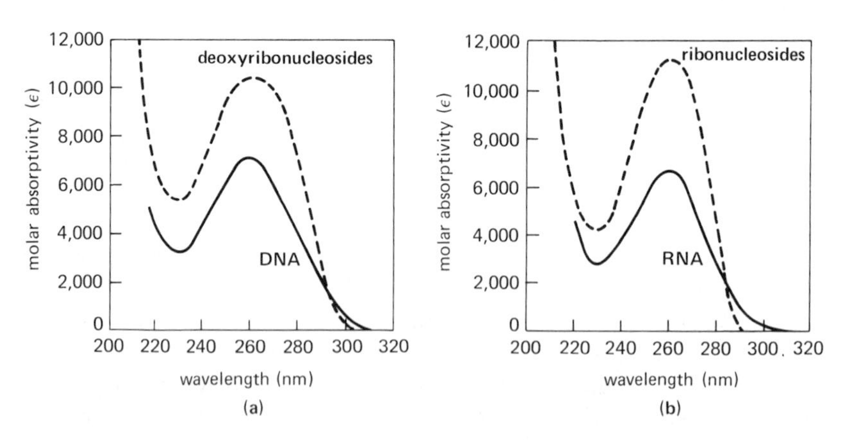

FIGURE 4-14 The molar absorptivity of double-stranded DNA and double-stranded RNA compared with their component mononucleosides. [*M. lysodeikticus* DNA data from Felsenfeld and Hirschman, *J. Mol. Biol.*, **13**, 407 (1965). Rice dwarf virus RNA data from Miura et al., *Virology*, **28**, 571 (1966). Nucleoside spectra calculated from the base composition (72% G + C for the DNA; 44% G + C for the RNA) and the absorption data of Voet et al., *Biopolymers*, **1**, 193 (1963).]

calculate a hypochromicity at 260 nm of about 40%. Although this phenomenon was first studied in double-stranded polynucleotides, it is also found in single strands and oligonucleotides. In single strands the magnitude of the hypochromicity is generally less. Single-stranded polynucleotides are 15–20% hypochromic and dinucleoside phosphates are 0–10% hypochromic at 260 nm. One can relate the magnitude of the hypochromicity to the amount of stacking of the bases. The qualitative explanation for the decrease in absorption is in terms of the dipole induced by the light in a typical base.[86–88] The magnitude of the dipole induced is proportional to the absorption. A base in a double strand has a smaller dipole induced by the light, because of the opposite sign dipoles induced by the bases on either side. The long wavelength hyperchromicity has been ascribed to $n \rightarrow \pi^*$ transitions which are polarized perpendicular to the plane of the bases,[89] but DeVoe[90] has been able to calculate this effect using only $\pi \rightarrow \pi^*$ transitions. The absorption parameter which is easiest to calculate theoretically is the hypochromism H. This is a measure of the integrated absorption.

$$H = 1 - \frac{\int \epsilon_{\text{polymer}}(\nu)\, d\nu}{\int \epsilon_{\text{monomers}}(\nu)\, d\nu} = 1 - \frac{\int \frac{\epsilon_{\text{polymer}}(\lambda)\, d\lambda}{\lambda^2}}{\int \frac{\epsilon_{\text{monomers}}(\lambda)\, d\lambda}{\lambda^2}}$$

It is important to specify the interval of integration when reporting an experimental hypochromism. If H becomes negative it is termed hyperchromism.

The circular dichroism spectra of double-stranded DNA and RNA are compared in Fig. 4-15 with the CD of their constituent nucleosides. The shapes of the nucleic acid CD curves are completely different from each other. The DNA curve is called conservative[91] with approximately equal positive and negative lobes on either side of 260 nm. It has the approximate shape of the derivative of the 260 nm absorption band. The negative peak near 300 nm is noteworthy because it may be correlated with the hyperchromic region in absorption. An approximate theory of CD[92] relates the shape of curves to the geometry of helixes. The DNA curve is consistent with B form geometry; the RNA form with RNA geometry. The main structural difference which leads to the CD difference is the tilting of the bases in RNA. The greater asymmetry of the tilted bases in RNA gives a nonconservative CD spectrum near 260 nm. The sign of the CD is definitely associated with right-handed helixes. Calculations of the CD for A form and C form DNA are not consistent with the measured CD of DNA in aqueous solution. However, the CD spectrum of DNA in 80% ethanol[93] is consistent with A form geometry.

As the CD of the nucleic acids tends to that of their constituents at high temperatures, one sees that melting of DNA causes mainly a decrease in

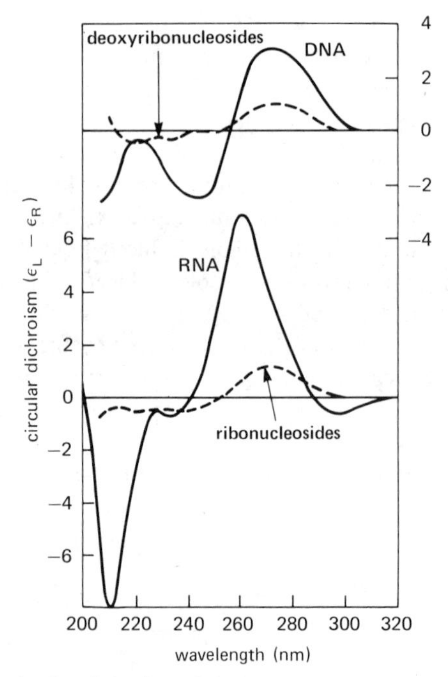

FIGURE 4-15 The circular dichroism of double-stranded DNA and RNA compared with their component mononucleosides. *M. lysodeikticus* DNA data from Allen et al., *Biopolymers,* **11**, 853 (1972). Rice dwarf virus RNA data from Samejima et al., *J. Mol. Biol.,* **34**, 39 (1968). Nucleoside spectra calculated from the base composition (72% G + C for the DNA; 44% G + C for the RNA) and the CD data of Cantor et al., *Biopolymers,* **9**, 1059, 1079 (1970).

the magnitude of the CD. Melting double-stranded RNA causes a red shift of the spectrum and a disappearance of the negative lobe near 300 nm.

The optical properties have mainly been used as an empirical measure of conformation as in absorbance vs. temperature melting curves. However, they can be used more quantitatively. The base composition of a double-stranded DNA or RNA can be calculated from the change of absorbance on heat denaturation.[94]

$$\Delta A(\lambda) = C\{f_{AT}\epsilon_{AA}(\lambda) + (1 - f_{AT}) \cdot \epsilon_{GG}(\lambda) + f_{AT} \cdot (1 - f_{AT}) \cdot [1 - 2K(\lambda)] \cdot [\epsilon_{AA}(\lambda) + \epsilon_{GG}(\lambda)]\}$$

$\Delta A(\lambda)$ is the increase of absorbance at wavelength λ on heat denaturation

C is concentration

f_{AT} is the mole fraction of A · T (or A · U) base pairs

$\epsilon_{AA}(\lambda)$, $\epsilon_{GG}(\lambda)$, $K(\lambda)$ are three parameters dependent on λ

The wavelength-dependent parameters have been tabulated for DNA denaturation,[94] but not yet for RNA. As the value of $K(\lambda)$ is near 0.5 for most wavelengths, the third term is negligible, and the equation is essentially linear in f_{AT} and can be easily solved. Data for at least two wavelengths must be measured to solve for the concentration and the mole fraction of A · T base pairs. A similar expression in terms of optical rotation has been described by Samejima and Yang[95] for DNA. It is assumed that the concentration is already known so only one wavelength suffices to determine the fraction of A · T base pairs.

$$f_{AT} = 1.321 - 3.774 \times 10^{-4} [\alpha]_{290}$$

$[\alpha]_{290}$ = specific rotation at 290 nm

The equation based on absorption is claimed to be accurate to ±0.02 mole fraction A · T; the optical rotation equation is probably less accurate.

It may be possible to obtain nearest-neighbor base frequencies (the mole fraction of each base pair present in both strands) from optical measurements. In a very long or circular double-stranded DNA or RNA (no end effects) there are only eight independent nearest-neighbor base frequencies. The ten independent frequencies allowed by Watson-Crick base pairing ($f_{AA} = f_{TT}$, f_{AT}, f_{TA}, $f_{AC} = f_{GU}$, $f_{CA} = f_{UG}$, $f_{AG} = f_{CU}$, $f_{GA} = f_{UC}$, $f_{GG} = f_{CC}$, f_{CG}, f_{GC}) are further constrained by two equations.[96]

$$f_{AT} + f_{AC} + f_{AG} = f_{TA} + f_{CA} + f_{GA}$$

$$f_{CG} + f_{CA} + f_{CT} = f_{GC} + f_{AC} + f_{TC}$$

The eight independent nearest-neighbor frequencies can in principle be obtained from measurements at eight or more wavelengths.

2. THEORY

The theory of the optical properties of polynucleotides is based on the general formalism of optical properties presented in the second chapter. However, approximations appropriate to polymers, or in particular, polynucleotides, will be discussed here. We begin with equations relating extinction coefficients to molecular wave functions [see Eqs. (2-6) and (2-16)]. For absorption:

$$\epsilon(\nu) = \frac{8\pi^3 \nu N_0}{6909\, hc} \sum_{A \neq 0}^{\text{transitions}} D_{OA}\, f(\nu - \nu_{OA}) \tag{4-5}$$

ν is the frequency in hertz; N_0 is Avogadro's number
h is Planck's constant; c is the speed of light
$D_{OA} = \text{Re}\, \boldsymbol{\mu}_{OA} \cdot \boldsymbol{\mu}_{OA}$ = dipole strength of transition from O to A. Re means real part.

$f(\nu - \nu_{OA})$ is a function which has the shape of an absorption band centered at frequency ν_{OA}. The integral of $f(\nu - \nu_{OA})\,d\nu = 1$. A gaussian is often used for $f(\nu - \nu_{OA})$.

For circular dichroism:

$$\epsilon_L(\nu) - \epsilon_R(\nu) = \frac{32\pi^3 \nu N_0}{6909\, hc} \sum_{A \neq 0}^{\text{transitions}} R_{OA}\, f(\nu - \nu_{OA}) \tag{4-6}$$

$R_{OA} = \text{Im}\, \boldsymbol{\mu}_{OA} \cdot \mathbf{m}_{AO}$ is rotational strength of transition from O to A. Im means imaginary part.

$f(\nu - \nu_{OA})$ is a function similar to that for absorption

If O→A is an allowed transition, we will use an experimentally determined shape which is identical for absorption and rotation. For forbidden transitions a shape must be assumed.

To calculate absorption and circular dichroism from these equations it is necessary to calculate dipole strengths, rotational strengths, and transition frequencies. This can be done by relating each of these quantities to the properties of subunits of the polymer.[86,87,97–99] The key assumption in this type of theory is that there is no electron exchange between subunits. This means that the subunits interact through their coulombic fields. The optical properties of the polymer will thus depend on the electronic properties (net charge, permanent dipole, polarizability and absorption spectrum) of the subunits and the geometry of the polymer. For polynucleotides this means base sequence and conformation (or primary, secondary, and tertiary structure). There are two different strategies used in calculating the optical properties of the polymer depending on whether the corresponding transition in the subunit is electrically allowed (such as $\pi \rightarrow \pi^*$) or magnetically allowed (such as $n \rightarrow \pi^*$). The occurrence of these transitions in the bases is discussed in Chapter 2.

Electrically allowed transitions / For electrically allowed transitions we ignore any effects of static electric fields (effects of net charges or permanent electric dipoles).[99] These effects are usually small compared to those of the transition charges or transition dipoles we do consider. For absorption, the easiest quantity to calculate is the hypochromism, which is a measure of the decrease in integrated absorption of the polymer compared to the monomers. The hypochromism H is:

$$H_{OA} = 1 - \frac{\int \epsilon(\text{polymer})\,d\nu}{\int \epsilon(\text{monomers})\,d\nu}$$

$$H_{OA} = \frac{4}{\sum_i h\nu_{ioa}\mu_{ioa}^2} \sum_b^{\text{transitions}} \sum_i \sum_{j \neq i} \frac{\nu_{ioa}\nu_{job}}{\nu_{job}^2 - \nu_{ioa}^2} V_{ioa;job}\, \boldsymbol{\mu}_{ioa} \cdot \boldsymbol{\mu}_{job} \tag{4-7}$$

$\int \epsilon(\text{polymer})d\nu$ or $\int \epsilon(\text{monomer})d\nu$ is the integral over an absorption band in the polymer (or monomer). The spectrum should be resolved into bands, or appropriate wavelengths for integration should be chosen.

ν_{ioa}, μ_{ioa} are the transition frequencies and transition electric dipole moments of the monomers which give rise to the integrated monomer absorption bands

ν_{job}, μ_{job} are the transition frequencies and transition electric dipole moments of all other bands in the monomers

$V_{ioa;job}$ is the potential energy of interaction between transition ($o \rightarrow a$) on monomer i and transition ($o \rightarrow b$) on monomer j

The sum over all transitions ($o \rightarrow b$) must be made, but the effect of the higher frequency, lower wavelength terms is diminished by the $(\nu_{ob}^2 - \nu_{oa}^2)$ denominator. The sign of the hypochromism is determined by the sign of this denominator and the sign of the geometric term, $V_{ioa;job}\, \mu_{ioa} \cdot \mu_{job}$. One speaks of bands borrowing and lending intensity. This terminology is consistent with the fact [easily verified from Eq. (4-7)] that the sum of intensity over all bands is conserved. This sum rule is not too helpful, because there always seem to be bands in the vacuum ultraviolet (below 180 nm) which can borrow intensity, but whose extinction coefficients cannot be measured easily.

The effect of the structure of the polymer is characterized by the term $V_{ioa;job}\, \mu_{ioa} \cdot \mu_{job}$. If a point dipole approximation is used for this term, it can be written as

$$V_{ioa;job}\, \mu_{ioa} \cdot \mu_{job} = \frac{\mu_{oa}^2 \mu_{job}^2}{R_{ij}^3} [\cos\theta_{ij} - 3\cos\theta_i \cos\theta_j] \cos\theta_{ij} \qquad (4\text{-}8)$$

where θ_i, θ_j, θ_{ij} are the angles [defined in Fig. 4-16(a)] specifying the transition dipole vectors and the line joining them. The advantage of this formulation is that the magnitudes and directions of the transition moments can be obtained from experiments on the monomers. The disadvantage is that a point dipole approximation is not a good approximation for the interaction of transitions in bases whose dimensions are as large as the distance between them. To avoid this problem one must know more about the transition charge distribution in each base. The best method is to get this detailed charge distribution from a molecular orbital wave function for the base which agrees with the experimental base absorption. One can then calculate bond transition dipole moments and use Eq. (4-8), or use transition monopoles at each nucleus.

$$V_{ioa;job} = \sum^{\text{nuclei}} \frac{\rho_{isoa}\rho_{jrob}}{R_{is,\, jr}} \qquad (4\text{-}9)$$

ρ_{iosa} is the transition monopole on nucleus s of base i

ρ_{jrob} is the transition monopole on nucleus r of base j

$R_{is,\, jr}$ is the distance between two transition monopoles

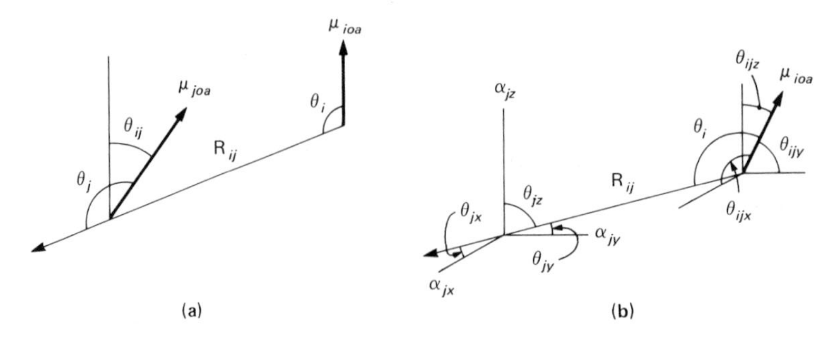

FIGURE 4-16 (a) The angles specifying the interaction between two point dipoles separated by the distance R_{ij}. θ_{ij} is the angle between $\mu_i oa$ and $\mu_j oa$; θ_i, θ_j are the angles each makes with the positive $\mathbf{R}_{ij} = \mathbf{R}_j - \mathbf{R}_i$ vector. (b) The angles specifying the interaction of a point dipole with a point polarizable group. The three principal components of the polarizability are α_{jx}, α_{jy}, α_{jz} along the three principal axes x, y, z. The dipole is oriented so as to make angles θ_{ijx}, θ_{ijy}, θ_{ijz} with respect to these three axes. The three principal axes make angles θ_{jx}, θ_{jy}, and θ_{jz} with the vector $\mathbf{R}_{ij}$; the dipole makes angle θ_i with this vector.

The sum over b in Eq. (4-7) can be replaced by the contribution of all these bands to the polarizability. If we make the point dipole approximation for $V_{ioa;job}$, the equation becomes very simple.[100]

$$H_{\mathrm{OA}} = 2 \sum_{i}^{\text{groups}} \sum_{j \neq i} \frac{\alpha_{jx}(\nu_{oa})}{R_{ij}^{3}} \left[\cos \theta_{ijx} - 3 \cos \theta_i \cos \theta_{jx}\right] \cos \theta_{ijx} \qquad (4\text{-}10)$$

plus identical terms for y and z

$\alpha_j(\nu_{oa})$ is the polarizability of subunit j at frequency ν_{oa}. Usually one uses the frequency at the maximum of the absorption band for ν_{oa}. The contribution of this band to the polarizability must be subtracted. There are three components to the polarizability (α_{jx}, α_{jy}, α_{jz}) along three perpendicular axes x, y, z.

The angles and distance are defined in Fig. 4-16(b). Equation (4-10) is the easiest one to use to determine the sign of the hypochromism. For positive polarizabilities, the sign of H depends only on the geometric term. It follows that a transition colinear with any component of the polarizability is hyperchromic. H is negative, because for example $\theta_{ijx} = 90°$, $\theta_{ijy} = 90°$, $\theta_{ijz} = 0°$, $\theta_j = 0°$, $\theta_{jz} = 0°$. Therefore we have ($\cos \theta_{ijz} - 3 \cos \theta_i \cos \theta_{jz}$); $\cos \theta_{ijz} = -2$ and $\cos \theta_{ijx} = \cos \theta_{ijy} = 0$. A transition dipole which is on one axis, but is parallel to another axis leads to hypochromism. Now H is positive, because $\theta_{ijx} = 90°$, $\theta_{ijy} = 0°$, $\theta_{ijz} = 90°$, $\theta_i = 90°$, $\theta_{jzy} = 90°$. Therefore $(\cos \theta_{ijzy} - 3 \cos \theta_i \cos \theta_{jzy}) \cos \theta_{ijzy} = 1$.

Equations (4-7) and (4-10) have been used to calculate the observed hypochromism of nucleic acids.[86,100] A reasonable magnitude and, of course, the correct sign of the hypochromism have been obtained. However, many aspects of the calculated results are not very satisfactory. The polarizability (or refractive index) of the solvent has a large effect which is difficult to calculate. The effect of base composition is not assessed correctly. Therefore, we can only claim qualitative understanding of the phenomenon. We can predict that usually the hypochromism of triple-strand helixes will be greater than that of double-strand helixes, which will be greater than single-strand helixes. Increased stacking of bases will increase the hypochromism. Calculated chain length dependence of the hypochromism is probably valid,[101,102] because the change of hypochromism with the number of bases will mainly depend on the inverse cube of the distance and not on the magnitudes and directions of transition moments and polarizability.

Similar equations can be written for the circular dichroism and rotational strength.

$$R_{\mathrm{OA}} = \frac{6909\,hc}{32\pi^3 N_0} \int \frac{(\epsilon_L - \epsilon_R) \cdot d\nu}{\nu} = \frac{6909\,hc}{32\pi^3 N_0} \int \frac{(\epsilon_L - \epsilon_R) \cdot d\lambda}{\lambda}$$

$$R_{\mathrm{OA}} = \sum_{i}^{\text{groups}} \operatorname{Im} \boldsymbol{\mu}_{iOa} \cdot \mathbf{m}_{iaO} - 2 \sum_{j \neq i}^{\text{groups}} \sum_{b \neq a}^{\text{transitions}} \frac{\operatorname{Im} V_{iOa;jOb}(\boldsymbol{\mu}_{iOa} \cdot \mathbf{m}_{jbO}\, \nu_{iOa} + \boldsymbol{\mu}_{jOb} \cdot \mathbf{m}_{iaO}\, \nu_{jOb})}{h(\nu_{jOb}^2 - \nu_{iOa}^2)}$$

$$+ \frac{2\pi}{c} \sum_{j \neq i}^{\text{groups}} \sum_{b \neq a}^{\text{transitions}} \frac{V_{iOa;jOb}\, \nu_{iOa}\, \nu_{jOb}\, (\mathbf{R}_{ij}) \cdot (\boldsymbol{\mu}_{iOa} \times \boldsymbol{\mu}_{jOb})}{h(\nu_{jOb}^2 - \nu_{iOa}^2)} \qquad (4\text{-}11)$$

The only new quantity in these equations is the magnetic dipole transition moment **m**. In general the discussion on hypochromism is applicable to rotational strength. The experimental determination of the rotational strength by integration of a circular dichroism band is, however, more difficult. The presence of both positive and negative bands makes it difficult to resolve the CD spectrum. There may be much cancellation of intensity. Furthermore, a CD curve with a maximum and a minimum may correspond to one monomer transition. The calculation of the rotational strength from monomer properties is also more involved. Unlike the electric dipole transitions, the magnetic dipole transitions cannot be obtained from the spectrum of the monomer. One tries therefore to minimize the effect of the magnetic dipole transitions by judicious choice of the subunits. Equation (4-11) has three types of terms whose sum gives the rotational strength. The choice of subunit determines the relative magnitudes of the three terms. It is obvious that if the subunit i is chosen to have a plane of symmetry, the first term (Im $_{iOa} \cdot \mathbf{m}_{iaO} = R_{Oai}$) must be zero. If we limit ourselves to $\pi \rightarrow \pi^*$ transitions in planar bases, then only the last term in Eq. (4-11) is nonzero. This term is exactly analogous to Eq. (4-7); the only difference is the replacement of $\mathbf{R}_{ij} \cdot \boldsymbol{\mu}_{iOa} \times \text{`}_{jOb}$ for $\boldsymbol{\mu}_{iOa} \cdot \boldsymbol{\mu}_{jOb}$. The contribution of $\pi \rightarrow \pi^*$ transitions in the bases to the CD of polynucleotides (which seems to be the dominant contribution) can thus be treated by methods identical to those for hypochromism. Replacing the sum over transitions ($O \rightarrow b$) by a polarizability leads to an equation analogous to Eq. (4-10).

$$R_{\mathrm{OA}} = -\left(\frac{\pi}{c}\right) \nu_{Oa}\, \mu_{Oa}{}^2 \sum_{i}^{\text{groups}} \sum_{j \neq i} \frac{\Delta\alpha_j(\nu_{Oa})}{R_{ij}{}^2}$$

$$[\cos\theta_{ij} - 3\cos\theta_i \cos\theta_j] \sin\theta_{ij} \sin\theta_i \sin\phi_{ij} \tag{4-12}$$

$\Delta\alpha_j(\nu_{Oa})$ is the difference in polarizability (at frequency ν_{Oa}) parallel and perpendicular to the symmetry axis of group j. We must choose groups that have at least an approximate axis of symmetry.

The angles are defined in Fig. 4-17. The sign and magnitude of the rotational strength depends on the inverse square of the distance and on many angles. We can draw general conclusions about the effect of structure. Spherically symmetrical groups will not contribute to the rotational strength ($\Delta\alpha_j = 0$). If bonds are chosen as the groups, $\Delta\alpha_j$ will be positive with symmetry axis along the bond. Colinear ($\sin\theta_i = 0$) or coplanar ($\sin\theta_{ij} = 0$) bonds will not contribute to the rotational strength. Two or more bonds that begin a right-handed helix will contribute a positive rotational strength; a left-handed helix of bonds will have a negative rotational strength. However, for the many bonds in a polynucleotide lying on helixes of different radii, one cannot simply equate helix sense and sign of rotational strength. Again, a good semiquantitative understanding of the relation between structure and rotational strength is obtained. However, to make a meaningful comparison with measured CD spectra it is necessary to calculate shapes of curves.

Each transition (O→a) in a monomer will split into N exciton transitions in a polymer of N monomer units. These exciton transitions determine the absorption and circular dichroism spectra of the polymer.[98,103–105] For a dimer such as dinucleoside phosphate, the calculated spectra are particularly simple.[105] Two transitions should occur at frequencies on either side of the monomer frequency, ν_{Oa}. The dipole strength and rotational strength of each exciton transition determine the shape of the spectrum. The splitting, that is the difference in frequency of the transitions, is usually small, so that the effect of the two transitions may not be apparent in

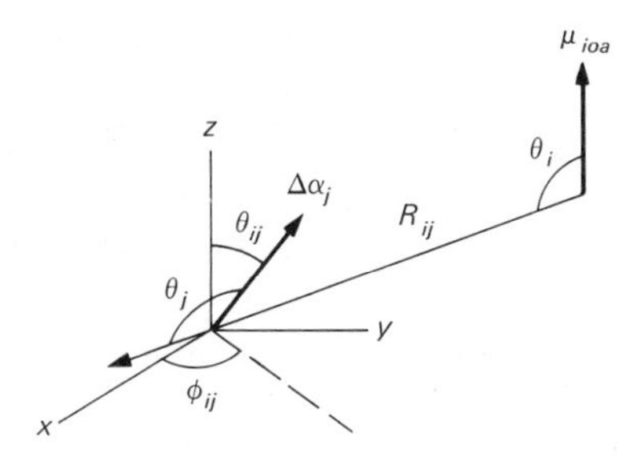

FIGURE 4-17 The angles specifying the interaction of a point dipole $\boldsymbol{\mu}_{iOa}$ with a polarizable group which has cylindrical symmetry. The axis of symmetry is labeled $\Delta\alpha_j$. The coordinate system x, y, z is chosen so that the vector R_{ij} lies in the yz plane and $\boldsymbol{\mu}_{iOa}$ is along z. The angles θ_{ij} and ϕ_{ij} specify the direction of the polarizability symmetry axis in spherical coordinates. θ_i and θ_j are the angles made by the dipole $\boldsymbol{\mu}_{iOa}$ and the polarizability symmetry axis $\Delta\alpha_j$ with the vector $\mathbf{R}_{ij}$;

absorption. However, as the two transitions have rotational strengths of opposite sign, the splitting will be very apparent in circular dichroism. This characteristic circular dichroism has been observed[106] and calculated[107,108] for dinucleoside phosphates. The CD spectrum of adenylyl-(3′-5′)-adenosine shown in Fig. 3-2 is an excellent example of the splitting of the 260 nm band of adenosine. This figure illustrates another aspect of this exciton splitting. If the rotational strength of the 260 nm band in ApA is obtained by integration of the CD band above 235 nm, one finds that its magnitude is essentially zero. This is called a conservative CD band.[91] Although conservative bands will mainly be seen in dimers or higher polymers, their presence is explained by a very general rule. This rule states that the sum of the rotational strengths of all transitions in a molecule is zero.

$$\sum_{A \neq O}^{\text{all transitions}} R_{OA} = 0$$

If the rotational strength for each polymer transition is considered to arise from interactions among monomer transitions as in Eq. (4-11), then the sum over all monomer transitions must also lead to zero. The polymer exciton transitions give rise to rotational strengths which mainly come from interactions among the same transition in many monomers. Summing over the exciton transitions from one monomer transition leads to a rotational strength which is approximately zero. For a dimer such as ApA the sum of the rotational strengths from the 260 nm adenine band is approximately zero. The sum is not exactly zero because of interaction with other bands. If interaction with other bands becomes dominant, the rotational strength will be nonzero and the band is termed nonconservative.[91] The nonconservative rotational strength which is just the sum over all exciton rotational strengths is the quantity calculated in Eqs. (4-11) and (4-12).

In a polymer of N monomer units each of the N exciton transitions will have a rotational strength and will occur at a frequency different from the monomer transition. A shape (such as a gaussian) can be assigned to each transition; therefore by use of Eq. (4-6) the circular dichroism spectrum can be calculated. Both conservative and nonconservative spectra are expected and seen.

To calculate the polymer exciton rotational strengths and frequencies one can expand the polymer wave functions as a linear combination of monomer wave functions. The coefficients in the expansion can be obtained by diagonalizing the matrix of the interaction energies or from first-order perturbation theory.[104,108,109] Semiquantitative agreement with the observed shape of CD curves has been obtained.[108] As agreement is not yet good enough to be able to draw quantitative structural conclusions from the calculations, we will not discuss them in detail, but instead discuss a simplifying assumption which still leads to qualitative agreement with experiment.

The N exciton frequencies for a polymer (ν_{OAK}; $K = 1, 2, \ldots N$) corresponding to one transition or a few similar transitions in a monomer will not be very different. It is reasonable to use a Taylor expansion around $\bar{\nu}$ using ν_{OAK} as essentially a continuous variable.[108,110] Equation (4-6) becomes

$$\frac{\epsilon_L(\nu) - \epsilon_R(\nu)}{\nu} = \frac{32\pi^3 N_0}{6909\,hc} f(\nu - \bar{\nu}) \sum_{K=1}^{N} R_{OAK} - \frac{\partial f(\nu - \bar{\nu})}{\partial \nu} \sum_{K=1}^{N} R_{OAK}(\nu_{OAK} - \bar{\nu}) + \ldots \tag{4-13}$$

$f(\nu - \bar{\nu})$ is a function which has the shape of an absorption band centered at an average frequency. The integral of $f(\nu - \bar{\nu})d\nu = 1$. We have assumed each transition in the base has approximately the same shape.

$\dfrac{\partial f(\nu - \bar{\nu})}{\partial \nu}$ is the derivative of the above function. Its integral over $\nu = 0$.

$\sum_{K=1}^{N} R_{OAK} = R_{OA}$. Equations (4-11, 4-12) related R_{OA} to monomer properties and polymer geometry.

$$\sum_{K=1}^{N} R_{OAK}(\nu_K - \bar{\nu}) = \frac{-\pi}{2c} \sum_{i}^{\text{groups}} \sum_{j \neq i} \sum_{a,b}^{\text{transitions}} \nu_{iOa} V_{iOa;jOb} \mathbf{R}_{ij} \cdot (\boldsymbol{\mu}_{iOa} \times \boldsymbol{\mu}_{jOb}) \tag{4-14}$$

The sums over transitions $(O \to a)$ and $(O \to b)$ include only the transitions in the monomers which occur in the polymer band labeled (OA). For the 260 nm band in polynucleotides they would be all $\pi \to \pi^*$ monomer bands above 225 nm.

In this approximation there are two main contributions to the CD spectrum. One has the shape of an absorption band and the magnitude of the rotational strength. The other has the shape of the derivative of an absorption band; its rotational strength is zero, but its magnitude can be easily calculated [Eq. (4-14)]. If $f(\nu - \bar{\nu})$ is approximated by a gaussian, the types of CD spectra (and ORD spectra) to be expected are illustrated in Fig. 4-18. The

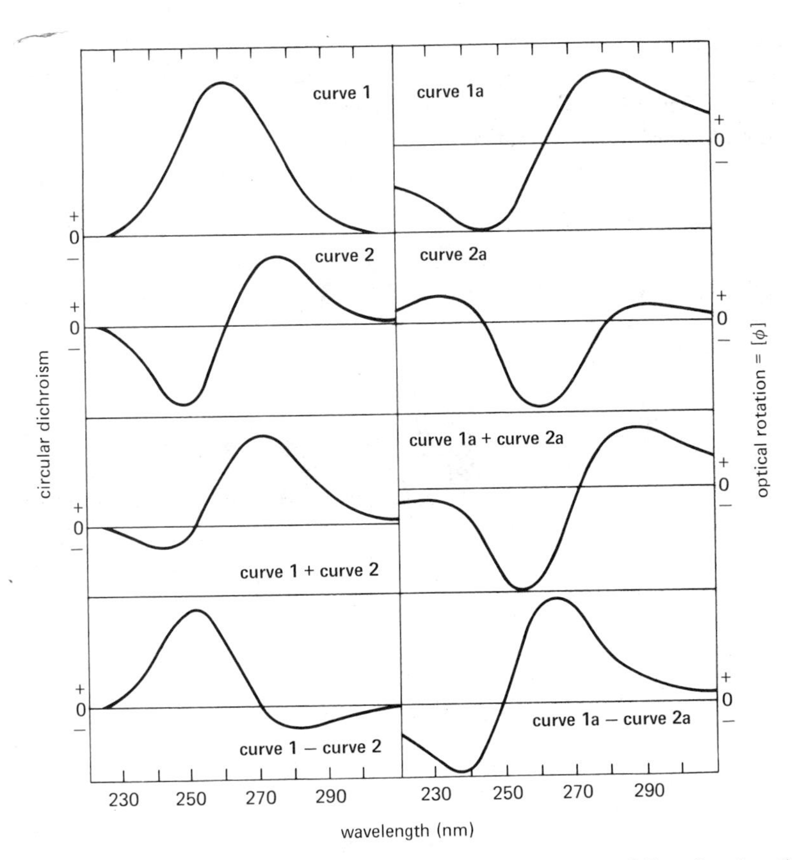

FIGURE 4-18 The shapes of circular dichroism curves (left side of figure) and optical rotatory dispersion curves (right side of figure) expected for polymers for each monomer band. The shapes were calculated from Eq. (4-15) and the Kronig-Kramers transform of Eq. (4-15). For each absorption band in a monomer there will be a polymer band which is a weighted sum of the top two curves. [From I. Tinoco, Jr., *J. Chim. Phys.*, **65**, 91 (1968). Reprinted with permission.]

curves in the figure were calculated from the following equations:

$$f(\nu - \bar{\nu}) = \frac{\sqrt{\pi}\, e^{-(\nu-\bar{\nu})^2/\Theta^2}}{\Theta}$$

$$\frac{\partial f(\nu - \bar{\nu})}{\partial \nu} = \frac{-2\sqrt{\pi}\,(\nu - \bar{\nu})e^{-(\nu-\bar{\nu})^2/\Theta^2}}{\Theta^3} \qquad (4\text{-}15)$$

and their Kronig-Kramers transforms.

Experimentally both kinds of CD spectra have been recognized. The derivative spectra with zero rotational strength were named conservative spectra and the absorption type spectra were named nonconservative.[91] Calculated and observed spectra for DNA and RNA are shown in Fig. 4-19. The agreement is considered fair. The reason for the different type of CD spectra in DNA and RNA seems to be the tilting and twisting of the bases. This increased asymmetry in RNA increases the rotational strength and leads to the observed nonconservative spectrum. Better agreement with experiment

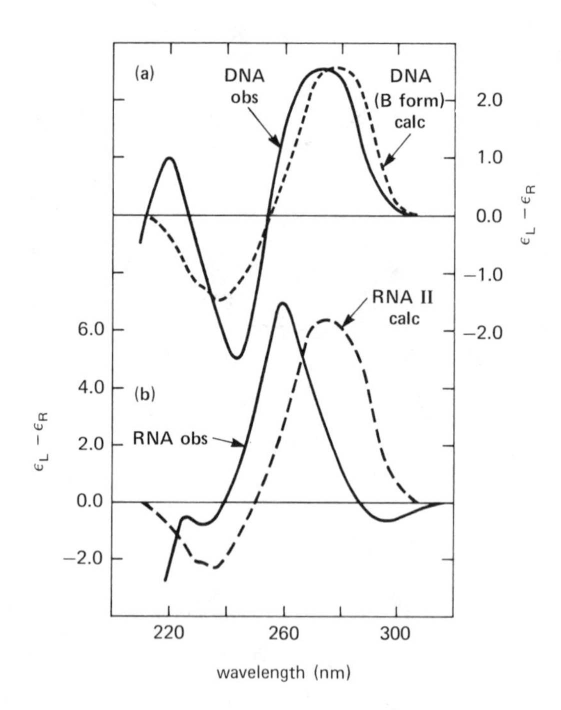

FIGURE 4-19 Observed and calculated circular dichroism curves for double-stranded DNA and RNA. The DNA is from calf thymus; the RNA is the double-stranded rice dwarf virus RNA. The calculated curves (dashed lines) were obtained by use of Eq. (4-13). [From W. C. Johnson, Jr. and I. Tinoco, Jr., *Biopolymers,* 7, 727 (1969). Reprinted with permission.]

is found for the DNA calculation than for the RNA calculation. The reason for this is not clear. It may be that $n \rightarrow \pi^*$ transitions which have not been considered in these calculations become significant for tilted bases. This might lead to the observed negative CD near 300 nm. The rotational strength of $n \rightarrow \pi^*$ transitions will be discussed in the next section.

A formally different approach to the calculation of the CD of polymers has been used by DeVoe.[88,111] He uses a classical model to derive expressions for the absorption and circular dichroism. This classical model leads to equations completely consistent with the most general quantum mechanical models.[112] The resulting equations are similar to those we have presented earlier, and reduce to them for the same level of approximation. However, they have the advantage of relating the measured spectra of the monomers in a more direct manner to the polymer CD spectra. The equations are:

$$\epsilon(\nu) = \frac{-8\pi^2 \nu N_0}{6909\, c} \sum_i \sum_j \operatorname{Im} A_{ij} \mathbf{e}_i \cdot \mathbf{e}_j$$

$$\epsilon_L - \epsilon_R(\nu) = \frac{16\pi^3 \nu^2 N_0}{6909\, c^2} \sum_i \sum_j \operatorname{Im} A_{ij} \mathbf{R}_{ij} \cdot \mathbf{e}_i \times \mathbf{e}_j \qquad (4\text{-}16)$$

$\mathbf{e}_i$, $\mathbf{e}_j$ are unit vectors which specify a direction of a component of the polarizability in a group. In general three will be needed for each group.

$\mathbf{R}_{ij}$ is a vector from group i to group j

The wavelength dependence of the spectrum is contained in A_{ij}. The A_{ij} are the coefficients of a complex matrix obtained by inverting a complex matrix.

$$A_{ij} = \left[\frac{\delta_{ij}}{\alpha_i(\nu)} + G_{ij}\right]^{-1} \qquad (4\text{-}17)$$

δ_{ij} = Kronecker delta = 1 if $i = j$ and 0 otherwise

$\alpha_i(\nu)$ is a complex polarizability for group i along the principal direction $\mathbf{e}_i$

$$G_{ij} = \frac{1}{R_{ij}^3}\left[\mathbf{e}_i \cdot \mathbf{e}_j - 3\,\frac{(\mathbf{R}_{ij} \cdot \mathbf{e}_i)(\mathbf{R}_{ij} \cdot \mathbf{e}_j)}{R_{ij}^2}\right]$$

$$G_{ij} = \frac{1}{R_{ij}^3}(\cos\theta_{ij} - 3\cos\theta_i \cos\theta_j)$$

The angles are defined in Fig. 4-17. The real and imaginary parts of the polarizability $\alpha_i(\nu)$ come from measurements of the refraction and absorption of group i. Both the magnitude and direction (three components) of the polarizability are needed. The imaginary part of the average polarizability of a monomer is

$$\text{Im}\,\alpha(\nu) = \frac{-2303\,c}{8\pi^2 \nu N_0}\,\epsilon(\nu) \tag{4-18}$$

$\alpha(\nu)$ is the average polarizability of the monomer $= \dfrac{(\alpha_x + \alpha_y + \alpha_z)}{3}$

The real part of the average polarizability of a monomer is related to its molar refraction $[R]$.

$$[R] = \frac{1}{\text{conc.}}\left[\frac{n^2 - 1}{n^2 + 2} - \frac{n_0{}^2 - 1}{n^2 + 1}\right] \tag{4-19}$$

$$\text{Re}\,\alpha(\nu) = \frac{3}{4\pi N_0}\,[R]$$

conc. is the concentration of monomers (moles/liter)
n is the refractive index of solution
n_0 is the refractive index of solvent
N_0 is Avogadro's number

The real and imaginary parts of the polarizability are also related by Kronig-Kramers transforms. Equation (4-16) has been used to calculate the absorption of B form DNA.[90] The agreement with experiment is not as good as seen in the quantum mechanical CD calculations. A point dipole approximation was used for each base, therefore better results should be obtained if point dipoles were used for each bond. Similar results would be expected for classical CD calculations. The one interesting result of this classical absorption calculation (based solely on $\pi{\rightarrow}\pi^*$ transitions) is that it gives the observed long wavelength hyperchromism around 290 nm. There is no need to postulate allowed $n{\rightarrow}\pi^*$ transitions polarized perpendicular to the planes of bases.

Electrically forbidden transitions / All the previous discussion has been about electrically allowed transitions such as $\pi{\rightarrow}\pi^*$ transitions and some $n{\rightarrow}\pi^*$ transitions (those like the heterocyclic nitrogens which have $s{\rightarrow}p_x$ atomiclike transitions). What about electrically forbidden magnetically allowed transitions such as the $n{\rightarrow}\pi^*$ in carbonyls which has $p_x{\rightarrow}p_z$ character? We cannot now ignore the effects of static fields, because they may make the dominant contribution to the rotational strength. These transitions have very weak intensity in absorption so only their rotational strengths will be discussed.[99,113]

$$R_A = -\sum_{i=1}^{N} 2\left[\sum_{j\neq i}\sum_{b\neq a}\frac{\operatorname{Im} V_{iOa;jOb}\boldsymbol{\mu}_{jOb}\cdot\mathbf{m}_{iaO}\nu_b}{h(\nu_b^2-\nu_a^2)}\right.$$

$$+\sum_{j\neq i}\sum_{b\neq a}\frac{\operatorname{Im} V_{iab;jOO}\boldsymbol{\mu}_{iOb}\cdot\mathbf{m}_{iaO}}{h(\nu_b-\nu_a)}$$

$$+\sum_{j\neq i}\sum_{b\neq a}\frac{\operatorname{Im} V_{iOb;jOO}\boldsymbol{\mu}_{iab}\cdot\mathbf{m}_{iaO}}{h\nu_b} \qquad (4\text{-}20)$$

$$+\sum_{j\neq i}\frac{\operatorname{Im} V_{iOa;OO}(\boldsymbol{\mu}_{iaa}-\boldsymbol{\mu}_{iOO})\cdot\mathbf{m}_{iaO}}{h\nu_a}$$

The terms involving the static fields are similar to those discussed earlier, except that now permanent dipoles ($\boldsymbol{\mu}_{aa}$, $\boldsymbol{\mu}_{OO}$) are present and transition dipoles between excited states ($\boldsymbol{\mu}_{ab}$) appear. An electrically forbidden transition in a monomer will not split into exciton transitions in the polymer, so only this nonconservative contribution to the CD will result. It will appear essentially with the position and band shape of the monomer band, but this may be hard to determine, because of the weakness of the absorption. No experimentally observed CD bands in polynucleotides have been assigned to electrically forbidden transitions. Which is not to say they are absent. In polypeptides the carbonyl $n\rightarrow\pi^*$ rotational strength is the longest wavelength CD band[113–115] and is very important in the visible ORD.

REFERENCES

1 R. Langridge, H. R. Wilson, C. W. Hooper, M. H. F. Wilkins, and L. D. Hamilton, *J. Mol. Biol.,* **2**, 19 (1960).
2 J. D. Watson, *The Double Helix*, Atheneum, New York (1968).
3 M. H. F. Wilkins, *Science,* **140**, 941 (1963).
4 L. D. Hamilton, *Nature,* **218**, 633 (1968).
5 A. Klug, *Nature,* **219**, 808, 843 (1968).
6 R. Signer, T. Caspersson, and E. Hammersten, *Nature,* **141**, 122 (1938).
7 T. Caspersson, *Chromosoma,* **4**, 605 (1940).
8 W. T. Astbury, Symp. Soc. Exp. Biol. I. *Nucleic Acids*, Cambridge University Press, (1947), p. 66.
9 J. M. Gulland and D. O. Jordan, Symp. Soc. Exp. Biol. I. *Nucleic Acids*, Cambridge University Press (1947).
10 J. M. Gulland, *Cold Spring Harbor Symp. Quant. Biol.,* **12**, 95 (1947).
11 M. H. F. Wilkins, R. G. Gosling, and W. E. Seeds, *Nature,* **167**, 759 (1951).
12 R. E. Franklin and R. G. Gosling, *Acta Cryst.,* **6**, 673 (1953).
13 L. Pauling and R. B. Corey, *Proc. Nat. Acad. Soc., U.S.,* **39**, 84 (1953).
14 E. Chargaff, *Experientia,* **6**, 201 (1950).

15 G. R. Wyatt and S. S. Cohen, *Biochem. J.*, **55**, 774 (1953).
16 J. D. Watson and F. H. C. Crick, *Nature,* **171**, 964 (1953).
17 H. R. Mahler and E. Cordes, *Biological Chemistry*, Harper & Row, New York (1971).
18 A. Kornberg, *The Enzymatic Synthesis of DNA*, Wiley, New York (1961).
19 J. Josse, A. D. Kaiser, and A. Kornberg, *J. Biol. Chem.*, **236**, 864 (1961).
20 K. C. Holmes and D. M. Blow, *The Use of X-ray Diffraction in the Study of Protein and Nucleic Acid Structure*, Interscience, New York (1966).
21 R. W. James, *The Optical Properties of the Diffraction of X-Rays*, G. Bell and Sons, London (1948).
22 W. Cochran, F. H. C. Crick, and V. Vand, *Acta Cryst.*, **5**, 581 (1952).
23 H. R. Wilson, *Diffraction of X-rays by Proteins, Nucleic Acids, and Viruses*, St. Martin's Press, New York (1966).
24 R. E. Dickerson in *The Proteins*, 2nd ed., Vol. II, H. Neurath, ed., Academic Press, New York (1964), p. 603.
25 C. Kittel, *Amer. J. Phys.*, **36**, 610 (1968).
26 P. Franklin, *An Introduction to Fourier Methods and the Laplace Transform*, Dover, New York (1949).
27 R. Langridge, D. A. Marvin, W. E. Seeds, H. R. Wilson, C. W. Hooper, and M. H. F. Wilkins, *J. Mol. Biol.*, **2**, 38 (1960).
28 A. Rich, *Revs. Mod. Phys.*, **31**, 191 (1959).
29 M. H. F. Wilkins, *Science,* **140**, 941 (1963).
30 J. Kraut, *Ann. Rev. Biochem.*, **34**, 247 (1965).
31 D. R. Davies, *Ann. Rev. Biochem.*, **36**, 321 (1967).
32 L. D. Hamilton, *Nature,* **218**, 633 (1968).
33 S. Arnott, S. D. Dover, and A. J. Wonacott, *Acta Cryst.*, **B25**, 2192 (1969).
34 M. H. F. Wilkins, W. E. Seeds, A. R. Stokes, and H. R. Wilson, *Nature,* **172**, 759 (1953).
35 R. E. Franklin and R. G. Gosling, *Nature,* **172**, 156 (1953).
36 R. Langridge, W. E. Seeds, H. R. Wilson, C. W. Hooper, M. H. F. Wilkins, and L. D. Hamilton, *J. Biophys. Biochem. Cytol.*, **3**, 767 (1957).
37 D. A. Marvin, M. Spencer, M. H. F. Wilkins, and L. D. Hamilton, *Nature,* **182**, 387 (1958).
38 D. R. Davies and R. L. Baldwin, *J. Mol. Biol.*, **6**, 251 (1963).
39 W. Fuller, M. H. F. Wilkins, H. R. Wilson, and L. D. Hamilton, *J. Mol. Biol.*, **12**, 60 (1965).
40 G. Milman, R. Langridge, and M. Chamberlain, *Proc. Nat. Acad. Sci., U.S.*, **57**, 1804 (1967).
41 M. J. Tunis and J. E. Hearst, *Biopolymers,* **6**, 1218 (1968).
42 D. A. Marvin, M. Spencer, M. H. F. Wilkins, and L. D. Hamilton, *J. Mol. Biol.*, **3**, 547 (1961).
43 M. H. F. Wilkins and J. T. Randall, *Biochim. Biophys. Acta,* **10**, 192 (1953).
44 R. Langridge, *Abstr.*, 7th Intl. Congr. Biochem., Tokyo, **1**, 57 (1967).
45 R. Langridge and P. J. Gomatos, *Science*, **141**, 694 (1963).
46 S. Arnott, F. Hutchinson, M. Spencer, M. H. F. Wilkins, W. Fuller, and R. Langridge, *Nature,* **211**, 227 (1966).
47 K. Tomita and A. Rich, *Nature,* **201**, 1160 (1964).
48 R. Langridge, M. A. Billeter, P. Borst, H. R. Burdon, and C. Weissman, *Proc. Nat. Acad. Sci., U.S.*, **52**, 114 (1964).
49 T. Sato, Y. Kyogoku, S. Higuchi, Y. Mitsui, Y. Iitaka, M. Tsuboi, and K. Miura, *J. Mol. Biol.*, **16**, 180 (1966).

50 S. Arnott, M. H. F. Wilkins, W. Fuller, and R. Langridge, *J. Mol. Biol.,* **27**, 525 (1967).
51 S. Arnott, M. H. F. Wilkins, W. Fuller, J. H. Venable, and R. Langridge, *J. Mol. Biol.,* **27**, 549 (1967).
52 V. Sasisekharan and P. B. Sigler, *J. Mol. Biol.,* **12**, 296 (1965).
53 A. Rich and J. D. Watson, *Proc. Nat. Acad. Sci., U.S.,* **40**, 759 (1954).
54 A. Rich and J. D. Watson, *Nature,* **173**, 995 (1954).
55 M. Spencer, W. Fuller, M. H. F. Wilkins, and G. L. Brown, *Nature,* **194**, 1014 (1962).
56 M. Spencer, *Cold Spring Harbor Symp. Quant. Biol.,* **28**, 77 (1963).
57 M. Spencer and F. Poole, *J. Mol. Biol.,* **11**, 314 (1965).
58 W. Fuller, F. Hutchinson, M. Spencer, and M. H. F. Wilkins, *J. Mol. Biol.,* **27**, 507 (1967).
59 V. Sasisekharan, D. R. Davies, and P. B. Sigler, unpublished observations, quoted in Davies (31).
60 A. Rich, *Nature,* **181**, 521 (1958).
61 D. R. Davies and A. Rich, *J. Amer. Chem. Soc.,* **80**, 1003 (1958).
62 D. R. Davies, Intern. Biophys. Congr., Stockholm, 1961.
63 R. Langridge and A. Rich, *Nature,* **198**, 725 (1963).
64 A. Rich, D. R. Davies, F. H. C. Crick, and J. D. Watson, *J. Mol. Biol.,* **3**, 71 (1961).
65 A. Klug and J. Finch, personal communication to D. R. Davies (31).
66 A. Rich, *Biochim. Biophys. Acta,* **29**, 502 (1958).
67 M. J. B. Tunis-Schneider and M. F. Maestre, *J. Mol. Biol.,* **52**, 521 (1970).
68 J. Brahms and W. H. F. Mommaerts, *J. Mol. Biol.,* **10**, 73 (1964).
69 G. Green and H. R. Mahler, *Biopolymers,* **6**, 1509 (1968).
70 M. Maestre and J. Wang, *Biopolymers,* **10**, 1021 (1971).
71 M. Maestre and I. Tinoco, Jr., *J. Mol. Biol.,* **23**, 323 (1967).
72 E. O'Brien, *J. Mol. Biol.,* **7**, 107 (1963).
73 H. M. Sobell, K. Tomita, and A. Rich, *Proc. Nat. Acad. Sci., U.S.,* **49**, 855 (1963).
74 K. Hoogsteen, *Acta Cryst.,* **12**, 822 (1959).
75 F. S. Matthews and A. Rich, *J. Mol. Biol.,* **8**, 89 (1964).
76 S. Arnott, M. H. F. Wilkins, L. D. Hamilton, and R. Langridge, *J. Mol. Biol.,* **11**, 391 (1965).
77 J. Donohue, *Proc. Nat. Acad. Sci., U.S.,* **42**, 60 (1956).
78 J. Donohue and K. Trueblood, *J. Mol. Biol.,* **2**, 363 (1960).
79 B. Pullman, ed., *Molecular Associations in Biology*, Academic Press, New York (1968), p. 21.
80 J. Donohue, *Science,* **165**, 1091 (1969).
81 G. B. B. M. Sutherland and M. Tsuboi, *Proc. Roy. Soc.,* **A239**, 446 (1957).
82 M. Tsuboi, *J. Amer. Chem. Soc.,* **79**, 1351 (1957).
83 T. Simanouchi, M. Tsuboi, and Y. Kyogoku, *Advan. Chem. Phys.,* **7**, 435 (1964).
84 M. Tsuboi, *Prog. Theor. Phys., Suppl.,* **17**, 99 (1961).
85 M. Tsuboi, *J. Polymer Sci.,* **69**, 139 (1962).
86 I. Tinoco, Jr., *J. Amer. Chem. Soc.,* **82**, 4785 (1960); **83**, 5047 (1961).
87 W. Rhodes, *J. Amer. Chem. Soc.,* **83**, 3609 (1961).
88 H. DeVoe, *J. Chem. Phys.,* **43**, 3199 (1965).
89 A. Rich and M. Kasha, *J. Amer. Chem. Soc.,* **82**, 6197 (1960).
90 H. DeVoe, *Ann. N.Y. Acad. Sci.,* **158**, 298 (1969).
91 C. A. Bush and J. Brahms, *J. Chem. Phys.,* **46**, 79 (1967).
92 W. C. Johnson and I. Tinoco, Jr., *Biopolymers,* **7**, 727 (1969).
93 J. Brahms and W. F. H. M. Mommaerts, *J. Mol. Biol.,* **10**, 73 (1964).

94 G. Felsenfeld and S. Z. Hirschman, *J. Mol. Biol.*, **13**, 407 (1965).
95 T. Samejima and J. T. Yang, *J. Biol. Chem.*, **240**, 2094 (1965).
96 D. Gray and I. Tinoco, Jr., *Biopolymers,* **9**, 223 (1970).
97 J. G. Kirkwood, *J. Chem. Phys.*, **5**, 479 (1937).
98 W. Moffitt, D. O. Fitts, and J. G. Kirkwood, *Proc. Nat. Acad. Sci., U.S.*, **43**, 723 (1957).
99 I. Tinoco, Jr., *Adv. Chem. Phys.*, **4**, 113 (1962).
100 H. DeVoe and I. Tinoco, Jr., *J. Mol. Biol.*, **4**, 518 (1962).
101 A. Rich and I. Tinoco, Jr., *J. Amer. Chem. Soc.*, **82**, 6409 (1960).
102 J. Applequist, *J. Amer. Chem. Soc.*, **83**, 3158 (1961).
103 W. Moffitt, *J. Chem. Phys.*, **25**, 467 (1956).
104 I. Tinoco, Jr., R. W. Woody, and D. F. Bradley, *J. Chem. Phys.*, **38**, 1317 (1963).
105 I. Tinoco, Jr., *Radiation Res.*, **20**, 133 (1963).
106 M. M. Warshaw and C. R. Cantor, *Biopolymers,* **9**, 1079 (1970).
107 C. A. Bush and I. Tinoco, Jr., *J. Mol. Biol.*, **23**, 601 (1967).
108 W. C. Johnson, Jr., and I. Tinoco, Jr., *Biopolymers,* 8, 715 (1969).
109 P. M. Bayley, E. B. Nielson, and J. A. Schellman, *J. Phys. Chem.*, 73, 228 (1969).
110 I. Tinoco, Jr., *J. Chem. Phys.*, **65**, 91 (1968).
111 H. DeVoe, *J. Chem. Phys.*, **41**, 393 (1964).
112 W. Rhodes and M. Chase, *Revs. Mod. Phys.*, **39**, 348 (1967).
113 J. A. Schellman and P. Oriel, *J. Chem. Phys.*, **37**, 2114 (1962).
114 I. Tinoco, Jr., and R. W. Woody, *J. Chem. Phys.*, **46**, 4927 (1967).
115 R. W. Woody, *J. Chem. Phys.*, **49**, 4797 (1968).
116 S. Arnott, *Prog. Biophys. Molec. Biol.*, **21**, 265 (1970).
117 J. M. Rosenberg, N. C. Seeman, J. P. Kim, F. L. Suddath, H. B. Nicholas, and A. Rich, *Nature*, **243**, 150 (1973).

chapter 5
molecular weight and long-range structure

I INTRODUCTION

In this chapter we discuss methods for determining the sizes and shapes of polynucleotides. Faced with a DNA or RNA of unknown properties, we wish to answer questions such as these: What is its molecular weight, and therefore how much genetic information does it contain? What is its topology? Is it linear, cyclic, or branched? How flexible is it? How much energy will be required to pack it into a virus or chromosome, or to bend it into a loop in some complex conformation?

Most of this chapter deals with methods used to answer these questions. These methods include direct visualization or counting, such as electron microscopy, autoradiography, and other techniques of radioactive labeling. They include the scattering of visible light and of X-rays, and the measurement of concentration distribution at sedimentation equilibrium in a density gradient. And they include the well-known transport methods for characterizing macromolecular structure, such as sedimentation, diffusion, viscosity, flow birefringence and dichroism, and fluorescence depolarization. The chapter begins with a discussion of conformational statistics of polynucleotide chains, which is basic to an understanding of the experimental methods. It ends with a brief survey of some of the results which have been obtained up to now on the molecular weights, flexibility, and topology of naturally occurring nucleic acids.

The reader is referred to review articles[1-3] which deal with many of the topics discussed here. We have drawn heavily on standard textbooks on polymer physical chemistry[4,5] for much of the material in this chapter, and these should be consulted for further details.

Throughout this chapter we have concentrated on the application of polymer physical chemistry techniques to rodlike or relatively long, flexible chain conformations which are characteristic of native and denatured DNA, synthetic polynucleotides, and most isolated RNA. We have largely neglected, particularly in the section on transport properties, the treatment of rigid, relatively impermeable, globular structures which contain nucleic acids, such as tRNA, ribosomes, and viruses. Because these matters are discussed in great detail elsewhere, we content ourselves with referring the reader to standard texts.[5]

II CONFORMATIONAL STATISTICS

A. Introduction

Many of the experimental techniques used to study nucleic acids, particularly scattering and hydrodynamic techniques, are sensitive both to the molecular weight and to the conformation in solution. In interpreting these measurements, it is therefore important to have a firm theoretical understanding of the way in which the polymer size depends both on the number of monomers contained in it – that is, on its molecular weight – and on its physical state in solution.

There are two limiting types of polymer chain conformation – the completely rigid rod and the completely flexible random coil. Single-stranded polynucleotides approach the latter of these limits, while native duplex DNA is intermediate between them, and therefore raises some difficult problems in conformational statistics due to its limited but not negligible flexibility. In this section we shall accordingly treat not only the limiting cases of the rigid rod and the flexible random coil, but also the so-called wormlike coil, which is of intermediate flexibility.

Native DNA molecules isolated from biological sources are of high molecular weight, often in the 10^7 to 10^9 molecular weight range. Since DNA segments have finite volume, it is clear that many conformations of the chain in solution will be excluded because of the prohibition of physical overlap of two segments in the same small volume of space. This effect, known as the "excluded volume effect," is one that is common to all polymers, not just to DNA. It has been the subject of very extensive and complicated theoretical and experimental study. We shall here be able to discuss only some of the more qualitative findings of these studies.

B. Measure of Average Dimensions

A solution of flexible polymers contains molecules, each of which may have a different conformation at any given instant. Further, due to the random bombardment of polymer molecules by solvent molecules (Brownian motion), the size and conformation of any given polymer molecule will change with time. It is, therefore, clear that we cannot specify a single dimensional parameter to characterize flexible polymers in solution, but must instead deal with average dimensions. These averages over internal conformations are denoted by angular brackets.

One commonly used average dimension is the mean square end-to-end distance, $\langle L^2 \rangle$. Figure 5-1 represents an instantaneous snapshot of a single polymer molecule in solution. This molecule contains $N+1$ repeating elements, or monomers, which are numbered from 0 to N, and which are represented by circles in the figure. A vector $\mathbf{L}$ is drawn between the beginning and end of the chain, which equals the sum of individual bond vectors $\mathbf{b}_i$ connecting monomer units $i-1$ and i:

$$\mathbf{L} = \sum_{i=1}^{N} \mathbf{b}_i$$

Thus

$$L^2 = \mathbf{L} \cdot \mathbf{L} = \left(\sum_{i=1}^{N} \mathbf{b}_i \right) \cdot \left(\sum_{j=1}^{N} \mathbf{b}_j \right) = \sum_{i=1}^{N} \sum_{j=1}^{N} \mathbf{b}_i \cdot \mathbf{b}_i$$

From vector analysis $\mathbf{b}_i \mathbf{b}_j = \mathbf{b}_i \mathbf{b}_j \cos \theta_{ij}$, where θ_{ij} is the angle between bonds i and j. Since the bond lengths will remain essentially constant,

$$\langle L^2 \rangle = \sum_{i=1}^{N} \sum_{j=1}^{N} b_i b_j \langle \cos \theta_{ij} \rangle \tag{5-1}$$

A second common measure of average dimensions is the mean square radius, $\langle R^2 \rangle$. For polymers in which each monomeric unit has the same molecular weight, $\langle R^2 \rangle$ is defined as the average distance of a unit from the center of mass of the chain:

$$\langle R^2 \rangle = (N+1)^{-1} \sum_{i=0}^{N} \langle r_i^2 \rangle$$

$\mathbf{r}_i$ is the vector from the ith element to the center of mass.

A useful relation has been obtained[6] between $\langle R^2 \rangle$ and the mean square distances $\langle L_{ij}^2 \rangle$ between all pairs of elements i and j in the chain, which we quote without proof. It is

$$\langle R^2 \rangle = \frac{1}{2(N+1)^2} \sum_{i=0}^{N} \sum_{j=0}^{N} \langle L_{ij}^2 \rangle \tag{5-2}$$

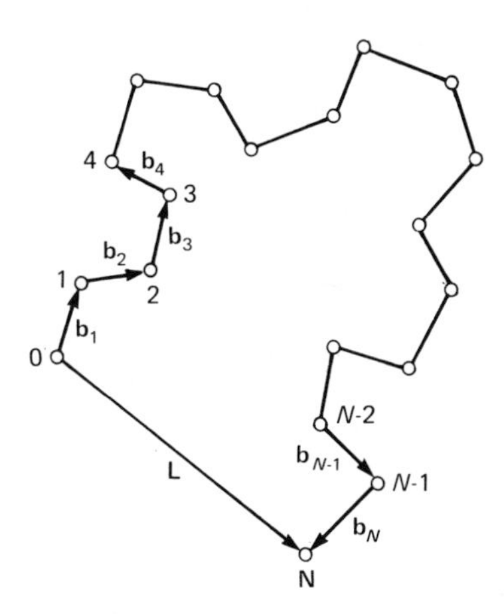

FIGURE 5-1 Two-dimensional projection of the instantaneous conformation of a polymer molecule in solution. The circles represent monomers, the arrows represent bond vectors $\mathbf{b}_i$.

Another important average dimension which we shall have occasion to consider is the mean reciprocal distance, $\langle L_{ij}^{-1} \rangle$, between elements i and j. If, in a particular conformation, these elements are separated by a distance L_{ij}, their reciprocal distance is L_{ij}^{-1}; and this quantity, averaged over all conformations, is $\langle L_{ij}^{-1} \rangle$. Knowledge of this quantity is necessary to calculate the hydrodynamic properties of polymer molecules, as will be seen in succeeding sections. At this point it suffices to remark that except for the rigid rod, $\langle L_{ij}^{-1} \rangle$ is not equal to $\langle L_{ij}^{2} \rangle^{-1/2}$.

C. Rigid Rod

We now proceed to apply the above considerations to particular models of polymer chains. A simple example, and one which is relevant to native DNA of low molecular weight (below 2×10^5), is the rigid rod. Since the rod is not flexible, no averaging over internal conformations is required. The distance between any two elements i and j is given by

$$L_{ij} = b \, |i - j| \tag{5-3}$$

where all bonds have constant length b, and therefore the distance between the elements at the ends of the chain is given by $L = bN = \mathcal{L}$. $\mathcal{L}$ is the contour length of the chain, as measured along its backbone.

The mean square radius is calculated by substituting Eq. (5-3) in Eq. (5-2). It is easily shown that the result, for long rods, is

$$R^2 = \frac{1}{12} \mathcal{L}^2 \tag{5-4}$$

Obviously, the mean reciprocal distance between elements i and j of the rigid rod is

$$L_{ij}^{-1} = \frac{1}{b\,|i-j|} \tag{5-5}$$

D. Flexible Linear Chains

At the opposite extreme, let us consider a completely flexible chain containing N bonds, each of length b. We imagine that the bonds are connected by universal joints. The conformation of the polymer is therefore that of a "random walk," in which successive steps are completely uncorrelated in direction. The central quantity to be determined for such a polymer chain is its distribution function; that is, the probability $W(L, N)dL$ that after N steps the end of the chain will be at a distance between L and $L + dL$ from the origin. The distribution function for a random walk is derived in many standard references[3,4]; it is

$$W(L, N)dL = 4\pi \left(\frac{3}{2\pi Nb^2}\right)^{3/2} \exp\left[-\frac{3L^2}{2Nb^2}\right] L^2 dL \tag{5-6}$$

Chains that conform to this distribution function are often called gaussian chains, because of the gaussian, or bell-shaped, character of the distribution.

This distribution function can be used to obtain many useful average dimensions of the polymer chain. First, we consider the mean square end-to-end distance, $\langle L^2 \rangle$. According to the general procedure for calculating averages with the distribution function, we write

$$\langle L^2 \rangle = \int_0^\infty L^2 W(L, N)dL \tag{5-7}$$

Substituting the expression for $W(L, N)$, Eq. (5-6), and integrating, we then obtain

$$\langle L^2 \rangle = 4\pi \left(\frac{3}{2\pi Nb^2}\right)^{3/2} \int_0^\infty L^4 \exp\left[-\frac{3L^2}{2Nb^2}\right] dL = Nb^2 \tag{5-8}$$

This result is remarkable in that the mean square distance between the ends of the polymer chain is proportional to the first power of N, and not, as in the case of the rigid rod, to N^2.

This result can also be obtained from Eq. (5-1) by noting that for $i = j$, $\langle \cos \theta_{ij} \rangle = 1$, while for a freely jointed chain, $\langle \cos \theta_{ij} \rangle = 0$ if $i \neq j$.

The mean square radius $\langle R^2 \rangle$ of the freely jointed chain may be obtained by use of Eqs. (5-2) and (5-8) which when combined give

$$\langle R^2 \rangle = \frac{b^2}{2(N+1)^2} \sum_{i=0}^{N} \sum_{j=0}^{N} |i-j|$$

Evaluation of the double summation for long chains ($N \gg 1$) yields

$$\langle R^2 \rangle = \frac{1}{6} b^2 N = \frac{1}{6} \langle L^2 \rangle \tag{5-9}$$

This relation will often hold with sufficient accuracy for linear chains, even if they are not freely jointed, and therefore $\langle L^2 \rangle$, which is usually more easily calculated, may be related to $\langle R^2 \rangle$, which is usually more easily measured.

The mean reciprocal distance between the ends of the chain, $\langle L^{-1} \rangle$, is obtained by use of the expression

$$\langle L^{-1} \rangle = \int_0^\infty L^{-1} W(L, N) dL \tag{5-10}$$

which embodies the usual prescription for taking averages using a distribution function. Using Eq. (5-6) and evaluating the integral, this becomes

$$\langle L^{-1} \rangle = 4\pi (3/2\pi N b^2)^{3/2} \int_0^\infty L \exp\,[-3L^2/2Nb^2]\; dL = \sqrt{6/\pi}(Nb^2)^{-1/2} \tag{5-11}$$

Comparison with Eq. (5-8) leads to the interesting result that

$$\langle L^{-1} \rangle = \sqrt{6/\pi}\; \langle L^2 \rangle^{-1/2} \tag{5-12}$$

It is important to note that $\langle L^{-1} \rangle$ is not equal to $\langle L^2 \rangle^{-1/2}$, but that these are related by a factor of $(6/\pi)^{1/2}$, which equals 1.382.

It should be remembered that Eqs. (5-9) and (5-12) have been derived only for polymer chains which obey the gaussian distribution function Eq. (5-6). Effects of chain stiffness and excluded volume, both of which are important for nucleic acids, will modify these relationships.

Although the freely jointed chain is very simple to treat theoretically, it obviously does not contain many of the structural features which are present in real polymer chains. The next most complicated type of polymeric chain structure which can be treated is that in which bond angles are held constant, although free rotation about bonds is allowed. If we refer back to Eq. (5-1) and let all the bonds lengths b and bond angles θ be the same, this equation becomes

$$\langle L^2 \rangle = Nb^2 + 2b^2 \sum_{i=2}^{N} \sum_{j=1}^{i} \langle \cos\theta_{ij} \rangle$$

It is easily shown that $\langle \cos\theta_{ij} \rangle = (\cos\theta)^{|i-j|}$. This then leads, after evaluation of the summations in the above equation, to

$$\langle L^2 \rangle = Nb^2 \frac{1 + \cos\theta}{1 - \cos\theta} - 2b^2 \cos\theta \frac{1 - \cos^N \theta}{(1 - \cos\theta)^2} \tag{5-13}$$

Most commonly with high polymers, it will be true that $N(1 - \cos\theta) \gg 1$. In this case Eq. (5-13) becomes

$$\langle L^2 \rangle = Nb^2 \frac{1 + \cos\theta}{1 - \cos\theta} \tag{5-14}$$

Comparing Eq. (5-14) with Eq. (5-18), we see that again $\langle L^2 \rangle$ is proportional to N, while the effective bond length is increased by the factor $[(1 + \cos\theta)/(1 - \cos\theta)]^{1/2}$.

A still more realistic type of polymer chain is one in which not only are bond angles fixed, but also internal rotation about bonds is hindered. It has been shown by a number of workers that, if ϕ is the angle of rotation about a bond, measured from $\phi = 0$ at the *trans* position, then

$$\langle L^2 \rangle = Nb^2 \left(\frac{1 + \cos\theta}{1 - \cos\theta} \right) \left(\frac{1 + \langle \cos\phi \rangle}{1 - \langle \cos\phi \rangle} \right)$$

We thus see that $\langle L^2 \rangle$ for the freely jointed chain, the chain with constant bond angles but free rotation, and the chain with constant bond angles and hindered rotation, can all be written in the form

$$\langle L^2 \rangle = \sigma b^2 N$$

This suggests that the actual chain can be replaced by a statistically equivalent chain which contains fewer segments, N_e, than the real chain, each with a longer effective bond length b_e; and that these segments are statistically independent so that the chain behaves as a succession of statistical segments which are freely jointed to one another.[7] Thus the distribution function for the equivalent chain is just that for the freely jointed chain, Eq. (5-6), with N replaced by N_e and b by b_e, respectively.

The mean square end-to-end distance of the equivalent chain is then

$$\langle L^2 \rangle = b_e{}^2 N_e \tag{5-15}$$

while its contour length, which is chosen to be equal to that of the real chain is

$$\mathcal{L} = Nb = N_e b_e \tag{5-16}$$

Comparing these equations, we see that

$$\sigma = \frac{b_e}{b} = \frac{N}{N_e}$$

b_e is called the statistical segment length, and is clearly equal to the root mean square end-to-end length of the bonds in a statistical segment:

$$b_e = \langle L_\sigma{}^2 \rangle^{1/2}$$

E. Wormlike Chains

The utility of the concept of the statistically equivalent chain requires that there be enough statistical segments that gaussian statistics are applicable. In other words, N_e must still be considerably greater than unity, even though it may be substantially less than N. For sufficiently short or inflexible polymer chains, this requirement may not be satisfied. Native DNA is, in fact, stiff enough to make gaussian statistics inapplicable for many purposes. This situation demands more elaborate and difficult treatment than that outlined above. Because relatively stiff chains may be envisioned to bend only gradually and smoothly in solution, somewhat like a worm, they are often called wormlike chains.[8]

In order to characterize the stiffness of polymer chains, we imagine that the first step in the polymer chain starts at the origin and is taken along the positive z-axis. We then ask, what is the average projection $\langle z \rangle$ of the end of the chain along the z-axis after N steps. It is obvious that if the polymer were a completely stiff rod, $\langle z \rangle$ would have its maximum value of Nb; and it is almost as obvious that for a freely jointed chain, $\langle z \rangle$ would equal b, since steps beyond the first would have equal probabilities of being taken in the positive or negative z-directions. Polymers of stiffness intermediate between the rigid rod and freely jointed chain will have intermediate values of $\langle z \rangle$.

In a more quantitative fashion, we may write

$$\langle z \rangle = \sum_{i=1}^{N} \langle z_i \rangle$$

We consider the projections of all succeeding bonds upon the first, and can therefore write

$$\langle z \rangle = b + b \langle \cos \theta_{1,2} \rangle + b \langle \cos \theta_{1,3} \rangle + \cdots + b \langle \cos \theta_{1,N} \rangle$$

For the chain with constant bond angles θ and free rotation, we use $\langle \cos \theta_{i,i+k} \rangle = (\cos \theta)^k$ to obtain

$$\langle z \rangle = b \sum_{i=0}^{N-1} \cos^i \theta = b \frac{1 - \cos^N \theta}{1 - \cos \theta}$$

where the second equality arises from summation of the geometric series.

If $\cos \theta$ is not too close to unity, $\langle z \rangle$ becomes $b/(1 - \cos \theta)$ as N goes to infinity, i.e., for very long chains. On the other hand, for weakly bending chains, θ may be very close to $0°$ and therefore $\cos\theta$ very close to unity, so that $\langle z \rangle$ can become very large. We therefore imagine passage to a continuous case in which both the segment length b and the bond angle θ are taken to the infinitesimal limit in such a way that the quantity

$$a = b/(1 - \cos\theta) \qquad (5\text{-}17)$$

remains finite. The quantity a is called the "persistence length" of the wormlike chain, and represents the average extension along the z-axis of an infinitely long polymer chain.

Let us now consider wormlike chains of finite contour length $\mathcal{L} = Nb$. For bond angles θ very close to zero, it is an adequate approximation to write

$$\cos\theta \approx 1 - \theta^2/2$$

so that

$$\frac{\theta^2}{2} = \frac{b}{a}$$

and

$$\cos^N \theta \approx (1 - \theta^2/2)^N \approx \exp[-N\theta^2/2]$$

Combining these results, we obtain

$$\cos^N \theta \approx e^{-\mathcal{L}/a} \qquad (5\text{-}18)$$

and finally

$$\langle z \rangle = a(1 - e^{-\mathcal{L}/a}) \qquad (5\text{-}19)$$

It is easily verified that this expression behaves properly at the limits of very long and very short chains. As $\mathcal{L}$ goes to ∞, $e^{-\mathcal{L}/a}$ goes to zero; so $\langle z \rangle$ goes to a, the persistence length, as discussed above. As $\mathcal{L}$ becomes much less than a, we may write

$$e^{-\mathcal{L}/a} \approx 1 - \mathcal{L}/a$$

so that $\langle z \rangle$ equals $\mathcal{L}$ for short chains. That is, wormlike chains of contour length much less than their persistence length behave like rigid rods.

It is also possible to obtain the mean square end-to-end distance of the wormlike chain by use of the above results. We return to Eq. (5-13), in which it is not now valid to drop the second term. Using Eqs. (5-17) and (5-18), and elsewhere setting $\cos\theta = 1$, we obtain

$$\langle L^2 \rangle = 2a(\mathcal{L} - a + ae^{-\mathcal{L}/a}) \qquad (5\text{-}20)$$

For very long chains, $\mathcal{L} \gg a$, and this becomes

$$\langle L^2 \rangle = 2a\mathcal{L}$$

Since $\mathcal{L}$ is proportional to the number of segments in the polymer chain, this equation shows that a very long wormlike chain also behaves like a gaussian chain in that $\langle L^2 \rangle$ is proportional to N. Further insight into the physical significance of a may be obtained by solving Eqs. (5-15) and (5-16) for b_e:

$$b_e = \frac{\langle L^2 \rangle}{\mathcal{L}}$$

Comparisons of the two above equations shows that

$$b_e = 2a$$

Thus the statistical segment length of the wormlike chain is twice its persistence length.

For very short chains, $\mathcal{L} \ll a$, and Eq. (5-20) becomes, after expansion of the exponential,

$$\langle L^2 \rangle = 2a\left[\mathcal{L} - a + a\left(1 - \frac{\mathcal{L}}{a} + \frac{\mathcal{L}^2}{2a^2} - \cdots\right)\right] = \mathcal{L}^2 + \cdots$$

This again demonstrates that the limiting behavior of a short wormlike chain is that of a rigid rod.

It is also straightforward, using Eqs. (5-2) and (5-20), to obtain an expression for the mean square radius of the wormlike chain. The result, after replacing sums by integrals and neglecting terms of order unity in comparison with N is

$$\langle R^2 \rangle = 2a\mathcal{L}\left[\frac{1}{6} - \frac{a}{2\mathcal{L}} + \frac{a^2}{\mathcal{L}^2} - \frac{a^3}{\mathcal{L}^3}(1 - e^{-\mathcal{L}/a})\right] \tag{5-21}$$

For long chains with $\mathcal{L} \gg a$, this expression coincides with Eq. (5-9) for gaussian random coils, while for short chains with $\mathcal{L} \ll a$, it agrees with Eq. (5-4) for rigid rods.

No exact distribution function for the wormlike chain has yet been obtained. This has made the calculation of dimensional averages other than $\langle z \rangle$, $\langle L^2 \rangle$, and $\langle R^2 \rangle$ very difficult. However, Hermans and Ullman[9] derived a differential equation which the distribution function must obey, and devised a method for computing averages directly from the differential equation. Additional exact results obtained in this way are

$$\langle L^4 \rangle = \frac{20}{3}\mathcal{L}^2 a^2 - \frac{208}{9}\mathcal{L}a^3 + \frac{8}{27}(e^{-3\mathcal{L}/a} - 1)a^4$$
$$+ 32(1 - e^{-\mathcal{L}/a})a^4 - 8e^{-\mathcal{L}/a}\mathcal{L}a^3 \tag{5-22}$$

and[10]

$$\langle L^6 \rangle = 48a^6 \left(\frac{35}{54} x^3 - \frac{259}{54} x^2 + \frac{452}{27} x - \frac{6143}{243} \right.$$

$$+ \frac{3162}{125} e^{-x} + \frac{213}{25} x e^{-x} + \frac{7}{10} x^2 e^{-x} - \frac{4}{243} e^{-3x} \tag{5-23}$$

$$\left. - \frac{1}{81} x e^{-3x} + \frac{1}{3375} e^{-6x} \right)$$

where $x = \mathcal{L}/a$.

Daniels[11] has obtained an approximate distribution function for the wormlike chain, which incorporates a first-order correction to the gaussian distribution. It is

$$W_D(L, \theta, \mathcal{L}) dL d\theta d\phi = \left(\frac{3}{4\pi \mathcal{L} a} \right)^{3/2} e^{-3L^2/4a\mathcal{L}} \left[1 - \frac{5}{4} \frac{a}{\mathcal{L}} + \frac{2L^2}{\mathcal{L}^2} - \frac{33}{80} \frac{L^4}{a\mathcal{L}^3} \right.$$

$$+ \left(\frac{3L}{2\mathcal{L}} - \frac{25}{8} \frac{La}{\mathcal{L}^2} + \frac{153}{40} \frac{L^3}{\mathcal{L}^3} - \frac{99}{160} \frac{L^5}{a\mathcal{L}^4} \right)$$

$$\left. P_1 (\cos \theta) + \frac{L^2}{2\mathcal{L}^2} P_2 (\cos \theta) \ldots \right] L^2 \sin \theta \, dL d\theta d\phi \tag{5-24}$$

θ and ϕ are the angular variables in a spherical polar coordinate system with its origin at the beginning of the chain and the polar (z) axis tangent to the chain at the origin. $P_1(\cos \theta)$ and $P_2(\cos \theta)$ are the first and second Legendre polynomials.

We can examine the adequacy of the Daniels distribution function by comparing its predictions with the known exact results. We have first of all that

$$\langle z \rangle_D = \int_0^\infty \int_0^\pi \int_0^{2\pi} L \cos \theta \, W_D(L, \theta, \mathcal{L}) dL d\theta d\phi = a$$

This agrees with the exact result, Eq. (5-19), for long chains.

By integrating Eq. (5-24) over all angles, we obtain the length distribution function

$$W_D(L, \mathcal{L}) dL = \int_0^\pi \int_0^{2\pi} W_D(L, \theta, \mathcal{L}) d\theta d\phi dL$$

$$= \sqrt{2\pi} (3/2 \, a \, \mathcal{L})^{3/2} e^{-3L^2/4a\mathcal{L}} \left(1 - \frac{5}{4} \frac{a}{\mathcal{L}} \right. \tag{5-25}$$

$$\left. + \frac{2L^2}{\mathcal{L}^2} - \frac{33}{80} \frac{L^4}{a\mathcal{L}^3} \right) L^2 dL$$

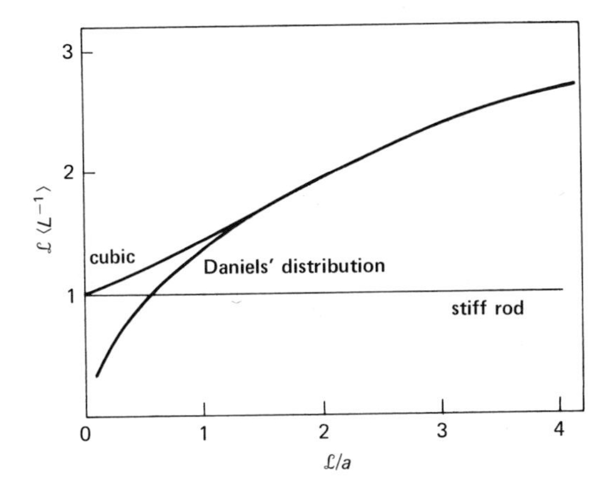

FIGURE 5-2 The behavior of the reciprocal end-to-end distance, $\langle L^{-1}\rangle$, of a wormlike chain, as a function of the number of persistence lengths in the chain. [Reprinted with permission from J. E. Hearst and W. H. Stockmayer, *J. Chem. Phys.*, **37**, 1425 (1962)].

Use of this distribution function yields

$$\langle L^2\rangle_D = 2a(\mathcal{L} - a)$$

and

$$\langle L^4\rangle_D = \frac{20}{3}\,\mathcal{L}^2 a^2 - \frac{208}{9}\,\mathcal{L} a^3$$

Comparison of these results with Eqs. (5-20) and (5-22) again indicates that the Daniels distribution gives good results for long chains, but becomes increasingly inadequate as $\mathcal{L}$ becomes comparable to or less than the persistence length a.

The distribution function Eq. (5-25) may be used to calculate $\langle L^{-1}\rangle$ for wormlike chains. The result is

$$\langle L^{-1}\rangle = \sqrt{6/\pi}(2a\mathcal{L})^{-1/2}\,[1 - a/20\mathcal{L}] \tag{5-26}$$

By comparison of Eqs. (5-26) and (5-12), it may be seen that the term in square brackets represents a correction to the simple proportionality between $\langle L^{-1}\rangle$ and $\langle L^2\rangle^{-1/2}$ which is introduced by chain stiffness.

The behavior of $\mathcal{L}\,\langle L^{-1}\rangle$ as a function of $\mathcal{L}/a$ is plotted in Fig. 5-2 according to Eq. (5-26). It is apparent that this expression fails as $\mathcal{L}/a$ approaches zero, since $\mathcal{L}\,\langle L^{-1}\rangle$ falls below unity at $\mathcal{L}/a = 0.573$; while it should instead approach the rigid rod limit of unity at $\mathcal{L}/a = 0$, and should never fall below unity. Hearst and Stockmayer[12] have shown that for a slightly flexible rod with $\mathcal{L}/a$ near zero,

$$\mathcal{L}\langle L^{-1}\rangle = 1 + \frac{1}{6}\frac{\mathcal{L}}{a}$$

They have further shown that it is adequate for most purposes to approximate the behavior of $\mathcal{L}\langle L^{-1}\rangle$ for small $\mathcal{L}/a$ by a cubic equation in $\mathcal{L}/a$ which has the proper slope and intercept at $\mathcal{L}/a = 0$ for a slightly flexible rod, and which joins smoothly with the curve predicted by the Daniels distribution. The point of intersection is at $\mathcal{L}/a = 4.40$.

It is of interest to examine the flexibility of wormlike chains from a different viewpoint, by inquiring into the connection between their persistence length and their elastic behavior. Qualitatively, it is clear that the lower the energy required to produce a bend through a given angle in the chain backbone, the smaller will be the persistence length. The following discussion will make the connection quantitative.[13]

Imagine at first that the chain is confined to a surface, so that bending can occur only in the surface plane. Generalization to three dimensions will be made later. Tangents drawn to the axis of the coil at points a distance ds apart along the contour make an angle $d\theta$ with each other. Then the free energy required to produce such a bend is

$$dG = \frac{\alpha}{2}\left(\frac{d\theta}{ds}\right)^2 ds$$

α is the bending force constant of the chain. It is necessary to use the Gibbs free energy G rather than the internal or potential energy E because the energy of bending will in general depend on temperature. By integration of this equation, the total bending free energy of a chain of contour length $\mathcal{L}$ is

$$\Delta G = \frac{\alpha}{2}\int_0^{\mathcal{L}}\left(\frac{d\theta}{ds}\right)^2 ds$$

For small displacements, as will be observed in short, stiff chains, θ may be expected to vary linearly with s:

$$\theta = cs$$

We also have the boundary conditions on the chain

$$\theta(s = 0) = 0$$
$$\theta(s = \mathcal{L}) = \theta_{\mathcal{L}} = c\mathcal{L}$$

Combining these expressions and performing the integration yields

$$\Delta G = \frac{\alpha c^2 \mathcal{L}}{2} = \frac{\alpha \theta_{\mathcal{L}}^{\,2}}{2\mathcal{L}}$$

Thus the total bending free energy is proportional to the square of the angle which the tangent to the end of the chain makes with the tangent to the beginning of the chain and is inversely proportional to the contour length of the chain. These relations will hold in general only for chains short enough that there is a linear variation of θ with contour length, but this suffices for our purposes. The bending force constant, α, is revealed as twice the free energy needed to produce a bend of one radian in a unit length of polymer chain.

Up to this point we have imagined that we somehow could bend the chain to our wishes through an angle θ, and ΔG represented the free energy we must expend to do this. In fact, of course, the polymer is immersed in a temperature bath, and all angles of bending can, in principle, occur spontaneously. The probability of a given bend is governed by a Boltzmann factor $e^{-\Delta G/kT}$. We are then able only to predict average quantities. For our present purposes, we wish to calculate the mean square angle of bending, $\langle \theta_{\mathcal{L}}^2 \rangle$. This is done according to the standard statistical mechanical prescription.

$$\langle \theta_{\mathcal{L}}^2 \rangle = \frac{\int_{-\infty}^{\infty} \exp\,[-\Delta G/kT]\,\theta_{\mathcal{L}}^2\, d\theta_{\mathcal{L}}}{\int_{-\infty}^{\infty} \exp\,[-\Delta G/kT]\, d\theta_{\mathcal{L}}}$$

which becomes

$$\langle \theta_{\mathcal{L}}^2 \rangle = \frac{\int_{-\infty}^{\infty} \exp\,[-\alpha\theta_{\mathcal{L}}^2/2\mathcal{L}kT]\,\theta_{\mathcal{L}}^2\, d\theta_{\mathcal{L}}}{\int_{-\infty}^{\infty} \exp\,[-\alpha\theta_{\mathcal{L}}^2/2\mathcal{L}kT]\, d\theta_{\mathcal{L}}} = \frac{\mathcal{L}kT}{\alpha}$$

This equation still embodies the initial restriction that bending can occur only on a surface; that is, only in a single direction perpendicular to the chain axis. In fact, of course, bending can occur in two mutually perpendicular directions each perpendicular to the chain backbone. On the average, these two bending modes are independent, so the final expression for the mean square bending angle is

$$\langle \theta_{\mathcal{L}}^2 \rangle = \frac{2\mathcal{L}kT}{\alpha}$$

The average projection of the position of the end of the chain along the z-axis can be written in integral form as

$$\langle z \rangle = \int_0^{\mathcal{L}} \langle \cos\theta_s \rangle\, ds$$

But for short, stiff chains of the type we have been considering, $\langle \cos\theta_s \rangle$ may be expanded in a Taylor's series.

$$\langle \cos\theta_s \rangle = 1 - \frac{1}{2}\langle \theta_s^2 \rangle + \cdots + = 1 - \frac{kTs}{\alpha} + \cdots \approx \exp\,[-kTs/\alpha]$$

We therefore obtain

$$\langle z \rangle = \frac{\alpha}{kT}\left(1 - \exp\left[-\mathcal{L}kT/\alpha\right]\right) \tag{5-27}$$

Comparison of Eqs. (5-27) and (5-19) leads to the identification

$$a = \frac{\alpha}{kT} \tag{5-28}$$

Thus, as is plausible physically, the persistence length is proportional to the bending force constant. It is also inversely proportional to the temperature, since the increased thermal buffeting of the chain at higher temperature will lead to increased bending.

F. Excluded Volume Effect

Chain stiffness is only one of the factors that prevents real polymers from behaving like ideal, gaussian chains. Another is the excluded volume effect, which may be imagined to be due to two causes.

The first cause of excluded volume effects is the fact that real polymer segments occupy volume, and thus overlaps among them are prohibited. On the contrary, the gaussian distribution, Eq. (5-6), is derived assuming there is no correlation between steps in the random walk. It is clear that the more compact configurations are likely to involve overlaps, which therefore must be excluded when calculating average dimensions. As a consequence, the average dimensions of a polymer chain are expanded by the presence of excluded volume effects.

The excluded volume effect is also influenced by differential thermodynamic interactions between solvent-solvent, polymer-polymer, and solvent-polymer pairs. It is a well-known experimental fact that polymers are difficult to dissolve in most solvents, and that one must look for "good solvents" in which a given polymer dissolves easily. On the molecular level, a good solvent is one which thoroughly solvates, i.e., surrounds, the polymer segments, thereby causing the segments to be on the average further apart than they would be in the bulk polymer. Thus good solvents cause chain expansion and contribute to a positive excluded volume effect.

Chain expansion will also occur in polyelectrolytes, because the repulsion between groups of like charge in the polymer will make the polymer swell to minimize its electrostatic free energy. This effect is of obvious importance in nucleic acids, due to the negatively charged phosphate groups in the chain backbone.

On the other hand, a poor solvent will tend to make the polymer segments cluster together, since polymer-solvent contacts are energetically unfavorable. This clustering will tend to counteract the expansion due to the physical volume of the segments. Under particular conditions, these two opposing influences may exactly balance, and the polymer will then display ideal gaussian dimensions. A solvent in which this occurs is known as a θ-solvent, and the temperature at which the polymer behaves ideally in the given solvent is known as the θ-temperature for the polymer-solvent system. As the solvent becomes poorer than a θ-solvent, say by lowering the temperature, polymer will tend to precipitate. In fact, the θ-temperature may be shown to be that temperature at which an infinitely long polymer will precipitate.

Experimentally, we measure intramolecular excluded volume through its effect on polymer dimensions. It is common to define an expansion parameter α_L such that

$$\alpha_L{}^2 = \frac{\langle L^2 \rangle}{\langle L^2 \rangle_0}$$

represents the ratio of mean square end-to-end distances in the given solvent and in a θ-solvent. The θ-condition is denoted by the subscript zero. A similar expansion factor can be defined for the mean square radius:

$$\alpha_R{}^2 = \frac{\langle R^2 \rangle}{\langle R^2 \rangle_0}$$

The theory of the excluded volume effect is one of the most difficult in all polymer physical chemistry, and no generally satisfactory solution to the problem has been obtained. An exact solution has been obtained when excluded volume effects are very small; that is, very close to the θ-point.[14] However, most experiments are carried out under conditions rather far from the θ-point, and thus the search for a theory for α_L valid over the whole range of polymer sizes and interactions has persisted.

One of the earliest theories of intramolecular excluded volume effects, that due to Flory and co-workers,[15-17] is still among the qualitatively most satisfactory. In a good solvent, the more highly expanded polymer configurations have a lower Gibbs free energy of mixing. However, elastic work must be done to expand the chain, so the elastic free energy of more highly expanded configurations is higher. Equilibrium occurs when the sum of these two free energies is a minimum. Quantitative implementation of this line of reasoning gives the equation

$$\alpha_L{}^5 - \alpha_L{}^3 = 2C_M \psi_1 (1 - \theta/T) M^{1/2} \tag{5-29}$$

where θ is the θ-temperature and $\alpha_L{}^N = (\alpha^2)^{N/2} = (\langle L^2 \rangle / \langle L^2 \rangle_0)^{N/2}$. C_M depends only on the unperturbed properties of the solvent and polymer segments, and not on molecular weight M. ψ_1 is an entropic parameter characterizing the dilution of polymer with solvent.

Equation (5-29) is not entirely adequate, since it predicts[18] that very near the θ-point, $\alpha_L{}^2$ will increase about twice as rapidly as is predicted by the exact theory. Moreover, it says that $(\alpha_L{}^5 - \alpha_L{}^3)/M^{1/2}$ is a constant, independent of M; while experimentally this function is observed to increase slowly with M. On the other hand, for the large excluded volume effects commonly encountered in practice, $\alpha_L{}^5 \gg \alpha_L{}^3$, so that Eq. (5-29) predicts that $\alpha_L{}^5$ will vary as $M^{1/2}$, or $\alpha_L{}^2$ as $M^{0.3}$. This limiting dependence seems to agree well with experimental results on most systems and with the Monte Carlo computer calculations described below.

In the absence of exact analytical theories of the excluded volume effect, computer simulation studies have been proved to be of considerable value.[19] The computer is programmed to generate random walks on a three-dimensional lattice, and to reject intersecting walks. Because of the random, gambling nature of the procedure, these calculations are often called Monte Carlo calculations.

For walks with reasonably large numbers of steps, it has been found that the results can be expressed in terms of a parameter ϵ such that

$$\langle L^2 \rangle = b^2 N^{1+\epsilon} \tag{5-30}$$

where N is the number of steps in the chain.

Monte Carlo calculations on lattices, and the Flory Eq. (5-29), show that ϵ reaches a limiting value of 1/5 for long chains; but some other theories[20-21] suggest the limiting value should be 1/3.

The use of ϵ to represent excluded volume has been extended to the calculation of dimensional averages other than $\langle L^2 \rangle$. In these calculations it is assumed that ϵ is invariant with N, which is not a good assumption for small N. Combination of Eqs. (5-2) and (5-30) leads to (5-31).

$$\langle R^2 \rangle = \frac{b^2 N^{1+\epsilon}}{(2+\epsilon)(3+\epsilon)} = \langle R^2 \rangle_0 \frac{N^\epsilon}{\left(1 + \frac{5}{6}\epsilon + \frac{1}{6}\epsilon^2\right)} \tag{5-31}$$

where $\langle R^2 \rangle_0$ is the mean square radius in the absence of excluded volume.

Equation (5-30) can be obtained from a distribution function of the perturbed gaussian form

$$W(L, N)dL = 4\pi \left(\frac{3}{2\pi b^2 N^{1+\epsilon}}\right)^{3/2} \exp[-3L^2/2b^2N^{1+\epsilon}]\, L^2 dL \tag{5-32}$$

When $\epsilon = 0$, this distribution function reduces to Eq. (5-6). If we use Eq. (5-32) to calculate $\langle L^{-1} \rangle$ we obtain

$$\langle L^{-1} \rangle = \sqrt{\frac{6}{\pi}} \frac{1}{bN^{(1+\epsilon)/2}}$$

This agrees with Eq. (5-12) if $\langle L^2 \rangle$ is given by Eq. (5-30).

High molecular weight DNA embodies both wormlike coil and excluded volume behavior, because it is both stiff and very long. It is therefore necessary to combine the treatments of the preceding two sections to obtain an adequate description of the configurational statistics of native DNA. Such combination has not as yet been possible in any rigorous manner, but some plausible expressions for the dimensional properties may be written down.[23,24]

The correspondence between the parameters L, N, a, and $\mathcal{L}$ is such that $N = \mathcal{L}/2a$ and $NL^2 = 2\mathcal{L}a$, where distances are measured in units of b. Thus the distribution function may be written

$$W(L, \mathcal{L})dL = 4\pi \left[\frac{3}{2^{1+\epsilon}(2a)^{1-\epsilon}}\right]^{3/2} \exp\left[-3L^2/2\mathcal{L}^{1+\epsilon}(2a)^{1+\epsilon}\right] L^2 dL \tag{5-34}$$

From this it may be calculated that

$$\langle L^2 \rangle = (2\mathcal{L}a)(\mathcal{L}/2a)^{\epsilon}$$

whereas comparison with Eq. (5-20) suggests that

$$\langle L^2 \rangle = (2\mathcal{L}a)(\mathcal{L}/2a)^{\epsilon}\left[1 - \frac{a}{\mathcal{L}} + \frac{a}{\mathcal{L}} e^{-\mathcal{L}/a}\right] \tag{5-35}$$

might be a more adequate expression. Similarly, a plausible expression for the mean square radius is

$$\langle R^2 \rangle = \frac{(2\mathcal{L}a)(\mathcal{L}/2a)^{\epsilon}}{6 + 5\epsilon + \epsilon^2}\left[1 - \frac{3a}{\mathcal{L}} + \frac{6a^2}{\mathcal{L}^2} - \frac{6a^2}{\mathcal{L}^2}(1 - e^{-\mathcal{L}/a})\right] \tag{5-36}$$

The mean reciprocal distance between chain segments is approximately

$$\mathcal{L}\langle L^{-1} \rangle = \sqrt{6/\pi}\,(\mathcal{L}/2a)^{(1-\epsilon)/2}[1 - a/20\mathcal{L}] \tag{5-37}$$

Methods for determining the structural parameters a, b, and ϵ which characterize polynucleotides under particular conditions in solution will be discussed later in this chapter.

G. Circular Polymer Chains

Circular molecules of DNA have been observed recently with increasing frequency. Very long circular synthetic polymers are formed with negligible probability at ordinary concentrations because of the entropy decrease associated with joining of the ends. However, circular DNAs, both double-stranded and single-stranded, have been identified in a wide variety of viruses, bacteria, and higher organisms. In many cases these circular DNA molecules have been observed directly by electron microscopy. With a few exceptions, all of these DNA molecules are probably circular at the completion of their synthesis, and thus the competition between cyclization and end-to-end aggregation, typical of synthetic condensation polymers, has no chance to occur.

The dimensional properties of flexible circular polymers can be obtained by the same sort of reasoning used above for linear molecules.[6] The distribution function for the distance between any two segments i and j in the circular polymer chain may be obtained by the following argument. These segments must be found at a distance L_{ij} from each other, regardless of whether the contour length between them is measured clockwise or counterclockwise around the circle. Because the probabilities that i and j will be found at a distance L_{ij} from one another going in either direction around the circle are independent, we can write the probability distribution function as a product of the individual probabilities corresponding to the two contour lengths,

$$W(L_{ij}) = W(L, |i-j|)W(L, N-|i-j|)$$

From Eq. (5-6) this may be written as

$$W(L_{ij})dL_{ij} = A \exp[-3L_{ij}^2/2|i-j|b^2] \exp[-3L_{ij}^2/2(N-|i-j|)b^2]\, L_{ij}^2 dL_{ij},$$

where A is a normalization constant. Evaluation of this constant finally gives

$$W(L_{ij}) = 4\pi \left[\frac{3}{2\pi b^2|i-j|\left(1-\dfrac{|i-j|}{N}\right)}\right]^{3/2} \times \exp\left[-3L_{ij}^2/2b^2|i-j|\left(1-\dfrac{|i-j|}{N}\right)\right]L_{ij}^2 dL_{ij} \qquad (5\text{-}38)$$

The average dimensions of the cyclic polymer chain may be obtained simply by use of this distribution function. It is obvious that $\langle L^2\rangle$, the mean square distance between the ends of the chain, is meaningless, since the ends are connected. However, using Eqs. (5-7) and (5-38), we find for the mean square distance between elements i, and j,

$$\langle L_{ij}^2\rangle = |i-j|(1-|i-j|/N)b^2 \qquad (5\text{-}39)$$

Comparing this with Eq. (5-8), we see that the average distances between segments in a circular molecule are smaller than those between the same segments in a linear molecule, because of the effective compression exerted by the requirement that the ends of the chain be joined.

The mean square radius of the circular polymer chain may be evaluated by using Eq. (5-2) in combination with Eq. (5-39). This leads after some manipulation to

$$\langle R^2 \rangle = \frac{1}{12} N b^2 \qquad (5\text{-}40)$$

Comparing this result with Eqs. (5-8) and (5-9), we see that for chains with the same number of segments,

$$\langle R^2 \rangle_{\text{cyclic}} = \frac{1}{2} \langle R^2 \rangle_{\text{linear}}$$

This result again emphasizes the fact that the average dimensions of the cyclic chain are smaller than those of the linear chain, due to the restriction that the ends of the chain be joined.

The mean reciprocal distance between elements i and j can also be evaluated using the distribution function Eq. (5-38) with Eq. (5-10). The result is

$$\langle L_{ij}^{-1} \rangle = \sqrt{\frac{6}{\pi}} \, \frac{1}{b} \left[\frac{1}{|i-j|(1-|i-j|/N)} \right]^{1/2} \qquad (5\text{-}41)$$

which by comparison with Eq. (5-39) leads to

$$\langle L_{ij}^{-1} \rangle = \sqrt{6/\pi} \, \langle L_{ij}^{2} \rangle^{-1/2}$$

Thus the same constant of proportionality, $\sqrt{6/\pi}$, between $\langle L^{-1} \rangle$ and $\langle L^2 \rangle^{-1/2}$ is found for linear and cyclic chains.

The above dimensional averages for cyclic chains can be modified for chain stiffness and excluded volume effects in the same way as was done for linear chains.

III DIRECT VISUALIZATION

After the preceding discussions of the lengths or molecular weights, configurations, and base sequences of the nucleic acids, we turn to methods for determining these quantities experimentally. The most obvious and naive way to obtain information on such properties is to look at the molecule directly. Such an approach is, at present, clearly out of the question for small molecules; but for large biopolymers, such as nucleic acids, it is indeed a feasible approach. The two major methods for determining molecular lengths and conformations by direct visualization are electron microscopy and tritium autoradiography. Several other methods for determining molecular weights by fairly direct counting procedures will be briefly discussed.

A. Electron Microscopy

For nucleic acids of molecular weight from several hundred thousand to 500×10^6, electron microscopy has become an extremely useful and versatile technique for size and structure determination. Several reviews of experimental techniques have recently appeared.[25,26] In both electron microscopy and autoradiography, there are two major experimental problems. The first is to spread the nucleic acids in such a way as to remove undue tangles and allow ready tracing of strand contour and continuity. The second is to magnify the diameter of the molecule beyond the 20 Å typical of double-stranded DNA so that it may be discerned.

In early work on the electron microscopy of DNA,[26] a solution of the nucleic acid was sprayed from an atomizer onto a freshly cleaved mica surface. The shear stresses set up in passing through the atomizer nozzle acted to stretch out and align the polymer chains so that they could be shadowed in a direction perpendicular to the direction of their deposition on the surface. However, it was found that these shear stresses, and also those involved in ordinary laboratory operations such as pipetting, stirring, pouring, and passing through a syringe, can break high molecular weight DNA.[28,29] Thus, if it is desired to examine the intact native molecule, more gentle spreading methods must be devised.

One method was devised in which DNA was transferred from solution to a glass slide covered with a basic copolymer, polystyrene-4-vinylpyridine, by streaking the slide along the surface of the solution.[30] The negatively charged DNA is adsorbed on the positively charged copolymer, and is disentangled and straightened by streaming forces. In a similar method, a few milliliters of DNA solution were placed on a collodion film mounted on a glass slide. When the slide was tilted, the excess liquid ran off and the DNA was oriented in the direction of flow.[31]

The technique most generally used now, however, is that developed by Kleinschmidt and co-workers.[32] In this method, the nucleic acid is adsorbed to an insoluble monomolecular film of denatured basic protein, often cytochrome *c*, at an air–water interface. The spreading procedure originally developed[32] is diagrammed at the top of Fig. 5-3. A solution containing both nucleic acid and protein is floated down a ramp onto a trough containing an aqueous solution. The film forms by spreading of the protein over the surface of the solution in the trough, and the nucleic acid is immobilized in it. Other procedures have been devised which are simpler and require less DNA. The diffusion method[33] is shown in the middle of Fig. 5-3. In this case, the nucleic acid is in solution in the trough, and a protein film is formed on top. The polynucleotide then diffuses to the underside of the film, at a rate governed by its size, and sticks. In a third method,[34] shown at the bottom of

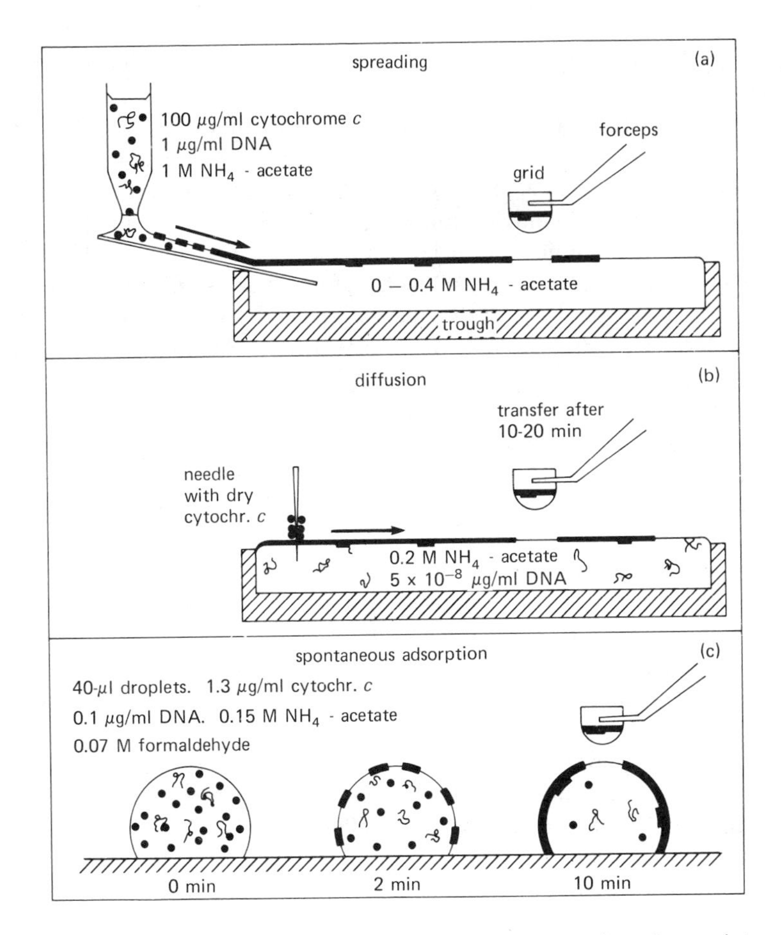

FIGURE 5-3 Scheme of electron microscopic preparation steps using a Langmuir trough. Top: spreading; middle: diffusion; bottom: spontaneous adsorption. [From D. Lang et al., *J. Mol. Biol.*, **23**, 163 (1967). Reprinted with permission.]

Fig. 5-3, formaldehyde and a critical amount of cytochrome *c* are added to a solution of DNA, and monolayer formation and adsorption occur spontaneously. If one droplet of DNA solution is used, only about a million molecules of a viral DNA are required by this technique. In working with very large DNA molecules, which are highly susceptible to shear breakage, a one-step release method has been devised in which DNA extraction takes place concurrently with protein film spreading. In the osmotic shock technique,[35,36] viral or bacterial protoplast particles are floated, in a solution of high ionic strength containing the protein for the film, down a ramp onto a trough containing pure water or low ionic strength solution. The shocked particles release their nucleic acid, which then diffuses to the film. It is also possible to use a protein denaturant, such as urea, in the trough, so that the virus or protoplast coat is denatured and the DNA extruded.[37]

The second problem, that of amplifying the diameter of the strands so that they may be seen in an electron microscope, is dealt with by picking up the monolayer on a grid and shadowing it, usually with a heavy atom such as Pt. (Of course, the complex of nucleic acid and basic protein is substantially thicker than nucleic acid alone.) When the atomic beam is directed at a low angle to the plane of the grid, the nucleic acid strand projects up from the plane and casts a wide shadow which may then be observed by contrast. Unidirectional shadowing gives a three-dimensional effect, but it is only possible to see strands lying perpendicular to the shadowing direction. It is thus necessary, if the continuity of the chain is to be observed, to stretch the polymer out along a single direction. This, however, leads to shear distortion and to the extension of a large molecule beyond the boundaries of a single electron microscope field, necessitating the taking of overlapping pictures.

A second important experimental development introduced by Kleinschmidt was therefore to shadow the sample from all directions by rotating it on a turntable.[32,35] It is possible to achieve similar results by shadowing a stationary sample sequentially in two or three directions.[38,39]. These techniques allow molecules to be successfully visualized in a rather small area. Very high contrast positive staining may also be employed, using uranyl acetate in organic solvents.[40–42]

After the sample has been shadowed or stained to give suitable contrast, it is photographed at a magnification of 5000 to 10,000-fold. The absolute size of the polynucleotide strand is then determined by comparison with a grating replica of known spacing. Contour lengths are measured by tracing enlarged photographs with a map measurer or thread.

Having measured the contour length of a polynucleotide strand, it is possible to determine its molecular weight if the mass per unit length is known. For double-stranded NaDNA, the average molecular weight of a base pair is 662, and the translation per residue is 3.46 Å in the B form. Thus the mass per unit length is 192 atomic mass units, or daltons, per angstrom. The lengths and molecular weights of DNA molecules from many sources have been measured in this way.[2,39] Agreement is usually good with results obtained by other methods. Electron microscopy has the advantage of being an absolute method (that is, not relying on calibration with samples of known molecular weight) and of having inherently high precision. Lang et al.[39] have estimated that the standard deviation of the contour length $\mathcal{L}$ due to all experimental sources of error was ±4%, and that this could be decreased to about 2% in favorable cases. By comparing the observed standard deviation of the sample with that estimated as due to experimental causes, the standard deviation in the natural length of DNA from T1, T3, and bovine papilloma viruses was found to be less than 1–2%.

The contour length of duplex DNA in electron microscopy has occasionally been found to vary, and in general is larger in solutions of lower ionic strength.[39–46] For example,[45] when DNA from λ bacteriophage is spread from 0.1 M NaCl onto a hypophase, or underlying solution, in the trough, also at 0.1 M NaCl, its length L is 10.8 μ; when spread onto pure water, L is 17.2 μ. When it is spread from 2 M ammonium acetate onto 0.1 M ammonium acetate, L is 13.6 μ; onto pure water, 17–18 μ. The lengthening of the DNA molecule may occur because of unwinding of the helix or tilting of the bases due to electrostatic repulsions between the phosphate groups at low ionic strength. There are also indications that, even under identical conditions, the mass per unit length of different DNAs may differ by a few percentage points.[47]

Single-stranded polynucleotides may also be studied by electron microscopy.[48–53] After denaturation of initially native duplex material with heat or alkali, it is necessary to react the strands with formamide[53] to prevent re-formation of base pairs.

For both single- and double-stranded polynucleotides, it is more accurate to measure molecular weights and lengths relative to an internal standard nucleic acid of known size on the same grid, than to attempt absolute length measurements.

Electron microscopy is immensely useful in determination of the topology of nucleic acids. Among the conformations which have been detected, in addition to the most commonly expected linear one, are branched[54,55], circular[56–59], supercoiled[60], and catenated[61] DNA. Figure 5-4 shows a typical circular DNA, a type which has been found as native double-stranded, single-stranded, and duplex RF (replicating form) in many viruses. Examples of supercoiled and catenated DNA are shown in Figures 5-47 and 5-48 at the end of this chapter.

Electron microscopy has also been used to assess the dimensional statistics of polynucleotide chains. In order to estimate the distribution and mean square value of the end-to-end distance L of the chain in three dimensions from observed quantities in two dimensions, some further theoretical development is required.[39]

We wish to calculate the probability $W(\rho)d\rho$ that the distance between the ends of the molecule, when it is projected on a plane (the monomolecular film or grid), will lie between ρ and $\rho + d\rho$. This is the product, integrated over all L, of two factors. The first is the probability that the end-to-end distance in three dimensions is between L and $L + dL$: $W(L)dL$. The second is the probability that, given a spacing of L in three dimensions, the ends are distanced ρ in two dimensions.

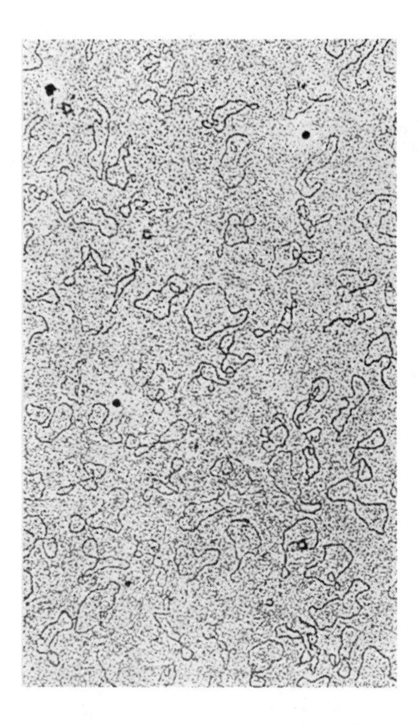

FIGURE 5-4 Electron micrograph of circular, double-stranded RF DNA from bacteriophage 174. [From A. K. Kleinschmidt, A. Burton, and R. L. Sinsheimer, *Science,* **142**, 961 (1963). Copyright 1963 by the American Association for the Advancement of Science. Reprinted with permission.]

The result is

$$W(\rho)d\rho = \frac{2}{\langle \rho^2 \rangle} \exp\left[-\rho^2/\langle \rho^2 \rangle\right] d\rho \tag{5-42}$$

where

$$\langle \rho^2 \rangle = 2\langle L^2 \rangle / 3 \tag{5-43}$$

is the mean square projection of the end-to-end distance in two dimensions.

By observing a large number of similar molecules in the electron microscope, it is possible to measure $\langle \rho^2 \rangle$ directly. Doing this for molecules of different contour lengths, one can then get $\langle L^2 \rangle$ as a function of $\mathcal{L}$. For long DNA molecules, containing many statistical segments, Lang et al.[33,39] found that $\langle L^2 \rangle$ fit, within sizable error limits, an equation of a form proposed by Katchalsky and Lifson[62] for polyelectrolyte expansion:

$$\langle L^2 \rangle^{1/2} = \frac{B\mathcal{L}}{2a} \left[1 + \sqrt{1 + (2a)^3/B^2\mathcal{L}}\right]$$

$2a$, as before, is the statistical segment length; while B is an electrostatic factor reflecting polymer charge, ionic strength and dielectric constant of the solution and temperature. For uncharged polymers, or at high ionic strengths, B is zero; and the equation becomes $\langle L^2 \rangle = 2a\,\mathcal{L}$, as obtained before for gaussian chains. From measurements on DNA from codfish, trout, holothuria, and T2 phage, at ionic strengths between 0.1 and 0.5, Lang et al.[33,39] obtained

$$\langle L^2 \rangle^{1/2} = 9.1 \times 10^{-2}\, \mathcal{L}\, [1 + \sqrt{1 + 1.84 \times 10^{-3}/\mathcal{L}}\,]$$

where dimensions are given in centimeters. This leads to a statistical segment length of 1850 Å. As will be seen later, this value is somewhat higher than that obtained from hydrodynamic or light-scattering measurements. This may reflect stretching out of the DNA on the electron microscope grid due to surface forces. The possible perturbations due to these forces indicate caution in assessing conformational distributions by electron microscopy.

The two-dimensional distribution function, Eq. (5-42), may be tested directly by observing the number of molecules with values of ρ in electron micrographs. The results for T3 DNA are shown in Fig. 5-5. At high ionic strength, agreement with theoretical expectations is surprisingly good. At lower ionic strengths, the distribution is shifted to higher values of ρ, presumably because of polyelectrolyte repulsions.

The high resolution of the electron microscope allows observation of the duplication of bacterial chromosomes in some detail. The first such observation was made autoradiographically, by Cairns.[63] This showed that the *E. coli* chromosome was a circular structure, consistent with the circular genetic map of this organism. In duplication, the circle was seen to split into a fork at one point, to proceed double for some distance, and then to rejoin at another fork (Fig. 5-6). Similar duplicating structures have been demonstrated by electron microscopy[64] in the very small bacterium *Mycoplasma hominis* H39, as shown in Fig. 5-7. The two branch points are located by the arrows, but the structure is otherwise unbranched. It can be seen that the newly replicated strands have the same thickness as the rest of the molecule. Therefore, they must be double-stranded. Furthermore, within experimental error, the branches between the forks are equally long. Both of these facts are consistent with the usual model for the duplication process. In both *E. coli* and *Mycoplasma hominis,* replication starts at a single point and continues in a unique direction. These techniques give resolution only down to about 10^3 base pairs, however; so they are not inconsistent with a mechanism in which duplication occurs from 5′ to 3′ on both strands, hence in opposite directions, for a length of several hundred base pairs.

It is intriguing to attempt to determine base sequences in nucleic acids by attaching a specific label to one of the nucleotides and observing the positions along the chain at which such labels appear in electron micrographs.[65]

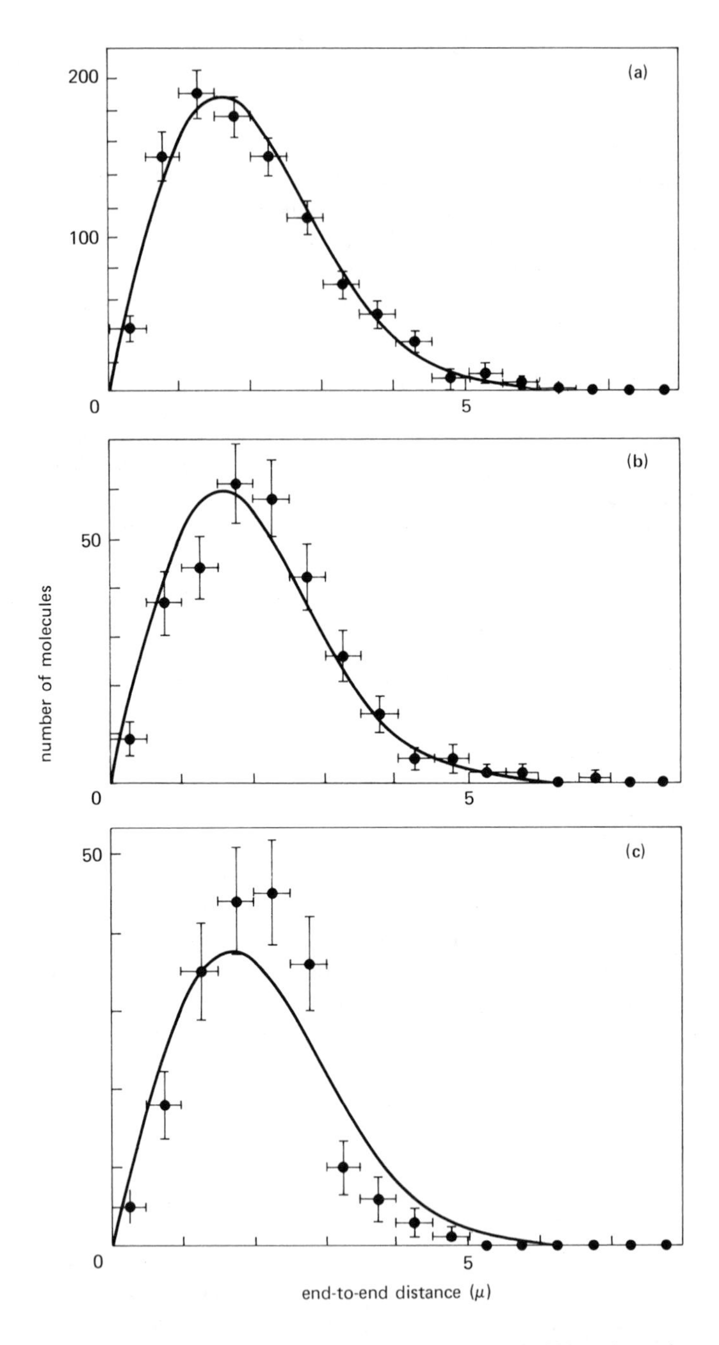

FIGURE 5-5 Distribution of end-to-end distances of T3 DNA after adsorption to cytochrome *c* film. Ionic strengths of the DNA solution in ammonium acetate are (a) 0.40 M, (b) 0.15 M, (c) 0.050 M. The line represents the theoretical distribution for random coils projected onto a plane. [From D. Lang et al., *J. Mol. Biol.*, **23**, 163 (1967). Reprinted with permission.]

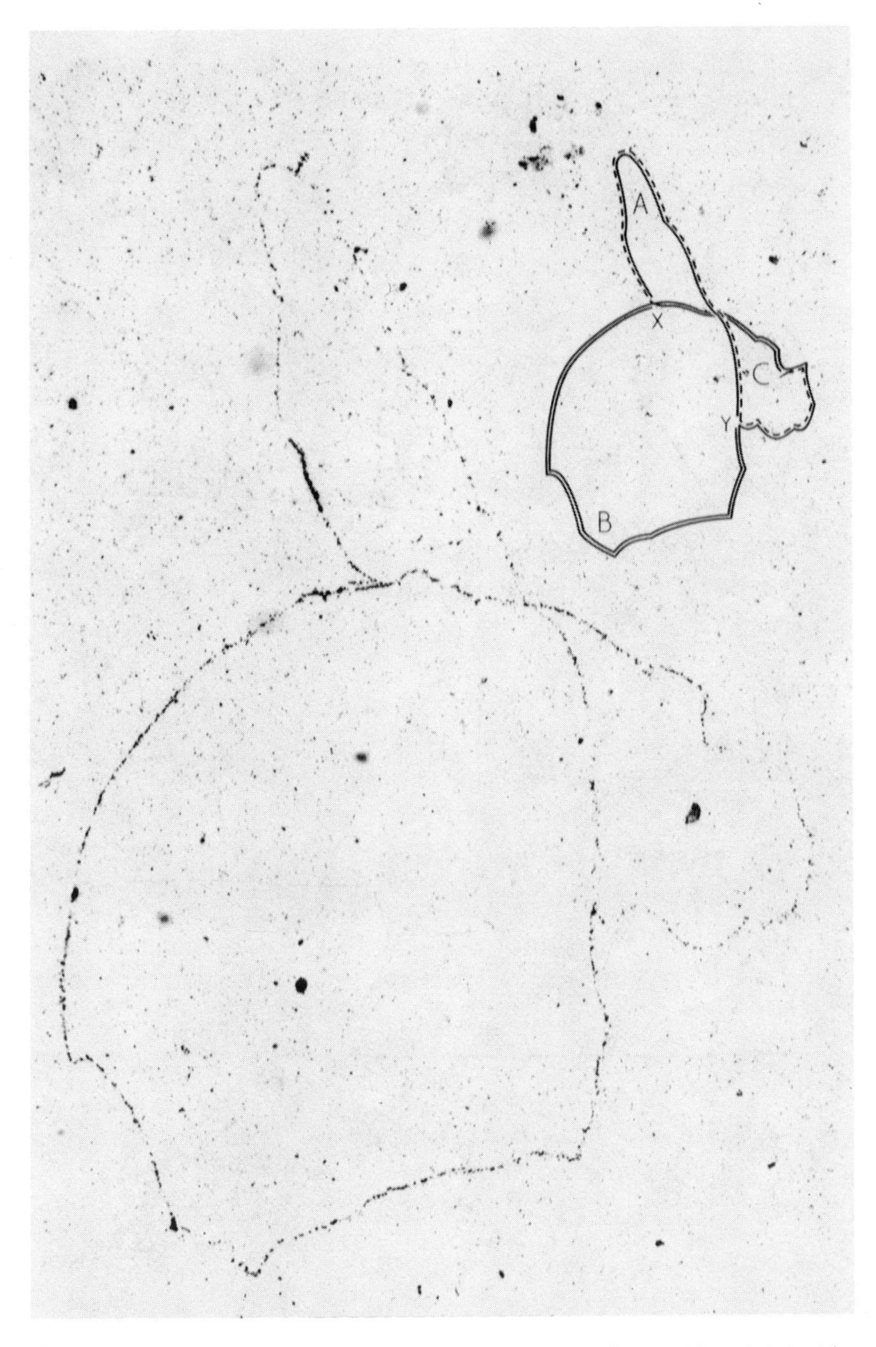

FIGURE 5-6 Autoradiograph of the chromosome of *E. coli* K12 Hfr, labeled with tritiated thymidine for two generations, extracted with lysozyme, and exposed to photographic film for 2 months. The scale is 100μ. [From J. Cairns, *Cold Spring Harbor Symp. Quant. Biol.*, 28, 44 (1963). Reprinted with permission.]

FIGURE 5-7 Tracing of a replicating chromosome. X 8000. The arrows indicate the two replicating forks. [From H. R. Bode and H. J. Morowitz, *J. Mol. Biol.*, **23**, 191 (1967). Reprinted with permission.]

It has been found[66] that diazotized 2-amino-*p*-benzenedisulfonic acid is a suitable reagent. At pH 9.2, this compound reacts with guanylic acid 60 times faster than with any other nucleotide. Once the reagent is attached to the G residues in the polynucleotide chain, its sulfonic acid groups bind two uranyl ions when the preparation is stained with uranyl acetate. These are visible in electron microscopy as pairs of dots 10–12 Å apart, arrayed perpendicular to the polynucleotide chain direction.[67] Pairs of such labels have been observed about 6 Å apart, a spacing which presumably corresponds to the distance between adjacent G residues along the single-stranded polyńucleotide chain. Only about 80% of the guanylic acid residues in RNA or denatured DNA are labeled. It is not clear, as yet, whether the residues not labeled are randomly located along the chain, or whether they are a reproducible set of hard-to-label nucleotides.

A different sort of sequence determination is embodied in the work of Inman,[44] who has obtained electron micrographs of partially denatured lambda DNA. The native material was heated, and the single-stranded regions were stabilized by reaction with formaldehyde. It was found that three zones of denaturation occurred with high frequency. The midpoints of these zones were located at 0.5, 0.7, and 0.98 of the total length of the DNA chain. These early melting regions may correspond to sequences rich in A and T (see Chapter 6). It is also possible to map regions of sequence non-homology between related nucleic acids, as for example DNA from wild-type and delation mutants, by renaturing complementary strands, one from each type. The nonhomologous regions are visible as single-stranded loops if they contain 50–100 bases or more.[53,68,69]

B. Autoradiography

The second major method for directly observing contour lengths and conformations of nucleic acids is autoradiography. This is a much more difficult technique than electron microscopy and consequently rarely used; but it has provided unique information on the very large chromosome from *E. coli*. Experimental methods are described in several sources.[3,70,72]

The problem of spreading DNA in sufficiently untangled form has been solved in two ways. For smaller DNA molecules, a procedure is used which is similar to that described earlier for electron microscopy. A glass microscope slide is drawn across the surface of a DNA solution, and shear forces align and untangle the molecules.[70] To study larger DNA, whole cells are put in a dialysis chamber. Dialysis is performed against a solution chosen to gently lyse the cells and liberate the DNA from the protein. The chamber is then drained and some untangled DNA adheres to the dialysis membrane.[72]

To be observed, the DNA must have been labeled with tritiated thymidine. The DNA is embedded in a photographic emulsion, and the decay of the tritium yields beta particles which leave tracks about 10^4 Å long in the emulsion. This is a 500-fold magnification of the 20 Å thickness of the double helix. The tracks emanating from the thymidines can be observed in the light microscope, after several weeks' exposure, as a line which defines the molecular contour. An autoradiograph of duplicating *E. coli* DNA is shown in Fig. 5-6.

The molecular weight of T2 DNA obtained from autoradiography[70] was about 110×10^6, in fair agreement with the value of 130×10^6 obtained from electron microscopy.[73] More spectacularly, the DNA from *E. coli* was found[63,74] to be circular and of molecular weight about 2.2×10^9. This is the largest DNA molecule yet measured.

C. Other Counting Methods

Several other methods, which involve basically the observation and counting of tracks or particles to determine molecular weights of nucleic acids, may be mentioned briefly at this point.[2,3]

The ^{32}P star method[75] gives the molecular weight independent of the state of purification or aggregation of the nucleic acid. DNA is uniformly labeled with radioactive phosphorus and embedded in a radiosensitive emulsion. Each decay leaves a track, and the tracks emanating from a single molecule form a star. By counting the tracks in the star, and knowing the specific activity of the phosphorus, one may calculate the molecular weight. This technique tends to give a lower limit to the molecular weight, since not all the tracks may be counted. However, when applied to T2 DNA,[76] it gave a molecular weight of 130×10^6, in good agreement with the value obtained by other techniques.

End-group analysis is a traditional way for determining the molecular weights of polymers. Application of this procedure to nucleic acids has been facilitated by the discovery of the enzyme polynucleotide kinase,[77-80] which may be used to transfer ^{32}P from highly labeled ATP to the free 5′-OH end(s) of polynucleotide chains. Knowing the specific activity of the ATP, and the total phosphorus content of the polynucleotide, one can obtain the number average molecular weight. This technique has been applied to T7 DNA,[81] in which one molecule of ^{32}P is incorporated for every 40,000 nucleotides. Since there are two 5′ ends of the duplex molecule, this leads to a molecular weight of 26 million.

One may determine the mass of DNA in a macroscopic sample of virus or bacteria by chemical means, and the number of virus particles or bacteria by plaque counting or microscopic observations. If the assumption is then made that there is but a single molecule of DNA per virus particle or bacterium, an assumption which has usually been borne out in practice, the mass of a DNA molecule may be found. Since not all virus particles in a preparation will always form plaques, this method tends to give an overestimate of the molecular weight.

Finally, one may label the DNA in a virus with radioactive phosphorus or tritium. Decay of these radioactive isotopes causes inactivation of the virus with a killing efficiency of about 10%. Knowing the fraction of the nucleotides labeled, and the lifetime of the isotope, one can use the "suicide rate" of the virus to get its total nucleic acid content. Again with the assumption of one DNA or RNA molecule per virus particle, this gives the molecular weight of the nucleic acid.

IV SCATTERING OF RADIATION

A. Light Scattering

Light scattering is one of the standard methods of obtaining information about molecular weight, size, and intermolecular interactions of polymers in solution. It has the advantage, as will be discussed below, of yielding such information independently of any assumptions about molecular conformation. Light scattering was used in the middle 1950s to determine molecular weights of DNA fragments in the range 0.5×10^6 to 3×10^6 daltons. Its extension to larger, native DNA molecules has been delayed until quite recently because of difficulty in studying scattering at sufficiently low angles. After a discussion of the basic theory of light scattering, this section surveys recent advances in low angle scattering and the use of scattering to obtain molecular weights and persistence lengths of native DNA chains.

The general theory of light scattering is discussed in a number of sources.[82-84] We first consider two-component systems in which the dimensions of the polymer are small (say, 1/20 or less) compared to the wavelength of light. For visible light, of wavelength 5000 Å, this restricts particle dimensions to 250 Å or less, substantially less than those characteristic of high molecular weight nucleic acids. Scattering particles of this small size may be regarded as points, and there is negligible angular dependence of scattering intensity due to intraparticle interference effects.

The equation for the intensity of light, i_θ, scattered through an angle θ with respect to the incident beam of intensity I_0 is

$$i_\theta = I_0 \frac{2\pi^2 n^2 (\partial n/\partial c)^2 (1 + \cos^2\theta) c}{N_A \lambda^4 r^2 \left(\dfrac{1}{M} + 2Bc + 2Cc^2 + \cdots\right)} \tag{5-44}$$

We see that i_0 is proportional both to I_0 and to c, the weight concentration of polymer. Thus for dilute solutions, when c is small, the incident intensity must be high if the scattered beam is to have adequate intensity. N_A is Avogadro's number. r is the distance from the scattering cell to the detector; and the r^{-2} dependence reflects the fact that while the total scattered intensity remains constant, it passes through successive imaginary spherical surfaces of area $4\pi r^2$ surrounding the cell, so the intensity per unit area arriving at the detector falls off as r^{-2}. M is the polymer molecular weight, B and C are virial coefficients, n is the solution refractive index, and $\partial n/\partial c$ is the refractive index increment, measured at constant temperature, pressure, and chemical potential of diffusible components.

The factor $(1 + \cos^2\theta)$ is appropriate for unpolarized light. If the beam is initially incident in the x-direction, it can only induce oscillating electric moments in a scattering center along the y- and z-axes, and scattered light is propagated perpendicular to these oscillating moments. Thus, looking at the scattered beam at $\theta = 90°$, say down the z-axis, one sees contributions only from the light whose electric vector is polarized in the y-direction. On the other hand, looking at $\theta = 0$, along the x-axis, one sees contributions from both y- and z-polarized beams. Thus the intensity is least at 90° and most at 0° and 180°. If scattering is carried out with plane-polarized light, $(1 + \cos^2\theta)$ is replaced by $2 \sin^2\theta'$, where θ' is the angle between the direction of observation and the direction of polarization of the electric vector of the incident beam. The $1/\lambda^4$ dependence, where λ is the wavelength of light, is characteristic of Rayleigh scattering. It is well known that this leads to the blue color of the sky, since blue light, having shorter wavelength than red light, is scattered more strongly into the observer's line of vision.

Any sort of scattering, be it of X-rays or visible light, occurs because of inhomogeneities in the sample from which scattering occurs. In X-ray diffraction from crystals, these inhomogeneities are the regular variations in electron density associated with the regular arrangement of atoms and molecules in the crystal lattice. In light scattering, the inhomogeneities are due to variations in the refractive index of various microscopic portions of the solution, about the mean value, n. The refractive index fluctuation is related to the refractive index change attendant on a given concentration fluctuation, which involves $n(\partial n/\partial c)$. The probability of a given concentration fluctuation is related to the bulk concentration, c, and to the solution nonideality – that is, the tendency of the macromolecules to cluster or to avoid each other – as reflected in the second, third, . . . virial coefficients B, C, . . . , respectively. These considerations give rise to the terms $n^2(\partial n/\partial c)^2 c/(1/M + 2Bc + 3Cc^2 + \cdots)$.

In polyelectrolyte solutions with added salt, we have (at least) three-component systems. Then a fluctuation in polymer concentration will in general be correlated with a fluctuation in salt concentration, leading to scattering which is severely ionic strength-dependent. It is therefore customary to work at relatively high salt concentrations (~0.1 M), where polymer-salt interactions are constant. An exhaustive review of the thermodynamics of multicomponent systems, with application to light scattering, is given by Casassa and Eisenberg.[85] They show that, to take multicomponent effects properly into account, one should dialyze the polymer solution against the solvent and subtract the scattering of the latter from that of the former. Refractive index increments of DNA-salt solutions have been accurately determined by Cohen and Eisenberg.[86]

It is convenient to write Eq. (5-44) in the form

$$R_\theta = \frac{i_\theta r^2}{I_0(1 + \cos^2\theta)} = \frac{Kc}{1/M + 2Bc + 3Cc^2 + \cdots} \tag{5-45}$$

R_θ is thus independent of geometrical factors. K is an optical constant:

$$K = 2\pi^2 n_0{}^2 (\partial n/\partial c)^2 / N_A \lambda^4$$

where we have substituted n_0, the solvent refractive index, for n, the solution refractive index. This is permissible in dilute solutions. According to Eq. (5-45), a plot of Kc/R_θ has an intercept of $1/M$ and a limiting slope of $2B$. Therefore, one can use light scattering to measure molecular weight and solution nonideality with no assumptions about molecular structure. If the solution is polydisperse, the intercept gives the reciprocal of the weight average molecular weight.

Our primary interest lies in molecules whose dimensions are not negligible relative to the wavelength of light. For these larger particles, the scattering intensity is decreased relative to the value predicted in Eq. (5-44), except at $\theta = 0$, due to destructive interference between waves scattered from different parts of the same molecule. We express this angular dependence of scattering by a factor $P(\theta)$, which represents the ratio of intensity scattered through an angle θ, to that scattered through zero angle (neglecting the trivial factor $1 + \cos^2\theta$). Thus Eq. (5-45) becomes

$$R_\theta = \frac{Kc}{1/M + 2Bc + 3Cc^2 + \cdots} P(\theta) \tag{5-46}$$

It is shown in the references cited above[5,82-84] that

$$P(\theta) = \frac{1}{N^2} \sum_{i=1}^{N} \sum_{j=1}^{N} \left\langle \frac{\sin \mu L_{ij}}{\mu L_{ij}} \right\rangle \tag{5-47}$$

$$\mu = \frac{4\pi}{\lambda} \sin\frac{\theta}{2}$$

A very valuable result can be obtained from Eq. (5-47) by analyzing $P(\theta)$ at low scattering angles. "Low" is defined here such that $\mu L_{ij} = (4\pi/\lambda)L_{ij} \sin(\theta/2) \ll 1$. Then, using the Taylor's series expansion, $\sin x = x - x^3/3! +$

$x^5/5! - \cdots$, valid for $x < 1$, we have

$$\lim_{\theta \to 0} P(\theta) = \frac{1}{N^2} \sum_{i=1}^{N} \sum_{j=1}^{N} \left\langle \frac{1}{\mu L_{ij}} \left[\mu L_{ij} - \frac{(\mu L_{ij})^3}{3!} + \cdots \right] \right\rangle$$

$$= 1 - \frac{\mu^2}{3! N^2} \sum_{i=1}^{N} \sum_{j=1}^{N} \langle L_{ij}{}^2 \rangle + \cdots$$

Comparing this expression with Eq. (5-2) for $\langle R^2 \rangle$, we find

$$\lim_{\theta \to 0} P(\theta) = 1 - \frac{1}{3} \mu^2 \langle R^2 \rangle = - \frac{16\pi^2}{3\lambda^2} \langle R^2 \rangle \sin^2 \left(\frac{\theta}{2} \right),$$

or, inverting,

$$\lim_{\theta \to 0} P(\theta)^{-1} = 1 + \frac{16\pi^2}{3\lambda^2} \langle R^2 \rangle \sin^2 \left(\frac{\theta}{2} \right) \tag{5-48}$$

Therefore, we see that the limiting slope of a plot of $1/P(\theta)$ vs. $\sin^2(\theta/2)$ will give $\langle R^2 \rangle$, with no assumptions about macromolecular structure.

In general, light scattering from solutions of large particles will show both concentration and angular dependence. Thus Kc/R_θ must be extrapolated both to $c = 0$ and $\theta = 0$ to get $\langle M \rangle_w$, $\langle R^2 \rangle$, and B. This simultaneous extrapolation may be done in a Zimm plot,[87] by plotting Kc/R_θ vs. $\sin^2(\theta/2) + kc$, where k is an arbitrarily chosen constant of convenient magnitude. Two such plots for scattering from solutions of ϕX 174 DNA are shown in Fig. 5-8.[88]

From Eq. (5-46), we see that

$$\frac{Kc}{R_\theta} = \left(\frac{1}{M} + 2Bc + 3Cc^2 + \cdots \right) P(\theta)^{-1} \tag{5-49}$$

Thus extrapolating points at a given concentration to $\theta = 0$, and then extrapolating these extrapolated points to $c = 0$, we get a line whose limiting slope is $2B/k$. In Fig. 5-8(a), at low salt concentration, we see $B > 0$, which indicates repulsion between the molecules due largely to polyelectrolyte effects. In Fig. 5.8(b), at high salt, $B < 0$, which indicates intermolecular attraction. If we extrapolate first to $c = 0$, and then to $\theta = 0$, we obtain a line whose limiting slope, from Eq. (5-48), is $(16\pi^2/3\lambda^2) \langle R^2 \rangle$. We see from comparison of Figs. 5-8(a) and 5-8(b) that this slope is greater in low salt, testifying to the expansion of the chain due to intramolecular polyelectrolyte repulsions. Regardless of the order of extrapolation, the limiting lines extrapolate to the same intercept, which is $1/M$ or $1/\langle M \rangle_w$.

The complete angular dependence of $P(\theta)$ has been worked out, in the case of linear polymers, rigorously only for random coils and rigid rods. For rods of length $\mathcal{L}$, the result is[89]

$$P(\theta) = \frac{2}{\mu\mathcal{L}} \int_0^{\mu\mathcal{L}} \frac{\sin x}{x}\, dx - \left[\frac{\sin(\mu\mathcal{L}/2)}{(\mu\mathcal{L}/2)}\right]^2$$

For flexible gaussian chains, the average required in Eq. (5-47) may be performed using the distribution function, Eq. (5-6), to give[90,91]

$$P(\theta) = (2/u^2)(e^{-u} + u - 1)$$

where

$$u = \mu^2 \langle R^2 \rangle$$

For wormlike coils, particularly with excluded volume, the distribution function is not known exactly. Several treatments,[83,92–96] mainly using the expansion of sin $\mu L/\mu L$ in powers of $\mu^2 L^2$ shown above with the exactly known moments $\langle L^2 \rangle$ and $\langle L^4 \rangle$ for the wormlike chain, Eqs. (5-20) and (5-22), have given approximate curves of $P(\theta)$ vs. $\sin^2(\theta/2)$. One of these,[96] using the Daniels distribution function, Eq. (5-24), to perform the average in Eq. (5-47), gave the results shown in Fig. 5-9. The quantity $x = \mathcal{L}/2a$, so $x = \infty$ corresponds to the random coil limit and $x = 0$ to the rigid rod. By appropriate modification of this distribution function for excluded volume effects, it was also possible to calculate approximate scattering curves for high molecular weight, native DNA, for which the excluded volume parameter ϵ is not zero.[96]

Calf thymus DNA, with molecular weight 17×10^6, has a contour length $\mathcal{L} = 17 \times 10^6/192 = 8.9 \times 10^4$ Å. The persistence length a of DNA is about 500 Å, so the mean square end-to-end distance $\langle L^2 \rangle = 2a\,\mathcal{L} = 8.9 \times 10^7$ Å^2, neglecting excluded volume effects. The mean square radius is $\langle R^2 \rangle = \frac{1}{6} \langle L^2 \rangle = 1.5 \times 10^7$ Å^2. Commercial light scattering instruments have a minimum detection angle $\theta_{min} = 30°$, so $\sin(\theta_{min}/2) = 0.259$. Thus for light with wavelength 5000 Å, $\mu_{min} = (4\pi/\lambda)\sin(\theta_{min}/2) = 6.5 \times 10^{-4}$, and $\mu_{min}^2 = \mu_{min}^2 \langle R^2 \rangle = 6.3$. For such a DNA sample, $x = \mathcal{L}/2a = 130$. Taking the curve in Fig. 5-9 with $x = 100$, and drawing the tangent to it at $u = 6.3$, we find $P(0)_{apparent}^{-1} \approx 2$. Thus, from Eq. (5-49), M will be underestimated by a factor of two. This has been pointed out previously[97,98] and a direct experimental demonstration is shown in Fig. 5-10.[99] In order to get true limiting values for native DNA molecular weights above 6×10^6, θ_{min} must be substantially below 30°.

Consequently, several light scattering photometers which permit measurements to angles as low as 6–16° have been constructed.[99–102] They all require intense, narrow, well-collimated incident beams. Although continuous lasers would appear ideally suited to this purpose, conventional high-pressure Hg lamps are adequate and have been the light sources used. Froelich et al.[99]

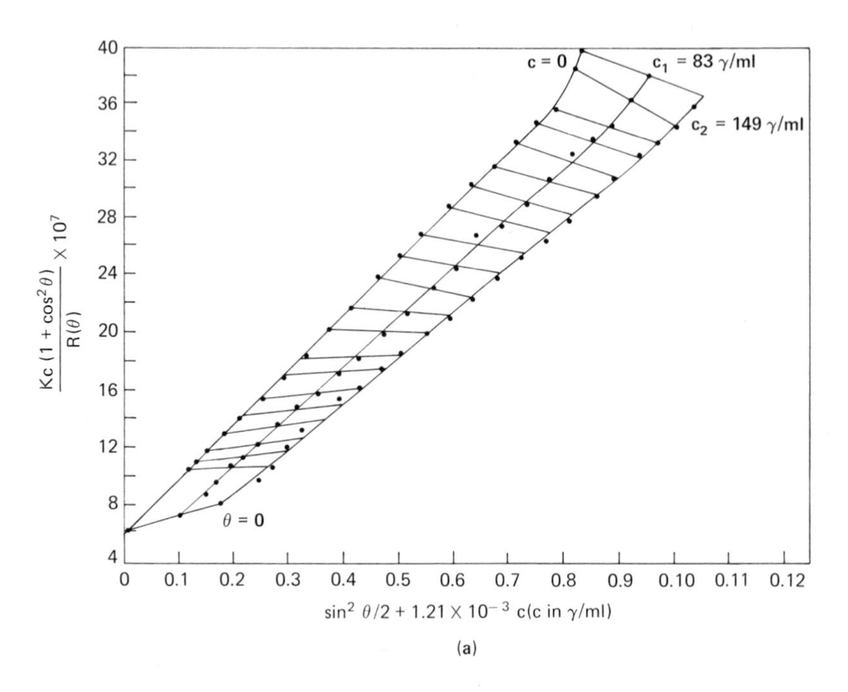

(a)

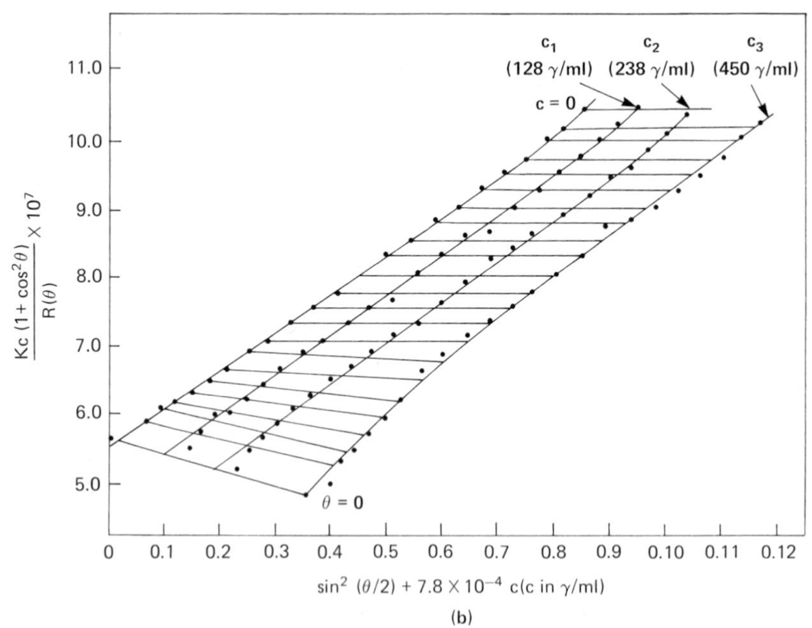

(b)

FIGURE 5-8 Light scattering of ϕX 174 DNA in NaCl + 10^{-3} M phosphate buffer, pH 7.5 at 37°C. (a) 0.02 M NaCl; (b) 0.2 M NaCl. [From R. L. Sinsheimer, *J. Mol. Biol.*, **1**, 37, 43 (1959). Reprinted with permission.]

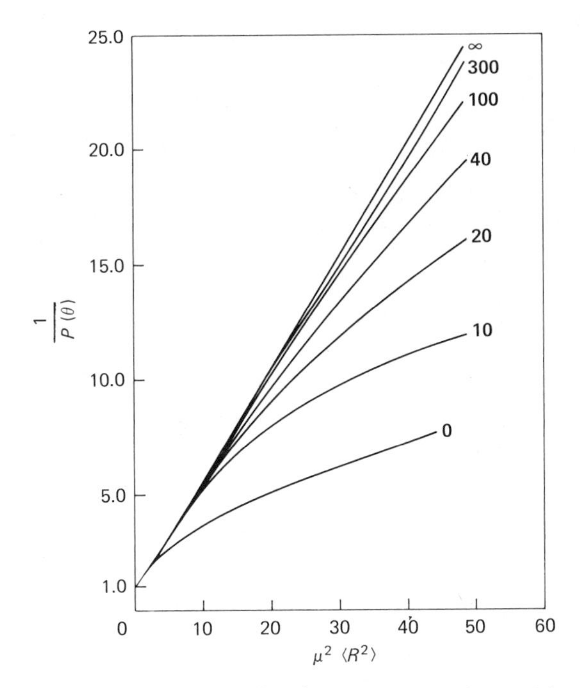

FIGURE 5-9 $1/P(\theta)$ as a function of $\mu^2 \langle R^2 \rangle$ for wormlike chains with no long-range excluded volume and containing the indicated number of statistical segment lengths. [From P. Sharp and V. A. Bloomfield, *Biopolymers*, **6**, 1205 (1968). Reprinted by permission of John Wiley & Sons, Inc.]

and Harpst et al.[100,101] used a long, rectangular sample cell in which the incident beam does not interfere with measurement of the scattered light down to low angles. Meyerhoff et al.[102] used a double prism to reflect the scattered light to the detector without reflecting the primary beam at angles as low as 6°.

At these low angles, scattering by dust particles becomes particularly severe. Methods have thus been devised for improved clarification of solutions. Froelich et al.[99] emulsify the DNA solution with a chloroform–isoamyl alcohol mixture, centrifuge the aqueous phase, and use the upper portion of the supernatant. Harpst et al.[101,102] filter solutions through Millipore filters, a procedure which has been shown not to break T7 DNA ($M = 25 \times 10^6$) or even T2 DNA ($M = 130 \times 10^6$).[103] This result is remarkable, since it would be thought that shear stresses in the narrow pores of the filter would break DNA; but since the flow rate in any single pore is very small, high shear stresses are not developed.

The only naturally occurring DNA whose molecular weight has been determined primarily by light scattering is the single-stranded DNA from

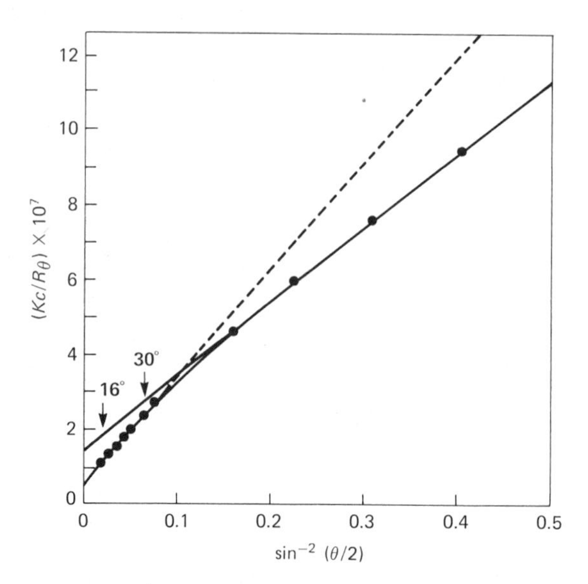

FIGURE 5-10 Plot of Kc/R_θ for calf thymus DNA, with data obtained in a low-angle light scattering photometer. – – –, initial slope; ————, slope extrapolated from 30° and above. [From D. Froelich et al., *Biophys. J.*, **3**, 115 (1963). Reprinted with permission.]

bacteriophage ϕX 174.[88] M was in fact obtained in two independent ways, thus reinforcing confidence in the result. The molecular weight of the entire virus was determined by light scattering as 6.2×10^6, while chemical analysis indicated a DNA content of 25.5%. Thus M for the DNA is 1.6×10^6. Light scattering measurements on the isolated DNA itself (Fig. 5-8) gave $M = 1.6$ to 1.8×10^6.

Early work by Doty and collaborators[104-108] used light scattering to determine the molecular weights of sonicated fragments of calf thymus DNA, and correlated these with sedimentation coefficients and intrinsic viscosities.[109,110] However, since a conventional light scattering photometer was used, the results were not reliable for molecular weights greater than $3–6 \times 10^6$.

With the low-angle photometers described here, it has been possible to measure reliably molecular weights of DNA from calf thymus ($M = 17$ to 18×10^6)[99,111] and T7 phage ($M = 25 \times 10^6$).[111] Theoretical calculations which take into account both the wormlike character and excluded volume of the DNA molecule[96] indicate that extrapolations to zero angle from the angular region 10–25° will slightly overestimate M, by perhaps 8% for $M = 25 \times 10^6$; but this is within current experimental error.

Light scattering can also be used to evaluate chain stiffness in DNA.

From low-angle scattering, the persistence length a can be obtained by measuring the limiting slope and intercept, thereby getting $\langle R^2 \rangle$ and M (or the contour length $\mathcal{L}$), and using an equation such as (5-21) or (5-36). It has been pointed out, however,[112] that small uncertainties in $\langle R^2 \rangle$ can lead to large errors in a by this procedure.

It is also possible to evaluate the persistence length from light scattering measurements at high angles.[113] It may be shown from the expressions for $P(\theta)$ given above that, at angles large enough such that $\mu \mathcal{L}$ or $\mu \langle R^2 \rangle^{1/2}$ are much greater than unity, $1/P(\theta)$ varies as $\sin(\theta/2)$ for rigid rods and as $\sin^2(\theta/2)$ for random coils. Sadron[114] found that for DNA at low concentrations,

$$Kc/R_\theta = [\langle M \rangle_w P(\theta)]^{-1} = A' \sin(\theta/2) + B'$$

Thus DNA behaves more like a rigid rod than a random coil at high angles. This is to be expected because, as noted in connection with Bragg's law for X-ray diffraction, large angles correspond to small distances; and over distances of a few hundred angstroms, duplex DNA is indeed rather stiff and rodlike.

According to a theory of scattering developed by Luzzati and Benoit,[115] A' in this equation depends only on λ and on M_L, the mass per unit length of the chain. B' depends on M_L, $\mathcal{L}$, and the average curvature of the chain, which is related to the persistence length. For wormlike chains, Ptitsyn and Fedorov[113] obtained

$$B' = \frac{2}{\pi^2 \langle M \rangle_n} - \frac{2}{3\pi^2 M_L a}$$

For high molecular weight DNA, $\langle M \rangle_n \gg M_L a$, so $B' < 0$, and the intercept of a plot of Kc/R_θ vs. $\sin(\theta/2)$, extrapolated from high angles will be negative. This is indeed observed.[114]

The values of the persistence length for native DNA obtained from light scattering will be compared with those obtained by other techniques at the end of this chapter.

High-angle light scattering can also be used to determine M_L, the mass per unit length. It has been shown[116] that for rigid rods, the expression for $P(\theta)$ gives

$$\lim_{\mu\mathcal{L} \to \infty} \left(\frac{R_\theta}{Kc}\right) = \frac{M}{\mu\mathcal{L}} - \frac{2M}{\mu^2\mathcal{L}^2}$$

One can thus obtain $M_L = M/\mathcal{L}$ by plotting $\mu R_\theta / Kc$ as a function of μ or $\sin(\theta/2)$. This procedure has been applied to sonicated DNA by Eisenberg and Cohen,[86,124] who found values for M_L of 203 daltons per angstrom for NaDNA and 270 daltons per angstrom for CsDNA. These compare well with the values of 192 and 263 daltons per angstrom for the Na and Cs salts, respectively, corresponding to the B form of DNA.

Mauss et al.[117] obtained similar results for sonicated DNA, and also studied M_L for sonicated DNA in the presence of proflavine dye. It is found that M_L decreases as the amount of dye bound increases. This is in accord with the intercalation hypothesis of Lerman,[118,119] in which the dye is inserted between the bases, thereby unwinding and lengthening the double helix. This topic is discussed in greater detail in Chapter 7.

B. Low-Angle X-Ray Scattering

As we have seen previously, X-ray diffraction from crystals or fibers can give structural information in the range of dimensions from about 1 to 20 Å. Light scattering, on the other hand, is useful for size and shape determination in the range 200–2000 Å. There is thus a large gap, in the region 20–200 Å, between the size ranges covered by these two techniques. Many interesting aspects of nucleic acid structure have dimensions which fall in the gap.

Although Bragg's law, $\lambda = 2d \sin \theta_b$, is exact only for crystals of infinite extent, it gives a qualitative indication of the range of dimensions d of single molecules that can be investigated sensitively for given values of wavelength λ and diffraction angle θ_b. "Low-angle" scattering is defined operationally as scattering through angles sufficiently small that $\sin \theta_b \approx \theta_b$, $\cos \theta_b \approx 1$, which is adequate for $\theta_b < 5°$. Then the range of d is

$$d \approx \frac{\lambda}{2\theta_b}$$

For CuK_α X-radiation, $\lambda = 1.54$ Å. Experimentally, one can work in the range of scattering angles $10' < 2\theta_b < 10°$. Then d will lie between 10 Å and 500 Å. As an example of the sort of data obtained, a composite scattering curve for NaDNA is shown in Fig. 5-11.

The basic theory of low angle X-ray scattering from solutions is conveniently developed in several references.[120–124] As is the case with X-ray diffraction and light scattering, observable scattering is obtained only from inhomogeneities in the sample. In X-ray diffraction from crystals, these inhomogeneities are the regular variation in electron density throughout the crystal. In light scattering, they are the random variations in refractive index associated with the dissolved solute particles. In low angle X-ray scattering from solutions, there are inhomogeneities in electron density due to the differences in electron density between solvent and solute.

The intensity of scattering of X-radiation by a single electron through an angle θ, which is twice the Bragg angle θ_b, is

$$i(\theta) = 7.9 \times 10^{-26} \left(\frac{1 + \cos^2 \theta}{2r^2}\right) I_0 = 7.9 \times 10^{-26}\ r^{-2} I_0$$

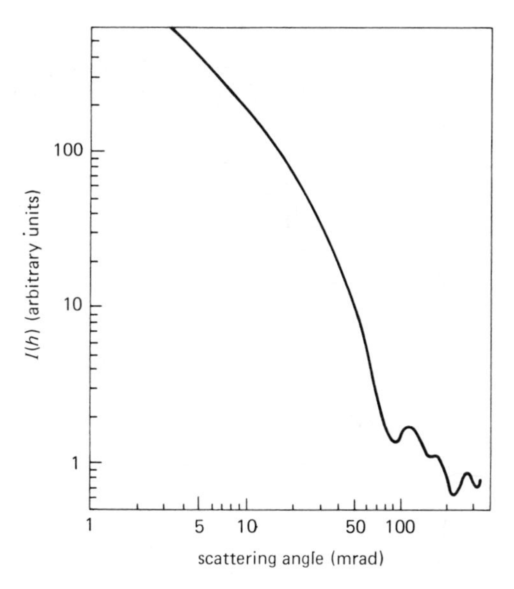

FIGURE 5-11 The complete low-angle X-ray scattering curve for NaDNA dialyzed against 0.05 M NaCl. The curve is a composite of several partial runs with different DNA concentrations and different instrumental parameters. [From S. Bram and W. W. Beeman, *J. Mol. Biol.*, **55**, 311 (1971). Reprinted with permission.]

I_0 is the incident intensity, r is the distance from the scattering center to the detector, and 7.9×10^{-26} is the scattering factor for a free electron. It is worth pointing out that the scattered intensity is independent of the X-ray wavelength, in contrast to Rayleigh scattering. This is so because the X-ray frequency ($\nu = c/\lambda = 3 \times 10^{10}/1.5 \times 10^{-8} = 2 \times 10^{18}$ sec^{-1}, where c is the speed of light) is much greater than the frequency of oscillation of an electron in an atom or molecule (about 10^{15} sec^{-1}). In the low-angle approximation, $(1 + \cos^2 \theta)/2 = 1$.

Now consider scattering in the forward direction ($\theta = 0$) by a gas of particles of molecular weight M_2, each containing $N_A z$ electrons per gram. Since scattering is being observed at zero angle, the radiation scattered from each electron will be in phase, there will be no interference effects, and the scattered amplitudes will add. Thus the amplitude of scattering from each particle is zM_2 times that from a single electron. The scattered intensity, which is proportional to the square of the amplitude, will be $(zM_2)^2$ that from a single electron. If the concentration of particles is c grams per cubic centimeter, and the particles are independent in the scattering volume V, so that they constitute an ideal gas, there are CN_A/M_2 particles per centimeter from which the scattered intensity is additive, so

$$i(0) = 7.9 \times 10^{-26}\, r^{-2} I_0 z^2 M_2 c N_A V$$

If instead of an ideal gas we had a binary solution, z would be interpreted as the excess (or deficiency) of electrons per gram in the solute, z_2, relative to the solvent, z_1, and $i(0)$ as the excess scattering of solution relative to pure solvent. In this case we replace $z^2 M_2 c N_A$ in the above equation by $(\partial\rho_e/\partial c)_{T,P}{}^2\ (\overline{\delta c^2})$, where ρ_e is the density of the solution in electrons per milliliter; $(\overline{\delta c^2})$ is the mean square concentration fluctuation, and is given by the same virial expansion as was used in the previous section on light scattering:

$$(\overline{\delta c^2}) = \frac{c}{VN_A}\left(\frac{1}{M} + 2Bc + \cdots\right)$$

It is easy to show that, for a two-component system,

$$\left(\frac{\partial\rho_e}{\partial c}\right)_{T,P} = N_A(z_2 - z_1\bar{v}_2\rho_1)$$

where $\bar{v}_2$ and ρ_1 are the partial specific volume (ml/g) of solute and density (g/ml) of solvent, respectively. Thus the intensity of forward scattering from a binary solution is

$$i(0) = 7.9 \times 10^{-26}\, r^{-2} I_0\ (z_2 - z_1\bar{v}_2\rho_1)^2\ (M^{-1} + 2Bc + \cdots)cN_A \quad (5\text{-}50)$$

Scattering experiments on nucleic acids are most often carried out in solutions containing polymer, water, and added salt. The modifications required for these three-component systems are described by Eisenberg and Cohen.[124] Briefly, the proper procedure, as in light scattering, is to dialyze the nucleic acid solution against the solvent and subtract the scattering of solvent from solution. With this procedure, one replaces $(\partial\rho_e/\partial c)_{T,P}$ for binary solutions, by $(\partial\rho_e/\partial c_2)_{T,\mu_1,\mu_3}$, the electron density increment at constant chemical potential of diffusible components. It may be shown that

$$\left(\frac{\partial\rho_e}{\partial c_2}\right)_{T,\mu_1,\mu_3} = N_A(z_2 - \xi_3 z_3) - \rho_{es}(\bar{v}_2 - \xi_3\bar{v}_3)$$

where ρ_{es} is the electron density (electrons/ml) of solvent mixture in the absence of polymer, $\bar{v}_3$ is the partial specific volume of salt, component 3, and ξ_3 is the number of grams of salt that must be removed per gram of polymer added, to maintain constant the chemical potential of the salt:

$$\xi_3 = -\left(\frac{\partial w_3}{\partial w_2}\right)_{T,\mu_3}$$

This correction arises because, in a three-component system, the solvent displaced from a scattering volume by the addition of nucleic acid will not in general have the same composition as bulk solvent, due to preferential interaction between the polymers and one or another of the solvent components.

The electron density increment can be related to the directly measured density increment[86] $(\partial\rho/\partial c_2)_{T,\mu_1,\mu_3}$ in the limit of vanishing polymer concentration c_2, by the expression

$$\frac{1}{N_A}\left(\frac{\partial\rho_e}{\partial c_2}\right)_{T,\mu_1,\mu_3} = z_2 + \frac{z_1 - z_3}{1 + 10^{-3}m_3M_3}\xi_3 - \frac{z_1 + 10^{-3}z_3m_3M_3}{1 + 10^{-3}m_3M_3}\left[1 - \left(\frac{\partial\rho}{\partial c_2}\right)_{T,\mu_2,\mu_3}\right]$$

where M_3 is the molecular weight, and m_3 the molality, in moles per kilogram of component 1, of the salt component 3. A plot of electron density increment as a function of c_3 for NaDNA in NaCl and CsDNA in CsCl is given in Fig. 5-12.

For scattering in other than the forward direction, there is interference between the rays scattered from different parts of the same particle. Thus, just as in light scattering, we can introduce the angular factor $P(\theta)$:

$$i(\theta) = i(0)P(\theta)$$

Combining all of these results, we finally obtain an equation for low-angle X-ray scattering identical to Eq. (5-49) for light scattering, with

$$R_\theta = \frac{r^2 i(\theta)}{I_0}$$

and

$$K = \frac{7.9 \times 10^{-26}}{N_A}\left(\frac{\partial\rho_e}{\partial c_2}\right)_{T,\mu_1,\mu_3}$$

These equations indicate that low-angle X-ray scattering, like light scattering, can be used to determine molecular weights, if the optical constant K and the absolute intensity (relative to the incident beam) can be measured. This is a difficult experimental task since $i(\theta)/I_0$ ranges from 10^{-4} to 10^{-6}. However, the primary beam can be attenuated by calibrated filters, or scattering from solutions of known properties can be used as a standard. Relative intensity measurements, alone, give $P(\theta)$ as a function of θ. This yields information on molecular shape, but not on molecular weight.

Several special cases of the angular dependence of scattering are worth mentioning in detail. First, in the section on light scattering we showed that, in general, at small angles

$$P(\theta) = 1 - \frac{1}{3}\mu^2 \langle R^2\rangle + \cdots$$

where $\mu = (4\pi/\lambda)\sin(\theta/2)$. Since the expansion of a gaussian function, e^{-x^2} for small values of the argument is $1 - x^2 + \cdots$, we will expect the scattered

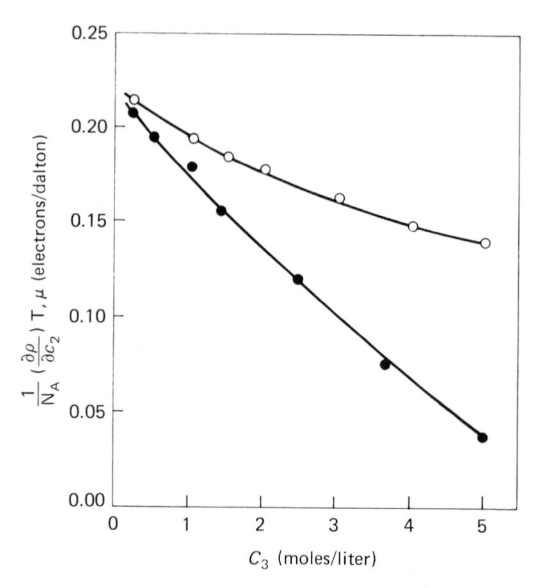

FIGURE 5-12 Density increments $N_A^{-1}(\partial\rho/\partial c_2)_{T,\,\mu_1,\,\mu_3}$, electrons per dalton, as a function of salt morality c_3: –○–○–, NaDNA in NaCl; –●–●–, CsDNA in CsCl. [From H. Eisenberg and G. Cohen, *J. Mol. Biol.*, 37, 355 (1968). Reprinted with permission.]

intensity to vary like a gaussian curve for small angles. Thus we have the approximation, due to Guinier,[125] that

$$i(\theta) = i(0) \exp\left[-\frac{1}{3}\mu^2 \langle R^2 \rangle\right]$$

Then the mean square radius may be determined from the slope of a plot of $\ln i(\theta)$ vs. μ^2:

$$\ln i(\theta) = \ln i(0) - \frac{1}{3}\mu^2 \langle R^2 \rangle$$

Although the longitudinal dimensions of long, rigid rods, such as those of duplex DNA, for which $\mu\mathcal{L} \gg 1$, are too long to be studied by low-angle X-ray scattering, two useful parameters may be obtained. We suppose that the radius of the rod is R_0. Then in the very low-angle range, for which $\mu\mathcal{L} \gg 1$ but $\mu R_0 \ll 1$, the scattering is sensitive only to the mass per unit length of the rod:

$$\frac{\mu i(\theta)}{Kc} = M_L\left(1 - \frac{1}{\mu\mathcal{L}}\right) \tag{5-51}$$

Thus a plot of $\mu i(\theta)$ vs. μ will asymptotically yield M_L, if K (i.e., the electron density increment) is known.

At somewhat higher values of μ, in the range $\mu R_0 \lesssim 1$, the scattering becomes sensitive to the cross-sectional area of the rod. For a cylinder of radius R_0, we define the cross-sectional radius of gyration R_c by

$$R_c^2 = \int_0^{R_0} r^2 \, \Delta\rho_e(r) \, dr$$

where $\Delta\rho_e(r)$ is the difference in electron density between solute and solvent a distance r from the axis of the rod. For a dilute solution of rods, uncorrelated in position and direction, it has been shown[121] that

$$\begin{aligned} \mu i(\theta) &= i(0)(1 - \tfrac{1}{2} R_c^2 \mu^2 + \cdots) \\ &\simeq i(0) \exp\left[-\mu^2 R_c^2/2\right] \end{aligned} \tag{5-52}$$

In more concentrated solutions, rods may align to give liquid crystalline gels. In that case, one observes additional diffraction effects due to the regular two-dimensional array (usually hexagonal) of rods with axes parallel.

The above equations have been derived for the scattering of a beam of point cross section. Because of the low scattered intensity from dilute solutions, it has been found useful to use a high, narrow slit for beam collimation.[120–122,126] Then the intensity scattered through a given angle is the superposition of scattering from all parts of the beam profile.

With this background in theory, we turn to results on particular systems. Applications of low-angle X-ray scattering to DNA solutions are well reviewed by Luzzati.[127] The partial molal volume $\bar{v}_2$ and mass per unit length M_L of various salts of double-stranded DNA have been obtained from measurements on liquid crystalline gels. Diffraction characteristic of the above-mentioned two-dimensional hexagonal lattice of rigid rods with intervening solvent is observed in these systems.[128] We let $s = \mu/2\pi = (2/\lambda)\sin(\theta/2)$. Then, if s_1 is the Bragg spacing of the first line in the diffraction pattern, in A^{-1},

$$s_1^{-2} \times 10^{-16} = \frac{1}{2}\sqrt{3} M_L \rho_1^{-1} \left[c^{-1} - 1 + \bar{v}_2 \rho_1\right]$$

ρ_1 is the solvent density, in grams per cubic centimeter. Thus, by plotting s_1^{-2} vs. c^{-1}, one obtains a straight line whose slope is proportional to M_L, and thus to p, the translation per residue in the double helix. The intercept on the c^{-1} axis is equal to $\frac{1}{2}\sqrt{3} M_L \rho_1^{-1} (1 - \bar{v}_2 \rho_1)$ and can thus be used to determine $\bar{v}_2$. One finds $\bar{v}_2 = 0.57\ \text{cm}^3/\text{g}$ for NaDNA, regardless of salt concentration.[129] M_L is found to be 192 daltons/Å, and p approximately 3.1 Å/per residue, in good agreement with the Watson-Crick model. Exact numerical agreement should not be taken too seriously, however, because of the neglect of preferential solvation effects in the above analysis.

The mass per unit length of the Na and Cs salts of DNA has also been determined by low-angle X-ray scattering in more dilute solutions,[124,128,130] using Eq. (5-51) and corrections for three-component effects. At low concentrations,[130] 2–12 mg/ml, a value of 200 ± 4 daltons/Å for NaDNA was found, compared with 192 daltons/Å expected for the B form. Figure 5-13 shows that the details of scattering from DNA at higher angles are also consistent with the existence of the B form, and not the A form, of NaDNA in dilute salt solutions. M_L measured for CsDNA is about 10% lower than the 263 daltons/Å expected for the B form, but this may be within experimental error. At higher concentrations, 10–56 mg/ml, values of M_L obtained[128] for both the Na and Cs salts of DNA are about 20–25% lower than those expected for the B form. (The large values of the concentrations used in these experiments, in order to obtain adequate scattering intensity, points up one of the difficulties of this method.)

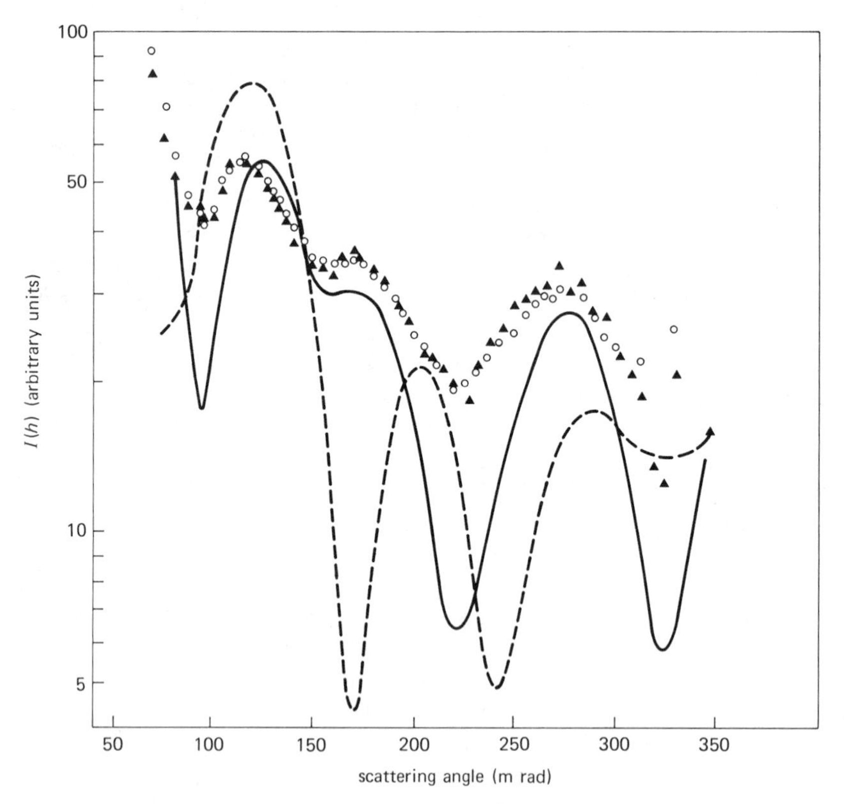

FIGURE 5-13 Observed scattering at larger angles from NaDNA dialyzed against 0.05 M NaCl, compared with theory for the B (solid line) and A (dashed line) forms. ▲, 107 mg DNA/ml, not sonicated (a gel); ○, 34 mg DNA/ml, sonicated. [From S. Bram and W. W. Beeman, *J. Mol. Biol.*, **55**, 311 (1971). Reprinted with permission.]

The axial radii of gyration R_c for several DNA salts have been determined using these techniques.[128,130] For NaDNA, a detailed study[130] at low DNA concentration suggests the existence of two radii of gyration (see Fig. 5-14) with values of about 7.8 Å and 9.6 Å. The smaller of these may be attributed to an electron-dense core of ionized DNA and some tightly associated Na^+ ions, and is in reasonable agreement with the R_c of 8.05 ± 0.20 Å for the model of the B form of NaDNA proposed by Langridge and co-workers.[131] The larger radius of gyration suggests a shell, about 20 Å thick, surrounding the core, which is much less electron dense relative to solvent, and has been attributed to a Debye-Hückel layer of counterions and perhaps electrostricted solvent.[130] This interpretation is complicated, however, by the theoretical observation[124,132] that in order for the assumption of independent electro-neutral density fluctuations, which was used in development of the scattering theory, to be accurately applicable to polyelectrolyte solutions with 0.1 M NaCl, the scattering angle should be less than 0.008 radians (0.45°). This is at the lower end of the range of angles at which scattering was measured.[130]

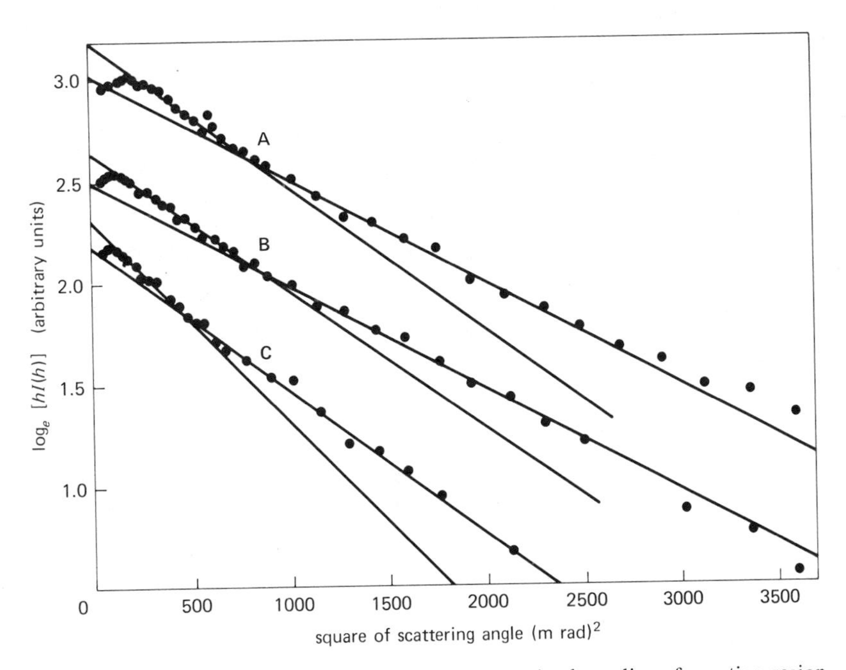

FIGURE 5-14 Some measured X-ray scattering curves in the radius of gyration region. Curve A: 12 mg NaDNA/ml, 0.05 M NaCl; curve B, 5.55 mg NaDNA/ml, 1.0 M NaCl; curve C, 7.83 mg CsDNA/ml, 0.05 M CsCl. The relative vertical positions are not significant. [From S. Bram and W. W. Beeman, *J. Mol. Biol.* **55**, 319 (1971). Reprinted with permission.]

Low-angle X-ray scattering has been used in a number of studies[133–136] to study structural transitions induced in nucleic acids by temperature, pH, and solvent changes (see Chapter 6). For example, polyribocytidylic acid has been studied in neutral aqueous NaCl solutions, and in water–alcohol solutions, at various temperatures, to elucidate changes in structure of the single-stranded molecule.[136] Measurements of optical absorbance, CD, and ORD were made of the same solution. Over a range of experimental conditions where the spectral properties changed markedly, the X-ray curves were all very similar and were interpretable in terms of a rodlike structure, with persistence length of about 200 Å and at least partial base stacking. Changes in M_L and R_c were modest, and appeared to rule out a transition from a stacked single-stranded helix to a completely unstacked random coil.

As mentioned previously, it has been proposed that some dyes, such as proflavine, interact with DNA by intercalation and partial unwinding of the helix.[118,119,137] If this is the case, M_L should decrease linearly as the degree of dye binding increases. This is very nicely confirmed by the results displayed in Fig. 5-15.[129]

The structure of RNA in solution may also be studied by low-angle X-ray scattering. High molecular weight RNA from a variety of sources – ribosomal RNA from ascites tumor cells, *E. coli*, and yeast,[138,139] and plant

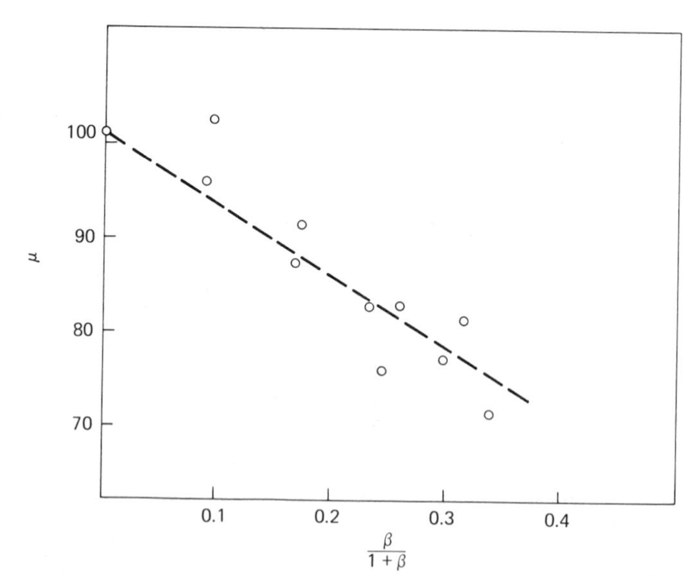

FIGURE 5-15 Mass per unit length of the proflavine-DNA complex; β is the ratio of bound proflavine molecules to base pairs. The dashed line corresponds to Lerman's intercalation model. [From V. Luzzati, *Prog. Nucleic Acid Res.*, **1**, 347 (1963). Reprinted with permission.]

virus RNA from tobacco mosaic virus and turnip yellow mosaic virus[140] – shows similar behavior, as depected in Fig. 5-16. The solid lines in the figure are those calculated for infinitely long rods, while the dotted lines are those expected theoretically for a gaussian distribution of finite rods with $L/R_c = 10$ or 12, connected by universal joints. Comparison of these theoretical curves with the experimental results suggests that more than 90% of the RNA in each chain is incorporated in short (50–150 Å) rigid helical segments which are connected by flexible regions that contain less than 10% of the bases. R_c and M_L of the rodlike portions are similar to those of duplex DNA, suggesting a double-stranded structure. It is well known that in TMV RNA this sort of structure is not the one that is biologically important. Therefore, the inference that, say, ribosomal RNA exists within the ribosome as an alternation of helical and random regions must be drawn with caution. It should be noted from Figure 5-16 that a distinction between the scattering from infinite and finite rods can be made only at very small values of the

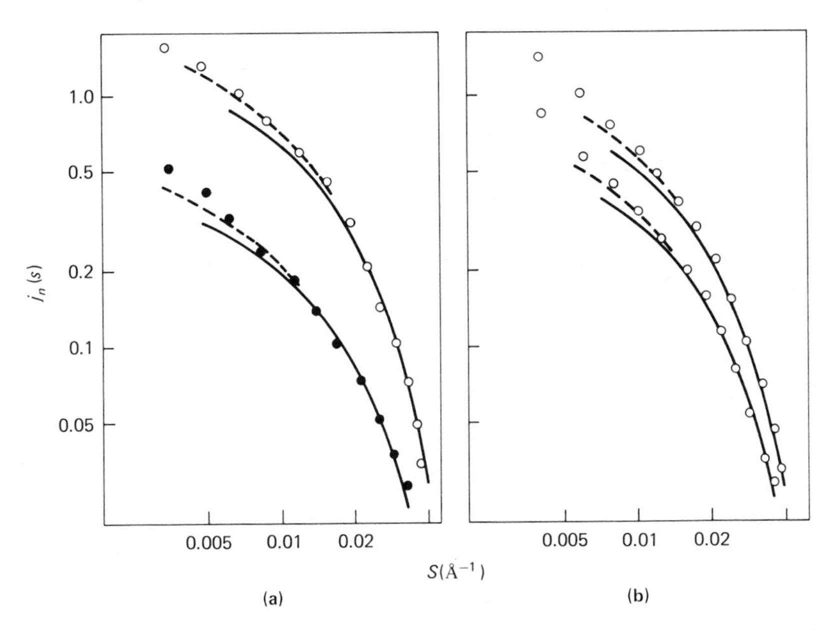

FIGURE 5-16 Low-angle X-ray scattering from viral RNA in aqueous solutions. (a) Tobacco mosaic virus RNA; –○–○–, 0.15 M NaCl, 0.015 M Na citrate pH 6.8, RNA conc. $c = 3.01\%$; –●–●–, ionic strength $<10^{-3}$ M, $c = 0.98\%$. (b) Turnip yellow mosaic virus RNA, 0.14 M NaCl, 0.01 M Na borate, pH 6.8, $c = 1.50\%$ and 2.49% for upper and lower curves. The solid lines represent the theoretical curves for a very long rigid rod with M_L and R_c giving best fit to experiment. The dashed lines are theoretical curves for broken rods of segment length $\bar{l}$, with $\bar{l}/R_c = 12$ in (a) and $\bar{l}/R_c = 10$ in (b). [From J. Witz, L. Hirth, and V. Luzzati, *J. Mol. Biol.*, **11**, 613 (1965). Reprinted with permission.]

scattering parameter s. At larger s, one is seeing only the cross-sectional structure, which of course is independent of length.

The use of low-angle X-ray scattering to study the structure of transfer RNA will be discussed in Chapter 8.

V SEDIMENTATION EQUILIBRIUM IN DENSITY GRADIENTS

The ultracentrifuge has been of great value and versatility in the physical chemical study of nucleic acids. Its uses may be broadly divided into preparative and analytical categories; and, in another dimension, into sedimentation equilibrium and sedimentation velocity. In these techniques the application of large centrifugal fields to a macromolecular solution causes a redistribution of species and the net motion of the macromolecule (if it is denser than solvent) away from the axis of rotation. The nonuniform concentration distribution produced leads to a diffusion flux tending to reestablish uniformity. In sedimentation velocity experiments, discussed in the next section, sedimentation prevails over diffusion and the macromolecule continues to sediment with finite velocity toward the bottom of the cell, though the concentration profile may be markedly influenced by diffusion. In sedimentation equilibrium experiments, on the other hand, centrifugal and diffusive forces finally balance and an equilibrium concentration distribution results which may be analyzed by thermodynamic methods.

We shall be particularly concerned with adaptations of the basic sedimentation equilibrium experiment to the special requirements of nucleic acid studies. The most notable of these adaptations is the density gradient technique,[141] in which the nucleic acid finds its position of neutral buoyancy in a density gradient established by the equilibrium redistribution of heavy salts in an aqueous solution in an ultracentrifugal field. This characterization of macromolecules by rather a novel property – buoyant density – has enabled distinctions between nucleic acids to be made on the basis of isotopic composition, G · C content, double- or single-strandedness, topological structure (in the presence of dyes), and other important features.

The presence of large amounts of added salts leads to the necessity of thermodynamic analysis of three- (and sometimes four-) component systems, because the differential interactions between the three components lead to correlations in their redistribution in the ultracentrifugal field. This effect can be turned around to investigate hydration of nucleic acids in aqueous salt solutions. However, we will begin our treatment of sedimentation equilibrium by considering two-component systems and will consider the more interesting but complicated multicomponent situation after a suitable background has been developed.

Several extensive and useful reviews of density gradient equilibrium sedimentation are available.[142–145] The reader is referred to these for more detailed coverage of theory, experimental procedures, and applications than we can present here.

A. Basic Thermodynamics

When sedimentation equilibrium is established, the total chemical potential of each component i, $\mu_{i,\,\mathrm{tot}}$ must be constant throughout the cell. $\mu_{i,\,\mathrm{tot}}$ depends on concentration (measured in molality m) and pressure P, both of which vary with distance r from the axis of rotation. It also depends on the work done on the system in bringing a mole of component i, of mass M_i, to r from $r = 0$ in the ultracentrifugal field. That is,

$$\mu_{i,\,\mathrm{tot}}(r) = \mu_i(r) - \frac{1}{2} M_i \omega^2 r^2 = \text{constant}$$

or

$$\frac{d\mu_{i,\,\mathrm{tot}}}{dr} = \sum_{j=1}^{\nu} \left(\frac{\partial \mu_i}{\partial m_j}\right)_{T,\,P,\,\mu_k \neq j} \frac{dm_j}{dr} + \left(\frac{\partial \mu_i}{\partial P}\right)_{T,\,M} \frac{dP}{dr} - M_i \omega^2 r = 0$$

In these equations μ_i is the ordinary chemical potential of component i. There are ν components. In solutions of charged molecules, such as nucleic acids in ionic solutions, the total chemical potential should also contain an electrostatic term. However, this can be eliminated by proper choice of neutral components. Many such choices are possible, as discussed by Schmid and Hearst.[146]

From standard thermodynamic relations we know that

$$\left(\frac{\partial \mu_i}{\partial P}\right)_{T,\,M} = \bar{V}_i = M_i \bar{v}_i$$

where $\bar{V}_i$ is the partial molal volume, and $\bar{v}_i$ the partial specific volume, of component i. The (dP/dr) can be evaluated by the following argument. Consider a cylindrical volume of solution with cross section A, with one end at r and the other at $r + dr$. The volume is Adr, and if the density of the solution is ρ, the mass of the volume is $\rho A dr$. Thus the total force acting on the volume at equilibrium is

$$AP(r) - AP(r + dr) + \rho A \omega^2 r dr = 0$$

Expanding $P(r + dr)$ in a Taylor series about r, and keeping only the first two terms, we obtain

$$\frac{dP}{dr} = \rho \omega^2 r$$

Thus the equation for sedimentation equilibrium of component i can be written

$$\left(\frac{\partial \mu_i}{\partial m_i}\right)_{T,\,P,\,m_{k \neq i}} \frac{dm_i}{dr} = M_i(1 - \bar{v}_i\rho)\omega^2 r - \sum_{\substack{j=1 \\ j \neq i}}^{\nu} \left(\frac{\partial \mu_i}{\partial m_j}\right)_{T,\,P,\,m_{k \neq j}} \frac{dm_j}{dr} \qquad (i = 1, 2, \cdots, \nu) \tag{5-53}$$

In this equation we have separated the concentration dependence of μ_i into a "self" term and a sum of terms representing the influence of the other components.

B. Two-Component Systems

In an ideal, two-component system, the chemical potential of the polymeric component is simply

$$\mu_p = \mu_p{}^0 + RT \ln m_p$$

where $\mu_p{}^0$ is the standard state chemical potential at one molal concentration, R is the gas constant, and T the absolute temperature. Therefore Eq. (5-53) becomes

$$\frac{1}{m_p}\frac{dm_p}{dr} = \frac{M_p(1 - \bar{v}_p\rho)\omega^2 r}{RT} \tag{5-54}$$

It may be that the solution density ρ changes with distance r down the cell, due to hydrostatic compression or, much more importantly, in three-component systems, as discussed below, due to redistribution of added small molecules. Suppose that at position r_0, the density is exactly equal to the reciprocal of the polymer partial specific volume: $\rho_0 = 1/\bar{v}_p$. Thus at r_0 there is no sedimenting force, since $(1 - \bar{v}_p\rho_0)$ is zero. If the density is a linear function of r near r_0, we may expand in a Taylor's series:

$$\rho = \rho_0 + \frac{d\rho}{dr}(r - r_0)$$

Substitution of this into Eq. (5-54) gives

$$-M_p\bar{v}_p \frac{d\rho}{dr}\omega^2 [(r - r_0) + r_0](r - r_0) = \frac{RT}{m_p}\frac{dm_p}{d(r - r_0)}$$

Neglect of $(r - r_0)$ with respect to r_0 in the square brackets in this equation and integration then yields

$$m_p(r) = m_p(r_0) \exp\left[-\frac{(r - r_0)^2}{2\sigma^2}\right] \tag{5-55}$$

where

$$\sigma^2 = \frac{RT}{M_p \bar{v}_p (d\rho/dr)\omega^2 r_0} \tag{5-56}$$

Thus we see that in the presence of a linear density gradient $d\rho/dr$, the polymer is distributed in a gaussian band centered on r_0, with standard deviation σ which is inversely proportional to the square root of the molecular weight M_p. Equation (5-56) suggests that determination of the band width will enable a determination of M_p, since the other quantities are all measurable. Further, since $(d\rho/dr)$ is proportional to ω^2, it would appear that by varying the angular velocity by one order of magnitude, it should be possible to determine molecular weights over four orders of magnitude.

C. Three-Component Systems[142,146–150]

The treatment developed in the previous section assumed that the solvent was a single component, and that solvent and polymer distributed themselves independently. In actual density gradient experiments, the solvent consists of two components (usually water and a heavy salt such as CsCl or Cs_2SO_4, etc.). Furthermore, the macromolecular component will in general interact preferentially with one or the other of the solvent components, thus leading to interdependence of the distribution of macromolecule and solvent at equilibrium. Qualitatively, if the polymer interacts preferentially with the denser of the solvent components, it will band at a buoyant density higher than the reciprocal of its partial specific volume; while if the polymer interacts more strongly with the less dense solvent component, it will band at a density lower than the reciprocal of its partial specific volume.

Nevertheless, it turns out that to very good approximation, the concentration profile of a polymer homogeneous in density and molecular weight follows a gaussian curve given by Eq. (5-55), if σ^2 is defined by the equation

$$\sigma^2 = \frac{RT}{M_{S,0}\bar{v}_{S,0}\left(\dfrac{d\rho}{dr}\right)_{\text{eff}}\omega^2 r_0} \tag{5-57}$$

Here M_S and $\bar{v}_S$ are the molecular weight and partial specific volume of the *solvated* macromolecule, as defined by Eqs. (5-62) and (5-63) below, and the subscript zero refers to quantities measured at r_0. The effective density gradient $(d\rho/dr)_{\text{eff}}$, as discussed below, is the resultant of composition and compression density gradients associated with both solvent and polymer.

In deriving Eq. (5-57) we will first consider the "physical density gradient" arising from the equilibrium distribution of solvent in the absence of polymer. The solvent components are labeled with subscripts 1 and 3, while polymer is component 2. Equation (5-53) can now be written for the salt component, neglecting the influence of the other components on μ_3, as

$$M_3(1 - \bar{v}_3\rho)\omega^2 r\,dr = \left(\frac{\partial \mu_3}{\partial m_3}\right)_P dm_3$$

where it is understood that all processes are isothermal. Neglecting small changes of ρ and μ_3 with pressure, this can be rearranged to give

$$\frac{dm_3}{dr} = \frac{M_3(1 - \bar{v}_3{}^0\rho^0)\omega^2 r}{(\partial\mu_3/\partial m_3)^0}$$

The superscript zero denotes quantities measured at atmospheric pressure.

The "composition density gradient" is defined as

$$\left(\frac{d\rho}{dr}\right)^0 = \left(\frac{d\rho}{dm_3}\right)^0 \frac{dm_3}{dr}$$

$$= M_3(1 - \bar{v}_3{}^0\rho^0)\omega^2 r \frac{(d\rho/dm_3)^0}{(\partial\mu_3/\partial m_3)^0}$$

$$= \frac{M_3(1 - \bar{v}_3{}^0\rho^0)\omega^2 r}{RT} \frac{d\rho}{d\ln a_3} = \frac{\omega^2 r}{\beta^0}$$

where the next to last equality follows from the general relation $\mu_3 = \mu_3{}^0 + RT\ln a_3$. a_3 is the activity of component 3. β^0 has values on the order of 10^9 to 10^{10} cgs units, so that a centrifuge operating at 60,000 rpm with a mean r of 6.5 cm will produce density gradients of 0.1 to 0.01 g/cm^4.

Since the solvent is also somewhat compressible at the high pressures of several hundred atmospheres produced in the ultracentrifuge, there will also be a contribution to the density gradient due to compression. This "compression density gradient" is

$$\left(\frac{d\rho}{dr}\right)_{\text{comp}} = \frac{d\rho}{dP}\frac{dP}{dr} \approx \kappa(\rho^0)^2\omega^2 r$$

where κ is the isothermal compressibility of the solvent. Typically, the compression density gradient is about 10% of the composition density gradient. It should be noted that although pressure increases the magnitude of the density gradient, it does not cause redistribution of salt.

The physical density gradient, $(d\rho/dr)$ is the sum of the composition and compression density gradients:

$$\frac{d\rho}{dr} = \left(\frac{d\rho}{dr}\right)^0 + \left(\frac{d\rho}{dr}\right)_{\text{comp}} = \left(\frac{1}{\beta^0} + \kappa(\rho^0)^2\right)\omega^2 r = \frac{\omega^2 r}{\beta}$$

As promised in the previous section, this equation shows that the density gradient is proportional to ω^2.

We now consider what happens to the polymer (component 2) in the density gradient. Equations (5-53) for solvent component 1 and polymer are, written out explicitly,

$$M_1(1 - \bar{v}_1 \rho)\omega^2 r dr = \left(\frac{\partial \mu_1}{\partial m_1}\right)_{m_2} dm_1 + \left(\frac{\partial \mu_1}{\partial m_2}\right)_{m_1} dm_2$$

$$M_2(1 - \bar{v}_2 \rho)\omega^2 r dr = \left(\frac{\partial \mu_2}{\partial m_2}\right)_{m_1} dm_2 + \left(\frac{\partial \mu_2}{\partial m_1}\right)_{m_2} dm_1 \tag{5-58}$$

It is unnecessary to write an equilibrium equation for m_3, since it is not an independent variable at constant T and P according to the Gibbs-Duhem equation.

It is convenient now to define a solvation parameter

$$\Gamma \equiv -\frac{(\partial \mu_1/\partial m_2)_{m_1}}{(\partial \mu_1/\partial m_1)_{m_2}} = \left(\frac{\partial m_1}{\partial m_2}\right)_{\mu_1} \tag{5-59}$$

where the second equality follows from the triple product rule. It is seen that Γ represents the number of moles of solvent 1 that must be added to a very large volume of solution upon addition of one mole of macromolecule in order to keep the chemical potential of solvent 1 constant. It thus represents, in rough physical terms, the amount of solvent which is "bound" by the macromolecule and which thus must be "replenished" in the bulk solution. If we use the further transformation

$$\left(\frac{\partial \mu_1}{\partial m_2}\right)_{m_1} = \frac{\partial^2 G}{\partial m_2 \partial m_1} = \left(\frac{\partial \mu_2}{\partial m_1}\right)_{m_2}$$

where G is the Gibbs free energy of the solution, Eq. (5-58) can be written

$$M_1(1 - \bar{v}_1 \rho)\omega^2 r dr = \left(\frac{\partial \mu_1}{\partial m_1}\right)_{m_2} dm_1 - \Gamma \left(\frac{\partial \mu_1}{\partial m_1}\right)_{m_2} dm_2$$

$$M_2(1 - \bar{v}_2 \rho)\omega^2 r dr = \left(\frac{\partial \mu_2}{\partial m_2}\right)_{m_1} dm_2 - \Gamma \left(\frac{\partial \mu_1}{\partial m_1}\right)_{m_2} dm_1$$

The buoyant density ρ_0 may now be found by eliminating dm_1 from these equations. We define the solvation parameter on a mass basis, as the number of grams of solvent "bound" per gram of macromolecule:

$$\Gamma' = (M_1/M_2)\Gamma \tag{5-60}$$

The experimental determination of Γ' is discussed in Chapter 7. We then obtain

$$M_2(1+\Gamma')\left[1-\left(\frac{\bar{v}_2+\Gamma'\bar{v}_1}{1+\Gamma'}\right)\rho\right]\omega^2 r dr$$

$$=\left(\frac{\partial\mu_2}{\partial m_2}\right)_{m_1}\left[1-\frac{\left(\frac{\partial\mu_1}{\partial m_2}\right)^2_{m_1}}{\left(\frac{\partial\mu_1}{\partial m_1}\right)_{m_2}\left(\frac{\partial\mu_3}{\partial m_2}\right)_{m_1}}\right]dm_2 \qquad (5\text{-}61)$$

The band center is the place where the polymer concentration is a maximum; that is, $(dm_2/dr)_{r=r_0} = 0$. This gives

$$\frac{1}{\rho_0}=\frac{\bar{v}_2+\Gamma'\bar{v}_1}{1+\Gamma'}=\bar{v}_S \qquad (5\text{-}62)$$

The experimentally determined buoyant density is thus seen to be the reciprocal of the partial specific volume of the solvated macromolecule, $\bar{v}_S$. Similarly, we may define the solvated polymer molecular weight M_S:

$$M_S = M_2(1+\Gamma') \qquad (5\text{-}63)$$

as the molecular weight of dry polymer plus that of associated solvent.

It may be shown, by thermodynamic transformations which need not be detailed here, that the right-hand side of Eq. (5-61) is equal to $(\partial\mu_2/\partial m_2)_{\mu_1}\, dm_2$. Now we make the assumption that the polymer is at such a low concentration that it behaves ideally, so that

$$\mu_2 = \mu_2{}^0 + RT \ln m_2$$

and therefore

$$\left(\frac{\partial\mu_2}{\partial m_2}\right)_{\mu_1}=\frac{RT}{m_2}$$

The partial derivative is taken at constant chemical potential μ_1 rather than at constant m_1 because solvation prohibits m_2 from varying independently of m_1. The assumption of ideal behavior of component 2 is inadequate for accurate molecular weight determinations, as will be discussed later.[151]

With the above substitutions, Eq. (5-61) becomes

$$M_S(1-\bar{v}_S\rho)\omega^2 r dr = \frac{RT}{m_2}\, dm_2$$

We note that M_S and $\bar{v}_S$ as well as ρ, vary with r because solvation changes with position in the density gradient. Expansion of M_S, $\bar{v}_S$, and ρ in Taylor series about $r = r_0$ and integration of the resulting differential equation finally leads to

$$m_2 = m_{2,0} \exp\left[-\frac{(r-r_0)^2}{2\sigma^2}\right] \tag{5-64}$$

where σ^2 is given by Eq. (5-57).

The effective density gradient $(d\rho/dr)_{\text{eff}}$ is

$$\left(\frac{d\rho}{dr}\right)_{\text{eff}} = \left(\frac{d\rho}{dr}\right) + \frac{\rho_0}{\bar{v}_{S,0}}\left(\frac{d\bar{v}_S}{dr}\right)$$

As shown by Hearst and Vinograd,[147] $(d\rho/dr)_{\text{eff}}$ may be written in the more physically clear form

$$\left(\frac{d\rho}{dr}\right)_{\text{eff}} = \left(\frac{d\rho}{dr}\right)^0 - \left(\frac{d\rho_S{}^0}{dr}\right)_P + (\kappa - \kappa_S)\rho_0{}^2\omega^2 r \tag{5-65}$$

where $\rho_S{}^0$ and κ_S are the density at 1 atm pressure and the isothermal compressibility, respectively, of the solvated macromolecule. The effective density gradient is thus seen to be the physical density gradient of the solvent diminished by that of the solvated macromolecule.

The effective density gradient is the sum of three effects.[146] The most important is the redistribution of the salt under the influence of the ultracentrifugal field to generate the composition density gradient. The solvation effect decreases $(d\rho/dr)_{\text{eff}}$. For example, consider polynucleotide at a position $r > r_0$. Since the salt molality increases with r, the water activity, and thus the solvation Γ', must decrease with increasing r. Thus the polynucleotide will be shifted to higher buoyant density, broadening the band and decreasing $(d\rho/dr)_{\text{eff}}$. The compressibility effect increases $(d\rho/dr)_{\text{eff}}$, because $\kappa_{\text{DNA}} < \kappa_{\text{solvent}}$. Thus again considering a position $r > r_0$, the solvent is compressed more than the polymer, raising the density of the former and thereby lowering the buoyant density of the latter. Thus the polynucleotide moves to smaller r, narrowing the band and increasing the effective density gradient.

The complete and accurate determination of all the parameters in Eq. (5-65) for the effective buoyant density gradient is formidable, although it has been undertaken in several cases.[151, 152] However, $(d\rho/dr)_{\text{eff}}$ can be measured directly and simply by an isotopic substitution technique.[86, 142, 146] If the ^{14}N in a polynucleotide is replaced by ^{15}N, a density shift $\Delta\rho$ occurs which is associated with a displacement Δr in the band maximum. For completely substituted *E. coli* DNA, $\Delta\rho = 0.014$ g/cm^3.[154] In the case of other polynucleotides it can be calculated from a knowledge of the base composition. Assuming that isotopic substitution changes only the polynucleotide mass and not its solvation or other thermodynamic properties, the fractional change in buoyant density equals the fractional change in mass. Thus

$$\frac{\Delta\rho}{\rho_0} = \frac{\Delta M_2}{M_2(1 + \Gamma')}$$

where ΔM_2 is the molecular weight change due to isotopic substitution. Then one has

$$\left(\frac{d\rho}{dr}\right)_{\text{eff}} = \frac{\Delta\rho}{\Delta r} = \frac{\rho_0 \Delta M_2}{M_2(1 + \Gamma')\Delta r} \qquad (5\text{-}66)$$

For many applications of density gradient equilibrium sedimentation, it suffices to know the composition density gradient. This depends on

$$\beta^0 = \frac{RT}{M_3(1 - \bar{v}_3{}^0\rho^0)} \frac{d \ln a_3}{d\rho}$$

The parameters in this equation are all accessible from other equilibrium measurements. Ifft et al.[148] have used literature values to get β^0 as a function of ρ for CsCl, RbCl, RbBr, KBr, and sucrose. For CsCl, β^0 varies by only about 1% between $\rho = 1.55$ g/cm^3, and $\rho = 1.80$ g/cm^3, a density range which encompasses the variation in buoyant densities of DNA encountered in CsCl. Thus, with an initially uniform CsCl concentration of initial density 1.70 g/cm^3, the equilibrium density gradient will be nearly linear across the cell, as shown in Fig. 5-17.

In order to know the density at all points within the cell, it suffices to know the density gradient and the density at one point within the cell. It is convenient to choose this latter as the isoconcentration point r_e, at which the equilibrium density ρ_e is equal to that of the initially uniform solution. It was originally assumed that the isoconcentration point could be taken at the midpoint of the cell: $r_e = (r_M + r_B)/2$, where r_B and r_M are the distances of the bottom and meniscus of the solution from the axis of rotation. More accurate estimates have since been made. For a sector-shaped cell, and with the assumption that β^0 is constant throughout the cell, Ifft et al.[148] used the requirement of conservation of mass of solute to show that

$$r_e = [(r_m{}^2 + r_B{}^2)/2]^{1/2}$$

If the buoyant density ρ_0^* is known for some standard nucleic acid, the buoyant density ρ_0 of an unknown sample can be determined by measuring the distance between the standard and unknown bands. It is most accurate to write

$$\rho_0 = \rho_0^* + \frac{\omega^2}{2\beta^0}(r_0{}^2 - r_0^{*2})^{1/2}$$

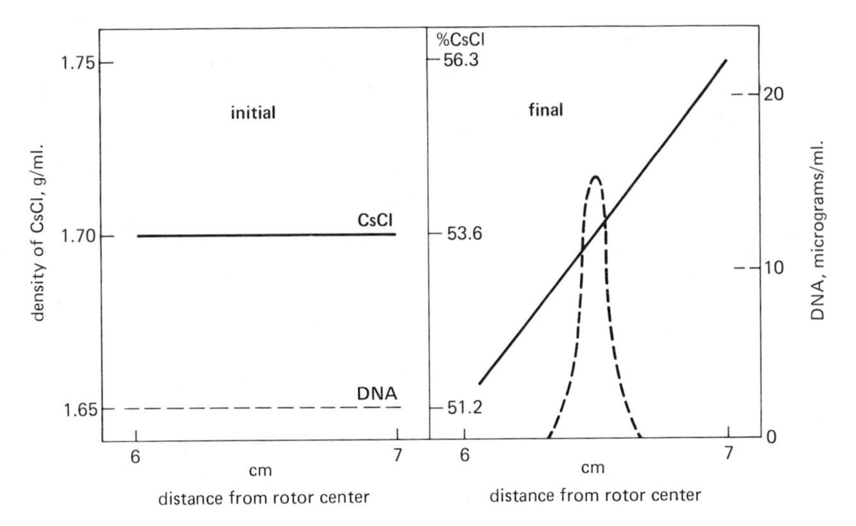

FIGURE 5-17 Concentration distribution of CsCl and DNA at the beginning and end of a density gradient sedimentation equilibrium experiment at about 45,000 rpm. [From J. Vinograd and J. E. Hearst, *Fortsch. Chem. Org. Naturstoffe,* **20**, 372 (1962). Reprinted with permission.]

where r_0^* and r_0 are the band centers of the unknown and standard samples, although the linear approximation

$$\rho_0 = \rho_0^* + (\omega^2/\beta)(r_0 - r_0^*)$$

is also useful.

D. Applications

There have been many clever applications of density gradient equilibrium centrifugation, of which we shall be able to mention only a few.

The first attempts to measure the molecular weights of homogeneous viral DNA from the band width in density gradient equilibrium, Eq. (5-57), led to results that were low by a factor of about two. This has been shown[151] to be due to neglect of virial coefficient corrections for the finite concentration of nucleic acid. In a nonideal solution, the concentration derivative of the chemical potential of component 2 appearing in Eq. (5-61) should be written

$$\left(\frac{\partial \mu_2}{\partial m_2}\right)_{m_1} = \frac{RT}{m_2}\,[1 + 2Bm_2 + 3Cm_2{}^2 + \cdots]$$

B and C are the second and third virial coefficients, respectively. This leads to determination of an apparent molecular weight from (5-57):

$$M_S^{(\mathrm{app})} = \frac{M_S}{1 + 2Bm_2 + 3Cm_2{}^2 + \cdots}$$

To determine the true hydrated molecular weight, M_S, $M_S^{(\mathrm{app})}$ must be extrapolated to zero concentration. Schmid and Hearst[151] found this extrapolation could best be carried out by plotting ln $M_S^{(\mathrm{app})}$ against $\langle m_2 \rangle$, the first moment of concentration distribution of DNA across the cell:

$$\langle m_2 \rangle = \frac{\int m_2{}^2\, d(r - r_0)}{\int m_2\, d(r - r_0)}$$

After M_S is determined in this fashion, M_2 may be found from Γ' (see Chapter 7) according to Eq. (5-63). This is the molecular weight of anhydrous polymer in the salt in which the sedimentation equilibrium was studied. Most commonly, this is a Cs salt, so M_2 is that for CsDNA. To get the molecular weight of another DNA salt, e.g., NaDNA, one multiplies M_2 by the ratio of the mean residue weight of that DNA to CsDNA. For NaDNA, this ratio is 0.75–0.77, depending on the base composition and extent of glucosylation of the DNA.

We next consider applications depending on the buoyant density of the nucleic acids and their reaction products.

In CsCl, the buoyant density of CsDNA is linear in G · C content over the range 20–80% GC,[155–158] as seen in Fig. 5-18(b). Using a value of $\rho = 1.710$ g/cm^3 for *E. coli* DNA, Schildkraut et al.[158] found

$$\rho = 1.660 + 0.098\, X_{\mathrm{GC}} \qquad (5\text{-}67)$$

where X_{GC} is the mole fraction of GC base pairs in the native DNA.

In Cs_2SO_4, the dependence of buoyant density on X_{GC} is nonlinear, and is much less than with CsCl,[159] as shown in Fig. 5-18(a). Thus Cs_2SO_4 density gradients are not well suited for routine determination of DNA base composition. However, since the density gradient in Cs_2SO_4 is roughly twice that in CsCl at the same rotor speed, the former is useful for the comparison or separation of two DNAs of widely different buoyant densities.

The buoyant density has been found to be sequence-dependent in synthetic polynucleotides of defined base sequence. Thus, poly(dA) · poly(dT) has $\rho = 1.638$ g/cm^3 in CsCl,[160] while poly(dAT) · poly(dAT) has $\rho =$ 1.672[160] or 1.679 g/cm^3.[158] The average of these is about 1.66 g/cm^3, in good agreement with Eq. (5-67). In naturally occurring DNAs, these sequence differences are averaged out. Further examples are given in Fig. 5-18 and by Szybalski.[154]

The buoyant density of RNAs in CsCl is quite high, and the solubility of CsCl in water is insufficient to band most RNAs. Double-stranded RNA, and DNA-RNA hybrids, form gaussian bands in Cs_2SO_4. However, most

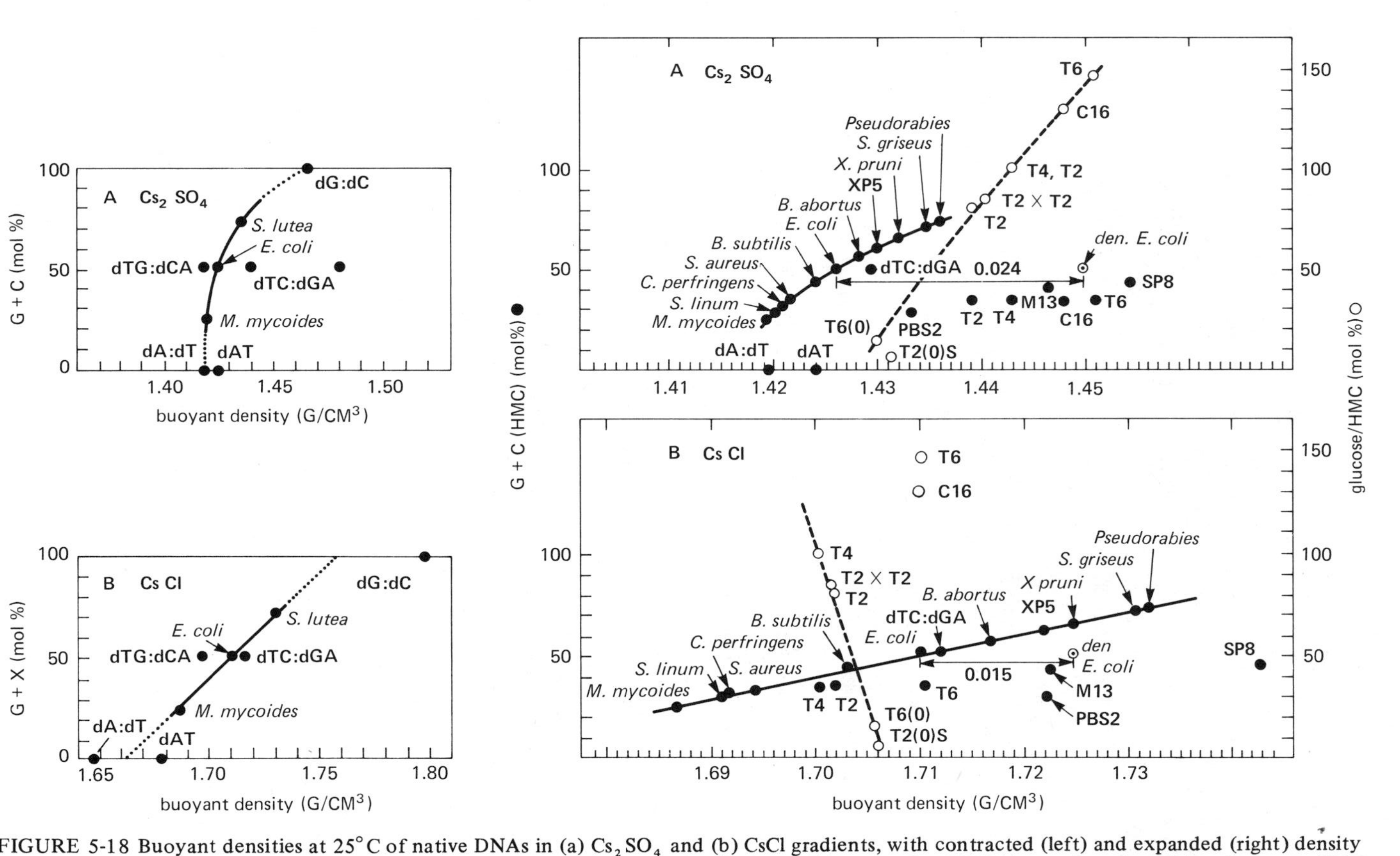

FIGURE 5-18 Buoyant densities at 25°C of native DNAs in (a) Cs_2SO_4 and (b) CsCl gradients, with contracted (left) and expanded (right) density scales, as a function of base composition (●, solid lines), expressed as mole percent of G + C or G + HMC, or as a function of the glucose to HMC ratio (mole %), the latter for T-even phage DNA (○, dashed lines). [From W. Szybalski in *Methods in Enzymology,* Vol. XXIB, L. Grossman and K. Moldave, eds., Academic Press, New York (1968), p. 330. Reprinted with permission.]

single-stranded RNAs precipitate in Cs_2SO_4. Density gradient studies on RNA in Cs_2SO_4 are surveyed by Erickson and Franklin[161] and Szybalski.[144] Several mixed-solvent systems have been found by Lozeron and Szybalski[162] that are adequate for all RNAs. However, for exact thermodynamic analysis, these solvents must be considered as four- or five-component systems.

As may be seen in Fig. 5-18, the buoyant density is a function of the extent of glucosylation of the hydroxymethylcytosine residues in DNA from T-even bacteriophages. As the percent glucosylation rises, the buoyant density of the DNA increases in Cs_2SO_4, and decreases in CsCl. This behavior is related to the buoyant density ρ_G of glucose in the two salts. In Cs_2SO_4, ρ_G is about 1.5 g/cm^3, which is greater than that of DNA, while in CsCl, ρ_G is about 1.6 g/cm^3, which is less than that of the DNA.

Alkylation decreases the buoyant density of DNA, because of the low density of the alkyl moiety. This can be seen by comparing, in CsCl, a buoyant density of 1.722 g/cm^3 for *B. subtilis* phage PBS2 DNA, which contains uracil instead of thymine (5-methyluracil) with the density of 1.688 g/cm^3 predicted for $X_{G \cdot C} = 0.28$ by Eq. (5-67).[144] Alkylation of DNA with nitrogen mustard (*N*-methyl-*bis*(2-chloroethyl) amine · HCl) also causes a large reduction in ρ.[163]

As discussed in more detail in Chapter 7, silver and mercury ions bind specifically to the heterocyclic bases, rather than to the phosphate groups, of DNA.[164,166] The initial strong binding has a limiting ratio of 1 metal ion/base pair, suggesting chelation between the bases. Hg^{2+} binds preferentially to A · T-rich DNA, and Ag^+ preferentially to G · C-rich DNA. In Cs_2SO_4, there is a large buoyant density increase attendant upon the ion binding, about 0.15 g/cm^3 for 1 Ag^+/base pair.[166] This has enabled the separation of the dAT-like fraction of crab DNA from the bulk of the crab DNA.[164] Mercury complexes have been used to separate A · T- and G · C-rich regions of lambda phage DNA, prepared by shear fragmentation.[165,167,168] Denatured DNA binds Ag^+ and Hg^{2+} much more strongly than duplex DNA. Binding in a density gradient can thus be used to separate native and denatured material.[169]

The two strands of many duplex DNAs have different affinities for G-rich synthetic polynucleotides, such as poly G or poly I,G.[170,171] The denatured and separated single strands, after complexing with G-rich polymers, will have different densities. That which complexes most extensively will have the higher ρ. This has been used as the basis for a method of preparative fractionation of the two strands of phage lambda[172] and other DNAs. Preferential binding of G-rich polymers has been taken to indicate the presence of dC-rich clusters.[173] There is evidence that transcription of genetic messages starts at pyrimidine clusters.

The initial use of density gradient equilibrium centrifugation was by

Meselson and Stahl[154] in demonstrating the semiconservative mode of replication of bacterial DNA. *Escherichia coli* bacteria were grown for 14 generations with $^{15}NH_4Cl$ as the only nitrogen source until the DNA contained essentially only ^{15}N, and was thus heavier than normal DNA. $^{14}NH_4Cl$ was then suddenly added to the medium, and the culture allowed to continue to grow. Examination of the buoyant density of the DNA one generation after this addition showed one peak, at a buoyant density midway between the ^{15}N and ^{14}N peaks. After two generations, two peaks of equal area were found, one corresponding to the buoyant density of the material observed in the preceding generation, the other corresponding to pure ^{14}N DNA. These results, and others like them, are consistent with the Watson-Crick scheme of semiconservative replication. At first both parental DNA strands contain ^{15}N. After one generation, each new duplex molecule has one ^{15}N parental strand and one ^{14}N daughter strand, and thus an intermediate density. After two generations, half of the duplex molecules have one ^{14}N and one ^{15}N strand each, while the other half have both parental and daughter strands containing only ^{14}N. Thus two bands of equal area are formed.

Similar work has since been done on the mode of duplication of DNA from higher organisms, of transcription, and of replication of duplex RNA. Other high-density labels, such as ^{2}H and ^{13}C, and the base analogs 5-bromouracil, 5-iodouracil, and 5-fluorouracil, have been used.

DNA denaturation and renaturation can be followed by changes in buoyant density. This is discussed in Chapter 6.

VI TRANSPORT PHENOMENA

A. Sedimentation

Sedimentation measures the translational motion of a polymer under the influence of a centrifugal force. The sedimentation coefficient, s, equals the centrifugal velocity of the polymer produced by unit centrifugal acceleration. Measurement of the sedimentation coefficient can give information on the number of segments in the polymer chain and on their average distance from one another. Thus information may be obtained about molecular weight and dimensional statistics of polynucleotides.

The discussion of viscous flow of suspensions, or macromolecular solutions, most frequently seen is based on the Navier-Stokes equations, which are treated in many standard texts.[174] An alternative treatment, more convenient in many ways for polymer applications, was developed by Oseen[175] and adapted for macromolecular solutions by Burgers.[176] This treatment considers the motion of the solvent produced by the application of forces to points in the fluid. These points may be identified with the positions of the polymer segments, and the force is due to the relative motion of solvent and segment. Thus the Oseen-Burgers formulation is obviously appropriate and convenient for the calculation of the hydrodynamic behavior of solutions of chain macromolecules.

One may write that the force $\mathbf{F}_i$ on the solvent, which is flowing with velocity $\mathbf{v}_i$ at the position of a segment moving with velocity $\mathbf{u}_i$, is

$$\mathbf{F}_i = -\zeta(\mathbf{v}_i - \mathbf{u}_i)$$

ζ is the translational frictional coefficient of a segment. For a sphere of diameter b in a solvent of viscosity η_0, this is $3\pi\eta_0 b$ according to Stokes' law.

The velocity $\mathbf{v}_i$ differs from the bulk velocity of the solvent, $\mathbf{v}^0$, because of the perturbation of solvent flow by the motion of the other segments in the chain. This effect is called "hydrodynamic interaction." Thus we may write the velocity at the position of the ith segment as a sum of $\mathbf{v}^0$ and the perturbations due to all the other segments:

$$\mathbf{v}_i = \mathbf{v}^0 + \sum_{j \neq i = 1}^{N} \mathbf{T}_{ij} \cdot \mathbf{F}_j$$

The hydrodynamic interaction tensor $\mathbf{T}_{ij}$ is an operator which converts the force exerted on the solvent by the jth segment of the polymer chain, into a velocity perturbation at the position of the ith segment. It is a function of the vector distances $\mathbf{r}_{ij}$ between pairs of chain segments. It is in this way, through hydrodynamic interaction, that the polymer dimensions are brought into the problem.

A further factor which must be taken into account is Brownian motion. If the shear rates in the system are sufficiently low with respect to the rotational diffusion coefficient of the polymer, $\mathbf{T}_{ij}$ may be averaged with equal probability over all orientations of the molecule with respect to a fixed external coordinate system. In addition, for flexible polymers, it is necessary to average over all internal configurations of the chain. The result is[177]

$$\langle T_{ij} \rangle = (1/6\pi\eta_0)\langle L_{ij}{}^{-1} \rangle$$

where the angular brackets denote averaging over internal configurations.

The concepts outlined above were integrated by Kirkwood[178-180] into a general theory of irreversible processes in solutions of macromolecules composed of identical subunits. Application to the calculation of the translational frictional coefficient f proceeds as follows. The total force $\mathbf{F}$ exerted by the macromolecule on the solvent is the sum of the forces exerted by the subunits:

$$\mathbf{F} = \sum_{i=1}^{N} \mathbf{F}_i$$

It is also equal to the product of f and the relative velocity of the macromolecular center of mass with respect to bulk solvent:

$$\mathbf{F} = -f(\mathbf{v}^0 - \mathbf{u})$$

This equation defines f.

We may solve these equations with $\mathbf{u}_i$ set equal to $\mathbf{u}$ for all i since, on the average, each segment moves with the velocity of the polymer center of mass. This yields

$$f = \frac{N\zeta}{1 + \dfrac{\zeta}{6\pi\eta_0 N} \sum\limits_{i=1}^{N} \sum\limits_{\substack{j=1 \\ i \neq j}}^{N} \langle L_{ij}^{-1} \rangle} \tag{5-68}$$

N, as before, is the number of segments in the chain. If the double summation were missing from the denominator of Eq. (5-68), the frictional coefficient of the polymer would be the sum of the monomeric frictional coefficients. This is the so-called "free-draining" limit. The presence of the double summation reflects hydrodynamic interaction, which lowers f because solvent is immobilized to a certain extent within the polymer domain and moves with the polymer through solution.

There are several approximations in the theory outlined above which have drawn attention. The first is the treatment of the solvent as a hydrodynamic continuum. However, frictional coefficients calculated according to Eq. (5-68) are surprisingly accurate even for short-chain hydrocarbons, except when the solvent molecules are substantially larger than the polymer segments.[181,182] Thus the approximation that the solvent behaves as a hydrodynamic continuum should be entirely adequate for aqueous solutions of high-molecular-weight polynucleotides.

Another approximation is the premature averaging of the Oseen interaction tensor employed in the above treatment. That is, the product $\mathbf{T}_{ij} \cdot \mathbf{F}_j$ should be averaged, not $\mathbf{T}_{ij}$ and $\mathbf{F}_j$ separately. However, a recalculation of the frictional coefficients of flexible chains, by methods which avoid this premature averaging,[183] leads to an increase in f of less than 2%.

A third approximation is the representation of monomers as point sources of friction. For a rigid, rodlike polymer, it has been shown[184] that this approximation can lead to singularities and negative values of the translational frictional coefficient for physically plausible values of the hydrodynamic interaction. However, this difficulty does not seem to afflict any of the calculations which are surveyed here.

Finally, it has been noticed that Kirkwood's derivation of Eq. (5-68) is not exact.[185,186] Among other things, this has the disturbing consequence that the frictional coefficients of a long rigid rod and a rigid ring of the same length are calculated to be equal (in the limit of very long rods). Physical intuition suggests, on the other hand, that the rod would have a larger f than the ring, because of the less compact shape of the rod. Even more disturbing is the fact that when the calculations are carried out exactly, using Oseen's theory but not Kirkwood's approximations, it is found that f(ring) is greater than f(rod).[186,187] This paradox may be related to the inadequacies in the Oseen approximation noted in the preceding paragraph. It should be noted that these results are derived for rigid rods and rings, and may have no bearing on the results reported later for (more or less flexible) linear and circular DNA.

Despite these difficulties, it appears clear that the theory outlined above is substantially correct when applied to rigid rod and random coil polymers, and that its predictions can be applied with confidence to the hydrodynamic properties of nucleic acids.

We may now apply the above theory to several different macromolecular models which will be of use in interpreting nucleic acid properties. We first consider a rigid rod of length $\mathcal{L}$ and diameter b. This can be modeled by N touching spheres of diameter b, such that $\mathcal{L} = Nb$. Using Eq. (5-5) for $1/L_{ij}$ for rods, and replacing the summations in Eq. (5-68) by integrations, we obtain[180]

$$f = \frac{N\zeta}{1 + (\zeta/3\pi\eta_0 b)(\ln N - 1)}$$

If the rod were free-draining, one would have $f = N$. On the other hand, if Stokes' law is used for ζ, one finds that $f \approx N\zeta/\ln N$. Other treatments[188] of the translational frictional coefficients of rigid rods all give results in the form

$$f = \frac{N\zeta}{\ln N - \gamma}$$

where γ varies according to the details of the calculation, but generally lies between 0 and 0.6.

For random coils, in the absence of excluded volume effects, one obtains[180] using Eq. (5-11) for $\langle L_{ij}^{-1} \rangle$ in Eq. (5-68)

$$f = N\zeta / \left[1 + \frac{8}{3} \zeta N^{1/2} / (\sqrt{6\pi^3}\, \eta_0 b) \right]$$

In the nondraining limit, with $\zeta = 3\pi\eta_0 b$ this gives

$$f = \frac{3\sqrt{\pi}}{8} 6\pi\eta_0 \left(\frac{b^2 N}{6} \right)^{1/2} = 0.665 \times 6\pi\eta_0 \langle R^2 \rangle^{1/2}$$

Thus a nondraining random coil behaves like a sphere with an effective radius $0.665\langle R^2 \rangle^{1/2}$. For flexible chains with excluded volume, a similar treatment, using Eq. (5-33) leads to[189]

$$f = K_1(\epsilon) b N^{(1+\epsilon)/2} = K_2(\epsilon) 6\pi\eta_0 \langle R^2 \rangle^{1/2} \tag{5-69}$$

where $K_1(\epsilon)$ and $K_2(\epsilon)$ are functions only of ϵ, in the nondraining limit, and vary rather slowly with ϵ.

Hearst and Stockmayer[190] used Eq. (5-68) to calculate the frictional coefficients of wormlike chains without excluded volume. This calculation has been extended[191] to wormlike chains with excluded volume and other long-range interaction effects, so that the theory of the translational motion of native, high-molecular weight DNA is now understood fairly completely. The resulting equation for s will be presented below.

For nucleic acids, the most important method of determining frictional coefficients is through measurement of the sedimentation coefficient, s. A simple derivation of the relation between f and s in two-component systems is as follows. Under conditions of steady flow of macromolecule from the meniscus toward the bottom of the cell, the centrifugal force must just balance the frictional force. That is

$$\omega^2 r(1 - \bar{v}_2\rho)M_2/N_A = fu$$

since there is no bulk movement of solvent. Then we define the sedimentation coefficient as the velocity produced by unit centrifugal acceleration:

$$s \equiv \frac{u}{\omega^2 r} = \frac{M_2(1 - \bar{v}_2\rho)}{N_A f} \tag{5-70}$$

Thus s has units of seconds, and typically has a magnitude of 10^{-13} seconds. In practice, one denotes a unit of 10^{-13} seconds as one Svedberg (S), in honor of the developer of the ultracentrifuge.

The above derivation implies that all the macromolecules move with the same velocity, u, so that an initially narrow band or sharp boundary would remain narrow. Actually, a certain amount of band spreading will occur due to diffusion. Also, a certain amount of skewing of the band will generally be observed, since s depends on the polymer concentration, c, as discussed below. These details are treated in a number of standard references.[150,192,193] However, in most applications of sedimentation velocity to nucleic acids, it suffices to follow the position of the band maximum with time.

Equation (5-70) holds in three-component systems, if component 2 again refers to the solvated macromolecule. In infinitely dilute solution (denoted by superscript zero), one then has[194-196]

$$s^0 = \frac{M_s(1 - \bar{v}_s\rho)}{N_A \eta_r f}$$

where M_s and $\bar{v}_s$ are defined in Eqs. (5-62) and (5-63). η_r is the relative viscosity of the solvent ($= \eta_{\text{solvent}}/\eta_{H_2O}$).

It is common to correct the measured sedimentation coefficient to standard conditions of 20°C and a solution having the viscosity and density of pure water:[192]

$$s_{20,w} = s\left(\frac{\eta_{H_2O,\,T}}{\eta_{H_2O,\,20°}}\right)\eta_r(T)\,\frac{(1-\bar{v}\rho)_{20,w}}{(1-\bar{v}\rho)_{T,\text{ solution}}}$$

These correction factors have been tabulated for a number of useful solvent systems by Studier.[197]

Bruner and Vinograd[192] have shown that for both T7 DNA and MS2 RNA, plots of $s^0\eta_r$ vs. ρ in fairly concentrated NaCl or CsCl solutions, ranging in concentration from 1 M to the buoyant density of the nucleic acids, are linear. These plots extrapolate at $\rho = 1$ to $s^0_{20,w}$ obtained as in the above equation by independent methods, and they also give the correct buoyant density when extrapolated to $s = 0$ (the buoyant condition).

Several experimental difficulties have had to be overcome before sedimentation velocity could be used reliably and routinely for the study of nucleic acids in solution. A major problem has been the severe concentration dependence of s of high molecular weight DNA, arising from its inordinately large molecular volume. In order to measure s^0, characteristic of a single isolated molecule, it is necessary to measure s at a series of finite concentrations and then extrapolate to infinite dilution. A linear extrapolation may be made according to the equation

$$1/s = (1/s^0)(1 + Kc)$$

It has been shown[198-200] that in a wide variety of polymer systems, K is proportional to the intrinsic viscosity $[\eta]$ (which must be expressed in the same units as c):

$$K = k[\eta]$$

For typical random coil polymers, k is about 1.6,[199] and is substantially less for more rigid, asymmetric molecules. For native DNA of molecular weight 0.3×10^6 to 16×10^6, obtained from many sources, $k = 0.80 \pm 0.10$ (standard deviation).[201,202] Studies on bacteriophage DNA and fragments with molecular weights up to 130×10^6 gave a slightly higher value,[203-207] which according to Aten and Cohen[206] is 0.97 ± 0.15.

These results indicate, for example, that the sedimentation coefficient of T2 phage DNA, with intrinsic viscosity $[\eta] = 326$ dl/g,[208] will be 25% lower measured at $c = 10\ \mu$g/ml, than its value at infinite dilution. It is therefore obvious that, in order to extrapolate reliably to infinite dilution, methods of measuring s of DNA at very low c must be used.

This has been accomplished by reviving the use of absorption optics as employed by Svedberg in his first studies with the ultracentrifuge.[192] Taking advantage of the strong nucleotide absorption around 260 nm, Shooter and Butler[209] and Schumaker and Schachman[210] incorporated ultraviolet absorption optics into the analytical ultracentrifuge. This lowered the minimal feasible concentration from 0.1% obtainable with schlieren optics, to 0.001%. For additional convenience, photoelectric scanning has been developed to replace the usual photographic-densitometric recording of the cell image. Details of these experimental advances are given by Schachman and Edelstein.[211]

A complication which has appeared in the studies of a number of workers is an apparent dependence of the sedimentation coefficient and boundary shape on the angular velocity of the rotor at high speeds.[204,206,212] This phenomenon becomes more pronounced at higher c and with higher molecular weight DNA. The rotor speed dependence of s is unlikely, on theoretical grounds, to be produced by hydrodynamic alignment of asymmetric molecules.[213,214] It may be due to convection,[206] to a loss of part of the DNA which pellets to the bottom of the rotor due to a speed-dependent reversible aggregation,[204] or to a combination of these. Whatever the cause, this phenomenon may be avoided simply by working at rotor speeds below 15,000 rpm. The boundary instability encountered at these low speeds by Eigner et al.[202] has not been noticed by other workers.

In order to avoid the speed dependence and strong concentration dependence of s, Burgi and Hershey[215] adapted the technique of zone centrifugation[216] to DNA solutions. In this technique a very low concentration, between 0.1 and 1.0 μg/ml, of radioactively labeled DNA is layered on a preformed sucrose density gradient in a preparative ultracentrifuge tube. The gradient is calibrated with a DNA of a known molecular weight and sedimentation coefficient. The DNA sediments in a zone or band, and the distance D traveled by the center of the band after a given duration of centrifugation are determined by assay of the radioactivity of drops collected after puncturing the bottom of the tube with a pin. It was found that the ratio of the distances traveled in a 5%–20% sucrose gradient of two DNAs was related to the ratios of their molecular weights and sedimentation coefficients by

$$\left(\frac{D_1}{D_2}\right) = \left(\frac{S_1}{S_2}\right) = \left(\frac{M_1}{M_2}\right)^{0.35}$$

A useful adaptation of the zone centrifugation idea to the analytical ultracentrifuge is the band centrifugation technique of Vinograd et al.[217] This involves layering a thin lamella of a solution of macromolecules onto a denser liquid in a rotating ultracentrifuge cell. The method differs from zone centrifugation in that the density gradient which stabilizes the system against convection is not formed in the bulk solution prior to application of the sample, but is generated during the experiment by the diffusion of small molecules due to the difference of their chemical potential in the sample solution and in the sedimentation solvent, and to the influence of the centrifugal field. Furthermore, large density and viscosity gradients are generally avoided. The sedimentation coefficient can usually be determined with sufficient accuracy by following the motion of the band maximum or center. Experimental and theoretical refinements of this technique are described in a number of papers.[218-225] Studier[197] has compiled a list of solvents which are useful in band centrifugation studies.

It is evident from Eq. (5-13) that before M or f can be evaluated from sedimentation studies, the partial specific volume $\bar{v}_2{}^0$ and solution density ρ^0 must be known. These quantities have been carefully determined for the Na and Cs salts of DNA by Cohen and Eisenberg.[86] Results for various concentrations of NaCl and CsCl, at infinite dilution of polymer, are given in Table 5-1. Hearst[226] has tabulated the specific volumes of the Cs, Rb, Na, Li, K, and NH_4 salts of DNA, as estimated using the crystal volumes for the salts. The most commonly used result is $\bar{v}_{\mathrm{NaDNA}} = 0.55_6$ ml/g. For tRNA from yeast, $\bar{v}$ has been measured as 0.531.[227]

We now turn to a consideration of some of the results obtained from sedimentation velocity studies of nucleic acids. A log-log plot of s vs. M is shown in Fig. 5-19 for native duplex DNA from many sources. Crothers and Zimm[205] fit these data to an equation of the form

$$s^0_{20,w} - b_s = K_s M^{a_s} \tag{5-71}$$

In $0.2\,M$ neutral buffer, $b_s = 2.7$, $a_s = 0.445$, and $\log K_s = -1.819$. These values are fairly sensitive to the molecular weights of the DNA samples, which must be measured independently.

Combining Eqs. (5-69) and (5-70), we see that for flexible chains,

$$s^0 = \text{const.} \times M^{(1-\epsilon)/2} = K_s M^{a_s} \tag{5-72}$$

where the constant depends upon both ϵ and segment length. Comparing this equation with Eq. (5-71), we see that the deviation of a_s from 0.500 reflects excluded volume or other long-range interaction effects, while K_s would be expected to give information about the statistical segment length in native DNA. The constant term b_s, which was introduced to "straighten out" log–log plots of s vs. M, was attributed to chain stiffness.[205] It will be noted that b_s becomes numerically important only for relatively low molecular weight DNA. At higher molecular weights, DNA behaves hydrodynamically like an expanded random coil.

Eigner and Doty,[207] in an earlier, widely used correlation of sedimentation coefficient and molecular weight of native DNA, used

$$S^0_{20,w} = 0.116 M^{0.325}$$

for M between 0.3×10^6 and 3×10^6, and

$$S^0_{20,w} = 0.034 M^{0.405}$$

for M between 3×10^6 and 130×10^6.

TABLE 5-1 PARTIAL SPECIFIC VOLUMES AND DENSITIES OF NaDNA IN NaCl AND CsDNA IN CsCl AT 25° C[a]

[NaCl] (M)	ρ^0 (g/ml)	$\bar{v}_2{}^0$ (ml/g)	[CsCl] (M)	ρ^0	$\bar{v}_2{}^0$
0	0.997	0.499	0	0.997	0.440
0.200	1.005	0.503	0.200	1.022	0.446
0.976	1.036	0.528	0.970	1.120	0.460
3.160	1.118	0.539	1.860	1.232	0.471
4.190	1.155	0.543	2.325	1.290	0.467
			3.440	1.428	0.471

[a]From G. Cohen and H. Eisenberg, *Biopolymers,* **6**, 1077 (1968).

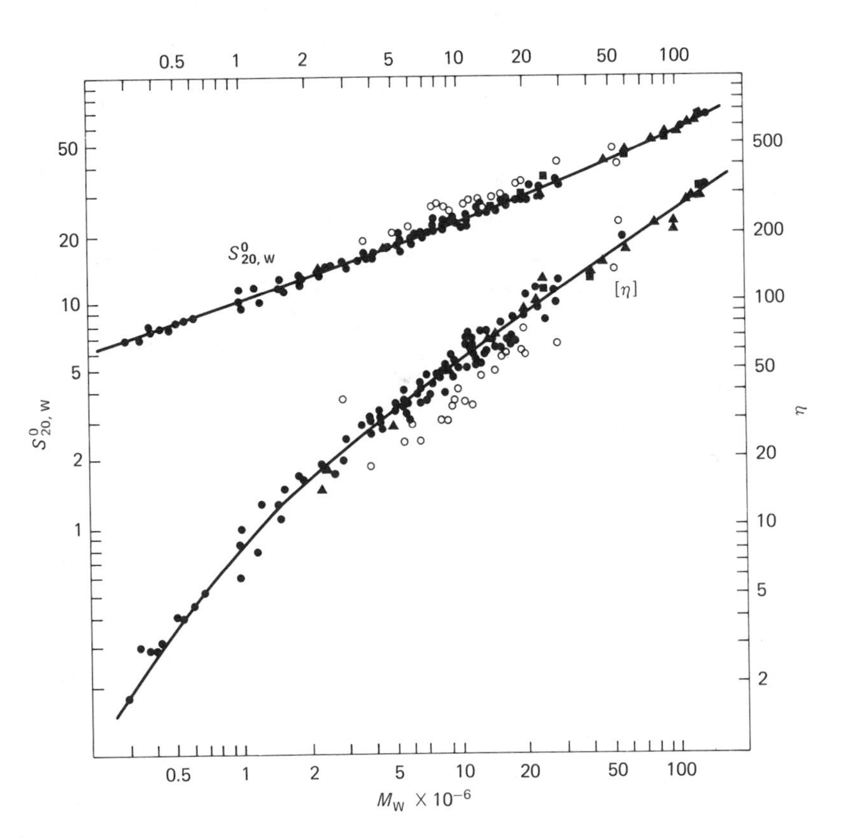

FIGURE 5-19 Sedimentation and viscosity data for native DNA as a function of molecular weight computed from S° and [η] data by the Scheraga-Mandelkern equation. [From J. Eigner and P. Doty, *J. Mol. Biol.,* **12**, 549 (1965). Reprinted with permission.]

The effect of excluded volume and chain stiffness on the sedimentation coefficients of polymers has been extensively studied.[228] Hearst and Stockmayer[190] calculated s for wormlike chains without excluded volume, using Eqs. (5-68) and (5-70) with the dimensional statistics discussed in Section I, E of this chapter. Gray et al.[191] generalized this treatment to wormlike chains with excluded volume effects. The result, using the modified gaussian distribution, Eq. (5-34), for large contour lengths, is

$$s^0 = \frac{9.178 \times 10^{-3} M_L{}^{(1+\epsilon)/2} (M/2a)^{(1-\epsilon)/2} (1 - \bar{v}\rho)}{(1-\epsilon)(3-\epsilon)3\pi\eta_0} + \frac{1.660 \times 10^{-3} M_L (1 - \bar{v}\rho)}{3\pi\eta_0}\left[1 - \ln\frac{b}{2a} + \ln\alpha - \left(\frac{2}{1-\epsilon}\right) \sqrt{6/\pi}\,\alpha^{(1-\epsilon)/2} + \frac{1}{3}\alpha + \frac{1}{2}c\alpha^2 + \frac{1}{3}d\alpha^3\right] \qquad (5\text{-}73)$$

where α, c, and d are tabulated functions of ϵ.

Equation (5-73) is of the same form as Eq. (5-71), and allows the determination of the unknown structural parameters ϵ, a, and b in terms of the experimentally determined constants a_s, K_s, and b_s. The results are discussed at the end of this chapter.

Reliable correlations of s vs. M for single-stranded polynucleotides have been difficult to achieve, because of a pronounced tendency for inter- and

TABLE 5-2 PARAMETERS IN $s^0_{20,w} = K_s M^{a_s}$ FOR SINGLE-STRANDED POLYNUCLEOTIDES

Solvent and conditions	a_s	K_s	*Ref.*[a]
DNA			
Phosphate buffer, $[Na^+]$ = 0.013 M, pH 6.8, 25°C			
39% GC	0.34	0.064	207
42% GC	0.35	0.060	207
50% GC	0.36	0.056	207
66% GC	0.41	0.034	207
Citrate buffer, $[Na^+]$ = 0.195 M, pH 7, 25°C	0.48	0.022	207
0.9 M NaCl – 0.1 M NaOH, 20°–25°	0.400	0.0528	197
1 M NaCl, 0.01 M Tris, pH 8, 20°–25°	0.549	0.0105	197
RNA			
1.1 M formaldehyde, 0.09 M Na_2HPO_4 + 0.01 M NaH_2PO_4, pH 8.5, 20°–25°	0.40	0.05	229
99% dimethylsulfoxide, 10^{-3} M EDTA, pH 7.1	0.31	0.052[b]	230

[a] 207 J. Eigner and P. Doty, *J. Mol. Biol.*, **12**, 549 (1965).
197 F. W. Studier, *J. Mol. Biol.*, **11**, 373 (1965).
229 H. Boedtker, *J. Mol. Biol.*, **35**, 61 (1968).
230 J. H. Strauss, Jr., R. B. Kelly, and R. L. Sinsheimer, *Biopolymers*, **6**, 793 (1968).

[b] Measured in DMSO at 25°C, and not corrected to 20,w.

intramolecular aggregation. This tendency has been overcome in various ways: by working at low ionic strength[207]; by denaturing and rapid quenching of dilute solutions at higher ionic strengths[207]; by working in alkaline solutions[197]; and, in the case of RNA, by reacting the RNA with formaldehyde[229] or denaturing it with dimethyl sulfoxide.[230]

The parameters a_s and K_s in Eq. (5-72) are given for single-stranded polynucleotides, treated in the ways listed above, in Table 5-2. It is to be noted that in low ionic strength, neutral buffer, a_s and K_s depend on the G · C content. As the percent G · C increases, a_s increases, and thus ϵ decreases, according to Eq. (5-72). This indicates greater base–base interactions at higher percent G · C. By properly adjusting the salt concentration and temperature, it is possible to find theta-solvent conditions for poly A, in which $a_s = 0.5$ and $\epsilon = 0$.[231] At high salt concentrations, polyelectrolyte repulsions may be so damped out, and base–base interactions predominant over physical excluded volume effects to such an extent, that ϵ is actually negative.[197]

The use of a_s and ϵ to determine statistical segment lengths of single-stranded polynucleotides will be discussed at the end of this chapter.

B. Viscosity and Shearing Phenomena

1. VISCOSITY

Viscometry has long been one of the most important techniques for characterizing polymers. It has maintained much of this prominence in the study of nucleic acids.

We first consider some basic definitions. Consider two parallel planes, each of area A; a distance h apart in the z-direction, with the fluid under study between them. The upper plane moves with velocity V in the x-direction with respect to the lower, and a linear velocity gradient is set up in the fluid:

$$v(z) = \frac{zV}{h} = z\,\frac{dv}{dz}$$

Then the force F needed to move the upper plane with respect to the lower is

$$F = \eta A\,\frac{dv}{dz}$$

This can be written as

$$P = \eta\dot{\epsilon}$$

where $P = F/A$ is the shear stress, and $\dot{\epsilon} = dv/dz$ is the shear rate. η is the viscosity of the fluid, which represents the stress necessary to establish unit shear rate. The units of η are dyne-sec/cm^2, or poise. Since water at 20°C has

$\eta = 0.01002$ poise, a commonly used unit of viscosity is the centipoise, or 10^{-2} poise.

An alternative definition of viscosity is instructive. Multiplying both sides of the above equation by $\dot{\epsilon}$, one has

$$P\dot{\epsilon} = \eta\dot{\epsilon}^2$$

The left-hand side of this equation has units of dyne/cm^2-sec, or ergs/cm^3-sec. Thus, η equals the energy dissipation per unit volume per unit time in a fluid deformed at unit rate of shear.

The viscosity of pure solvent is denoted η_0. Addition of macromolecules raises the viscosity to a new value, η. This occurs because the large polymers, extending across the stream lines, greatly enhance resistance to flow. The fractional increase in viscosity is called the specific viscosity, η_{sp}:

$$\eta_{sp} = \frac{\eta - \eta_0}{\eta_0} = \eta_{\text{rel}} - 1$$

where the relative viscosity is

$$\eta_{\text{rel}} = \frac{\eta}{\eta_0}$$

In the limit of low concentration, c, η_{sp} is proportional to c. Thus, one defines the intrinsic viscosity $[\eta]$ as

$$[\eta] = \lim_{c \to 0} \frac{\eta - \eta_0}{\eta_0 c} \tag{5-74}$$

An equivalent definition is

$$[\eta] = \lim_{c \to 0} \frac{\ln \eta_{\text{rel}}}{c} \tag{5-75}$$

It is evident from Eq. (5-74) that $[\eta]$ is proportional to the fractional increment in viscosity of the solution due to the addition of a single macromolecule. Measurement of $[\eta]$ may thus be expected to give information about the properties of individual polymer molecules.

The fundamental equation for $[\eta]$ in terms of molecular structure and solution forces can be obtained as follows.[180] $\eta_0\dot{\epsilon}^2$ is the energy dissipation in pure solvent, while $\eta\dot{\epsilon}^2$ is the energy dissipation in the polymer solution. Thus, $(\eta - \eta_0)\dot{\epsilon}^2$ is the energy dissipation due to dissolved polymer. But this arises from the frictional interactions between the monomeric elements of the polymer and the solvent. If $\mathbf{F}_l$ is the force exerted on the solvent by the lth monomer, and the velocity of the solvent at the position of the lth monomer is $\mathbf{u}_l^0$, the energy dissipation per second due to the lth monomer is $-\mathbf{F}_l \cdot \mathbf{u}_l^0$. This must be summed over all N monomers in the chain, averaged over all configurations of the polymer, and multiplied by the number of polymer molecules per cubic centimeter of solution, to get the energy dissipation per cubic centimeter per second.

$$(\eta - \eta_0)\dot{\epsilon}^2 = -\left(\frac{N_A c}{M}\right) \sum_{l=1}^{N} \langle \mathbf{F}_l \cdot \mathbf{u}_l{}^0 \rangle$$

With a velocity gradient in the z-direction, $\mathbf{u}_l{}^0 = \dot{\epsilon} z_l \hat{\mathbf{e}}_x$, where z_l is the z-coordinate of the lth monomer, measured relative to the center of mass (or of frictional resistance) of the polymer chain, and $\hat{\mathbf{e}}_x$ is a unit vector in the x direction. Also, since the flow is in the x-direction, $\mathbf{F}_l = \mathbf{F}_{lx} \hat{\mathbf{e}}_x$. Thus one obtains[232]

$$[\eta] = -\left(\frac{N_A}{\eta_0 M \dot{\epsilon}}\right) \sum_{l=1}^{N} \langle \mathbf{F}_{lx} z_l \rangle \tag{5-76}$$

In this equation, $[\eta]$ is in cubic centimeters per gram. It is also common to measure c in weight percent, that is, in grams per 100 ml of solution. In this case, the units of $[\eta]$ are deciliters per gram, and the rhs of Eq. (5-76) must be divided by 100.

This equation has been applied to rigid rods as follows. In the absence of hydrodynamic interaction, F_{lx} would simply be $-\zeta \dot{\epsilon} z_l$. Further, $z_l = (N/2 - l)\, b \cos\theta$ for a rod of N spheres of diameter b whose origin is at the middle of the rod, and whose length is $\mathcal{L} = Nb$. The angle that the long axis of the rod makes with the z-axis is θ. Thus, Eq. (5-76) becomes, after evaluation of the summation and replacement of ζ by $3\pi\eta_0 b$,

$$[\eta] = \frac{\pi N_A \mathcal{L}^2 b \langle \cos^2\theta \rangle}{4M_0}$$

If all orientations of the rod were equally probable, one would have $\langle \cos^2\theta \rangle = 1/3$. In fact, the molecule will tend to spend more time parallel to the flow lines than perpendicular to them, so $\langle \cos^2\theta \rangle$ will be less than $1/3$. Also, we have neglected hydrodynamic interaction between the beads. When these two factors are taken into account, we get[180,233]

$$[\eta] = \frac{2\pi N_A \mathcal{L}^2 b}{45 M_0 \ln(\mathcal{L}/b)} \tag{5-77}$$

For large axial ratios, $\mathcal{L}/b \gg 1$, $[\eta]$ will vary as $\mathcal{L}^{1.8}$ according to Eq. (5-77).

A different molecular weight dependence is to be expected for random coil polymers. Einstein[234] showed that for solid spheres,

$$\eta_{sp} = (5/2)\phi = (5/2)cv_2$$

where ϕ is the volume fraction of polymer, and v_2 is the polymer specific volume. For spheres of radius R, $v_2 = (4\pi/3)R^3 N_A/M$, so using Eq. (5-74) we find

$$[\eta] = (10\pi/3)N_A(R^3/M)$$

As was mentioned earlier in the discussion of translational motion, the hydrodynamic behavior of a flexible macromolecule is in many ways closely related to that of a solid sphere, with an effective radius proportional to the root mean square radius $\langle R^2\rangle^{1/2}$. Thus, Flory[4] proposed for random coils an equation analogous to the one derived for spheres,

$$[\eta] = 6^{3/2}\Phi\langle R^2\rangle^{3/2}/M = \Phi\langle L^2\rangle^{3/2}/M \tag{5-78}$$

Φ is a parameter which is independent of M.

The form of Eq. (5-78) has been justified by Zimm,[235] employing the familiar device of normal coordinates.[236,237] The beads in the polymer chain move under the forces discussed earlier: direct pulls exerted by the nearest neighboring beads along the chain, Brownian motion, and frictional resistance of the solvent as modified by hydrodynamic interaction. The first few normal modes of motion of the polymer, under the influence of these forces, are shown in Fig. 5-20. The forces F_{xl} and coordinates z_l needed in Eq. (5-76) can be expressed in terms of these normal coordinates, and $[\eta]$ evaluated. Φ is found to be 2.84×10^{23} cgs units for a polymer in a theta-solvent. Subsequent refinement of this result,[238] avoiding the premature averaging of the hydrodynamic interaction tensor $\mathbf{T}_{ij}$ discussed in the preceding section and used by Zimm, gives a slightly lower value, $\Phi = 2.69 \times 10^{23}$.

The Flory parameter Φ in Eq. (5-78) is a decreasing function of the excluded volume parameter ϵ.[181,239-244] It has been found[245] that a close representation of its dependence on ϵ is

$$\Phi(\epsilon) = 2.84 \times 10^{23}(1 - 2.68\epsilon + 2.74\epsilon^2) \tag{5-79}$$

It will be noticed that Eqs. (5-78) and (5-79) provide a means for determining the effective segment length b for random coils.[245] Intrinsic

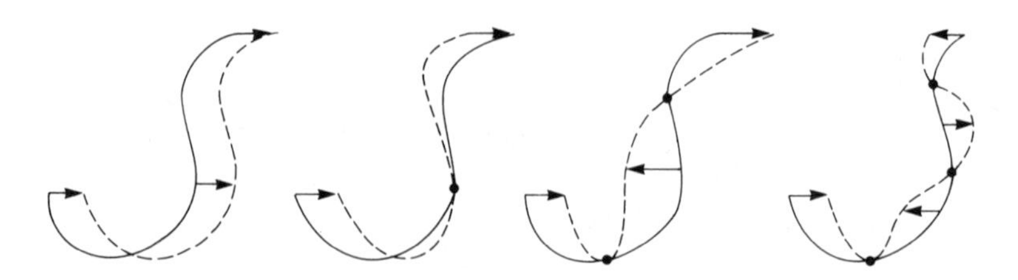

FIGURE 5-20 Schematic representation of the first four normal modes of a chain molecule. [From B. H. Zimm, in *Rheology*, Vol. 3, F. Eirich, ed., Academic Press, New York (1960), p. 1.]

TABLE 5-3 a_η AND K_η FOR SINGLE-STRANDED POLYNUCLEOTIDES OF VARIOUS COMPOSITION IN VARIOUS SOLVENTS, WITH EFFECTIVE SEGMENT LENGTH b[a]

Solvent	Composition	a_η	$K_\eta(dl/g)$	b,Å	Ref.[c]
$[Na^+]$ = 0.013 M, pH 6.8, 25°C	DNA, 39% GC	0.97	2.1×10^{-5}	14.7	207
$[Na^+]$ = 0.013 M, pH 6.8, 25°C	42%	0.94	2.6×10^{-5}	14.4	207
$[Na^+]$ = 0.013 M, pH 6.8, 25°C	50%	0.91	3.1×10^{-5}	14.0	207
$[Na^+]$ = 0.013 M, pH 6.8, 25°C	66%	0.88	2.9×10^{-5}	12.5	207
$[Na^+]$ = 0.195 M, pH 7, 25°C	DNA, all samples	0.55	49.0×10^{-5}	12.1	207
1 M NaCl, pH 7.5, 0°C	poly A	0.713	0.483^b	14.5	231
7.8°C	poly A	0.680	0.446^b	13.7	231
20°C	poly A	0.569	0.533^b	13.2	231

[a] From V. A. Bloomfield, *Biochem. Biophys. Res. Commun.*, **34**, 765 (1969).

[b] K_η in this case is defined by $[\eta] = K_\eta z^{a_\eta}$, where z is the average degree of polymerization and $[\eta]$ is measured in equivalents per liter. To transform to grams per milliliter multiply by 0.351, since M_0 = 351 for Na(poly A). Also, $z = M/351$.

[c] 207 J. Eigner and P. Doty, *J. Mol. Biol.*, **12**, 549 (1965).
231 H. Eisenberg and G. Felsenfeld, *J. Mol. Biol.*, **30**, 17 (1967).

viscosity data is usually reported in the form

$$[\eta] = K_\eta M^{a_\eta} \tag{5-80}$$

From the above equations, and Eq. (5-31), we find

$$a_\eta = \frac{1 + 3\epsilon}{2} \tag{5-81}$$

$$K_\eta = \Phi(\epsilon) b^3 \left[\left(1 + \frac{5}{6}\epsilon + \frac{1}{6}\epsilon^2 \right) M_0^{1+\epsilon} \right]^{-3/2} \tag{5-82}$$

Thus, by determining ϵ and K_η from a log–log plot of $[\eta]$ vs. M and Eq. (5-81), we can get b from Eq. (5-82).

In Table 5-3 are shown values of the parameters K_η and a_η for single-stranded DNAs of different composition,[207] and of poly A,[231] in various solvents. It may be noted that at low ionic strength and low GC composition, $a_\eta = 0.97$, which indicates large polyelectrolyte repulsions in addition to the usual excluded volume effects. Also given in Table 5-3 are values of the effective segment length b, calculated using Eqs. (5-79)–(5-82). These results will be discussed in more detail in Section VII of this chapter.

For stiff-chain molecules such as wormlike coils, it is necessary to take bending modes into account in calculation of intrinsic viscosity. Harris and Hearst[246,247] have made such a calculation, using Green's function techniques to calculate the normal modes and associated relaxation times in the free-draining limit, and using perturbation theory to pass from these results to the viscosity in the presence of hydrodynamic interaction and excluded volume.[248]

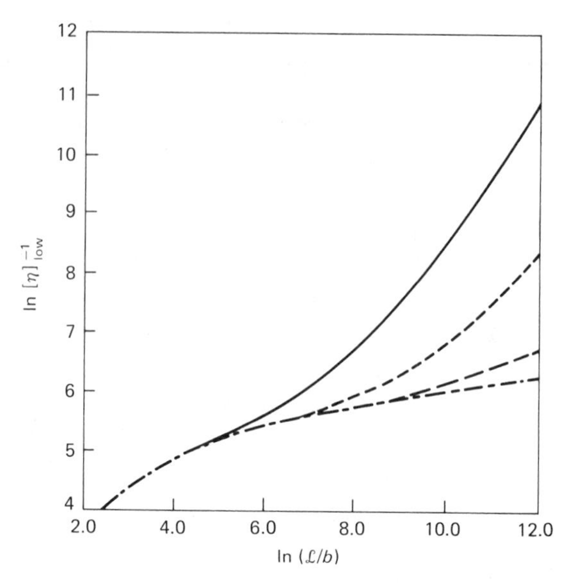

FIGURE 5-21 1/ln $[\eta]$ vs. ln $(\mathcal{L}/b)$ displaying the deviation from rigid-rod behavior for small $\mathcal{L}/a$ and constant b/a. – • –, rigid rod, $b/a = 0$; ——— ———, $b/a = 10^{-5}$, – – –, $b/a = 10^{-4}$; ————, $b/a = 10^{-3}$. [From J. E. Hearst, E. Beals, and R. A. Harris, *J. Chem. Phys.*, **48**, 5371 (1968). Reprinted with permission.]

In Fig. 5-21 is plotted $1/\ln[\eta]$ vs. $\ln(\mathcal{L}/b)$ for short wormlike chains, with varying values of $b/2a$, the ratio of coil diameter to statistical segment length. The coincidence of these plots for very small contour lengths shows that all chains behave as rigid rods ($b/2a = 0$), regardless of flexibility, so long as they are sufficiently short. At larger $\mathcal{L}$, deviations from rigid rod behavior are noted in the expected direction.

The behavior of $[\eta]$ for longer wormlike chains is exemplified in Fig. 5-22 for $\epsilon = 0.1$, the value found for native DNA in 0.195 M Na^+ neutral buffers. The dependence of Φ on $b/2a$ and ϵ for very long chains is given in Table 5-4. For native DNA, $b/2a$ is approximately 3×10^{-2}.[23]

Other theoretical treatments of the intrinsic viscosity of wormlike chains with excluded volume effects have been surveyed.[245]

Application of viscometry to nucleic acid solutions has been very productive, but here also, as in the case of sedimentation, technical difficulties have had to be overcome. The difficulties arise from the great sensitivity of high-molecular weight DNA to shearing stresses. These distort the molecule and make its intrinsic viscosity at shear rates commonly found in capillary viscometers (10^3 sec^{-1}) much lower than that appropriate to the molecule in the absence of shearing stress. For DNA of molecular weight 16×10^6 or less, a multibulb capillary viscometer with a minimum average

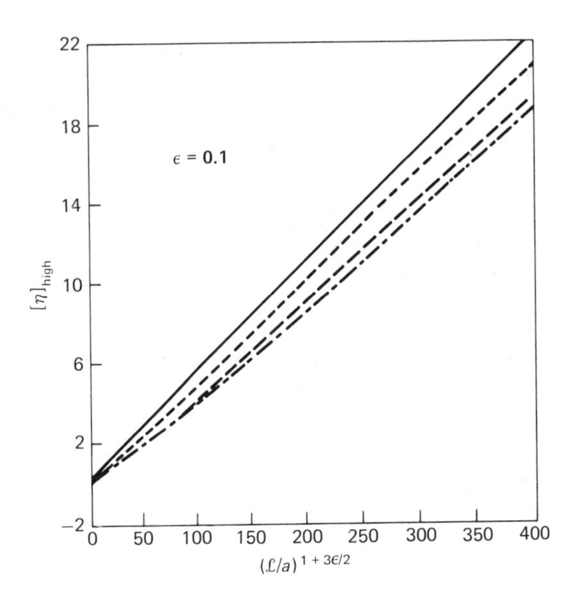

FIGURE 5-22 $[\eta]$ vs $(\mathcal{L}/a)^{0.65}$ for chains with large $\mathcal{L}/a$ and excluded volume parameter $\epsilon = 0.1$ ———, $b/a = 0.3$; – – – –, $b/a = 3 \times 10^{-2}$; —— ——, $b/a = 3 \times 10^{-3}$; —— · ——, $b/a = 3 \times 10^{-4}$. [From J. E. Hearst, E. Beals, and R. A. Harris, *J. Chem. Phys.*, **48**, 5371 (1968). Reprinted with permission.]

gradient of 30 $\sec^{-1}$ provides an adequate extrapolation to zero rate of shear.[249] For calf thymus DNA, of molecular weight about 7×10^6, it is necessary to go to shear stresses of 0.1 dyne/cm^2, corresponding to shear rates of about 10 $\sec^{-1}$ in aqueous solution, to obtain measured viscosities essentially equal to their zero shear values.[250]

Several studies[205,251] indicate that the shear dependence of viscosity of DNA solutions becomes less severe as the concentration decreases, but that it persists to infinite dilution. These studies also demonstrate that the limiting slopes of plots of $[\eta]$ vs. $\dot{\epsilon}$ are horizontal, in accord with most theories of non-Newtonian viscosity. These theories attribute non-Newtonian behavior to a variety of effects, including finite chain extensibility, internal friction due to barriers to internal rotation, excluded volume effects, anisotropy of hydrodynamic interaction, and variation of hydrodynamic interaction with distance between polymer chain elements. They all agree, however, in predicting a dependence of $[\eta]$ on $\dot{\epsilon}$ of the form

$$[\eta] = [\eta]_0(1 - \alpha^2 \dot{\epsilon}^2 + \cdots)$$

For flexible chains,

$$\alpha = \alpha' \tau_1 = \frac{\alpha'' M [\eta]_0 \eta_0}{RT}$$

TABLE 5-4 LIMITING VALUE OF $\Phi(\epsilon)$ AS FUNCTION OF $b/2a$ ($=\lambda b$)[a]

λb	$\epsilon = 0$	$\epsilon = 0.1$	$\epsilon = 0.2$	$\epsilon = 0.3$
3×10^{-1}	2.83	2.09	1.56	1.16
3×10^{-2}	2.80	2.04	1.50	1.11
3×10^{-3}	2.77	2.00	1.46	1.06
3×10^{-4}	2.73	1.96	1.41	1.02

[a] From J. E. Hearst, E. Beals, R. A. Harris, *J. Chem. Phys.*, **48**, 5371 (1968).

where τ_1 is the lowest relaxation time of the chain, corresponding to the first normal mode.[235] α' and α'' are constants. One calculation,[252] assuming for T4 DNA a molecular weight of 120×10^6, gave $\alpha^2 = 0.75$, in good agreement with experiment.

To achieve satisfactorily low rates of shear for the study of high-molecular weight DNA, it has been necessary to resort to rotating cylinder, or Couette, viscometers. Early designs[208,250,251,254] had drawbacks of mechanical complexity and contamination of the nucleic acid solutions with polyvalent metal ions from the machined viscometer surfaces.

These disadvantages have been obviated by the all-glass rotating cylinder viscometer designed by Zimm and Crothers.[253] This employs an inner rotating cylinder which is supported by its buoyancy in solution, centered by surface tension, and driven by magnetic interaction between a small iron plug in the bottom of the rotor and an external rotating magnetic field. A disadvantage of the Zimm-Crothers viscometer is that different rotors must be used if different shear rates are desired in a given solution in order to extrapolate to zero shear. Consequently, modifications have been developed[255-257] which permit variation of rotor speed by use of a variable magnetic field.

Other difficulties associated with the above viscometer are a need for extreme surface cleanliness and for relatively large volumes of solution (about 3 ml). These difficulties have been overcome in a rotating Cartesian-diver viscometer designed by Gill and Thompson.[258] The rotor is driven by the interaction of a rotating magnetic field with a second magnetic field caused by eddy currents induced in a conducting ring in the bottom of the rotor. It is kept submerged at constant depth by a servomechanism-controlled pressure-regulating system. Despite some complexities of construction, this viscometer is versatile and easy to use, and will probably become the instrument of choice for DNA work.

The concentration dependence of η_{sp} of nucleic acid solutions, particularly for native DNA, is also quite strong, and increases with increasing $[\eta]$ or M according to the relation

$$\frac{\eta_{sp}}{c} = [\eta] + k'[\eta]^2 c + \cdots \tag{5-83}$$

The Huggins constant, k, is 0.5 ± 0.2 for a wide variety of DNA samples.[205,249] A viscosity-concentration plot for DNA which is linear over a much larger range of c than is Eq. (5-83), thus easing the extrapolation, is based on Eq. (5-75).[205] It is easy to show

$$\frac{\ln \eta_{\mathrm{rel}}}{c} = [\eta] + k\,[\eta]^2 c + \cdots$$

where

$$k'' = k' - 0.5$$

With k' nearly equal to 0.5, a plot according to this equation will be almost horizontal.

The major use of intrinsic viscosity measurements is to determine molecular weights. Since viscosity does not provide an absolute measure of M, it must be calibrated with samples of known molecular weight (determined, for example, by light scattering). This calibration gives the necessary constants in the $[\eta] - M$ relationship, such as K_η and a_η in Eq. (5-80).

An early useful viscosity-molecular correlation for low-molecular weight native DNA ($M < 3 \times 10^6$), due to Doty et al.[259] is

$$[\eta] = 1.45 \times 10^{-6} \langle M \rangle_w^{1.12}$$

Another commonly used correlation equation, proposed by Eigner and Doty,[207] has

$$[\eta] = 1.05 \times 10^{-7} M^{1.32}$$

for M between 0.3×10^6 and 3×10^6, and

$$[\eta] = 6.9 \times 10^{-4} M^{0.70}$$

for M between 3×10^6 and 130×10^6.

Figure 5-19 shows a plot of $\log[\eta]$ vs. $\log M$ for native DNA which covers a very wide range of molecular weights. Solvent conditions are the standard ones of 0.2 M Na^+, pH 7, 22°C. Crothers and Zimm[205] fit this curve with the empirical equation

$$\log([\eta] + 5) = -2.863 + 0.665 \log M \tag{5-84}$$

The form of Eq. (5-84) has been shown to be expected for wormlike chains with excluded volume effects.[260]

2. SCHERAGA-MANDELKERN EQUATION

By combining results from sedimentation and viscosity studies several useful pieces of information can be obtained. First of all, it is possible to obtain approximate values of molecular weight from measurements of s and $[\eta]$. When it is remembered that several of the usual methods of determining

M – sedimentation and diffusion, and light scattering – are difficult or impossible to apply to large native DNA molecules, this possibility becomes of great importance. Combination of Eqs. (5-69), (5-70), and (5-78) leads to an expression for M in which the geometrical factors characterizing the polymer chain have been eliminated. One obtains the Scheraga-Mandelkern equation[261,262]

$$M = \left[\frac{s^0 [\eta]^{1/3} \eta_0 N_A}{10^{13} \beta (1 - \bar{v}\rho)} \right]^{3/2} \tag{5-85}$$

β is an empirical parameter which has the value 2.5×10^6 for synthetic flexible polymers, and which is relatively insensitive to molecular shape. Because of the stiffness of the native DNA backbone, there has been doubt about using this value of β to obtain molecular weights. However, it has been shown[207] that duplex DNA has a value of β close to 2.5×10^6 at both high and low molecular weights. It becomes somewhat larger at intermediate molecular weights, but virtually all of the data can be accommodated by $\beta = (2.5 \pm 0.3) \times 10^6$. Hearst et al.,[248] combining their treatment of $[\eta]$ with that of Gray et al.[191] of s, get somewhat higher values of β. For example, with $\epsilon = 0$ and $b/2a = 3 \times 10^{-2}$, $\beta \times 10^{-6}$ is calculated to vary from 2.67 at $\mathcal{L}/2a = 10$ to 2.74 at $\mathcal{L}/2a = 10^4$; with $\epsilon = 0.1$, it varies from 2.79 to 2.86.

Crothers and Zimm[205] combined Eqs. (5-71) and (5-84) to obtain a modified version of Eq. (5-85),

$$M = \left[\frac{(s^0 - 2.7)([\eta] + 5)^{1/3} \eta_0 N_A}{10^{13} \beta' (1 - \bar{v}\rho)} \right]^{3/2}$$

which is applicable to native DNA in 0.2 M salt. β' has the value 2.27×10^6 over the entire molecular weight range studied.

For single-stranded DNA from T7 phage, β has been found to be $(2.51 \pm 0.1) \times 10^6$ for neutral DNA and $(2.45 \pm 0.1) \times 10^6$ for alkaline DNA, over a wide range of ionic strengths.[263]

Another important result of the extensive collection of sedimentation and viscosity data which has been built up for native DNA[207] is the recognition that s and $[\eta]$ are monotonic functions of each other, within experimental accuracy. This is seen in Fig. 5-23. This plot demonstrates convincingly that duplex DNA molecules of different molecular weight, regardless of biological source, base composition, or unusual or glucosylated bases, form a homologous series. There appears to be no unusual stiffness or flexibility, and no branched or otherwise nonlinear structures (except for certain well-defined circular species discussed below) in any of the preparations examined thus far.

3. SHEAR DEGRADATION[228]

Not only does hydrodynamic shear decrease the viscosity of polynucleotides in solution, but it can also cause breakage of high molecular

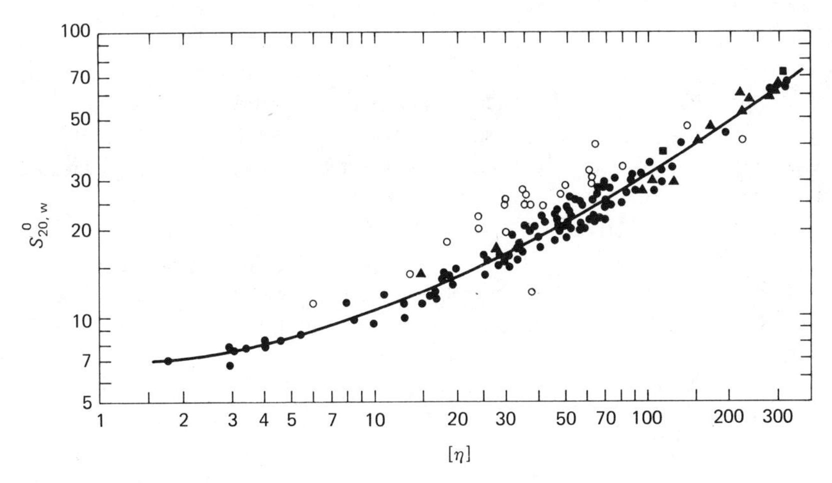

FIGURE 5-23 Sedimentation and viscosity data for native DNA in solvents of counterion concentration 0.1 to 1.0 M. [From J. Eigner and P. Doty, *J. Mol. Biol.*, **12**, 549 (1965). Reprinted with permission.]

weight DNA. Although the possibility of shear degradation of high polymers was apparently first pointed out by Frenkel[264] in 1944, the importance of this phenomenon for the characterization of native DNA was first recognized in 1959. Davison[265] studied the distribution of sedimentation coefficients of DNA, from T2 and T4 phage, which had been passed through a hypodermic needle at various speeds. He found substantial lowering of s after high shear, and emphasized that common laboratory operations such as stirring, pipetting, and filling ultracentrifuge cells can cause breakage of large DNA molecules. To this list may be added shaking during deproteinization,[266] treatment in a homogenizer,[267] passage through a glass spray atomizer,[268] and ultrasonic irradiation.[259, 269–271] On the other hand, it has been shown[97, 272] that intact native DNA from T2 and T7 bacteriophages can be passed through nitrocellulose or Millipore filters without significant breakage. This is apparently because the shear stress developed in any given capillary is below that needed to rupture the molecule.

It has also been shown[273] that single-stranded DNA in alkaline solution is even more sensitive to shear breakage than native DNA.

Before the possibility of shear breakage of DNA was realized, the maximum molecular weights of DNAs isolated from a variety of sources were in the range 6 to 10×10^6. After the necessity of avoiding undue shear was realized, isolation of much larger DNA molecules was achieved.[274–276] It has been found essential in such isolations to work at fairly high DNA concentrations (400–500 μg/ml) during steps involving shear.[277] This tactic is rationalized by the discovery[278] that the critical shear stress needed to break DNA increases with increasing concentration. This "self-protection" effect apparently disappears below a critical concentration, which for T2 DNA is 0.1–0.2 μg/ml.[279] The self-protection effect may involve suppression of turbulence, thus reducing the maximum shear stresses in the system,[279] nonspecific aggregation,[280] or suppression of maximum elongation of the molecule due to intermolecular excluded volume effects.[281] A useful review of the isolation of intact DNA molecules from viruses, bacteria, and higher organisms has been written by Josse and Eigner.[1]

Several studies have been directed toward an understanding of the physical processes leading to shear breakage, but the phenomenon remains somewhat obscure. Levinthal and Davison[282] studied the breakage of T2 DNA by shearing in capillaries. In order to calculate the tensile force acting to rupture the double helix, they assumed as a model that DNA behaved as a rigid, impermeable rod oriented at 45° to the flow lines. The critical tensile force was estimated to be 1.1×10^{-3} dynes, in close agreement with the forces calculated to be necessary to disrupt a C–O or C–C bond (8.9×10^{-4} dynes and 8.1×10^{-4} dynes, respectively). Although the model for DNA as a rigid rod is not in accord with current understanding of the zero-shear configuration, it may represent high-shear configurations fairly well. A more serious assumption is that the rod is oriented at 45° to the streamlines. Studies[283] on flexible fibers have shown that both bending and orientation, much more nearly parallel to the flow lines, may occur at high stresses.

Harrington and Zimm[284] have studied DNA degradation in a variety of high-shear devices. In order to calculate the critical stress for breakage, they used a semiempirical equation which depends on the ratio of the viscosity increment produced by addition of polymer to solvent, to the mean extension of the molecule normal to the stream lines. These quantities were evaluated for the polymer at rest, but it may be argued that this ratio will not be greatly different at high shears. For T2 DNA in phosphate buffer, the critical tensile force was estimated to be only 3×10^{-5} dynes, some thirty times less than that found by Levinthal and Davison.[282] An intermediate value of 4.3×10^{-4} dynes was estimated[285] for the critical shear stress for DNA breakage in a high-speed rotary homogenizer with specially sharpened blades.

These latter studies indicate that the experimentally determined critical shear stress is substantially below that needed to break the covalent bonds in the DNA backbone. Thus, it may be proposed that the mechanical stress serves to lower the activation energy for a chain scission reaction in which participation of solvent is essential. One piece of evidence in support of this interpretation is the finding[286] that chain scission in DNA occurs with 90% C–O bond rupture, 10% P–O bond rupture, and no C–C bond rupture, whereas the C–C bond is probably the weakest in the DNA backbone.

The conclusion that the shear breakage mechanism is largely chemical is buttressed by the study by Yew and Davidson[281] of shear breakage of bonds between cohered ends of phage λ DNA molecules (see Section VII; C). They found that the rate constant for breakage had a very large temperature coefficient, with an activation energy of 130 ± 20 kcal/mole. This is nearly the same as the activation energy for the pure thermal reaction, 114 ± 17 kcal/mole.[287]

According to transition-state theory, the rate constant of the breakage reaction is

$$k_b = (kT/h)\exp[-\Delta G^{\ddagger}/RT]$$

where h is Planck's constant and $\Delta G^{\ddagger}$ is the free energy of activation. The free energy profile for the breakage reaction is shown in Fig. 5-24. r^* is the distance which the bond must be stretched to reach the transition state. The shear stress is assumed to lower $\Delta G_t^{\ddagger}$, the thermal free energy of activation, by an amount Δu:

$$\Delta G_{sh}^{\ddagger} = \Delta G_t^{\ddagger} - \Delta u$$

$\Delta G_{sh}^{\ddagger}$ is the free energy of activation in the presence of the shear stress. Δu is taken to be proportional to the shear stress, $\eta\dot{\epsilon}$, and to r^*

$$\Delta u = A\eta\dot{\epsilon}r^*$$

where A may be supposed to depend on the size, shape, and frictional properties of the polymer. Thus,

$$k_{b,sh} = k_{b,t}\exp[A\eta\dot{\epsilon}r^*/RT]$$

Taking $\dot{\epsilon}$ as the maximum shear rate at the capillary wall, it was estimated[281] that $A\eta\dot{\epsilon}_{\max}r^* = 10.9$ kcal/mole, from studies of the variation of $k_{b,sh}$ with

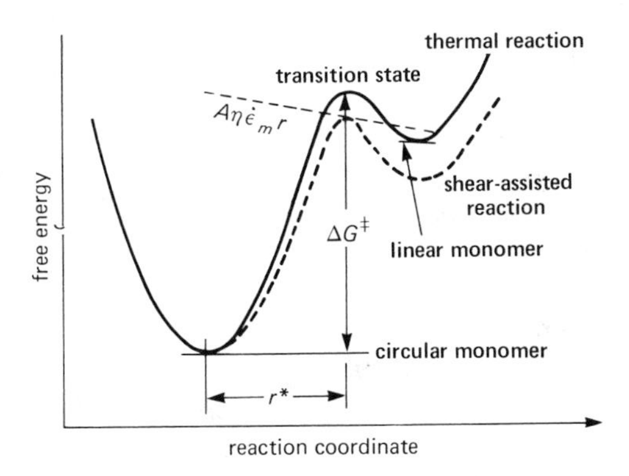

FIGURE 5-24 Free energy profile for shear breakage reaction. [From F. F. H. Yew and N. Davidson, *Biopolymers,* **6**, 659 (1968). Reproduced by permission of John Wiley & Sons, Inc.]

$\dot{\epsilon}_{max}$. This is not inconsistent with the -16 ± 37 kcal/mole obtained directly from thermal measurements.

It was further observed that

$$\left(\frac{\partial \ln k_{b,sh}}{\partial \dot{\epsilon}_{max}}\right)_T = \frac{A\eta r^*}{RT}$$

is the same for shear breakage of circular monomers, joined half-molecules, and joined quarter-molecules. This quantity is, therefore, experimentally independent of molecular length. However, it is also proportional to A. Since A is related to the maximum tension on the polymer, it should vary like $\mathcal{L}^2$ for rods or $\varphi^{3/2}$ for random coils. This leads to the suggestion that it is only the exceptional molecule, greatly extended due to a rare fluctuation, which is subjected to enough tension to cause breakage.[281] As remarked above, this necessity for unusual extension may also explain the self-protection effect.

C. Rotational Motion

Studies of rotational motion of nucleic acids can give information about their molecular weight, size, asymmetry, flexibility, and optical anisotropy. Rotational motion may be observed by measurements of birefringence or dichroism in orienting flow or electric fields, or of fluorescence depolarization. The sudden release of an orienting field leads to a relaxation of the polymer distribution from an ordered to a random state; the relaxation time for this process may also be related to size and structure.

1. ORIENTATION IN FLOW

Orientation in a flow field is a dynamic phenomenon. The polymer molecule is continually rotating, but tends to speed up or slow down at various angles with respect to the flow lines, thus biasing the distribution of polymer orientations. A quantitative treatment of this phenomenon is given in several reviews.[5,288,289] Here we shall content ourselves with a qualitative picture and a summary of results.

We first consider rigid particles, and examine for simplicity the two-dimensional situation depicted in Fig. 5-25. The hydrodynamic forces tend to rotate the rod clockwise, and it is evident that the torque will be greatest at an angle $\theta = 90°$ with respect to the flow lines, and least at $\theta = 0°$. Thus the rod will move most quickly through the former angle, and will linger longest at the latter. This biased distribution will be counteracted by Brownian motion, which will tend to randomize the distribution. A steady distribution is achieved when these two forces balance. The balance point is determined by the ratio $\dot{\epsilon}/D_r'$, where $\dot{\epsilon}$ is the shear rate and D_r is the rotational diffusion coefficient. D_r is the proportionality constant between

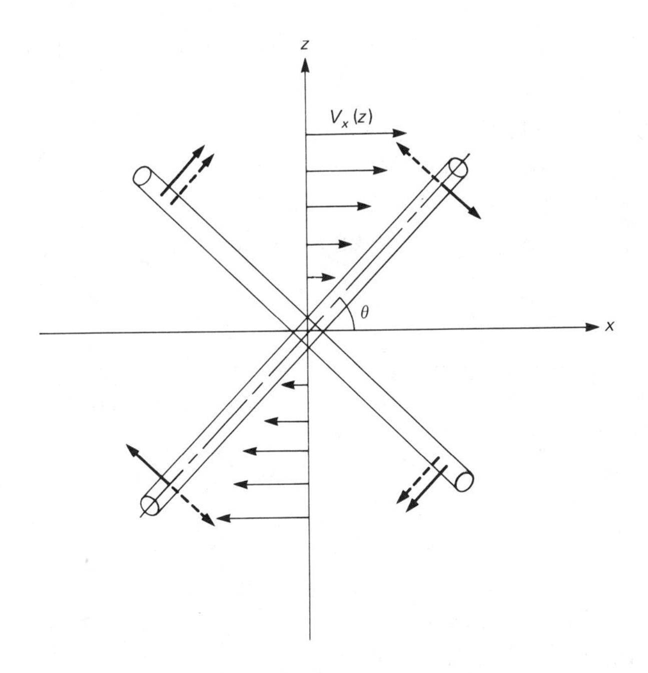

FIGURE 5-25 Forces acting on a rigid rod in a velocity gradient ———→, tangential component of velocity gradient; —— —— —→, Brownian motion forces.

the net flux of particles passing through angle θ per second and the angular gradient in the fraction of particles oriented at θ. Both of these quantities have units of $\sec^{-1}$. When $\dot{\epsilon}/D_r \ll 1$, it is clear from Fig. 5-25 that the most probable orientation is $\theta = 45^\circ$, where the hydrodynamic and Brownian motion forces are equal and opposite. The least probable orientation is $\theta = 135^\circ$, where the two forces add. On the other hand, when $\dot{\epsilon}/D_r \gg 1$, the hydrodynamic force will greatly outweigh the Brownian motion randomizing force save when all the particle axes are concentrated near $\theta = 0^\circ$. In these circumstances, obviously, the minimum in the angular distribution will be at $\theta = 90^\circ$.

D_r is related to the rotational frictional coefficient f_r by

$$D_r = \frac{kT}{f_r} \tag{5-86}$$

f_r is the torque needed to rotate the axis of the particle through solution with unit angular velocity. For spheres of radius R,[290]

$$f_r = 8\pi\eta_0 R^3$$

while for elongated prolate ellipsoids (or approximately for rods) of length a and diameter b,[291]

$$f_r = \frac{16\pi\eta_0 a^3}{3[2\ln(2a/b) - 1]}$$

for rotation about the minor axis.

D_r and f_r may be related to the relaxation time τ, which is defined by

$$\frac{\langle\cos\theta\rangle_t}{\langle\cos\theta\rangle_{t=0}} = e^{-t/\tau} \qquad (5\text{-}87)$$

That is, τ is the time required for $\langle\cos\theta\rangle$ to fall to $1/e$ of its initial value. In three dimensions, the orientation of the long axis of a rod can relax by rotation about both short axes, and is therefore twice as fast as it would be in two dimensions. The three-dimensional result is[5,289]

$$\tau = \frac{1}{2}D_r = \frac{f_r}{2kT}$$

The above discussion has been for two dimensions, save for the last equation. In reality, rotational orientation and relaxation will occur in three dimensions. This problem has been rigorously formulated, and solved in series form for small $\dot{\epsilon}/D_r$, by Peterlin and Stuart[292–294] for ellipsoids of revolution. Their qualitative conclusions are much the same as those arrived at in two dimensions. The distribution has a broad maximum near 45° for $\dot{\epsilon}/D_r \ll 1$ and a narrow maximum near 0° for $\dot{\epsilon}/D_r \gg 1$.

Finally, it is convenient to express the diffusion coefficient in terms of the molecular weight and intrinsic viscosity of the polymer. Collecting previous results, one obtains

$$\frac{1}{D_r} = A\beta$$

where

$$\beta = \frac{M[\eta]\eta_0}{RT} \qquad (5\text{-}88)$$

A is a function of axial ratio and, for prolate ellipsoids of revolution, or rigid rods, varies from 12/5 for spheres ($\mathcal{L}/b = 1$) to 15/2 for $\mathcal{L}/b = \infty$. β is proportional to the relaxation time.

We now turn to flexible polymers. The same dynamical considerations arise for flexible chains as for rigid polymers, with the additional complication that deformation of the polymer will occur in the shearing field. It may be seen from Fig. 5-26 that the chain will be extended when its ends are in the first and third quadrants, and compressed in the second and fourth quadrants. If the polymer is very flexible, or soft, it will readily deform in

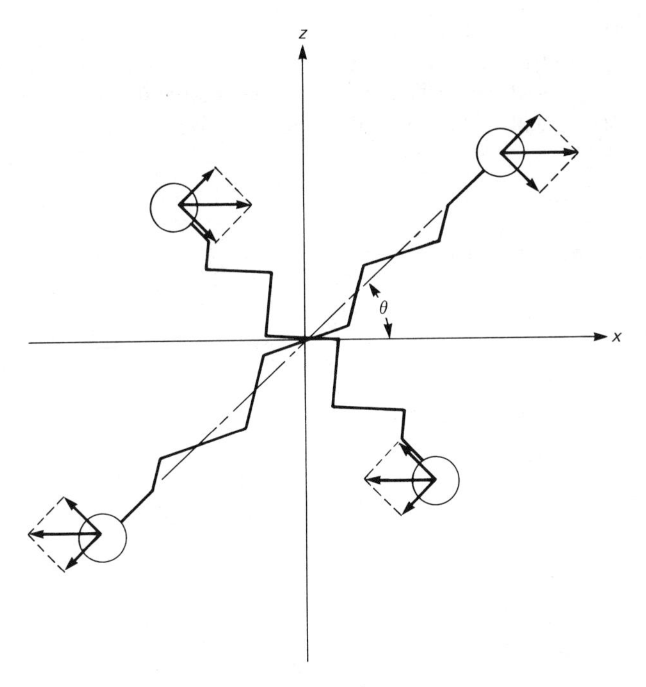

FIGURE 5-26 Alternating extensive and compressive stresses act on a flexible polymer chain as it rotates in a velocity gradient.

phase with its rotational motion. In this case, the normal coordinate theory of Zimm[235] is applicable for low shear gradients, which corresponds to small deformations of the chain. Zimm found that for nondraining gaussian coils, the relaxation time τ_1 for the first normal mode, which corresponds to an overall rotation of the chain, is

$$\tau_1 = 0.423\,\beta \tag{5-89}$$

Several approximate treatments have been developed for large shear rates and deformations.[288,289]

2. FLOW BIREFRINGENCE

Now let us consider the optical consequences of the orientation of polymer molecules in a field. Nucleic acid molecules have a refractive index n different from that of usual aqueous solutions. Thus when the nucleic acid molecules are oriented in a shearing field, the solution becomes birefringent. That is, it has different refractive indices parallel and perpendicular to the flow lines. This is known as "form birefringence," and arises in all solutions of nonspherical molecules whose refractive index differs from that of the solvent. However, even if the refractive index of the polymer equals that of the solvent, birefringence will arise when the polymer itself is optically anisotropic, with different refractive indices parallel and perpendicular to its long axis. This is called "intrinsic birefringence." We define the birefringence as

$$\Delta n = n_{\parallel} - n_{\perp}$$

where the average refractive index, n, is

$$n = \frac{1}{3}\,(n_{\parallel} + 2n_{\perp})$$

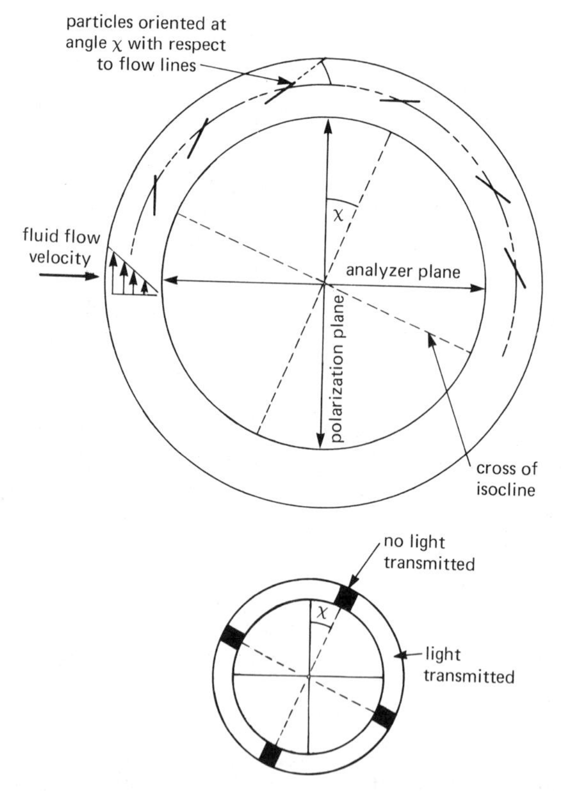

FIGURE 5-27 Schematic illustration of flow birefringence when all solute particles have the same orientation angle χ with respect to the flow lines. The upper diagram shows the *cross of isocline*, which indicates the four locations where the optic axes of solute particles are exactly parallel to the analyzer or polarizer plane. The lower diagram shows that no light is transmitted through the annular ring between the rotor and stator of the apparatus, at the locations of the cross of isocline. [From C. Tanford, *Physical Chemistry of Macromolecules* (1961). Reproduced by permission of John Wiley & Sons, Inc.]

Flow birefringence measurements are generally made in a concentric cylinder device, shown schematically in Fig. 5-27. The solution to be studied is confined between an inner stator and an outer rotor, and viewed through an analyzer crossed at right angles to a polarizer. If the solution is birefringent, light polarized in the z-direction will be deviated, due to separation into ordinary and extraordinary rays passing parallel and perpendicular to the principal optic axis of the solution. Thus light will be observed through the crossed analyzer. The only exception will occur when the plane of polarization of the light coincides with the principal optic axes. As shown in Fig. 5-27, this will happen only at four points, which define a dark "cross of isocline." The extinction angle χ varies like $\langle\theta\rangle$ in the above discussion. Harrington[295] has described an instrument which can measure very low birefringence ($\Delta n = 10^{-11}$) and operate at very low shear rates ($\dot{\epsilon} = 0.1$ to 3 $\sec^{-1}$). Both of these capabilities are necessary for work with high molecular weight DNA.

For rigid ellipsoids of revolution, Peterlin and Stuart[292-294] found

$$\chi = \frac{\pi}{4} - \frac{\dot{\epsilon}}{12D_r}\,[1 - 0(\dot{\epsilon}^2/D_r^{\,2})]$$

where 0(Z) denotes "terms of the order of z", so that

$$-\left(\frac{d\chi}{d\dot{\epsilon}}\right)_{\dot{\epsilon},\ c\to 0} = \frac{1}{12D_r}$$

at $\dot{\epsilon}/D_r < 1$. It is seen that χ tends to 45° as $\dot{\epsilon}/D_r$ goes to zero. For high $\dot{\epsilon}/D_r$, χ has been evaluated numerically.[289,296]

For nondraining random chains with no excluded volume, the Zimm[235] normal coordinate theory gives

$$\lim_{\substack{c\to 0\\ \dot{\epsilon}\to 0}} \cot 2\chi = 0.206\,\beta\dot{\epsilon}$$

where β is defined by Eq. (5-88) or

$$-\left(\frac{d\chi}{d\dot{\epsilon}}\right)_{\dot{\epsilon},\ c\to 0} = 0.103\,\beta \qquad (5\text{-}90)$$

The numerical coefficient in this equation varies with excluded volume parameter ϵ, rising to 0.115 at $\epsilon = 0.1$ and 0.126 at $\epsilon = 0.2$.[243,297]

The birefringence itself is the other quantity that can be measured in flow birefringence experiments. Tsvetkov[288,298] has observed that, unlike the extinction angle, the birefringence is largely independent of the mechanical properties of the polymer and can be calculated in the same way as for a rigid particle of appropriate structure. The characteristic birefringence $[n]$, which is the observed birefringence extrapolated to zero concentration and shear rate, and normalized with respect to solvent viscosity, can be written as the sum of three terms:

$$[n] = \lim_{\substack{c \to 0 \\ \dot{\epsilon} \to 0}} \left(\frac{\Delta n}{\dot{\epsilon} \eta_0 c} \right) = [n]_i + [n]_f + [n]_{fs} \tag{5-91}$$

$[\eta]_i$ is related to the intrinsic anisotropy of the segments, and is given by

$$[n]_i = \frac{4\pi}{45kT} \frac{(n_s^2 + 2)^2}{n_s} [\eta]_0 (\alpha_{\parallel} - \alpha_{\perp}) \tag{5-92}$$

n_s is the refractive index of the solvent, $[\eta]_0$ the polymer intrinsic viscosity at zero shear, and $\alpha_{\parallel}$ and $\alpha_{\perp}$ the principal polarizabilities of the segment, which is assumed to be cylindrically symmetrical. The second term in Eq. (5-91), $[n]_f$, is the characteristic birefringence due to the overall shape of the polymer domain, and corresponds to the form birefringence. $[n]_f$ has been shown to be negligible for DNA.[299]

Finally $[n]_{fs}$ is the birefringence due to local anisotropy resulting from the asymmetric shape of polymer segments. This term is large for rigid polymers with large persistence lengths. It is given by

$$[n]_{fs} = \frac{(n_s^2 + 2)^2 (n_k^2 - n_s^2)^2}{180\pi RT n_s^3 \beta \bar{v}} [\eta]_0 M_0 S(L_2 - L_1)_s \tag{5-93}$$

n_k is the refractive index of dissolved polymer; $\bar{v}$ is the polymer partial specific volume; S is the number of monomer units per segment; and $(L_2 - L_1)_s$ is a function of the segmental axial ratio. For native DNA, $(L_2 - L_1)_s = 2\pi$.

Figure 5-28 shows the extinction angle χ and birefringence Δn for T2 DNA[297] plotted as a function of shear rate $\dot{\epsilon}$. Both χ and Δn are strong functions of shear rate, and extrapolations to zero shear are required if meaningful results pertaining to the undistorted molecule are to be obtained. χ and Δn are also strongly concentration-dependent, because of the large hydrodynamic volume of high-molecular weight T2 DNA, so extrapolation to zero concentration is also necessary.

For T2 DNA, the limiting orientation time $(-d\chi/d\dot{\epsilon})_{\dot{\epsilon}, c \to 0}$ in Eq. (5-89) is 0.168 seconds, corresponding to a value for β of 1.63 seconds and for τ_1 of 0.69 seconds. This agrees well with the value of $\tau_1 = 0.64$ seconds computed from Eqs. (5-88) and (5-89) with $M = 120 \times 10^6$ and $[\eta] = 310\ cm^3/g$. Direct measurement of τ_1, by rapidly stopping the rotor in a Couette flow birefringence apparatus, or by stopping the flow in a flow dichroism device,[301] gives $\tau_1 = 0.5$ seconds. Agreement here seems reasonable.

It is evident that for an unknown DNA measurement of the extinction angle and intrinsic viscosity will give, by Eqs. (5-88) and (5-89), the molecular weight.

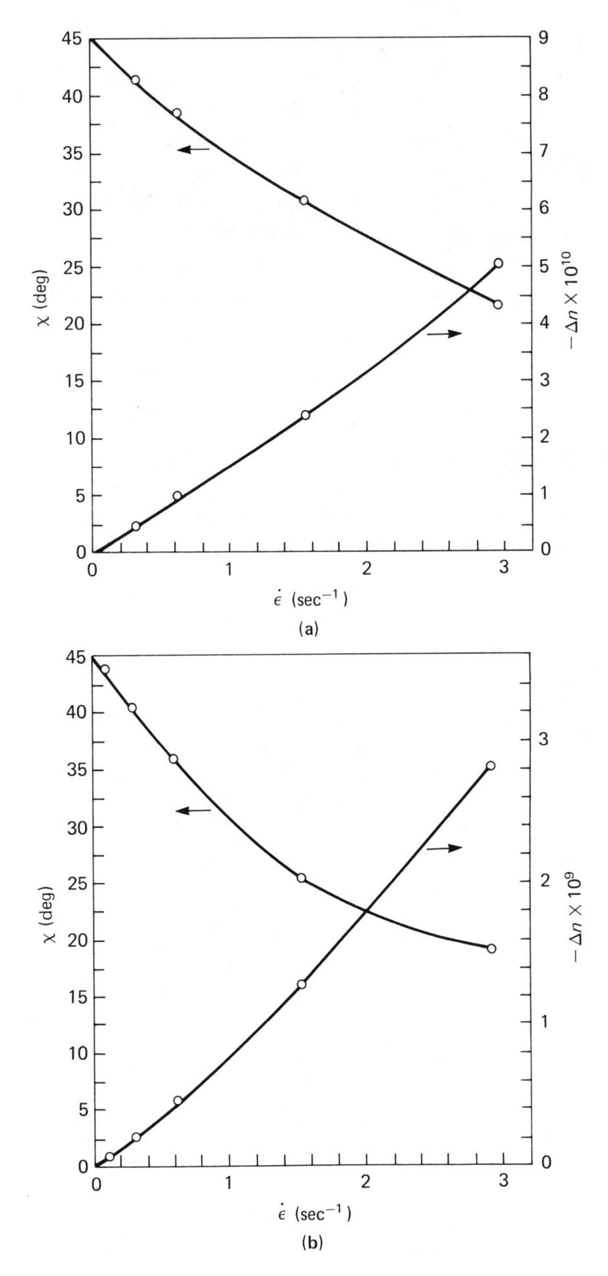

FIGURE 5-28 Extinction angle χ and birefringence Δn as a function of shear rate $\dot{\epsilon}$ for T2 DNA. [From R. E. Harrington, *Biopolymers,* 9, 159 (1970). Reproduced by permission of John Wiley & Sons, Inc.]

We observe from Fig. 5-28 that the birefringence is negative, a fact which was noted in the earliest investigations[302] and which was even then interpreted as evidence that the base pairs are perpendicular to the long axis of the DNA molecule. That is, the polarizability of the bases is greater in the plane of the heterocyclic ring than perpendicular to the plane.

The statistical segment length of DNA can be determined from flow birefringence measurements if the anisotropy of polarizability, $\alpha_{\parallel} - \alpha_{\perp}$, is known. Neglecting the very small form birefringence, Eqs. (5-91)–(5-93) can be written

$$\frac{[n]}{[\eta]_0} = \frac{4\pi}{45kT} \frac{(n_s{}^2 + 2)^2}{n_s} \cdot S \cdot \left[(\alpha_{\parallel} - \alpha_{\perp}) + \frac{M_0(dn/dc)^2}{2\pi N_A \bar{v}} \right]$$

Unfortunately, estimates of $\alpha_{\parallel} - \alpha_{\perp}$ vary widely, from -12.5 Å^3 to -19 Å^3,[299,303-305] Uncertainty in the anisotropy, or in the refractive index increment (which enters as the square), leads to equivalent uncertainties in S, the number of base pairs per statistical segment.

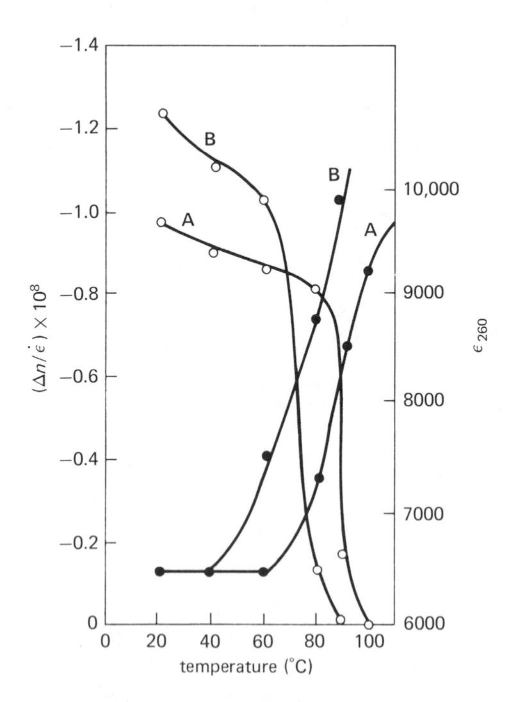

FIGURE 5-29 Dependence of flow birefringence (○) and molecular absorption coefficient at 260 nm, ϵ_{260} (●), upon temperature for a calf thymus DNA sample (c = 0.165 g/liter) at two salt concentrations. (a) 0.15 M NaCl; (b) 0.015 M NaCl. [From R. E. Harrington, *Encyclopedia of Polymer Science and Technology*, Vol. 7, p. 100. Reproduced by permission of John Wiley & Sons, Inc.]

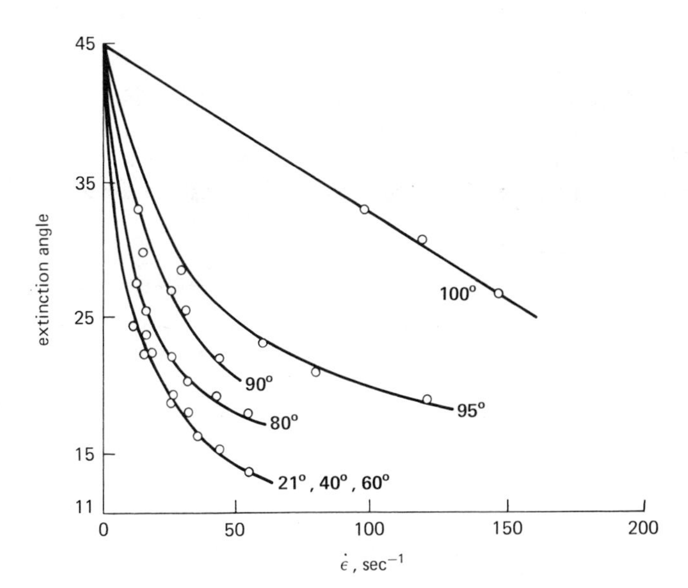

FIGURE 5-30 Extinction angle as a function of shear rate for calf thymus DNA ($c = 0.155$ g/liter) in 0.15 M NaCl buffer after denaturation at the various indicated temperatures. [From R. E. Harrington, *Encyclopedia of Polymer Science and Technology*, Vol. 7, p. 100. Reproduced by permission of John Wiley & Sons, Inc.]

Direct measurement of $\alpha_{\parallel} - \alpha_{\perp}$ would be possible for highly oriented solutions of very short, rodlike DNA. In this case the appropriate hydrodynamic treatment would be that for long prolate ellipsoids in the limit $\dot{\epsilon}/D_r \gg 1$ as discussed above. However, problems of polydispersity in the fragmented DNA sample, and of the need to include the form birefringence in the optical treatment, make this approach a difficult one.

Single-stranded polynucleotides, as remarked above, and shown in Fig. 5-29, have in general a substantially smaller birefringence than native DNA, due to their greater randomness. Figure 5-30 shows that denatured, single-stranded chains are also less orientable and deformable in shear than native DNA, as evidenced by the larger extinction angle for a given $\dot{\epsilon}$. This is presumably attributable to their greater compactness. However, it has been found[304] that high molecular weight Na poly A at low ionic strength has a substantial negative birefringence, indicative of base stacking perpendicular to the long axis of the molecule. This birefringence is decreased by the addition of Mg^{2+} or Mn^{2+}, which may serve to neutralize charge or to intramolecularly cross-link the poly A by coordination to the phosphate groups.

The few flow-optical studies that have been made to date on RNA have been reviewed by Tsvetkov[288] and Harrington.[289] The results are not such as to warrant firm statements about RNA macromolecular structure. It is of interest that the observed birefringence is positive, probably due to the predominance of the form birefringence contribution in Eq. (5-91).

It is evident from Fig. 5-29 that flow birefringence is a sensitive measure of thermal denaturation phenomena in DNA. It has also been found useful in determining the orientation of dye molecules bound to native DNA. These phenomena are discussed in greater detail in Chapters 6 and 7.

3. FLOW DICHROISM

Just as nucleic acid molecules have different refractive indices in directions parallel or perpendicular to the plane of the bases, so do they have different extinction coefficients. If we define the extinction coefficient for light with its electric vector polarized parallel to the helix axis in native DNA as $\epsilon_{\|}$, and the extinction coefficient for light polarized perpendicular as $\epsilon_{\perp}$, the molecular dichroism is

$$\Delta\epsilon = \epsilon_{\|} - \epsilon_{\perp}$$

while the ordinary extinction coefficient, for isotropic orientation of the molecular axes, is

$$\epsilon = \frac{1}{3}(\epsilon_{\|} + 2\epsilon_{\perp})$$

Under conditions of incomplete orientation of the rodlike molecules in flow, the observed extinction coefficient ϵ_α for light polarized along a particular axis α is

$$\epsilon_\alpha = \epsilon_{\|} \langle\cos^2 \theta_\alpha\rangle + \epsilon_{\perp} \langle\sin^2 \theta_\alpha\rangle$$

where θ_α is the angle between the rod axis and the α axis. Combination of these equations then gives the differences between ϵ_α and the isotropic extinction coefficient ϵ observed in a particular experiment, and $\Delta\epsilon$:

$$\epsilon_\alpha - \epsilon = \Delta\epsilon \left(\langle\cos^2 \theta_\alpha\rangle - \frac{1}{3}\right)$$

Flow dichroism and flow birefringence are connected by Kronig-Kramers transforms, just as are optical absorption and refractive index, or optical rotation and circular dichroism (see Chapter 2).

Flow dichroism of nucleic acids has been studied in devices with a variety of geometries. In one apparatus,[306] solution is flowed in the x-direction through a narrow rectangular channel between quartz plates, and light propagates perpendicular to the channel in the z-direction. A velocity gradient is established in the z-direction, and the light may be polarized in either the x- or y-directions. Another apparatus[307,308] employs transparent concentric cylinders, the inner rotating with respect to the outer, with the light beam directed radially, again parallel to the velocity gradient and perpendicular to the stream lines. These devices, because of the narrow gap of solution across which the light is propagated and absorbed, require high nucleic acid concentrations. Two other devices require lower concentrations. One[309] is a conventional Couette apparatus, with light propagated axially perpendicular to the velocity gradient and flow lines. The other[301] has the solution flowing down a long narrow channel, with an unpolarized beam propagating along the flow lines. Since in DNA the transition moments in the bases are perpendicular to the helix axis, orientation of the molecule in the velocity gradient makes the planes of the bases move toward parallelism with the electric vector of the light, and thus the absorption is enhanced.

The extinction angle has the same meaning in flow dichroism as in flow birefringence, but is measured by rotating the analyzer with respect to the polarizer so as to find the minimum (or maximum) in absorbance.

The effect of orientation and distortion of a polymer chain in a velocity gradient on its optical properties has been derived by Kühn and Grün.[310] There are N segments each of length b in the freely jointed chain, whose undistorted mean square length is $\langle L^2 \rangle_0 = b^2 N$, whose mean square length in the flow field is $\langle L^2 \rangle$, and whose contour length is $\mathcal{L} = Nb$. If $\epsilon_\parallel$ and $\epsilon_\perp$ are the extinction coefficients of each segment parallel and perpendicular to the local chain axis, and ϵ_1 and ϵ_2 are the extinction coefficients for the entire polymer in directions parallel and perpendicular to the end-to-end vector $\mathbf{L}$, one finds

$$\epsilon_1 = N\epsilon + \frac{2}{5} \Delta\epsilon \frac{\langle L^2 \rangle}{\langle L^2 \rangle_0}$$

$$\epsilon_2 = N\epsilon - \frac{1}{5} \Delta\epsilon \frac{\langle L^2 \rangle}{\langle L^2 \rangle_0}$$

ϵ and $\Delta\epsilon$ in these equations pertain to the individual segments. Clearly the first terms in these equations represent the additive, isotropic contribution of each segment to ϵ_1 and ϵ_2. The second terms reflect the fact that an elongation and orientation of the entire chain in the direction of its end points is accompanied by an orientation of the individual segments.

These equations have been combined[301] with the normal coordinate treatment[295] of polymer dynamics to eliminate $\langle L^2 \rangle$ and get an explicit expression for the dichroism in terms of the molecular length and segmental length and asymmetry. In the final dichroism apparatus described above, A_f and A_s are the absorbancies of the flowing and stationary solutions, respectively. With flow in the x-direction and the velocity gradient in the z-direction across the channel, one finds

$$\frac{\Delta A}{A} = \frac{A_f - A_s}{A_s} = \frac{\epsilon_y + \epsilon_z - 2\epsilon}{2\epsilon} = -\frac{\Delta\epsilon}{2\epsilon}\left(\langle \cos^2 \theta_x \rangle - \frac{1}{3}\right)$$

$$= \frac{3}{5}\frac{\Delta\epsilon}{\epsilon}\left\{\left[0.0303\left(\frac{\eta_0 \dot{\epsilon}}{kT}\right)^2 \mathcal{L}^2 b^4\right]^2 + \left[0.314\left(\frac{\eta_0 \dot{\epsilon}}{kT}\right)\mathcal{L}^{1/2} b^{5/2}\right]^2\right\}^{1/2} \quad (5\text{-}94)$$

The third equality in Eq. (5-94) shows that at very high shear rates, in which the polymer is perfectly aligned in the x-direction, so that $\langle \cos^2 \theta_x \rangle = 1$, $\Delta A/A = -\Delta\epsilon/3\epsilon$. In DNA, essentially all of the absorption at 260 nm is in the plane of the bases, perpendicular to the helix axis, so $\epsilon_{\parallel} \approx 0$ and $\Delta\epsilon/\epsilon \approx -3/2$. Thus the maximum value of $\Delta A/A$ is 0.5. The last equality suggests that, at high shear gradients, the dichroism should increase quadratically with $\dot{\epsilon}$.

The UV dichroism of native DNA, like its birefringence, is found to be large and negative.[306,308,309] This is consistent with the assignment of the UV absorbance to $\pi \rightarrow \pi^*$ transitions in the plane of the bases. Several studies[308,311] have indicated, however, that the dichroism decreases below 240 nm. For example, in Fig. 5-31 is plotted the dichroic ratio, $A_{\parallel}/A_{\perp}$, for T2 phage DNA, at rest and in a moderately large shear gradient.[311] It is observed that the dichroic ratio reaches a maximum at around 220 nm. This is consistent with the presence of an $n \rightarrow \pi^*$ transition, involving an electron on

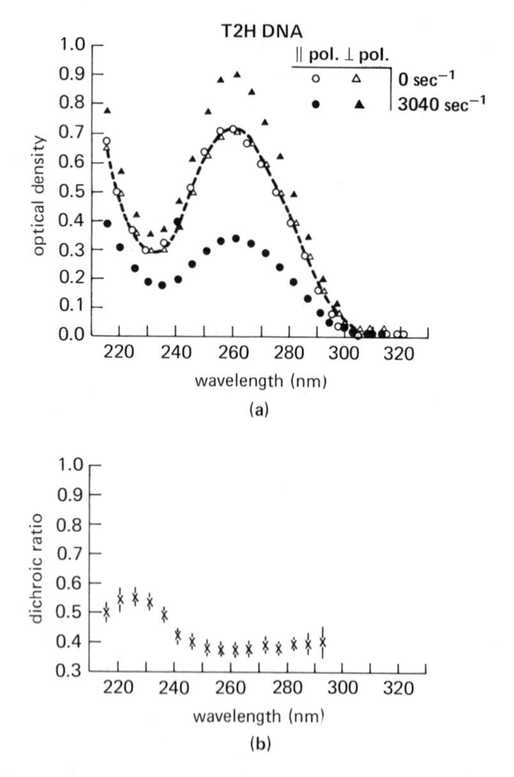

FIGURE 5-31 Polarized absorption spectra (a) and dichroic ratio (b) for a T2 DNA solution, 380 μg/ml in aqueous buffer. [From D. M. Gray and I. Rubenstein, *Biopolymers*, 6, 1605 (1968). Reproduced by permission of John Wiley & Sons, Inc.]

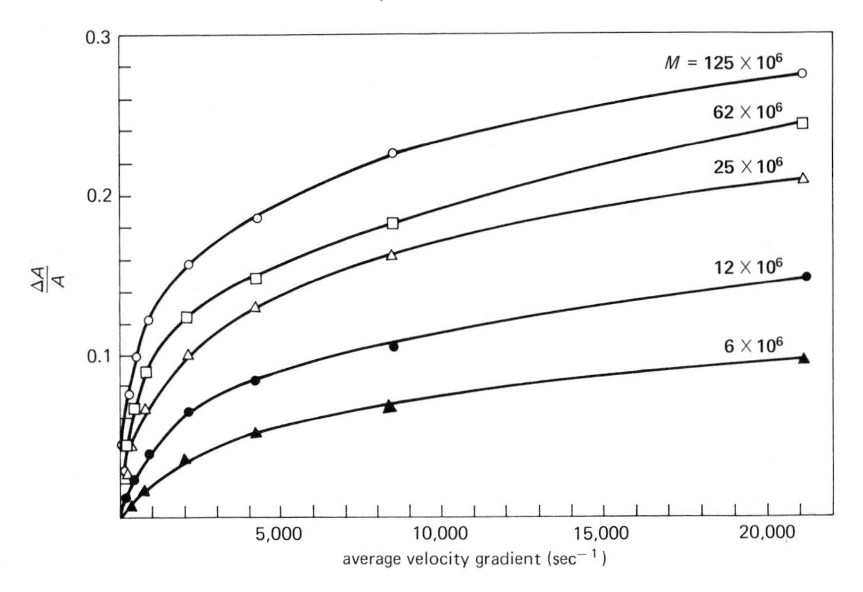

FIGURE 5-32 Velocity gradient dependence of dichroism of native DNA solutions ($c = 8\ \mu g/ml$) at 25°C in 0.1 M NaCl, phosphate buffer, pH 7.8. [From P. Callis and N. Davidson, *Biopolymers*, 7, 335 (1969). Reproduced by permission of John Wiley & Sons, Inc.]

the aza nitrogen, which would be polarized perpendicular to the ring. The presence of this transition at such a low wavelength is surprising. Similar dichroic spectra were obtained from oriented films of DNA.[311]

A linear dependence of flow dichroism on $\dot{\epsilon}$ is observed at low shear rates.[309] This is consistent with Eq. (5-94) since the second term in curly brackets will predominate at low $\dot{\epsilon}$. Figure 5-32 shows that at high shear rates, the dichroism becomes saturated.[301] This indicates that the DNA is becoming deformed and oriented to the maximum extent, although $\Delta A/A$ is still substantially below the theoretical maximum of 0.5 cited above. The behavior of $\Delta A/A$ vs. $\dot{\epsilon}$ shown in Fig. 5-32 is very different from the quadratic increase predicted from the normal coordinate theory in Eq. (5-94). This discrepancy is undoubtedly due to the fact that the theory is valid only for small deformations, while the chain is actually substantially deformed. Indeed, it has been calculated[309] that for T2 DNA with $\dot{\epsilon} = 160\ sec^{-1}$ in 1 M NaCl, $\langle L^2\rangle^{1/2}/\mathcal{L} = 0.50$ while $\langle L^2\rangle_0{}^{1/2}/\mathcal{L} = 0.036$, so that the molecule is stretched more than tenfold by the shear stresses.

Equation (5-94) also predicts a quadratic dependence of dichroism on molecular weight (or $\mathcal{L}$) in the high shear region. Figure 5-33 shows that a linear dependence is observed instead, although there are hints of a quadratic dependence at $\dot{\epsilon} = 50\ sec^{-1}$. At higher shear stresses, curves of $\Delta A/A$ vs. M

also level off,[301] again indicating substantial distortion of the molecular coil.

Another method of orienting polynucleotides is by use of electric fields. The orienting torque is provided by the coupling between the electric field and permanent and induced dipoles in the molecule. The magnitude of the torque depends on the dipole moment and electronic and ionic polarizability of a polyelectrolyte, and the dielectric constant and conductivity of the solvent. For a flexible polynucleotide in aqueous solution, the electrical orientation phenomenon is still not quantitatively understood. For a rigid molecule the orientation depends on the square of the electric field at low fields and saturates at high fields. The optical properties usually measured are either the difference in refractive index parallel and perpendicular to the field (electric birefringence), or the difference in extinction coefficient parallel and perpendicular to the field (electric dichroism).

Once the molecules are (partially) oriented by the electric field, removal of the field allows the relaxation of the orientation to be measured. This is a much simpler process to understand, because the electrical properties of the molecule are only involved in determining the original orientation. For a rigid rod the relaxation is simply

$$S = S_0 \exp[-6D_r t]$$

S is the signal at any time t
S_0 is the signal at zero time
D_r is the rotatory diffusion coefficient of the long axis of the rod

For a flexible nucleic acid, a sum of exponentials with different relaxation times is expected. The coefficient of each exponential will depend on the electrical and optical anisotropy of the polymer segments involved in each relaxation mechanism.

Few studies of this method have been made on nucleic acids, presumably because of the difficulty of interpretation.[228]

In summary, flow birefringence and dichroism are potentially very useful techniques for elucidating polynucleotide size and structure. Measurements of flow birefringence and intrinsic viscosity can be combined, using Eq. (5-88) to get the molecular weight. It should be noted that since $[\eta]$ varies as $M^{1/2}$ for random coils with no excluded volume, β varies as $M^{3/2}$ and so is a very sensitive measure of molecular weight. Flow birefringence and dichroism should also be sensitive indicators of statistical segment lengths, although the optical anisotropy of base pairs must be known with precision. Nonsteady state measurements also should give information on the internal dynamics of the chain.

4. FLUORESCENCE DEPOLARIZATION

Fluorescence depolarization has been a very useful technique for studying the rotational motion of proteins,[312] and should become useful with small or flexible nucleic acids. The experimental setup is as follows. A

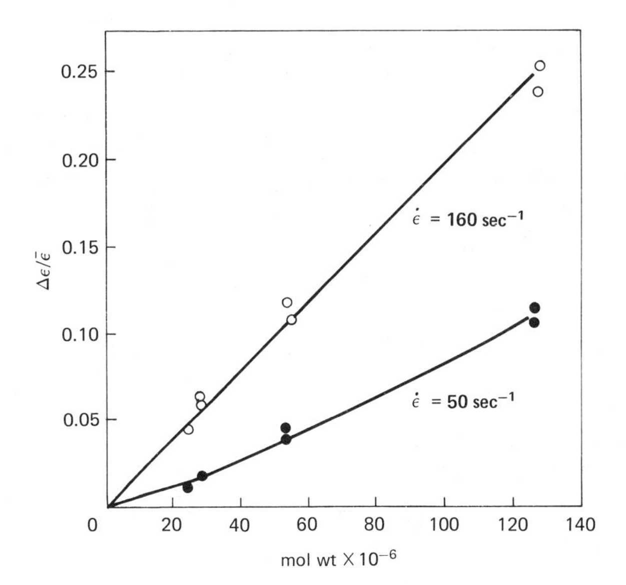

FIGURE 5-33 Fractional dichroism as a function of weight-average molecular weight at shear rates 50 sec^{-1} and 160 sec^{-1}, for various native viral DNAs. [From C. S. Lee and N. Davidson, *Biopolymers,* **6**, 531 (1968). Reproduced by permission of John Wiley & Sons, Inc.]

solution of macromolecules, with a fluorescing group attached, is irradiated by plane polarized light. The intensities of fluorescent light emitted parallel ($I_{\parallel}$) and perpendicular ($I_{\perp}$) to the plane of the exciting light are measured. The polarization of fluorescence is

$$p = \frac{I_{\parallel} - I_{\perp}}{I_{\parallel} + I_{\perp}}$$

During the fluorescence lifetime, τ_f, in which the chromophore exists in its excited state before emitting, the macromolecule may rotate, with a rotational relaxation time τ, defined by Eq. (5-87). If $\tau_f \ll \tau$, emission will occur before rotation is perceptible, so the light will be completely polarized. (Actually, in this case $p = 1/2$, since there will be an unavoidable depolarization due to the initially random orientation of chromophores in the solution.) If $\tau_f \gg \tau$, the molecules can rotate many times before fluorescence, so $p = 0$ and the emitted light is completely depolarized. However, if $\tau_f \approx \tau$, only partial depolarization will occur, and τ may be measured if τ_f is known. The equation describing this for steady-state measurement is

$$\left(\frac{1}{p} - \frac{1}{3}\right) = \left(\frac{1}{p_0} - \frac{1}{3}\right)\left(1 + \frac{3\tau_f}{\tau}\right)$$

p_0 is the polarization for an infinitely viscous solution ($\tau \gg \tau_f$). If the molecule is not spherical, but instead has several different rotational relaxation times about different axes, τ is the harmonic mean of the relaxation times.

Time-dependent fluorescence depolarization measurements may also be made in the nanosecond region,[313,314] enabling separate determination of several relaxation times. If we define the polarization anisotropy as a function of time by the equation

$$r(t) = \frac{I_{\parallel}(t) - I_{\perp}(t)}{I_{\parallel}(t) + 2I_{\perp}(t)}$$

then for an ellipsoid of revolution, $r(t)$ is the sum of three exponentials:

$$r(t) = r_0 \, [a_1(\theta)e^{-t/\tau_1} + a_2(\theta)e^{-t/\tau_2} + a_3(\theta)e^{-t/\tau_3}]$$

r_0 is the intrinsic polarization anisotropy of the fluorescing label or group, and θ is the angle between the transition moment of the fluorescent group and the symmetry axis of the ellipsoid. The weighting factors are functions only of this angle:

$$a_1(\theta) = \left(\frac{3}{2}\cos^2\theta - \frac{1}{2}\right)^2$$

$$a_2(\theta) = 3\cos^2\theta \sin^2\theta$$

$$a_3(\theta) = \frac{3}{4}\sin^4\theta$$

The rotational relaxation times are related to the diffusion coefficients $D_{\parallel}$ and $D_{\perp}$ for rotation about the symmetry axis and perpendicular to the symmetry axis by the equations

$$\tau_1 = \frac{1}{6D_{\perp}}$$

$$\tau_2 = \frac{1}{5D_{\perp} + D_{\parallel}}$$

$$\tau_3 = \frac{1}{2D_{\perp} + 4D_{\parallel}}$$

Using the relations between τ, f_r, and molecular radius R given in Section VI, C, 1, we find that for spheres $\tau = 4\pi\eta_0 R^3/kT$. Since typically $\tau_f = 10^{-9}$ to 10^{-8} second, R must be about 30 Å or less for the fluorescence depolarization technique to be applicable. Only tRNA, among the naturally occurring nucleic acids, is small enough to meet this criterion.

Nanosecond depolarization studies have been made of the fluorescence of ethidium bromide bound to yeast tRNA.[314] The dye is presumably bound rigidly to the tRNA, and has a lifetime $\tau_f = 26$ nsec when bound. In the presence of 0.003 M Mg^+, only one relaxation time of 24.8 nsec was observed, indicating a roughly spherical particle with a Stokes radius of 30.1 Å, axial ratio of 2, and hydration $\Gamma' \approx 1.2$ g/g. In the absence of Mg^{2+}, two relaxation times (15.5 nsec and 36.4 nsec) were observed, indicating a more asymmetric particle. The dimensions suggested for the long and short semiaxes were 70 Å and 15 Å, respectively.

VII RESULTS

In this section we survey some of the results on molecular weights and chain flexibility that have been obtained by the methods discussed in this chapter. In a final section, we discuss DNA with an unusual topology, cyclic DNA.

A. Molecular Weights

We have discussed a large number of methods for determining the molecular weights of nucleic acids. Some of these – electron microscopy, tritium autoradiography, ^{32}P star counting, end group labeling, sedimentation equilibrium, light scattering, and viral or bacterial molecular weight determination combined with measurement of percent nucleic acid content – are "absolute" methods. That is, they yield the molecular weight independent of theoretical assumptions or calibrations, requiring only knowledge of a few independently measurable parameters. Others, notably the Scheraga-Mandelkern relation between $[\eta]$, S, and M and the similar relation between $[\eta]$, τ_1, and M, are based on well-founded if not rigorous theory, and may have achieved the status of "semiabsolute" methods. Others, the correlations of S with M or $[\eta]$ with M, are frankly empirical, or "secondary," and require calibration with samples of known molecular weight.

As a practical matter, it is often easier to measure a sedimentation coefficient or intrinsic viscosity, and then to relate this to molecular weight by an equation such as (5-71) or (5-84). However, as these equations rely for determination of their constants on absolute molecular weight measurements, it is good to review briefly the reliability of these methods. In each case a homogeneous sample of nucleic acid molecules is assumed.

Electron microscopy allows lengths to be measured with considerable precision. The mass per unit length of 192 atomic mass units per Å, characteristic of the B form of NaDNA, is usually assumed to convert length into molecular weight. However, the mass per unit length may vary considerably, depending on the ionic strength at which the microscopy is done. Usually duplex DNA stretches at low ionic strength, leading to overestimates of molecular weight. In consequence, the molecular weight of an unknown nucleic acid is best measured by electron microscopy by comparing the measured length with that of a known nucleic acid on the same grid. This, unfortunately, converts electron microscopy to a secondary method. The same comments hold for tritium autoradiography.

^{32}P star counting is generally held to yield an underestimate of molecular weight, since tracks may be missed and not counted. However, exclusion of stars with a small number of tracks, as being due to background, may in some cases lead to an overestimate of M. End group labeling by kinase most likely leads to an overestimate, due to incomplete labeling; although extra ends, due for example to single-strand breaks or fragmented molecules, could have the opposite effect.

Sedimentation equilibrium tends to underestimate molecular weight, unless virial coefficient corrections are carefully made. The determination of M_2, the molecular weight of anhydrous polynucleotide, also depends on accurate knowledge of the hydration, Γ'. Light scattering on high molecular weight DNA also tends to lead to low values of M, since it is difficult to measure at low enough angles to obtain an accurate extrapolation to zero angle (see Fig. 5-10). Light scattering determination of M also depends sensitively on the value taken for the refractive index increment (dn/dc).

Plaque-count determinations of virus molecular weight tend, as remarked earlier, to lead to overestimates, with consequent overestimation of M for the viral nucleic acid. Other methods of determining viral molecular weight, such as by a combination of sedimentation and diffusion, should be relatively reliable, however.

Use of the Scheraga-Mandelkern equation to determine M depends on an accurate value for β. Too high an estimate for this quantity will lead to low values of M.

In Table 5-5 we see the values obtained for the molecular weights of several coliphage DNAs, using a number of these methods. These values have been critically reviewed in several places.[47,146,315,316] We see that in most cases all of these methods of molecular weight determination give reasonably concordant values. ^{32}P star counting gives results significantly higher than those obtained by other techniques for the larger DNAs from T5$^+$ and T2. The Scheraga-Mandelkern equation, on the other hand, with $\beta = 2.81 \times 10^6$ gives results that are consistently lower than those obtained by other means. It is apparent that the molecular weights of the larger DNAs, such as T5 and T2 (T4) are not known to better than 5%. It is important that more, and more accurate, values of M be obtained in this range, since calibration of S vs. M and $[\eta]$ vs. M plots depends crucially on the high-molecular weight values.

TABLE 5-5 MOLECULAR WEIGHTS ($\times 10^{-6}$) OF SOME COLIPHAGE DNAs

Method	ϕX174 RF	T7	T5st(0)	T5+	T4[a]	T2[a]
Electron microscopic length	3.1 ± 0.2[b]	24.0 ± 1.2[b]	74.5[b]	74.9[321]	119[321]	108 ± 4[b] 116[321]
^{32}P Star counting[315]		25 ± 2	77 ± 2	83 ± 4		132 ± 12
End group labeling[81]		26				
Sedimentation equilibrium[146]		24.8 ± 0.4		68.7 ± 6	113 ± 6	
Light scattering	3.2[88]	25 ± 2[101]				
% Virus mol. wt.	3.4 ± 0.2[88]	21.3 ± 3[317] 25.3 ± 2[318,319]		67.3 ± 3.1[319]	102[320] 105.7 ± 3.8[319]	
S and [η], Eq. (5-85)						
$\beta = 2.5 \times 10^6$	3.4	26.0	68	72.8	109	117
$\beta = 2.81 \times 10^6$	2.9	22.1	57.2	61.1	91.5	96.4

[a] Apart from minor differences in glucosylation, it is generally considered that T2 and T4 DNAs are essentially identical in size. Superscript numbers are references.

[b] Based on M_L = 192 atomic mass units/Å. Molecular lengths as tabulated in Thomas and MacHattie.[2]

The molecular weights and physical properties of many other viral DNA molecules have been tabulated.[2]

B. Polynucleotide Chain Stiffness

1. DOUBLE-STRANDED DNA

Assuming the validity of the wormlike chain model for duplex DNA, the stiffness of DNA is characterized by its persistence length a. The exact magnitude of this parameter has been the subject of much critical discussion.[112,322–324] There are essentially two ways of determining a. Perhaps the most reliable is by hydrodynamics. In this case the empirical data for S or $[\eta]$ as a function of M, such as summarized by Eq. (5-71) is compared with a theoretical relation such as Eq. (5-73). The constants in the empirical equation are then related to the structural parameters (ϵ, a, b) of the wormlike chain. The exact results obtained depend on the details of the theoretical calculations and on the molecular weights used in the empirical relation. However, despite variations in both of these aspects, three groups of workers[321,323,324] have found that for DNA in 0.2 M NaCl, 25°C, pH 7, the persistence length is 600 ± 100 Å. This value was obtained assuming no excluded volume ($\epsilon = 0$). ϵ most likely lies in the range 0.05–0.1 in this solvent, which would reduce estimated values of a by 50–100 Å. A similar estimate of a (415 Å, with $\epsilon = 0.1$) has been made from flow birefringence measurements.[297]

Light scattering is the other method for determining a. The limiting slope of a $Kc/R(\theta)$ vs. $\sin^2(\theta/2)$ plot gives the mean square radius $\langle R^2 \rangle$, according to Eq. (5-48). This in turn can be interpreted in terms of a persistence length according to Eq. (5-21). Although this approach does not rely on imperfectly understood theory, as does that based on hydrodynamics, the practical difficulties are greater. For high-molecular weight DNA, such as T7 DNA, measurements cannot be made at low enough angles to provide true limiting slopes.[96] The apparent limiting slopes are too high. For low-molecular weight DNA, polydispersity produces complications and anisotropy corrections to scattering should be applied. Further, a small uncertainty in limiting slope can lead to a large uncertainty in $\langle R^2 \rangle$ and thus in a.[112] Therefore it appears that the value of $a = 900 \pm 200$ Å adduced from light scattering[322,323] is likely to be high.

It is a testimony to the strength of base-stacking interactions, and other noncovalent forces stabilizing the double helix, as well as restrictions on rotation about backbone bonds, that fairly large numbers of single-strand breaks do not increase the flexibility of the DNA at room temperature, as measured by S and $[\eta]$.[325,326] This is true even in samples with more than one break per persistence length. At temperatures near the melting region, on the other hand, nicked chains are relatively more flexible than intact ones.[326]

As shown in Section II, the persistence length a is related to the bending force constant α, which is twice the free energy needed to produce unit curvature in unit length of rod, by $\alpha = 2kTa$. From standard thermodynamics, we also have

$$\Delta G = \frac{\alpha}{2}\frac{\theta_{\mathcal{L}}{}^2}{\mathcal{L}} = \Delta H - T\Delta S$$

Therefore,

$$a = \frac{\Delta H}{kT}\frac{\mathcal{L}}{\theta_{\mathcal{L}}{}^2} - \frac{\Delta S}{k}\frac{\mathcal{L}}{\theta_{\mathcal{L}}{}^2}$$

Thus by studying s vs. M as a function of T, one can determine a as a function of T and thus ΔH and ΔS, the enthalpy and entropy of bending unit length of rod through unit angle. This has been done by Gray and Hearst,[327] who found that $2a$ varied from 918 ± 32 Å at 5°C to 824 ± 33 Å at 49°C. The resulting analysis gives $\Delta H = 77 \pm 25$ kcal-Å/rad^2-mole, and $\Delta S = -0.17 \pm 0.09$ kcal-Å/rad^2-mole-deg.

The enthalpy of bending is rather small, indicating that not much energy is required to bend native DNA to pack into a phage head, for example. A rough calculation for one model of packing indicates that about 10 kcal – roughly the energy of two hydrogen bonds – would be needed to bend 1000 Å of DNA into the proper configuration.[327] The negative ΔS, if it is real, may indicate a restriction on the conformational mobility of DNA attendant on bending, such as might arise if there were special flexible regions in the molecule.

In principle, a combination of Eqs. (5-71) and (5-73) can also be used to obtain the hydrodynamic diameter b of the double helix. However, the theoretical approximations and experimental uncertainties make this procedure unreliable. The hydrodynamic calculations are not very sensitive to the value chosen for b. A value of 25 Å for the hydrated diameter gives good agreement with experiment[191,248,324] and with X-ray measurements of double-helix spacings.

2. SINGLE-STRANDED POLYNUCLEOTIDES

The values given in Table 5-2 can be used to obtain estimates of the effective segment length b in single-stranded polynucleotides. It has been shown, using Eqs. (5-68) and (5-70) with the chain statistics appropriate to flexible chains with excluded volume Eq. (5-33), that[189]

$$s^0 = \frac{C_D(\epsilon)(1 - \bar{v}_2\rho)M_0^{(1+\epsilon)/2} \times 10^{13}}{N_A \eta_0 b} M^{(1-\epsilon)/2}$$

where s is measured in svedbergs. Clearly, the coefficient of $M^{(1-\epsilon)/2}$ in this equation is K_s, while $a_s = (1 - \epsilon)/2$. $C_D(\epsilon)$ can be adequately approximated by the polynomial

$$C_D(\epsilon) = 0.1955(1 + 1.20\epsilon + 2.5\epsilon^2)$$

for linear chains.[328] As typical values from Table 5-2, we may take $a_s = 0.40$ and $K_S = 0.050$. The former gives $\epsilon = 0.2$, so $C_D\ (0.2) = 0.262$. Taking M_0, the monomer molecular weight, as 330 for the Na salt, $(1 - \bar{v}_2 \rho) = 0.45$, $\eta_0 = 10^{-2}$ poise, we find $b = 12.7$ Å.

A somewhat smaller value comes from light scattering on single-stranded polynucleotides in θ-solvents.[329,330] For example, 2 M NaCl is a θ-solvent for poly rU at 18°C. Under these conditions, light scattering yields[330]

$$\langle L^2 \rangle / n \langle l^2 \rangle = 17.6$$

where $\langle l^2 \rangle$ is the mean square bond length in the polynucleotide backbone. The six bonds in the backbone have lengths 1.56 Å for the bond between $C_3{}'$ and $C_4{}'$; 1.52 Å between $C_4{}'$ and $C_5{}'$; 1.47 Å between $C_5{}'$ and 0; 1.52 Å between 0 and P; 1.60 Å between P and 0; and 1.43 Å between 0 and $C_3{}'$. The total contour length of a nucleotide is therefore 9.10 Å, and $\langle l^2 \rangle = 2.31$ Å^2, so $b^2 = (17.6)(2.31) = 40.7$ Å^2, and $b = 6.4$ Å.

In either case, it is clear that the chain is considerably more extended than would be the case if free rotation were allowed. If the polynucleotide chain behaved like a hydrocarbon chain with tetrahedral bond angles, $\cos\theta = 1/3$ in Eq. (5-14), so $\langle L^2 \rangle / n \langle b^2 \rangle$ would be 2.0. Scott[331] has calculated $\langle L^2 \rangle / n \langle b^2 \rangle$ assuming free rotation about all bonds save that connecting $C_3{}'$ and $C_4{}'$ in the pentose ring, but taking into account the different bond lengths and angles along the chain. He obtained $\langle L^2 \rangle / nb^2 = 3.09$. For polyphosphates, the experimental value[332,333] is 6.6 to 7.2. The much larger value obtained for poly rU is apparently the result of highly restricted rotation about the backbone bonds, as suggested by Sundaralingam[334] and Arnott[335] on the basis of crystallographic analysis.

C. Circular DNA

Circular DNA molecules have, since 1962, been isolated from a wide variety of organisms.[336] These include bacterial viruses, tumor and other animal viruses, bacterial plasmids and sex factors, bacteria, mitochondria, kinetoplasts, and sperm. Circularity has been suggested or demonstrated by genetic linkage maps and by a host of physical techniques including electron microscopy, autoradiography, sedimentation velocity, and buoyant density titrations using alkali or dye molecules.

These circular molecules occur in a variety of forms. One finds single-stranded circles, double-stranded circles in which the ends of the strands are cohesive and joined by noncovalent bonds ("cohered circles"), double-stranded circles in which one strand is covalently cyclized and the other held in a cyclic configuration by base-pairing to the first ("nicked circles"), and double-stranded circles in which both strands are covalently cyclized ("closed circles"). Catenated forms, in which several circles are linked, as in a chain, have also been found. These forms are in many cases interconvertible. By heating or shearing, cohesive circles can be converted to linear molecules. By treatment with enzymes, X-radiation, or chemical agents, closed circles can be converted to nicked circles. Conversely, cohesive circles or nicked circles can be treated with a joining enzyme, polynucleotide ligase, to give closed circles. These transformations will be discussed in greater detail below.

The first circular DNA to be found was the single-stranded DNA from bacteriophage ϕX 174.[337] The circularity of this molecule was demonstrated in several ways. It was found that *E. coli* phosphodiesterase, which attacks single polynucleotide strands with free 3′–OH ends, did not hydrolyze ϕX 174 DNA. Nor did spleen phosphodiesterase, which attacks free 5′–OH termini. Thus ϕX 174 DNA either has no ends, or they are blocked. Sedimentation velocity studies of this DNA indicated two forms, of essentially the same molecular weight, one of which sedimented 10% slower than the other, even in alkakine solvents where hydrogen bonding could not occur. Limited treatment of the faster form with pancreatic DNase, an endonuclease, or with elevated temperatures, at first converted the faster form into the slower. More extensive degradation produced low molecular weight fragments. The more rapidly sedimenting form is the one without free ends. It was therefore concluded that this form was a covalently bonded circle, and that a single break converted it to a linear molecule of the same molecular weight. Because the cyclic chain is more compact, due to its ends being joined, it would be expected to sediment more rapidly. Shortly thereafter, the circular form of ϕX 174 DNA was observed in electron micrographs.[338]

Subsequently, several other single-stranded viral DNAs have been discovered. These have all been shown to be circular. Single-stranded circular DNA can also be produced by denaturation of nicked-circular DNA, as discussed below.

Circular duplex DNA was first produced by Hershey et al.,[339] who found that λ DNA aggregated due to the presence of cohesive sites at the ends of the molecule. By heating to 75°C at low DNA concentration (5 μg/ml) and high salt (0.6 M NaCl) briefly, to dissociated linear aggregates, and then cooling slowly back to room temperature, it was found that molecules were produced that sedimented 13% faster than linear λ DNA. Since a linear dimer (two DNA molecules joined end to end) would be expected to sediment about 40% faster than a linear monomer, since the 13% faster sedimenting molecules could be reversibly converted back to linear by heating or controlled shear, and since their formation occurred at low concentrations designed to enhance intramolecular reactions, it was concluded these molecules were "intramolecularly folded" – or, as they are now called, circular. Formation of these cohered circles implies that the cohesive sites at the ends are complementary in base sequence, so that Watson-Crick base pairing can occur between them.

Several other duplex DNA molecules, some related to λ DNA, have been shown to be cyclizable by annealing.[340–346] Some of these, indeed, will form interspecific dimers (such as λ-P21[340] and λ-ϕ80[341]), indicating essentially identical base sequences in the cohesive ends. Other pairs, such as λ and 186[340,342,343] will not mutually cohere. With these systems, it is also

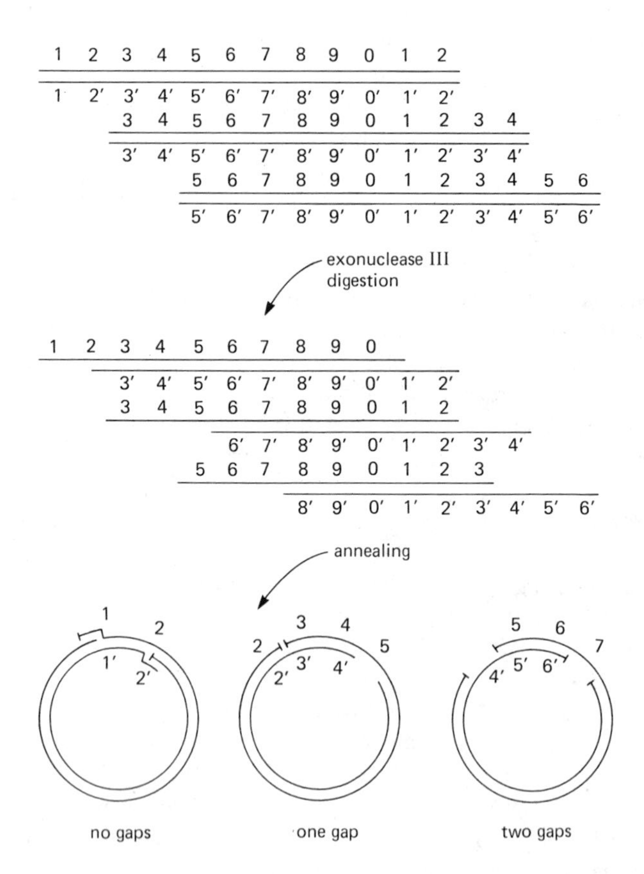

FIGURE 5-34 Scheme of terminal redundancy experiment, as described in text. The primed and unprimed numbers along each line correspond to complementary bases along the two strands of the duplex. [From C. A. Thomas, Jr., *J. Cell. Physiol.*, **70** (Supp. 1) 13 (1967). Reprinted with permission.]

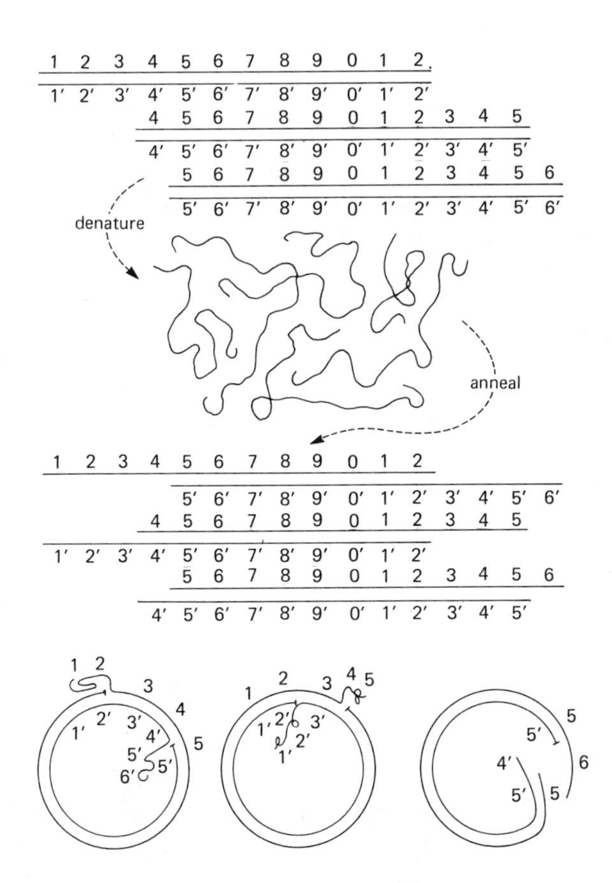

FIGURE 5-35 Circle formation by denaturation and annealing of a permuted collection of duplex DNA molecules. [From C. A. Thomas, Jr., *J. Cell. Physiol.*, **70** (Supp. 1) 13 (1967). Reprinted with permission.]

possible to form linear and circular dimers, linear trimers, and concatamers (interlocked rings).[343]

Cohered circles may be artificially produced in two other ways. Many DNA are terminally redundant.[347] That is, as shown in Fig. 5-34, the beginning sequence is repeated at the end. Degradation of these duplexes with exonuclease III exposes 5′-termini at both ends, and these being complementary, can cohere. This has been done with DNA from a nonglucosylated mutant of T2,[348] T3,[349], and P22.[350] Circles were found in all cases, after 260 to 2000–6000 nucleotides were exposed. It is not clear, however, why several hundred to several thousand bases must be exposed before circles can form in these systems; when single-stranded the cohesive ends of λ DNA contain only 12 nucleotides each.[351]

DNA from bacteriophages T2, T4, and P22 has also been shown to be circularly permuted.[347,350] As shown in Fig. 5-35, cohesive circles may be formed by denaturing and annealing of a population of these double-stranded molecules.

Nicked circles, in which one strand is covalently cyclized, may be produced by introducing one single-strand break in a closed circle.

Each of the above types of circular DNA molecules – single-stranded, cohesive, and nicked – has at least one single-stranded scission in its backbone, about which rotation can occur. Thus these molecules should all have substantial configurational flexibility, and should obey the gaussian or wormlike coil configuration statistics discussed earlier in this chapter. The only restraint on their mobility is that their ends must remain joined. Qualitatively, this has the effect of decreasing their average dimensions.

The angular dependence of light scattering has been calculated[352] for the cyclic chain without excluded volume. The result is

$$P(\theta) = 2u^{-1/2}\, e^{-u/4}\, E(u^{1/2}/2)$$

where u is defined as $(16\pi^2/\lambda^2)\sin^2(\theta/2)\langle R^2\rangle$, and $E(x)$ is the tabulated integral

$$E(x) = \int_0^x e^{t^2}\, dt$$

The reciprocal scattering function, $P(\theta)^{-1}$, is shown in Fig. 5-36 compared with linear and branched chains, the latter with four branches of equal length. It is clear that, in interpreting light scattering from cyclic chains, it is even more necessary than with linear chains to work at low angles to achieve adequate extrapolation to zero angle.

The hydrodynamic behavior of circular DNA is expected to differ from that of linear material of the same molecular weight. Because of its more compact structure, cyclic DNA should have a higher sedimentation coefficient, smaller intrinsic viscosity, and larger extinction angle for given shear gradient. These qualitative predictions are borne out experimentally and in detailed calculations.

For example, it is found that the ratio of sedimentation coefficients for linear and nicked circular DNA, S_L/S_C, is 14.5/16.0 = 0.906 for polyoma virus DNA[353] and 32.0/36.2 = 0.884 for λ DNA.[339] Calculated values,[191] taking into account chain stiffness and a slight difference between excluded volume parameters ϵ_L and ϵ_C, are 0.895 for polyoma DNA and 0.889 for λ DNA.

An empirical relation between $S_C{}^0$ and M for nicked circular DNA is[354]

$$S_C{}^0 - 2.7 = 0.01759\, M^{0.445}$$

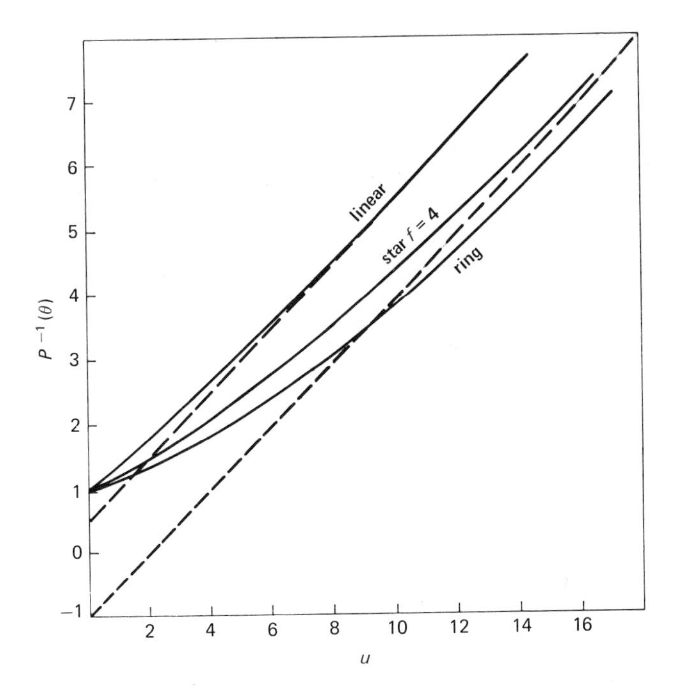

FIGURE 5-36 $1/P(\theta)$ as a function of the variable u defined in the text, for linear chains, star molecules with four equal arms, and cyclic chains. The upper dashed line is the asymptote for the linear model; the lower one, for both stars and rings. [From E. F. Casassa, *J. Polymer Sci.*, **A** 3, 605 (1965). Reproduced by permission of John Wiley & Sons, Inc.]

This is of the form predicted theoretically[191] and should be compared with Eqs. (5-71) and (5-73) for linear chains.

Similar calculations have been made of the intrinsic viscosities of the linear and cyclic forms of flexible polymer chains with excluded volume,[189,355] using the Zimm[235] normal coordinate theory. One may write for cyclic chains,

$$[\eta]_C = 12^{3/2}\,\Phi_C(\epsilon)\,\langle R^2\rangle^{3/2}/M$$

in parallel with Eq. (5-78) for linear chains. In Fig. 5-37 are plotted Φ_L and Φ_C, and the ratio $[\eta]_L/[\eta]_C$, as functions of ϵ, assuming that $\epsilon_L = \epsilon_C$.[189] It is observed that both Φ_L and Φ_C are fairly steeply decreasing functions of ϵ, while the ratio of intrinsic viscosities is predicted to start at 1.55 for $\epsilon = 0$ and rise steadily with ϵ.

In Fig. 5-38 is plotted the ratio $[\eta]_L/[\eta]_C$ for λ DNA, in which the parameter ϵ was varied by varying the ionic strength.[356] The ratio extrapolated to $\epsilon = 0$ is 1.6, in good agreement with the theoretical value

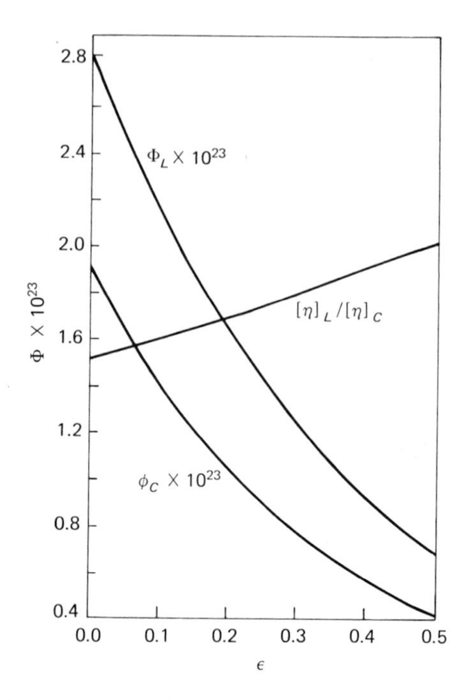

FIGURE 5-37 Intrinsic viscosity of cyclic and linear polymers as function of chain expansion. The Flory parameters for the straight chain, Φ_L, and ring, Φ_C, and the ratio of intrinsic viscosities of the two forms are plotted vs. ϵ. [From V. Bloomfield and B. H. Zimm, *J. Chem. Phys.*, **44**, 315 (1966). Reprinted with permission.]

cited above. The measurements were made at a salt concentration such that $\epsilon = 0.11$; the theoretical ratio is 1.61.

This figure illustrates another interesting point, which is that $[\eta]_L/[\eta]_C$ does indeed rise with increasing ϵ, whether one takes $\epsilon_L = \epsilon_C$ or uses a more sophisticated treatment. It has been argued[355] that because the first-order perturbation theory of excluded volume indicates that the circular chain expands more than the linear, $[\eta]_L/[\eta]_C$ should decrease with increasing excluded volume. This behavior is indicated by the line labeled F–K. The observed upward trend indicates that, although the cyclic chain may initially expand more rapidly under the influence of long-range interactions, the fact that the ends of the linear chain are free enables it to expand to a greater extent than the circular DNA at large values of ϵ.

The flow dichroism of linear and circular λ DNA has been measured by Lee and Davidson.[309] The ratio of dichroisms, $(\Delta A)_L/(\Delta A)_C$, which is equal to the ratio of birefringences. $(\Delta n)_L/(\Delta n)_C$, was found to be 1.85. This is substantially lower than the predicted value of 5.8.[189] The discrepancy is

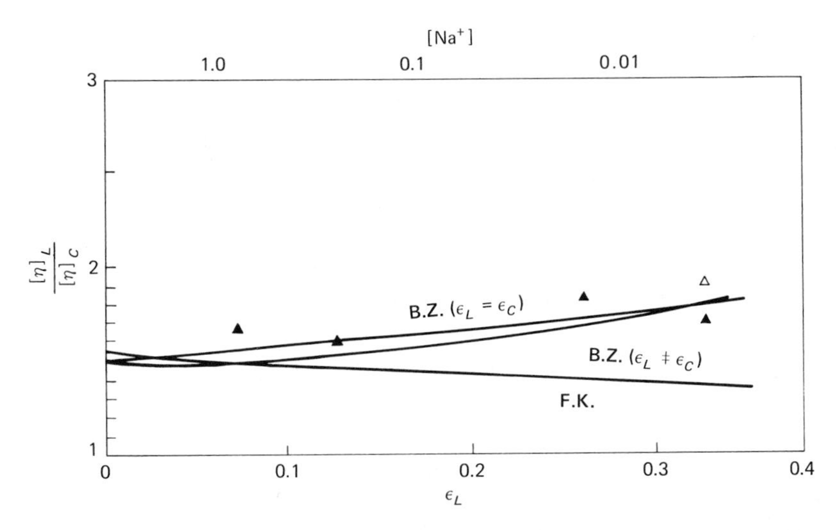

FIGURE 5-38 Intrinsic viscosity quotient for linear and circular λ bacteriophage DNA, as a function of ϵ_L and $[Na^+]$. The lines correspond to various theoretical predictions discussed in the text. [From R. J. Douthart and V. A. Bloomfield, *Biopolymers*, **6**, 1291 (1968). Reproduced by permission of John Wiley & Sons, Inc.]

probably due to the large distortion of the DNA molecules even at $\dot{\epsilon} = 5$ sec^{-1}, the lowest measured shear rate, since the theory is valid only for small deformations.

Cohered circular DNA, particularly from bacteriophage λ, has been studied in considerable detail. It has been postulated[357,358] that the cohesive ends of λ DNA are 5′-terminated single strands projecting from the ends of an otherwise duplex molecule, and that the base sequences on these strands are complementary. It has been shown that left halves of sheared λ DNA cohere only to right halves, indicating complementary ends.[357] Further, cohesive circles are converted to linear monomers, and linear aggregates are disrupted, by exposure to heat, in support of the model of cohesion by base pairing.[339] Each single-stranded end contains 12 nucleotides whose composition and sequence are known.[351]

By measuring the areas under the peaks in a sedimentation velocity experiment with a mixture of linear and circular DNA, it is possible to obtain the relative amounts of the two components. This technique has been used[359] to measure the equilibrium constant K for the reaction

$$L(\text{linear monomer}) \rightleftharpoons C(\text{circular monomer})$$

for DNA from the deletion mutant $\lambda b_2 b_5 c$ of λ phage DNA which has a molecular weight of 25.8×10^6. By obtaining the equilibrium constant as a function of temperature, and using the standard thermodynamic equations

$$\frac{d \ln K}{d(1/T)} = -\frac{\Delta H^0}{R}$$

$$\Delta S^0 = R \ln K + \frac{\Delta H^0}{T} = \frac{\Delta H^0}{T_m}$$

it was found that in 0.13 M Na^+, $\Delta H^0 = -91 \pm 10$ kcal/mole, $\Delta S^0 = -280 \pm 30$ cal/mole-deg; and in 2 M Na^+, $\Delta H^0 = -85 \pm 10$ kcal/mole, ΔS^0 -250 ± 30 cal/mole-deg. The midpoints T_m of the transitions, in which half of L is converted to C, are 50.6°C and 63.6°C in low and high salt, respectively. Coliphage 186 DNA, which has a molecular weight of 19.7×10^6 and cohesive ends that do not cohere with those of λ DNA, has a T_m of 63°C in 0.13 M Na^+.[342]

The cyclization reaction has been proposed to take place in two steps, as shown in Fig. 5-39. In the first step, it is supposed that the ends diffuse into each other's vicinity, within a small volume δV, but sufficiently far apart that there are no electrostatic contributions to ΔH_1 and ΔS_1. These contributions are all contained in the second step, in which base pairing occurs. Then $\Delta H_1 = 0$, and ΔS_1 can be calculated from the well-known Boltzmann equation

$$S = k \ln \Omega$$

where Ω is the number of configurations of equal energy accessible to the molecule. For if $\Omega(\delta V)$ is the number of configurations accessible when both ends are in the same small volume δV, and $\Omega(L)$ is the total number of configurations accessible to the linear chain, then

$$\Delta S_1 = S(\delta V) - S(L) = k \ln \frac{\Omega(\delta V)}{\Omega(L)}$$

According to Eq. (5-6) when the end of a chain of length $\mathcal{L}$ and segment length b is within a small distance l of the beginning of the chain,

$$\frac{\Omega(\delta V)}{\Omega(L)} = \int_0^l W(l', \mathcal{L})\, dl' = \left(\frac{3}{2\pi \mathcal{L} b}\right)^{3/2} \int_0^l \exp\,[-3l'^2/2\mathcal{L}b]\, 4\pi l'^2\, dl'$$

When $l^2 \ll \mathcal{L}b$, the exponential term will be essentially unity and can be taken outside the integral. Also

$$\int_0^l 4\pi l'^2\, dl' = \frac{4\pi l^3}{3} = \delta V$$

Thus, on a per mole basis,

$$\Delta S_1 = R \ln \left[\left(\frac{3}{2\pi \mathcal{L} b}\right)^{3/2} \delta V\right] = R \ln (j \delta V)$$

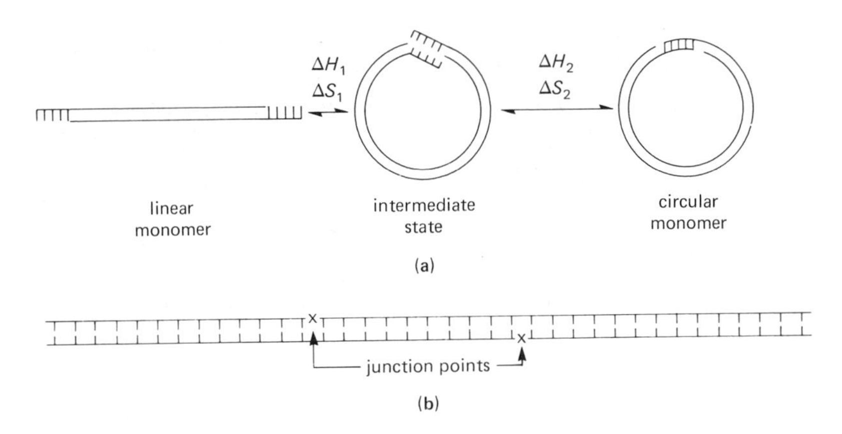

FIGURE 5-39 (a) Two-step model for the cyclization of DNA with cohesive ends. (b) Model for the structure of the helical duplex at the cohesive ends. [From J. C. Wang and N. Davidson, *J. Mol. Biol.*, **15**, 111 (1966). Reprinted with permission.]

j is the Jacobson-Stockmayer[360] factor. It has units of reciprocal volume, and represents the concentration of one end of the molecule in the neighborhood of the other.

Using $\mathcal{L} = 13.2\ \mu$ for $\lambda b_2 b_5 c$ DNA, taking $b(= 2a$, the statistical segment length for DNA) as 717 Å,[12] and guessing $\delta V = 500$ Å^3, Wang and Davidson[359] found $\Delta S_1 = -43 \pm 4.6$ cal/mole-deg. Thus, in 0.13 M Na^+, $\Delta H_2 = \Delta H = \Delta H_1 = -91 \pm 10$ kcal/mole, and $\Delta S_2 = -237 \pm 30$ cal/mole-deg.

These values of ΔH_2 and ΔS_2 may be used to estimate the number of base pairs, n, formed in the joining of the cohesive regions.[359] One has

$$n = \frac{\Delta S_2}{\Delta s}$$

and

$$n = \frac{\Delta H_2 + 7000}{\Delta h}$$

Δs is the entropy change upon incorporating one mole of base pairs into double helical regions. If it is assumed that the single-stranded cohesive ends have the same degree of order as poly A and poly U under similar conditions (50–60°C, 0.1 M salt), and that steric interactions at the junction points marked by "x" in Fig. 5-39 contribute negligibly to ΔS_2, then $\Delta s = -23$ cal/mole-deg.[361] Δh is the enthalpy change for renaturing one mole of base pairs. With assumptions similar to those for Δs, $\Delta h = -7.7$ kcal/mole.[359,361] The extra factor of 7000 cal arises from an additional stacking interaction at the junction points.

Use of these values gives $n = 10 \pm 2$ base pairs. A similar conclusion was arrived at[362] using a more detailed statistical mechanical analysis of the biological melting curve for λ DNA,[363] in which the presence of free ends is measured by the infectivity of the DNA. As noted above, chemical analysis indicates that $n = 12$.

The mechanism corresponding to Fig. 5-39 is

$$L \underset{k_2}{\overset{k_1}{\rightleftharpoons}} I \underset{k_4}{\overset{k_3}{\rightleftharpoons}} C \tag{5-96}$$

where I is the intermediate state. The concentration of I may be assumed small, and the steady-state approximation applied:

$$\frac{d[I]}{dt} = 0 = k_1[L] - (k_2 + k_3)[I] + k_4[C]$$

Along with the rate equation

$$-\frac{d[L]}{dt} = \frac{d[C]}{dt} = k_1[L] - k_2[I]$$

and the conservation of mass restriction, this leads to

$$-\frac{d[L]}{dt} = k[L]$$

where the apparent first-order rate constant k is

$$k = \frac{k_1 k_3 + k_2 k_4}{k_2 + k_3}$$

Measured values of $k(\text{min}^{-1})$ are

$$\log k = 14.9 - \frac{5200}{T}$$

in 2.0 M Na^+, and

$$\log k = 13.6 - \frac{5000}{T}$$

in 0.13 M Na^+. It may be calculated from this last equation that the half-time for cyclization at 37°C is 174 minutes. Since it has been found[364] that DNA injected by λ phage is "instantaneously" converted into a noninfectious, probably circular form, it appears likely that cyclization is accomplished *in vivo* by some specific, nonrandom mechanism. This has been discussed briefly by Wang and Davidson.[365] In terms of the previous equilibrium parameters, we have

$$K = \frac{[C]_{t=\infty}}{[L]_{t=\infty}} = \frac{k_1 k_3}{k_2 k_4}$$

and

$$K' = \frac{k_1}{k_2} = e^{\Delta S_1/R} \approx 4 \times 10^{-10}$$

The question now arises, which of the steps in Eq. (5-96) is rate-limiting? If diffusion-controlled formation of I is rate-limiting, $k_2 \ll k_3$, and

$$k = k_1 + \frac{k_2 k_4}{k_3} = (1 + K^{-1})k_1$$

If, on the other hand, rearrangement of I to give C is rate-limiting, $k_2 \gg k_3$, and

$$k = \frac{k_1 k_3}{k_2} + k_4 = (K + 1)k_4$$

These possibilities may be evaluated as follows. If the reaction is diffusion controlled, k_1, and hence k, should vary as T/η. Thus $k\eta/T$ should be approximately constant. It is found, however,[359] that this quantity increases markedly with T in 2.0 M NaCl. Further, the activation energy is 24 kcal, which is substantially larger than the few kilocalories (~5) expected either for a viscosity-limited diffusion-controlled reaction or for one in which the diffusion rate is governed by rates of internal rotation about chemical bonds along the polymer chain. Thus it appears that formation of base-pairs in the $I \rightleftharpoons C$ conversion is rate-limiting. Since an activation energy of 24 kcal/mole is rather high even for the formation of hydrogen-bonded base pairs, this buttresses the hypothesis that the single-stranded cohesive ends have significant structure that must be disrupted before joining of the ends can occur.

Further evidence that the rate of the cyclization reaction is not diffusion controlled comes from kinetic studies on 186 DNA.[342] If the process were controlled by diffusion together of the ends, due to segmental motion,[359] 186 DNA should cyclize 1.5 times as fast as $\lambda b_2 b_5 c$ DNA. In fact, it cyclizes 10 times as fast. The energy of activation is the same, 24 kcal/mole, as it is for $\lambda b_2 b_5 c$ so the difference must lie in the preexponential factor for the $I \rightleftharpoons C$ process.

It is worth noting, finally, that it has been demonstrated directly that joining of the cohesive ends is the same process for cyclization and for association of left and right ends of sheared half-molecules of $\lambda b_2 b_5 c$ DNA.[366] Measurement of ΔH^0 by studying the variation of equilibrium constant with temperature for association of half-molecules gave -108 ± 10 kcal/mole in 0.13 M Na^+, and -88 ± 10 kcal/mole in 2.0 M Na^+. This agrees within experimental error with the ΔH^0 values for the intramolecular cyclization of -91 ± 10 kcal/mole and -85 ± 10 kcal cited above. Furthermore, the Jacobson-Stockmayer[360] j-factor, defined in Eq. (5-95), is essentially the same theoretically and experimentally. Theoretically, with $\mathcal{L} = 13.2\ \mu$ and $b = 717$ Å, $j = 3.6 \times 10^{11}$ molecules per cubic centimeter. Experimentally, j can be measured both by equilibrium and kinetic methods. If the equilibrium constant for the formation of joined halves (JH) from left halves (LH) and right halves (RH) is

$$K_L = \frac{[JH]}{[LH]\,[RH]}$$

while the equilibrium constant for cyclization is K, then[352]

$$K = jK_L$$

Direct measurement of K and K_L gave[366] $j = (3.4 \pm 1.0) \times 10^{11}$ molecules per cubic centimeter. Agreement with theory is good. Similarly, if the observed rate constant k_L for the joining of right and left halves is defined by

$$\frac{d[JH]}{dt} = k_L\,[LH]\ [RH]$$

while the cyclization rate constant is k, then

$$k = jk_L$$

The kinetically measured value of j is $(2.8 \pm 0.5) \times 10^{11}$ molecules per cubic centimeter,[360] again in reasonable agreement with theory and equilibrium experiments.

Closed circular DNA may be recognized by a variety of properties, prominent among which is its greater compactness in solution as reflected in an elevated sedimentation coefficient.[353] This compactness was attributed by Vinograd et al.[353] to a twisted, or supercoiled, configuration of the closed circular DNA. This is shown in Fig. 5-40. As will become apparent from our subsequent discussion, however, closed circular DNA is not necessarily twisted. The closed circular DNA, component I, has a sedimentation coefficient of 20 S in neutral 1 M NaCl. This species may undergo various transformations, which change its sedimentation coefficient and other measurable properties. A single cleavage, or a small number of them, can be introduced into I in a variety of ways. This relaxes the twists, and gives rise to a more expanded, nicked circular molecule II, with sedimentation coefficient of 16 S. Denaturation of II gives separated, single-stranded DNA chains, one linear (16 S) and one circular (18 S). Introduction of a double-strand cleavage into I, using *E. coli* endonuclease I, yields linear species III with a sedimentation coefficient of 14 S. Alkaline denaturation of I leads to an unwinding of the supercoil twists until a closed circular species I′, with the same 16 S sedimentation as species II. Further denaturation leads to imposition of twists of opposite sense, and ultimately to a highly compact, 53 S structure. The experimental sedimentation velocity data reflecting these

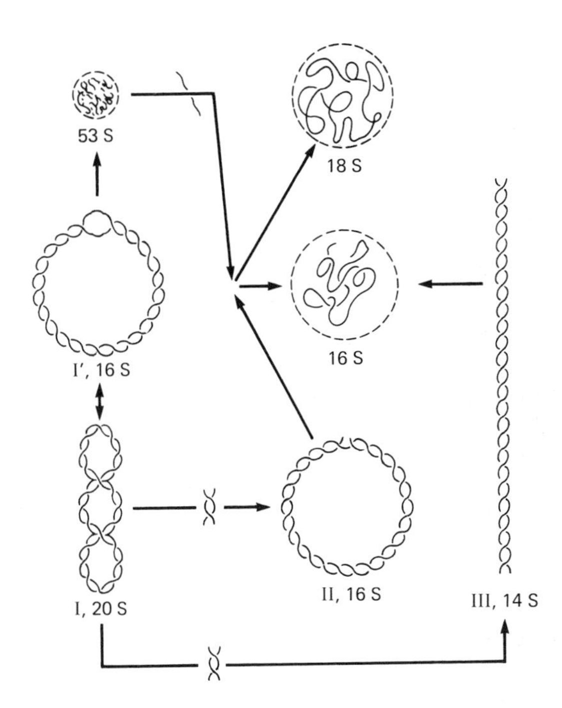

FIGURE 5-40 The several circular and linear forms of polyoma DNA. The dashed circles around the denatured forms indicate the relative hydrodynamic diameters. The sedimentation coefficients were measured in neutral and alkaline NaCl solutions. [From J. Vinograd et al., *Proc. Nat. Acad. Sci., U.S.,* **53**, 1104 (1965). Reprinted with permission.]

transformations, and equivalent ones in acid solution, are shown in Fig. 5-41.[336,353]

The nature of the twists in closed circular DNA has been the subject of much investigation. The fundamental fact is that once the strands of the duplex molecule are covalently cyclized, the total number of turns of one strand about the other remains a topological constant. Thus, changes in the number of Watson-Crick turns in the closed molecule will be uniquely related to the number of supercoil turns. The mathematical statement of this fact is[367,368]

$$\tau = \alpha - \beta \tag{5-97}$$

α is the "topological winding number," and equals the number of 360° revolutions made by one of the strands about the duplex axis when that axis is forced to lie in a plane. β is the "duplex winding number," and equals the number of revolutions of one strand about the duplex axis when the molecule is not constrained to lie in a plane. It, therefore, represents the number of

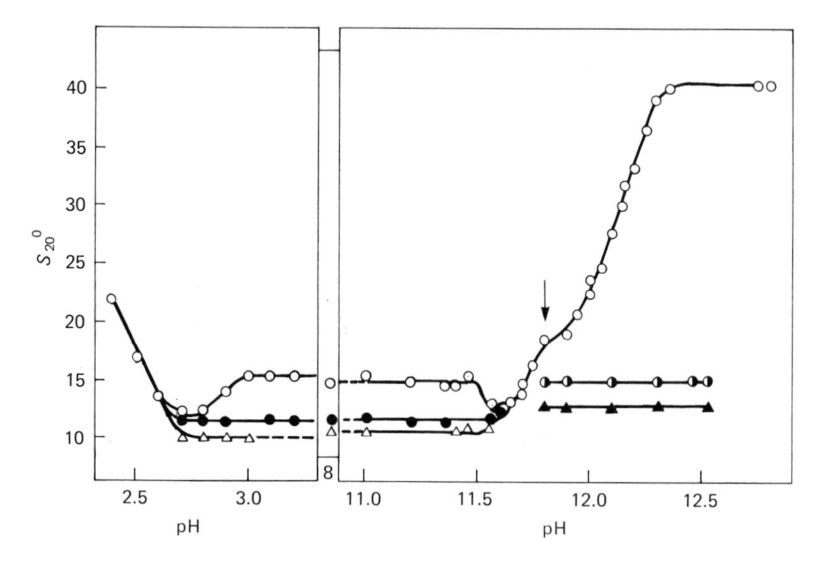

FIGURE 5-41 Sedimentation velocity – pH titration of the closed circular (○), nicked circular (●), and linear (△) duplex forms of polyoma DNA. [From J. Vinograd and J. Lebowitz, *J. Gen. Physiol.*, **49**, 103 (1966). Reprinted with permission.]

Watson-Crick turns in the molecule in solution. The difference between these two is τ, the "superhelix winding number," which equals the number of revolutions made by the duplex around the superhelix axis.[367]

These relations are illustrated in Fig. 5-42 for a closed circular "molecule" whose strands are not interwound. Thus $\alpha = 0$. The convention is adopted that right-handed duplex turns are positive. Two types of superhelices may be formed, as indicated in Figs. 5-42(c) and 5-42(d). The first is a *toroidal* superhelix. The sign convention is that right-handed toroidal superhelical turns are positive. In order to form the three left-handed toroidal superhelical turns as shown in Fig. 5-42(c), it is necessary to introduce three right-handed duplex turns. Thus, $\alpha = 0$, $\beta = +3$, $\tau = -3$. In Fig. 5-42(d) is shown an *interwound* superhelix. This is the type that probably exists in closed circular DNA, at least at moderate or high salt. Here, in order to introduce three right-handed interwound turns, three right-handed duplex turns must also be introduced. However, the sign convention for interwound turns is opposite that for toroidal turns – right-handed turns being counted as negative – because of a 90° change in the viewing axis. Thus, again $\alpha = 0$, $\beta = +3$, and $\tau = -3$.

Superhelical turns may be envisioned to arise in the following way. Consider a linear duplex in which each strand contains 1000 nucleotides. Imagine that only 900 of the possible base-pairs are formed before the duplex is covalently cyclized. Thus, since there are ten base-pairs per turn of the

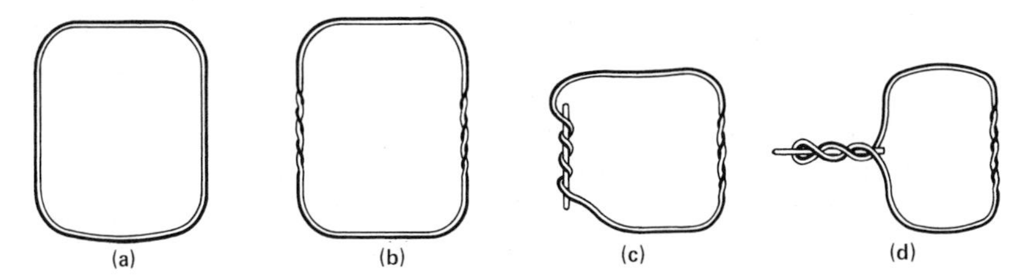

FIGURE 5-42 An illustration of the formation of superhelical turns in a closed duplex consisting of two noninterwound strands. The topological winding number, α, in this figure has a value of zero. (a) Two noninterwound strands. The duplex winding number, β, and the superhelical winding number, τ, have values of zero. (b) Three right-handed duplex turns and three left-handed duplex turns have been wound into the duplex in (a). The right- and left-handed duplexes cancel each other; $\alpha = \beta = \tau = 0$. (c) Three right-handed duplex turns and three left-handed superhelical turns have been introduced into the duplex in (a). The separate strands in the superhelical region rotate about each other. In this arrangement $\alpha = 0$, $\beta = +3$, and $\tau = -3$. The superhelical turns may be regarded as forming a section of a toroidal superhelix. (d) Three right-handed duplex turns and three right-handed interwound superhelical turns have been introduced into the duplex in (a). The paired strands in the interwound superhelix do not rotate around each other. In this arrangement $\alpha = 0$, $\beta = +3$, and $\tau = -3$. See text for the explanation of the sign convention for τ in interwound superhelixes. [From J. Vinograd, J. Lebowitz, and R. Watson, *J. Mol. Biol.*, **33**, 173 (1968). Reprinted with permission.]

duplex, $\alpha = 900/10 = 90$. Now suppose that the remaining 100 base pairs were formed, so that 10 additional right-handed duplex turns are formed, and $\beta = 100$. Thus, according to Eq. (5-97), $\tau = -10$. If the resulting superhelix is an interwound one, the sign convention indicates that these are right-handed superhelical turns. The reader is encouraged to experiment with flexible rubber tubing or telephone cords to verify these relationships.

It is evident from Fig. 5-42(d), and worth noting, that a superhelix with τ turns will, when projected onto a plane, cross itself $|\tau|$ times and exhibit $|\tau| + 1$ loops.

The sedimentation velocity behavior of closed circular DNA has been extensively studied. Particularly interesting is a sedimentation velocity-pH titration, as shown in Fig. 5-41. It is observed that component I between pH 3 and 11.4 has a sedimentation coefficient substantially higher than that of component II. At pH 11.5 (and 3.0 on the acid side) there is an early melting transition, occurring before the alkaline denaturation transition for components II and III, in which the sedimentation coefficient of I decreases until it equals that of II. This is the $\mathrm{I} \rightarrow \mathrm{I}'$ transition in Fig. 5-40. This early helix-coil transition occurs because there is a positive free energy of superhelix formation.[367] This free energy is stored in the superhelix as reduced conformational entropy and as bending energy. Since the free energy of the superhelix is higher than that of the nicked circle, it is less stable and, therefore, more susceptible to denaturation, up to the point at which all the right-handed twists are removed. Subsequent denaturation is more difficult, however, because the invariance of α imposes constraints on unwinding. Therefore, the later helix-coil transition, above pH 11.7 in Fig. 5-41, takes place at higher pH for component I than for II or III, and is broader because the topological restrictions make it somewhat anticooperative.

The fact that, in the early helix-coil transition region, S decreases, proves that the twists in the superhelix are right-handed. The duplex winding number, β, in Eq. (5-97), decreases on denaturation. Thus, in order for α to remain constant, τ must increase. That is, the number of right-handed twists must decrease, or the number of left-handed twists must increase. If the latter were the case, the molecule would be wound up even more tightly, and S would increase. This is contrary to observation, so the twists are right-handed.

An alkaline pH-sedimentation velocity titration, such as that shown in Fig. 5-41, can be used to estimate the superhelix winding number. If β_0 is the number of duplex turns in the untitrated superhelical molecule, and p is the degree of titration, then

$$\beta = \beta_0(1 - p)$$

This assumes that titration of ten base pairs unwinds one duplex turn. The quantity of interest is τ_0, the superhelix number in the untitrated molecule:

$$\tau_0 = \alpha - \beta_0$$

At the minimum in the titration curve, $\tau = 0$ and $p = p_c$, the critical degree of titration. These relations give

$$\tau_0 = -\beta_0 p_c \qquad (5\text{-}98)$$

It is mentioned below that p_c, as determined by titration in alkaline buoyant density gradients, is 0.032 ± 0.002 for polyoma DNA, and β_0 is 470 based on a molecular weight of 3.1×10^6 and an average residue molecular weight of 332. Thus τ_0 for polyoma DNA is -15 ± 1.[367] This analysis is based on the assumption that the average rotation of the duplex is independent of τ. However, the tendency to release the superhelical turns will lead to a decrease in the average rotation of the duplex.[369] That is, the number of base pairs per turn of the helix will be increased, β_0 will be decreased below the number calculated from the molecular weight, and thus the true $|\tau_0|$ will be lower than that calculated from the above equation. The importance of this correction is, as yet, uncertain.

As discussed in Chapter 6, as duplex DNA is titrated in alkaline CsCl, the guanine and thymine residues, freed from their hydrogen bondage, are deprotonated. The resulting negative charge is partially neutralized by Cs ions which, being heavy, increase the buoyant density of the DNA. Such a buoyant density titration for covalent circles and nicked circles of polyoma DNA is shown in Fig. 5-43.[367] The early helix-coil transition for component I is evident from its increased buoyant density between pH 11.4 and 11.8. At

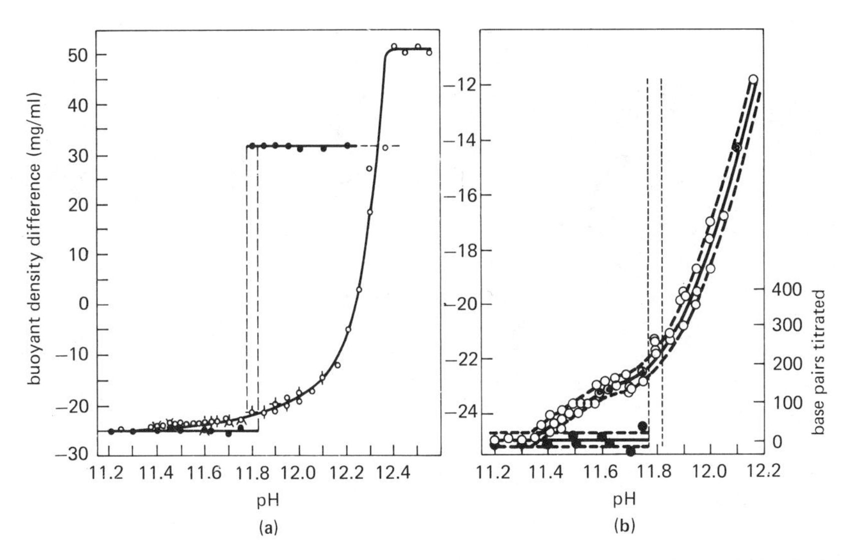

FIGURE 5-43 (a) The buoyant density titration in alkaline CsCl of polyoma DNA, both closed circular form I (○) and singly nicked form II (●). The dashed lines enclose the region in which the abrupt titration of II occurs. (b) An enlargement of the central region of (a). The right-handed ordinate gives the number of base pairs that have been titrated. [From J. Vinograd, J. Lebowitz, and R. Watson, *J. Mol. Biol.,* **33**, 173 (1968). Reprinted with permission.]

the latter pH, the degree of titration (p) is $p_c = 0.032$, so that 3.2% of the bases are titrated when all the initial supertwists have relaxed. As calculated above, this leads to $\tau_0 = -15 \pm 1$, so that there is about one twist in 1000 Å of contour length (or 300 base pairs). The breadth of the late transition, and the enhanced stability of component I to denaturation, is again apparent in Fig. 5-43.

Another useful method for determining the degree of supercoiling is titration with ethidium bromide.[370] This drug, as discussed in Chapter 7, apparently intercalates between the base pairs in duplex DNA, causing local unwinding. It is therefore evident that ethidium bromide can be used just like acid or alkali to estimate τ_0.

Equations (5-97) and (5-98) need to be modified slightly to reflect the geometry of the intercalation. If r is the number of dye molecules bound per nucleotide, and ϕ is the angle in degrees through which the duplex is unwound upon binding, then p must be replaced by $10(\phi/360) \cdot 2r$. Equation (5-98) is then replaced by

$$\tau_0 = -\frac{\phi r_c \beta_0}{18}$$

which, if $\phi = 12°$, becomes $\tau_0 = -(2/3) r_c \beta_0$.

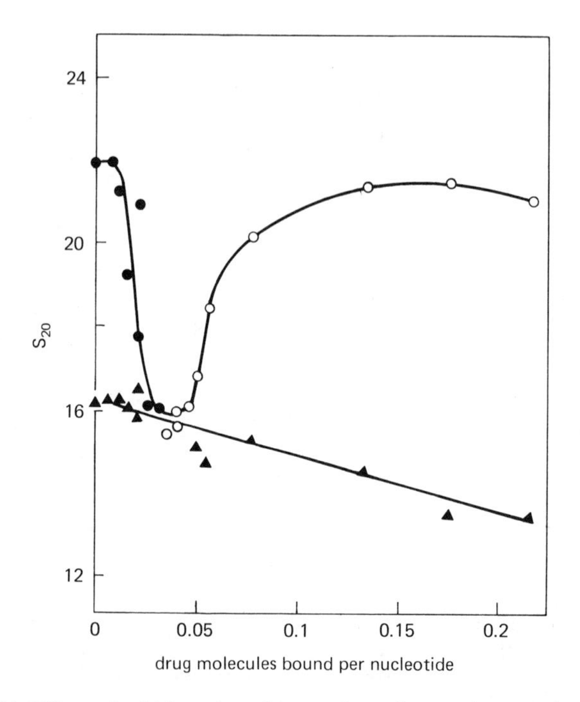

FIGURE 5-44 Effect of ethidium bromide on the sedimentation velocity of polyoma virus DNA. Both closed circular (○) and nicked circular (Δ) forms were present. (●) indicates that only a single sedimenting boundary was formed in the mixture. [From L. V. Crawford and M. J. Waring, *J. Mol. Biol.*, **25**, 23 (1967). Reprinted with permission.]

Figure 5-44[370] shows the effect of ethidium bromide on the sedimentation coefficient of polyoma virus DNA. As seen above in the alkaline titration, S_{I} first decreases until it equals S_{II} and then increases again with greater binding. This behavior again indicates that the superhelical turns are right-handed. The critical number of drug molecules bound per nucleotide, r_c, is 0.037, which, in connection with Eq. (5-99) and the value for β_0 given above, gives $\tau_0 = -12 \pm 3$. More recent studies[371,372] give similar results.

Buoyant density titrations can also be performed with ethidium bromide. Figure 5-45[371] shows such a titration with SV 40 DNA, a viral DNA very similar to polyoma DNA. Binding of ethidium bromide lowers the buoyant density of DNA. Since closed circular DNA initially unwinds its duplex turns more readily than nicked circular DNA, it will bind the dye more readily, and thus its buoyant density will decrease faster at first. When $\tau = 0$, further unwinding is more difficult for I than for II, so the buoyant density of the former rises above that of the latter. From analysis of curves of this type, a value for τ_0 of -12.7 ± 1.5 has been determined, in good agreement with the previous determinations.

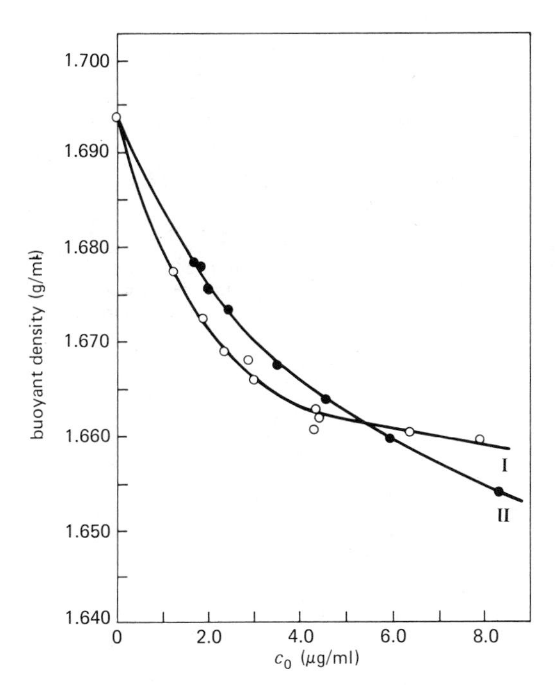

FIGURE 5-45 The buoyant densities of SV 40 DNA forms I (○) and II (●) in CsCl containing low concentrations c_0 of ethidium bromide. [From W. Bauer and J. Vinograd, *J. Mol. Biol.*, **47**, 419 (1970); **54**, 281 (1970). Reprinted with permission.]

Finally, one may hope to determine τ_0 simply by counting crossovers in the electron microscope. This technique presents some difficulties, because a certain number of accidental crossovers are bound to occur by the molecule settling on top of itself. To a certain extent, this may be alleviated by spreading forces which operate upon formation of the protein film. In favorable instances, at any rate, the number of twists seen in electron microscopy is comparable to that deduced from other techniques. This is shown in Fig. 5-46, which shows histograms of the number of crossovers in DNA from mitochondria and ϕX 174.[373] Under favorable circumstances, it is even possible to show directly from electron microscopy that the superhelix turns are right-handed. Such micrographs are shown in Fig. 5-47 for P22 viral DNA.[350]

The question now arises: what is the origin of these superhelical turns? Two suggestions immediately present themselves. The first is that a "blister" of unwound DNA is formed when the duplex is covalently cyclized. If the mechanism of closure were the same for all DNAs, the size of the blister should be the same, and thus τ_0 should be constant, regardless of the size of the DNA. The second suggestion is that the pitch of the Watson-Crick helix is greater at the time of closing than it is under the conditions of observation.

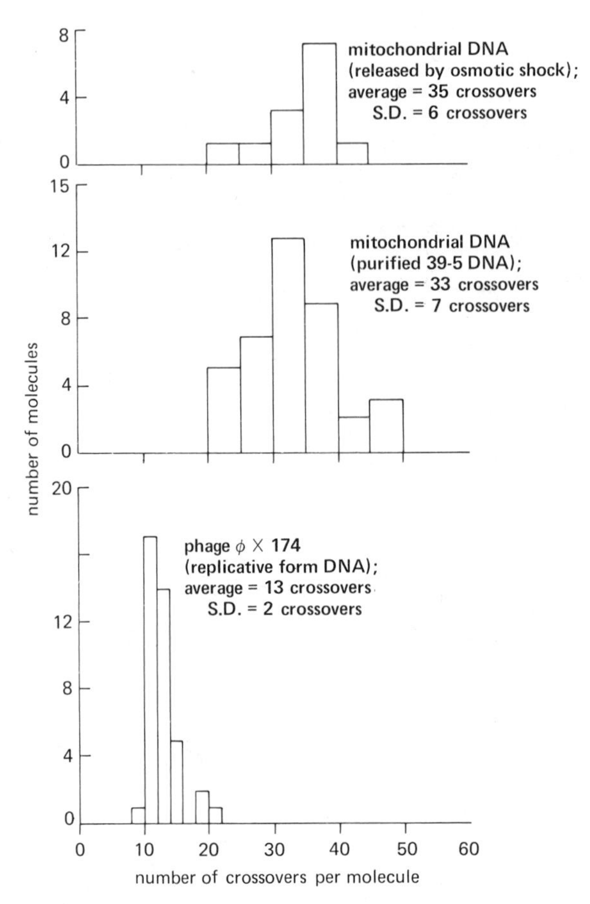

FIGURE 5-46 Histogram of the number of crossovers in twisted circular DNA from various sources. [From E. F. J. van Bruggen et al., *Biochem. Biophys. Acta,* **161**, 407 (Fig. 4) (1968). Reprinted with permission.]

Thus if all DNAs were closed under similar conditions, τ_0/β_0, the superhelix density, should be a constant.

The available evidence supports the latter of these postulates. For example, τ_0 has been measured by sedimentation velocity in ethidium bromide for four covalently closed cyclic DNAs from *E. coli.*[374] Although there was about a 20-fold range in molecular weights of these DNAs (from 1.45×10^6 to 33×10^6), this was accompanied by a comparable variation in τ_0 (from -8.5 to -136) so that the superhelix density was roughly constant.

It may be, in fact, that DNA molecules which are observed to be supercoiled in 1 *M* NaCl and 20°C in the ultracentrifuge, are not supercoiled in the cell. For example, it has been shown[375–377] that molecules closed *in vitro* by ligase do not exhibit supercoiling in media similar to that in which

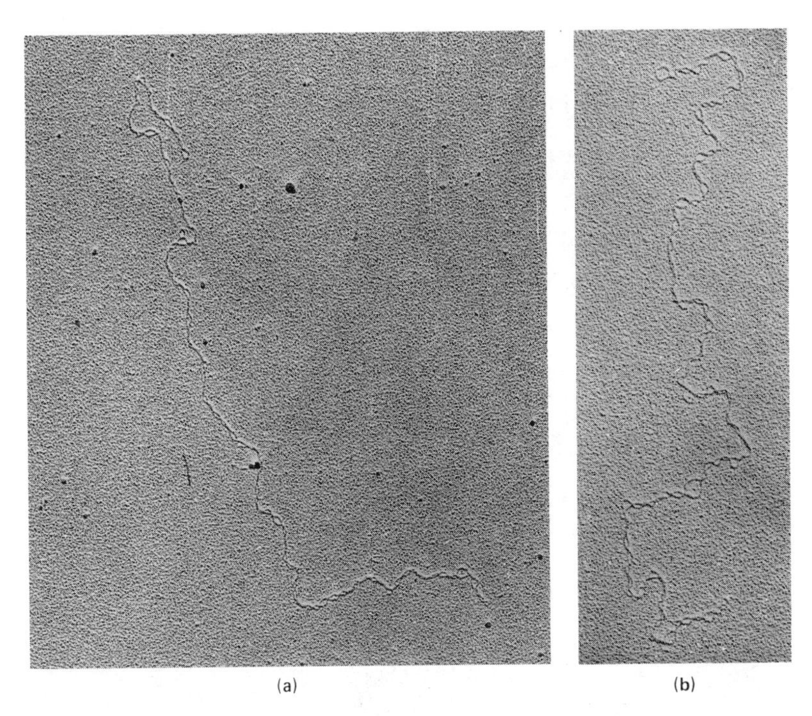

(a) (b)

FIGURE 5-47 Two examples of P22 superhelical DNA in which the right-handed screw sense of the tertiary turns can be seen. [From M. Rhoades and C. A. Thomas, Jr., *J. Mol. Biol.*, **37**, 41 (1968). Reprinted with permission.]

they are cyclized. Further, it has been demonstrated by sedimentation analysis[369,377,378] that both ionic strength and temperature can drastically change the number of superhelical turns. It therefore indeed seems that most supercoiling arises from tilting the bases, or changing the pitch, of the Watson-Crick helix. However, the evidence so far is insufficient to rule out small but significant contributions to supercoiling from other mechanisms.[374]

Correlations have been proposed between S and M for covalently closed DNA.[354] However, because it is apparent from the above discussion that the sedimentation coefficient of a closed circle will be a complex function of its degree of supercoiling as well as its molecular weight, such correlations should be used with caution in determining M. Conversion of the closed circle to a nicked circle and use of the correlation between $S_c{}^0$ and M given above is the procedure of choice.

There have been several theoretical calculations of the sedimentation coefficients of supercoiled DNA.[355,379,380] These all employ the hydrodynamic theory discussed in Section VI,A, but use different chain statistics. For polyoma DNA, which is rather short, a rigid model[380] gives good results.

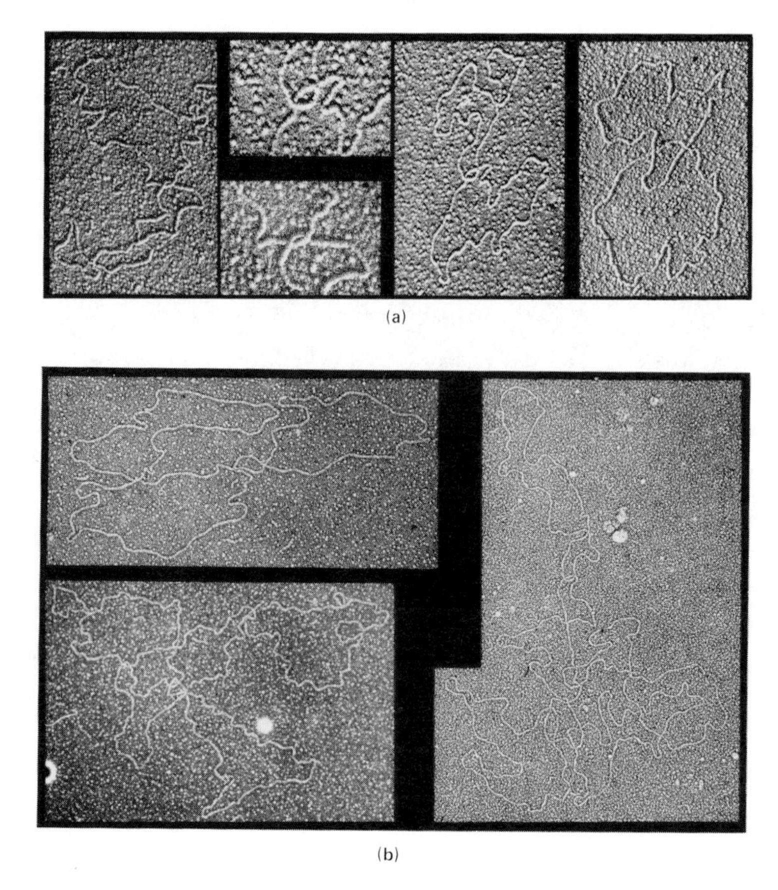

(a)

(b)

FIGURE 5-48 Dimeric (a) and oligomeric (b) forms of catenated mitochondrial DNA. [From B. Hudson and J. Vinograd, *Nature,* **216**, 650 (1967). Reprinted with permission.]

In general, however, the combination of chain stiffness, excluded volume, and supercoiling leads to complications too great for the calculated hydrodynamic behavior to be a reliable guide to degree of supercoiling.

The intrinsic viscosity of closed circular ϕX 174 has been measured.[381] It is 6 dl/g, substantially below the 15 dl/g found for nicked circular ϕX 174 DNA. This is to be expected for a tightly coiled molecule.

Finally, we briefly consider catenated DNA, in which two or more circular molecules are "topologically" bonded, as are the links of a chain. DNA molecules of this type were first prepared by cyclizing λ-DNA in the presence of high concentrations of cyclic 186 DNA.[343] It was shown that the base sequences in the cohesive ends of these molecules were uncomplementary, and thus λ DNA cyclization was uncomplicated by binding to 186 DNA. Proof that catenated structures were formed was obtained by using 5-bromouracil-labeled λ DNA, and observing a peak of intermediate buoyant density in a density gradient sedimentation equilibrium experiment.

Catenated DNA molecules have also been found in mitochondrial DNA from HeLa cells[382] and human leukemic leucocytes.[383] In both cases, the existence of catenanes was shown by density gradient binding experiments with ethidium bromide. Since at high drug concentrations nicked circles bind more ethidium bromide than closed circles, and therefore have a lower buoyant density, one would expect a mixture of these species to form two bands in a density gradient. Since ethidium bromide is fluorescent when bound to DNA, the bands can be simply detected by fluorescence. In fact, a third band of intermediate density was observed, which in HeLaDNA constituted about 10% of the DNA.[382] The identification of this intermediate density species as a catenane has been confirmed by electron microscopy, as shown in Fig. 5-48. Also shown in this figure are some higher oligomers, which can in principle exist in several topologically isomeric forms. It has been hypothesized that these catenanes are formed by recombination, or crossing over, of circular DNA molecules.[382] Further information on the occurrence and properties of these complex mitochondrial DNAs may be found in reviews.[354,384,385]

REFERENCES

1 J. Josse and J. Eigner, *Ann. Rev. Biochem.*, **35**, 789 (1966).
2 C. A. Thomas, Jr., and L. A. McHattie, *Ann. Rev. Biochem.*, **36**, 485 (1967).
3 J. Eigner in L. Grossman and K. Moldave, eds., *Methods in Enzymology*, Vol. 12B, Academic Press, New York (1968), p. 386.
4 P. J. Flory, *Principles of Polymer Chemistry*, Cornell University Press, Ithaca (1953).
5 C. Tanford, *Physical Chemistry of Macromolecules*, Wiley, New York (1961).
6 B. H. Zimm and W. H. Stockmayer, *J. Chem. Phys.*, **17**, 1301 (1949).
7 W. Kühn, *Kolloid Z.*, **76**, 258 (1936).
8 O. Kratky and G. Porod, *Rec. Trav. Chim.*, **68**, 1106 (1949).
9 J. J. Hermans and R. Ullman, *Physica*, **18**, 951 (1952).
10 S. Heine, O. Kratky, and G. Porod, *Makromol. Chem.*, **44–46**, 682 (1961).
11 H. E. Daniels, *Proc. Roy. Soc. (Edinburgh)*, **A63**, 290 (1952).
12 J. E. Hearst and W. H. Stockmayer, *J. Chem. Phys.*, **37**, 1425 (1962).
13 L. D. Landau and E. M. Lifshitz, *Statistical Physics*, Pergamon Press, London (1958), p. 478.
14 M. Fixman, *J. Chem. Phys.*, **23**, 1656 (1955).
15 P. J. Flory, *J. Chem. Phys.*, **17**, 303 (1949).
16 P. J. Flory and T. G. Fox, Jr., *J. Amer. Chem. Soc.*, **73**, 1904 (1951).

17 P. J. Flory and T. G. Fox, Jr., *J. Polymer Sci.*, **5**, 745 (1950).
18 W. H. Stockmayer, *Makromol. Chem.*, **35**, 54 (1960).
19 F. T. Wall, S. Windwer, and P. J. Gans, *Methods Computational Phys.*, **1**, 217 (1963).
20 W. Stockmayer, M. Kurata, and A. Roig, *J. Chem. Phys.*, **33**, 151 (1960).
21 M. Fixman and W. Stockmayer, *J. Polymer Sci.*, **C1**, 137 (1963).
22 A. Peterlin, *J. Chem. Phys.*, **23**, 2464 (1955).
23 H. B. Gray, Jr., V. A. Bloomfield, and J. E. Hearst, *J. Chem. Phys.*, **46**, 1493 (1967).
24 P. Sharp and V. A. Bloomfield, *J. Chem. Phys.*, **48**, 2149 (1968).
25 A. Kleinschmidt in *Methods in Enzymology*, Vol. 12B, L. Grossman and K. Moldave, eds., Academic Press, New York (1968), p. 361.
26 M. Beer in *Methods in Enzymology*, Vol. 12B, L. Grossman and K. Moldave, eds., Academic Press, New York (1968), p. 377.
27 C. E. Hall and M. Litt, *J. Biophys. Biochem. Cytol.*, **4**, 1 (1958).
28 P. F. Davison, *Proc. Nat. Acad. Sci., U.S.*, **45**, 1560 (1959).
29 C. Levinthal and P. F. Davison, *J. Mol. Biol.*, **3**, 674 (1961).
30 M. Beer, *J. Mol. Biol.*, **3**, 263 (1961).
31 I. Bendet, E. Schachter, and M. A. Lauffer, *J. Mol. Biol.*, **5**, 76 (1962).
32 A. Kleinschmidt and R. Zahn, *Z. Naturforsch.*, **14b**, 770 (1959).
33 D. Lang, A. K. Kleinschmidt, and R. K. Zahn, *Biochim. Biophys. Acta*, **88**, 142 (1964).
34 D. Lang and M. Mitani, *Biopolymers*, **9**, 373 (1970).
35 A. Kleinschmidt, D. Lang, and R. K. Zahn, *Z. Naturforsch.*, **16b**, 730 (1961).
36 A. K. Kleinschmidt, D. Lang, and R. K. Zahn, *Biochim. Biophys. Acta*, **61**, 857 (1962).
37 T. H. Dunnebacke and A. K. Kleinschmidt, *Z. Naturforsch.*, **22b**, 159 (1967).
38 L. A. MacHattie and C. A. Thomas, Jr., *Science*, **144**, 1142 (1964).
39 D. Lang, H. Bujard, B. Wolff, and D. Russell, *J. Mol. Biol.*, **23**, 163 (1967).
40 J. G. Wetmur, N. Davidson, and J. V. Scaletti, *Biochem. Biophys. Res. Commun.*, **25**, 684 (1966).
41 K. B. Easterbrook, *J. Virology*, **1**, 643 (1967).
42 C. N. Gordon and A. K. Kleinschmidt, *Biochim. Biophys. Acta*, **155**, 305 (1968).
43 A. K. Kleinschmidt, S. J. Kass, R. C. Williams, and C. A. Knight, *J. Mol. Biol.*, **13**, 749 (1965).
44 R. B. Inman, *J. Mol. Biol.*, **18**, 464 (1966).
45 R. B. Inman, *J. Mol. Biol.*, **25**, 209 (1967).
46 Y. Becker and I. Sarov, *J. Mol. Biol.*, **34**, 655 (1968).
47 D. Freifelder, *J. Mol. Biol.*, **54**, 567 (1970).
48 P. J. Highton and M. Beer, *J. Mol. Biol.*, **7**, 70 (1961).
49 D. Lang, A. K. Kleinschmidt, and R. K. Zahn, *Biophysik*, **2**, 73 (1964).
50 D. Freifelder, A. K. Kleinschmidt, and R. L. Sinsheimer, *Science*, **146**, 254 (1964).
51 D. Freifelder and A. K. Kleinschmidt, *J. Mol. Biol.*, **14**, 271 (1965).
52 A. M. Fiskin and M. Beer, *Science*, **159**, 1111 (1968).
53 B. C. Westmoreland, W. Szybalski, and H. Ris, *Science*, **163**, 1343 (1969).
54 C. C. Richardson, C. L. Schildkraut, and A. Kornberg, *Cold Spring Harbor Symp. Quant. Biol.*, **28**, 9 (1963).
55 R. B. Inman and R. C. Baldwin, *J. Mol. Biol.*, **5**, 172 (1962).
56 H. Ris and B. L. Chandler, *Cold Spring Harbor Symp. Quant. Biol.*, **28**, 1 (1963).
57 W. Stoeckenius, Appendix to R. Weil and J. Vinograd, *Proc. Nat. Acad. Sci., U.S.*, **50**, 730 (1963).

58 A. K. Kleinschmidt, A. Burton, and R. L. Sinsheimer, *Science,* **142**, 961 (1963).
59 A. K. Kaiser and R. B. Inman, *J. Mol. Biol.,* **13**, 78 (1965).
60 J. Vinograd, J. Lebowitz, R. Radloff, R. Watson, and P. Laipis, *Proc. Nat. Acad. Sci., U.S.,* **53**, 1104 (1965).
61 B. Hudson and J. Vinograd, *Nature,* **216**, 647 (1967).
62 A. Katchalsky and S. Lifson, *J. Polymer Sci.,* **11**, 409 (1953).
63 J. Cairns, *Cold Spring Harbor Symp. Quant. Biol.*, **28**, 43 (1963).
64 H. R. Bode and H. J. Morowitz, *J. Mol. Biol.,* **23**, 191 (1967).
65 M. Beer and E. N. Moudrianakis, *Proc. Nat. Acad. Sci., U.S.,* **48**, 409 (1962).
66 E. N. Moudrianakis and M. Beer, *Biochim. Biophys. Acta,* **95**, 22 (1965).
67 E. N. Moudrianakis and M. Beer, *Proc. Nat. Acad. Sci., U.S.,* **53**, 564 (1965).
68 R. W. Davis and N. Davidson, *Proc. Nat. Acad. Sci., U.S.,* **60**, 243 (1968).
69 C. S. Lee, R. W. Davis, and N. Davidson, *J. Mol. Biol.*, **48**, 1 (1970).
70 J. Cairns, *J. Mol. Biol.*, **3**, 756 (1961).
71 J. Cairns, *Cold Spring Harbor Symp. Quant. Biol.*, **27**, 311 (1962).
72 J. Cairns, *J. Mol. Biol.*, **4**, 407 (1962).
73 C. A. Thomas, Jr., and L. A. MacHattie, *Proc. Nat. Acad. Sci., U.S.,* **52**, 1297 (1964).
74 J. Cairns, *J. Mol. Biol.*, **6**, 208 (1963).
75 C. Levinthal and C. A. Thomas, Jr., *Biochim. Biophys. Acta*, **23**, 453 (1957).
76 I. Rubenstein, C. A. Thomas, Jr., and A. D. Hershey, *Proc. Nat. Acad. Sci., U.S.,* **47**, 1113 (1961).
77 A. Novogrodsky and J. Hurwitz, *Fed. Proc.,* **24**, 602 (1965).
78 A. Novogrodsky and J. Hurwitz, *J. Biol. Chem.,* **241**, 2923 (1966).
79 A. Novogrodsky, M. Tal, A. Traub, and J. Hurwitz, *J. Biol. Chem.*, **241**, 2933 (1966).
80 C. C. Richardson, *Proc. Nat. Acad. Sci., U.S.,* **54**, 158 (1965).
81 C. C. Richardson, *J. Mol. Biol.*, **15**, 49 (1966).
82 K. A. Stacey, *Light Scattering in Physical Chemistry*, Academic Press, New York (1956).
83 E. P. Geiduschek and A. Holtzer, *Advan. Biol. Med. Phys.,* **6**, 431 (1958).
84 D. McIntyre and F. Gornick, eds., *Light Scattering from Dilute Polymer Solutions*, Gordon and Breach, New York (1964).
85 E. F. Casassa and H. Eisenberg, *Adv. Protein Chem.,* **19** (1964).
86 G. Cohen and H. Eisenberg, *Biopolymers*, **6**, 1077 (1968).
87 B. H. Zimm, *J. Chem. Phys.*, **16**, 1093, 1099 (1948).
88 R. L. Sinsheimer, *J. Mol. Biol.,* **1**, 37, 43 (1959).
89 T. Neugebauer, *Ann. Physik* [5], **42**, 509 (1943).
90 P. Debye, *J. Appl. Phys.,* **15**, 338 (1944).
91 P. Debye, *J. Phys. Colloid Chem.,* **51**, 18 (1947).
92 A. Peterlin, *Nature,* **171**, 259 (1953).
93 A. Peterlin, *Macromol. Chem.*, **9**, 244 (1953).
94 A. Peterlin, *J. Polymer Sci.*, **10**, 425 (1953).
95 A. Peterlin, in *Proceedings of the Interdisciplinary Conference on Electromagnetic Scattering*, M. Kerker, ed., Pergamon Press, London (1963), p. 357.
96 P. Sharp and V. A. Bloomfield, *Biopolymers,* **6**, 1201 (1968).
97 J. A. V. Butler, D. J. R. Laurence, A. B. Robins, and K. V. Shooter, *Proc. Roy. Soc.*, **A 250**, 1 (1959).
98 C. L. Sadron, in *The Nucleic Acids*, Vol. 3, E. Chargaff and J. N. Davidson, eds., Academic Press, New York (1960), p. 1.
99 D. Froelich, C. Strazielle, G. Bernardi, and H. Benoit, *Biophys. J.*, **3**, 115 (1963).

100 J. A. Harpst, A. I. Krasna, and B. H. Zimm, *Fed. Proc.*, **24**, 538 (1965).
101 J. A. Harpst, A. I. Krasna, and B. H. Zimm, *Biopolymers*, **6**, 585 (1968).
102 G. Meyerhoff, U. Moritz, and R. L. Darskus, *Polymer Letters*, **6** 207 (1968).
103 A. I. Krasna and J. A. Harpst, *Proc. Nat. Acad. Sci., U.S.*, **51**, 36 (1964).
104 P. Doty and B. H. Bunce, *J. Amer. Chem. Soc.*, **74**, 5029 (1952).
105 M. E. Reichmann, B. H. Bunce, and P. Doty, *J. Polymer Sci.*, **10**, 109 (1953).
106 M. E. Reichmann, S. A. Rice, C. A. Thomas, and P. Doty, *J. Amer. Chem. Soc.*, **76**, 3047 (1954).
107 S. A. Rice and P. Doty, *J. Amer. Chem. Soc.*, **79**, 3937 (1957).
108 P. Ehrlich and P. Doty, *J. Amer. Chem. Soc.*, **80**, 4251 (1958).
109 P. Doty, B. B. McGill, and S. A. Rice, *Proc. Nat. Acad. Sci., U.S.*, **44**, 432 (1958).
110 J. Eigner and P. Doty, *J. Mol. Biol.*, **12**, 549 (1965).
111 J. A. Harpst, A. I. Krasna, and B. H. Zimm, *Biopolymers*, **6**, 595 (1968).
112 C. W. Schmid, F. P. Rinehart, and J. E. Hearst, *Biopolymers*, **10**, 883 (1971).
113 O. B. Ptitsyn and B. A. Fedorov, *Biofizika*, 8, 659 (1963).
114 C. Sadron, *J. Chim. Phys. Phys.-chim. Biol.*, **58**, 877 (1961).
115 V. Luzzati and H. Benoit, *Acta Cryst.*, **14**, 297 (1961).
116 A. Holtzer, *J. Polymer Sci.*, **17**, 432 (1953).
117 Y. Mauss, J. Chambron, M. Daune, and H. Benoit, *J. Mol. Biol.*, **27**, 579 (1967).
118 L. S. Lerman, *J. Mol. Biol.*, **3**, 18 (1961).
119 L. S. Lerman, *Proc. Nat. Acad. Sci., U.S.*, **49**, 94 (1963).
120 A. Guinier and G. Fournet, *Small-Angle Scattering of X-Rays*, Wiley, New York (1955).
121 V. Luzzati, *Acta Cryst.*, **13**, 939 (1960).
122 V. Luzzati in *X-Ray Optics and X-Ray Microanalysis*, Academic Press, New York (1963), p. 133.
123 O. Kratky, *Prog. Biophys.*, **13**, 105 (1963).
124 H. Eisenberg and G. Cohen, *J. Mol. Biol.*, **37**, 355 (1968).
125 A. Guinier, *C. R. Hebd. Seances, Acad. Sci.*, **204**, 1115 (1937).
126 V. Luzzati, *Acta Cryst.*, **10**, 139 (1957).
127 V. Luzzati, *Prog. Nucleic Acid Res.*, **1**, 347 (1963).
128 V. Luzzati, F. Masson, A. Mathis, and P. Saludjian, *Biopolymers*, **5**, 491 (1967).
129 V. Luzzati, A. Nicolaieff, and F. Masson, *J. Mol. Biol.*, **3**, 185 (1961).
130 S. Bram and W. W. Beeman, *J. Mol. Biol.*, **55**, 311 (1971).
131 R. Langridge, D. A. Marvin, W. E. Seeds, H. R. Wilson, C. W. Hooper, M. H. F. Wilkins, and L. D. Hamilton, *J. Mol. Biol.*, **2**, 38 (1960).
132 A. Vrij and J. Th. G. Overbeek, *J. Coll. Sci.*, **17**, 570 (1962).
133 V. Luzzati, A. Mathis, F. Masson, and J. Witz, *J. Mol. Biol.*, **10**, 28 (1964).
134 A. Mathis, *J. Chim. Phys.*, **65**, 46 (1968).
135 J. Witz and V. Luzzati, *J. Mol. Biol.*, **11**, 620 (1965).
136 A. Gulik, H. Inoue, and V. Luzzati, *J. Mol. Biol.*, **53**, 221 (1970).
137 V. Luzzati, F. Masson, and L. S. Lerman, *J. Mol. Biol.*, **3**, 634 (1961).
138 S. N. Timasheff, J. Witz, and V. Luzzati, *Biophys. J.*, **1**, 525 (1961).
139 V. Luzzati, J. Witz, and S. N. Timasheff, in *Acides Ribonucléiques et Polyphosphates*, Coloque CNRS, Strasbourg (1962), p. 123.
140 J. Witz, L. Hirth, and V. Luzzati, *J. Mol. Biol.*, **11**, 613 (1965).
141 M. Meselson, F. W. Stahl, and J. Vinograd, *Proc. Nat. Acad. Sci., U.S.*, **43**, 581 (1957).
142 J. Vinograd and J. E. Hearst, *Fortsch. Chem. Org. Naturstoffe*, **20**, 372 (1962).
143 J. Vinograd in *Methods in Enzymology*, Vol. VI, S. P. Colowick and N. O. Kaplan, eds., Academic Press, New York (1963), p. 854.

144 W. Szybalski in *Methods in Enzymology*, Vol. XII B, L. Grossman and K. Moldave, eds., Academic Press, New York (1968), p. 330.
145 J. J. Hermans and H. A. Ende, in *Newer Methods of Polymer Characterization*, B. Ke, ed., Interscience, New York (1964), p. 525.
146 C. W. Schmid and J. E. Hearst, *J. Mol. Biol.*, **44**, 143 (1969).
147 J. E. Hearst and J. Vinograd, *Proc. Nat. Acad. Sci., U.S.*, **47**, 999 (1961).
148 J. B. Ifft, D. H. Voet, and J. Vinograd, *J. Phys. Chem.*, **65**, 1138 (1961).
149 J. E. Hearst, J. B. Ifft, and J. Vinograd, *Proc. Nat. Acad. Sci., U.S.*, **47**, 1015 (1961).
150 H. Fujita, *Mathematical Theory of Sedimentation Analysis*, Academic Press, New York (1962), p. 258.
151 C. W. Schmid and J. E. Hearst, *J. Mol. Biol.*, **44**, 143 (1969).
152 J. E. Hearst and J. Vinograd, *Proc. Nat. Acad. Sci., U.S.*, **47**, 1005 (1961).
153 H. Eisenberg, *Biopolymers*, **5**, 681 (1967).
154 M. Meselson and F. W. Stahl, *Proc. Nat. Acad. Sci., U.S.*, **44**, 671 (1958).
155 N. J. Sueoka, J. Marmur, and P. Doty, *Nature*, **183**, 1429 (1959).
156 R. Rolfe and M. Meselson, *Proc. Nat. Acad. Sci., U.S.*, **45**, 1039 (1959).
157 N. Sueoka, *J. Mol. Biol.*, **3**, 31 (1961).
158 C. L. Schildkraut, J. Marmur, and P. Doty, *J. Mol. Biol.*, **4**, 430 (1962).
159 R. L. Erickson and W. Szybalski, *Virology*, **22**, 111 (1964).
160 R. D. Wells and J. E. Blair, *J. Mol. Biol.*, **27**, 273 (1967).
161 R. L. Erickson and R. M. Franklin, *Bacteriol. Revs.*, **30**, 267 (1966).
162 H. A. Lozeron and W. Szybalski, *Biochem. Biophys. Res. Commun.*, **23**, 612 (1966).
163 K. W. Kohn and C. L. Spears, *Biochim. Biophys. Acta*, **145**, 720 (1967).
164 N. Davidson, J. Widholm, U. S. Nandi, R. Jensen, B. M. Olivera, and J. C. Wang, *Proc. Nat. Acad. Sci., U.S.*, **53**, 111 (1965).
165 U. S. Nandi, J. C. Wang, and N. Davidson, *Biochemistry*, **4**, 1687 (1965).
166 R. H. Jensen and N. Davidson, *Biopolymers*, **4**, 17 (1966).
167 S. N. Cohen, U. Maitra, and J. Hurwitz, *J. Mol. Biol.*, **26**, 19 (1967).
168 A. Skalka, E. Burgi, and A. D. Hershey, *J. Mol. Biol.*, **34**, 1 (1968).
169 W. C. Summers and W. Szybalski, *J. Mol. Biol.*, **26**, 107, 227 (1967).
170 Z. Opara-Kubinska, H. Kubinski, and W. Szybalski, *Proc. Nat. Acad. Sci., U.S.*, **52**, 923 (1964).
171 H. Kubinski, Z. Opara-Kubinska, and W. Szybalski, *J. Mol. Biol.*, **20**, 313 (1966).
172 Z. Hradencna and W. Szybalski, *Virology*, **32**, 633 (1967).
173 W. Szybalski, H. Kubinski, and P. Sheldrick, *Cold Spring Harbor Symp. Quant. Biol.*, **31**, 123 (1966).
174 L. D. Landau and E. M. Lifshitz, *Fluid Mechanics*, Pergamon Press, London (1959).
175 C. W. Oseen, *Hydrodynamik*, Akademische Verlag, Leipzig (1927).
176 J. M. Burgers, in *2nd Report on Viscosity and Plasticity*, Amsterdam Acad. Sci., Nordemann, New York, Chap. 3 (1938).
177 J. G. Kirkwood and J. Riseman, *J. Chem. Phys.*, **16**, 565 (1948).
178 J. G. Kirkwood, *Rec. Trav. Chim.*, **68**, 649 (1949).
179 J. G. Kirkwood, *J. Polymer Sci.*, **12**, 1 (1954).
180 J. G. Kirkwood and J. Riseman in *Rheology*, Vol. 1, F. Eirich, ed., Academic Press, New York (1956), p. 495.
181 R. K. Dewan and K. E. Van Holde, *J. Chem. Phys.*, **39**, 1820 (1963).
182 R. D. Burkhart and J. C. Merrill, *J. Chem. Phys.*, **46**, 4985 (1967).
183 A. Horta and M. Fixman, *J. Amer. Chem. Soc.*, **90**, 3048 (1968).
184 R. Zwanzig, J. Kiefer, and G. H. Weiss, *Proc. Nat. Acad. Sci., U.S.*, **60**, 381 (1968).
185 J. J. Erpenbeck and J. G. Kirkwood, *J. Chem. Phys.*, **38**, 1023 (1963).

186 R. Zwanzig, *J. Chem. Phys.*, **45**, 1858 (1966).
187 C. M. Tchen, *J. Appl. Phys.*, **25**, 463 (1954).
188 V. A. Bloomfield, W. O. Dalton, and K. E. Van Holde, *Biopolymers,* **5**, 135 (1967).
189 V. Bloomfield and B. H. Zimm, *J. Chem. Phys.*, **44**, 315 (1966).
190 J. E. Hearst and W. H. Stockmayer, *J. Chem. Phys.,* **37**, 1425 (1962).
191 H. B. Gray, Jr., V. A. Bloomfield, and J. E. Hearst, *J. Chem. Phys.,* **46**, 1493 (1967).
192 T. Svedberg and K. O. Pedersen, *The Ultracentrifuge*, Oxford University Press, London (1940).
193 H. K. Schachman, *Ultracentrifugation in Biochemistry*, Academic Press, New York (1959).
194 R. J. Goldberg, *J. Phys. Chem.*, **57**, 194 (1953).
195 S. Katz and H. K. Schachman, *Biochim. Biophys. Acta,* **18**, 28 (1955).
196 R. Bruner and J. Vinograd, *Biochim. Biophys. Acta*, **108**, 18 (1965).
197 F. W. Studier, *J. Mol. Biol.,* **11**, 373 (1965).
198 S. Newman and F. Eirich, *J. Colloid Sci.*, **5**, 541 (1950).
199 M. Wales and K. E. Van Holde, *J. Polymer Sci.,* **14**, 81 (1954).
200 J. M. Creeth and C. G. Knight, *Biochim. Biophys. Acta,* **102**, 549 (1965).
201 Y. Kawade and I. Watanabe, *Biochim. Biophys. Acta,* **19**, 513 (1956).
202 J. Eigner, C. Schildkraut, and P. Doty, *Biochim. Biophys. Acta,* **55**, 13 (1962).
203 P. F. Davison and D. Freifelder, *J. Mol. Biol.,* **5**, 643 (1962).
204 J. Rosenbloom and V. N. Schumaker, *Biochemistry,* **2**, 1206 (1963).
205 D. M. Crothers and B. H. Zimm, *J. Mol. Biol.,* **12**, 525 (1965).
206 J. B. T. Aten and J. A. Cohen, *J. Mol. Biol.,* **12**, 537 (1965).
207 J. Eigner and P. Doty, *J. Mol. Biol.,* **12**, 549 (1965).
208 V. N. Schumaker and C. Bennett, *J. Mol. Biol.,* **5**, 384 (1962).
209 K. V. Shooter and J. A. V. Butler, *Trans. Faraday Soc.,* **52**, 734 (1956).
210 V. N. Schumaker and H. K. Schachman, *Biochim. Biophys. Acta,* **23**, 628 (1957).
211 H. K. Schachman and S. J. Edelstein, *Biochemistry,* **5**, 2681 (1966).
212 J. E. Hearst and J. Vinograd, *Archives Biochemistry Biophysics,* **92**, 206 (1961).
213 S. J. Singer, *J. Polymer Sci.,* **2**, 290 (1947).
214 J. M. Peterson, *J. Chem. Phys.,* **40**, 2680 (1964).
215 E. Burgi and A. D. Hershey, *Biophys. J.,* **3**, 309 (1963).
216 R. J. Britten and R. B. Roberts, *Science,* **131**, 32 (1960).
217 J. Vinograd, R. Bruner, R. Kent, and J. Weigle, *Proc. Nat. Acad. Sci., U.S.,* **49**, 902 (1963).
218 G. Kegeles, *J. Amer. Chem. Soc.*, **74**, 5532 (1952).
219 J. Vinograd, R. Radloff, and R. Bruner, *Biopolymers,* **3**, 481 (1965).
220 M. Gehatia and E. Katchalski, *J. Chem. Phys.,* **30**, 1334 (1959).
221 V. N. Schumaker and J. Rosenbloom, *Biochemistry,* **4**, 1005 (1965).
222 J. Vinograd and R. Bruner, *Biopolymers,* **4**, 131, 1055 (1966).
223 M. M. Rubin and A. Katchalsky, *Biopolymers,* **4**, 579 (1966).
224 J. Vinograd and R. Bruner, *Biopolymers,* **4**, 157 (1966).
225 H. Fujita and V. J. MacCosham, *J. Chem. Phys.,* **30**, 291 (1959).
226 J. Hearst, *J. Mol. Biol.,* **4**, 415 (1962).
227 T. Lindahl, D. D. Henley, and J. R. Fresco, *J. Amer. Chem. Soc.,* **87**, 4961 (1965).
228 V. A. Bloomfield, *Macromol. Rev.,* **3**, 255 (1968).
229 H. Boedtker, *J. Mol. Biol.,* **35**, 61 (1968).
230 J. H. Strauss, Jr., R. B. Kelly, and R. L. Sinsheimer, *Biopolymers,* **6**, 793 (1968).
231 H. Eisenberg and G. Felsenfeld, *J. Mol. Biol.,* **30**, 17 (1967).
232 H. A. Kramers, *J. Chem. Phys.,* **14**, 415 (1946).

233 J. G. Kirkwood and P. L. Auer, *J. Chem. Phys.*, **19**, 281 (1951).
234 A. Einstein, *Ann. Physik* [4] **19**, 289 (1906); **34**, 591 (1911).
235 B. H. Zimm, *J. Chem. Phys.*, **24**, 269 (1956).
236 P. E. Rouse, Jr., *J. Chem. Phys.*, **21**, 1272 (1953).
237 F. Bueche, *J. Chem. Phys.*, **22**, 603 (1954).
238 C. Y. Pyun and M. Fixman, *J. Chem. Phys.*, **42**, 3838 (1965).
239 P. J. Flory and T. G. Fox, *J. Amer. Chem. Soc.*, **73**, 1904 (1951).
240 H. Yamakawa and M. Kurata, *J. Phys. Soc., Japan*, **13**, 94 (1958).
241 M. Kurata and Y. Yamakawa, *J. Chem. Phys.*, **29**, 311 (1958).
242 O. B. Ptitsyn and Y. E. Eizner, *Zh. Fiz. Khim.*, **32**, 2464 (1958); *Zh. Tekhn. Fiz.*, **29**, 1117 (1959).
243 N. W. Tschoegl, *J. Chem. Phys.*, **39**, 149 (1963); **44**, 4615 (1966).
244 R. Ullman, *J. Chem. Phys.*, **40**, 2193 (1964).
245 P. A. Sharp and V. A. Bloomfield, *Macromolecules*, **1**, 380 (1968).
246 R. A. Harris and J. E. Hearst, *J. Chem. Phys.*, **44**, 2595 (1966).
247 J. E. Hearst, R. A. Harris, and E. Beals, *J. Chem. Phys.*, **45**, 3106 (1966).
248 J. E. Hearst, E. Beals, and R. A. Harris, *J. Chem. Phys.*, **48**, 5371 (1968).
249 J. Eigner, C. Schildkraut, and P. Doty, *Biochim. Biophys. Acta*, **55**, 13 (1962).
250 H. Eisenberg, *J. Polymer Sci.*, **25**, 257 (1957).
251 H. Eisenberg and E. H. Frei, *J. Polymer Sci.*, **14**, 417 (1954).
252 J. E. Hearst and Y. Tagami, *J. Chem. Phys.*, **42**, 4149 (1965).
253 B. H. Zimm and D. M. Crothers, *Proc. Nat. Acad. Sci., U.S.*, **48**, 905 (1962).
254 E. H. Frei, D. Treves, and H. Eisenberg, *J. Polymer Sci.*, **25**, 273 (1957).
255 E. V. Frisman, L. V. Shchagina, and V. I. Vorobiev, *Kolloidn. Zh.*, **27**, 130 (1965).
256 G. C. Berry, *J. Chem. Phys.*, **46**, 1388 (1967).
257 A. R. Sloniewsky, G. T. Evans, and P. Ander, *J. Polymer Sci.*, A-2 **6**, 1555 (1968).
258 S. J. Gill and D. S. Thompson, *Proc. Nat. Acad. Sci., U.S.*, **57**, 562 (1967).
259 P. Doty, B. B. McGill, and S. A. Rice, *Proc. Nat. Acad. Sci., U.S.*, **44**, 432 (1958).
260 P. A. Sharp and V. A. Bloomfield, *J. Chem. Phys.*, **48**, 2149 (1968).
261 L. Mandelkern and P. J. Flory, *J. Chem. Phys.*, **20**, 212 (1952).
262 H. A. Scheraga and L. Mandelkern, *J. Amer. Chem. Soc.*, **75**, 179 (1953).
263 A. H. Rosenberg and F. W. Studier, *Biopolymers*, **7**, 765 (1969).
264 Ya. Frenkel, *Acta Physiochim. URSS*, **19**, 51 (1944).
265 P. F. Davison, *Proc. Nat. Acad. Sci., U.S.*, **45**, 1560 (1959).
266 C. A. Thomas, Jr., *J. Gen. Physiol.*, **42**, 503 (1959).
267 H. S. Rosenkranz and A. Bendich, *J. Amer. Chem. Soc.*, **82**, 3198 (1960).
268 L. F. Cavalieri and B. H. Rosenberg, *J. Amer. Chem. Soc.*, **81**, 5136 (1959).
269 D. Freifelder and P. F. Davison, *Biophys. J.*, **2**, 235 (1962).
270 D. E. Hughes and W. L. Nyborg, *Science*, **138**, 108 (1962).
271 N. J. Pritchard, D. E. Hughes, and A. R. Peacocke, *Biopolymers*, **4**, 259 (1966).
272 A. P. Nygaard and B. D. Hall, *Biochem. Biophys. Res. Commun.*, **12**, 98 (1963).
273 P. F. Davison and D. Freifelder, *J. Mol. Biol.*, **16**, 490 (1966).
274 P. F. Davison, *Nature*, **185**, 918 (1960).
275 I. Rubenstein, C. A. Thomas, Jr., and A. D. Hershey, *Proc. Nat. Acad. Sci., U.S.*, **47**, 1113 (1961).
276 P. F. Davison, D. Freifelder, R. Ikde, and C. Levinthal, *Proc. Nat. Acad. Sci., U.S.*, **47**, 1123 (1961).
277 J. D. Mandell and A. D. Hershey, *Anal. Biochem.*, **1**, 66 (1960).
278 A. D. Hershey and E. Burgi, *J. Mol. Biol.*, **2**, 143 (1960).
279 E. Burgi and A. D. Hershey, *J. Mol. Biol.*, **4**, 313 (1962).
280 L. Cavalieri and B. Rosenberg, *Biophys. J.*, **1**, 317 (1961).

281 F. F. H. Yew and N. Davidson, *Biopolymers,* **6**, 659 (1968).
282 C. Levinthal and P. F. Davison, *J. Mol. Biol.,* **3**, 674 (1961).
283 O. L. Forgacs and S. G. Mason, *J. Colloid Sci.,* **14**, 473 (1959).
284 R. E. Harrington and B. H. Zimm, *J. Phys. Chem.,* **69**, 161 (1965).
285 R. F. Harrington, *J. Polymer Sci.,* **A.1 4,** 489 (1966).
286 O. Richards and P. D. Boyer, *J. Mol. Biol.,* **11**, 327 (1965).
287 J. C. Wang and N. Davidson, *J. Mol. Biol.,* **15**, 111 (1966); **19**, 469 (1966).
288 V. N. Tsvetkov in *Newer Methods of Polymer Characterization*, B. Ke, ed., Interscience, New York (1964), p. 536.
289 R. E. Harrington, *Encyclopedia of Polymer Science and Technology*, Vol. 7, p. 100.
290 G. G. Stokes, *Mathematical and Physical Papers*, Cambridge Univ. Press, London (1890).
291 F. Perrin, *J. Phys. Raduin,* **5**, 497 (1934).
292 A. Peterlin, *Z. Physik,* **111**, 232 (1938).
293 A. Peterlin, *Kolloid Z.,* **86**, 230 (1939).
294 A. Peterlin and H. A. Stuart, *Z. Physik,* **112**, 1, 129 (1939).
295 R. E. Harrington, *Biopolymers,* **9**, 141 (1970).
296 H. A. Scheraga, J. T. Edsall, and J. O. Gadd, Jr., *Ann. Comp. Lab. Harvard Univ.,* **26**, 219 (1951).
297 R. E. Harrington, *Biopolymers,* **9**, 159 (1970).
298 V. N. Tsvetkov, *Vysokomolekul. Soedin.,* **5**, 740 (1963).
299 V. N. Tsvetkov, *Vysokomolekul. Soedin.,* **5**, 747 (1963); *Polymer Sci. USSR,* **4**, 1456 (1963).
300 D. S. Thompson and S. J. Gill, *J. Chem. Phys.,* **47**, 5008 (1967).
301 P. Callis and N. Davidson, *Biopolymers,* **7**, 335 (1969).
302 R. Signer, T. Caspersson, and E. Hammersten, *Nature,* **141**, 122 (1938).
303 G. Weill, C. Resnick, and S. Stoylov, *J. Chim. Phys.,* **65**, 182 (1968).
304 S. Takashima, *Biopolymers,* **6**, 1437 (1968).
305 J. L. Sarquis and R. E. Harrington, *J. Phys. Chem.,* **73**, 1685 (1969).
306 L. F. Cavalieri, B. Rosenberg, and M. Rosoff, *J. Amer. Chem. Soc.,* **78**, 5235 (1956).
307 A. Wada and S. Kozawa, *J. Polymer Sci.,* A **2**, 853 (1964).
308 A. Wada, *Biopolymers,* **2**, 361 (1964).
309 C. S. Lee and N. Davidson, *Biopolymers,* **6**, 531 (1968).
310 W. Kühn and F. Grün, *Kolloid Z.,* **101**, 248 (1942).
311 D. M. Gray and I. Rubenstein, *Biopolymers,* **6**, 1605 (1968).
312 G. Weber, *Adv. Protein Chem.,* **8**, 415 (1953).
313 T. Tao, *Biopolymers,* **8**, 609 (1969).
314 T. Tao, J. H. Nelson, and C. R. Cantor, *Biochemistry,* **9**, 3514 (1970).
315 S. B. Leighton and I. Rubenstein, *J. Mol. Biol.,* **46**, 313 (1969).
316 J. E. Hearst, C. W. Schmid, and F. P. Rinehart, *Macromolecules,* **1**, 491 (1968).
317 P. Davison and D. Freifelder, *J. Mol. Biol.,* **5**, 635 (1962).
318 F. C. Bancroft and D. Freifelder, *J. Mol. Biol.,* **54**, 537 (1970).
319 S. B. Dubin, G. B. Benedek, F. C. Bancroft, and D. Freifelder, *J. Mol. Biol.,* **54**, 547 (1970).
320 D. J. Cummings and L. M. Kozloff, *Biochim. Biophys. Acta,* **44**, 445 (1960).
321 D. Lang, *J. Mol. Biol.,* **54**, 557 (1970).
322 J. B. Hays, M. E. Magar, and B. H. Zimm, *Biopolymers,* **8**, 531 (1969).
323 H. Eisenberg, *Biopolymers,* **8**, 545 (1969).
324 R. Ullman, *J. Chem. Phys.* **49**, 5466 (1968).
325 C. A. Thomas, *J. Amer. Chem. Soc.,* **78**, 1861 (1956).

326 J. B. Hays and B. H. Zimm, *J. Mol. Biol.,* **48**, 297 (1970).
327 H. B. Gray, Jr. and J. E. Hearst, *J. Mol. Biol.,* **35**, 111 (1968).
328 V. A. Bloomfield, *Biochem. Biophys. Res. Commun.,* **34**, 765 (1969).
329 H. Eisenberg and G. Felsenfeld, *J. Mol. Biol.,* **30**, 17 (1967).
330 L. D. Inners and G. Felsenfeld, *J. Mol. Biol.,* **50**, 373 (1970).
331 R. A. Scott III, *Biopolymers,* **6**, 625 (1968).
332 U. Strauss and P. Ander, *J. Phys. Chem.,* **66**, 2635 (1962).
333 U. Strauss and P. L. Wineman, *J. Amer. Chem. Soc.,* **80**, 2366 (1958).
334 M. Sundaralingam, *Biopolymers,* **7**, 821 (1969).
335 S. Arnott, *Prog. Biophys. Mol. Biol.,* **21**, 265 (1970).
336 J. Vinograd and J. Lebowitz, *J. Gen. Physiol.,* **49**, 103 (1966).
337 W. Fiers and R. L. Sinsheimer, *J. Mol. Biol.,* **5**, 408, 420, 424 (1962).
338 D. Freifelder, A. K. Kleinschmidt, and R. L. Sinsheimer, *Science,* **146**, 254 (1964).
339 A. D. Hershey, E. Burgi, and L. Ingraham, *Proc. Nat. Acad. Sci., U.S.,* **49**, 748 (1963).
340 R. L. Baldwin, P. Barrand, A. Fritsch, D. A. Goldthwait, and F. Jacob, *J. Mol. Biol.,* **17**, 343 (1966).
341 H. Yamagishi, K. Nakamura, and H. Ozeki, *Biochem. Biophys. Res. Commun.,* **20**, 727 (1965).
342 J. C. Wang, *J. Mol. Biol.,* **28**, 403 (1967).
343 J. C. Wang and H. Schwartz, *Biopolymers,* **5**, 953 (1967).
344 J. G. Wetmur, N.Davidson, and J. V. Scaletti, *Biochem. Biophys. Res. Commun.,* **25**, 684 (1966).
345 C. S. Lee and N. Davidson, *Biochem. Biophys. Res. Commun.,* **32**, 757 (1968).
346 C. S. Lee, N. Davidson, and J. V. Scaletti, *Biochem. Biophys. Res. Commun.,* **32**, 752 (1968).
347 C. A. Thomas, Jr., *J. Cell. Physiol.,* **70** (Supp. 1) 13 (1967).
348 L. A. MacHattie, D. A. Ritchie, C. C. Richardson, and C. A. Thomas, Jr., *J. Mol. Biol.,* **23**, 355 (1967).
349 D. A. Ritchie, C. A. Thomas, Jr., L. A. MacHattie, and P. C. Wensink, *J. Mol. Biol.,* **23**, 365 (1967).
350 M. Rhoades and C. A. Thomas, Jr., *J. Mol. Biol.,* **37**, 41 (1968).
351 R. Wu and E. Taylor, *J. Mol. Biol.,* **57**, 491 (1971).
352 E. F. Casassa, *J. Polymer Sci.,* **A 3**, 605 (1965).
353 J. Vinograd, J. Lebowitz, R. Radloff, and P. Laipis, *Proc. Nat. Acad. Sci., U.S.,* **53**, 1104 (1965).
354 B. Hudson, D. A. Clayton, and J. Vinograd, *Cold Spring Harbor Symp. Quant. Biol.,* **33**, 435 (1968).
355 M. Fukatsu and M. Kurata, *J. Chem. Phys.,* **4**, 4539 (1966).
356 R. J. Douthart and V. A. Bloomfield, *Biopolymers,* **6**, 1297 (1968).
357 A. D. Hershey and E. Burgi, *Proc. Nat. Acad. Sci., U.S.,* **53**, 325 (1965).
358 H. B. Strack and A. D. Kaiser, *J. Mol. Biol.,* **12**, 36 (1965).
359 J. C. Wang and N. Davidson, *J. Mol. Biol.,* **15**, 111 (1966).
360 H. Jacobson and W. H. Stockmayer, *J. Chem. Phys.,* **18**, 1600 (1950).
361 M. A. Rawitscher, P. D. Ross, and J. M. Sturtevant, *J. Amer. Chem. Soc.,* **85**, 1915 (1963).
362 N. R. Kallenbach and D. M. Crothers, *Proc. Nat. Acad. Sci., U.S.,* **56**, 1018 (1966).
363 A. D. Kaiser and R. B. Inman, *J. Mol. Biol.,* **13**, 78 (1965).
364 W. F. Dove and J. J. Weigle, *J. Mol. Biol.,* **12**, 620 (1965).
365 J. C. Wang and N. Davidson, *Cold Spring Harbor Symp. Quant. Biol.,* **33**, 409 (1968).
366 J. C. Wang and N. Davidson, *J. Mol. Biol.,* **19**, 469 (1966).

367 J. Vinograd, J. Lebowitz, and R. Watson, *J. Mol. Biol.*, **33**, 173 (1968).
368 D. Glaubiger and J. E. Hearst, *Biopolymers*, **5**, 691 (1967).
369 J. C. Wang, *J. Mol. Biol.*, **43**, 25 (1969).
370 L. V. Crawford and M. J. Waring, *J. Mol. Biol.*, **25**, 23 (1967).
371 W. Bauer and J. Vinograd, *J. Mol. Biol.*, **33**, 141 (1968); **47**, 419 (1970); **54**, 281 (1970).
372 B. Hudson, W. Upholt, J. Devinny, and J. Vinograd, *Proc. Nat. Acad. Sci., U.S.*, **62**, 813 (1969).
373 E. F. J. van Bruggen, C. M. Runnert, P. Borst, G. J. C. M. Ruttenberg, A. M. Kroon, and F. M. A. H. Schuurmans Stekhoven, *Biochim. Biophys. Acta*, **161**, 402 (1968).
374 J. C. Wang, *J. Mol. Biol.*, **43**, 263 (1969).
375 M. Gellert, *Proc. Nat. Acad. Sci., U.S.*, **57**, 148 (1967).
376 M. Gefter, A. Becker, and J. Hurwitz, *Proc. Nat. Acad. Sci., U.S.*, **58**, 240 (1967).
377 J. C. Wang, D. Baumgarten, and B. M. Olivera, *Proc. Nat. Acad. Sci., U.S.*, **58**, 1852 (1967).
378 J. A. Kiger, Jr., E. T. Young II, and R. L. Sinsheimer, *J. Mol. Biol.*, **33**, 395 (1968).
379 V. A. Bloomfield, *Proc. Nat. Acad. Sci., U.S.*, **55**, 717 (1966).
380 H. B. Gray, Jr., *Biopolymers*, **5**, 1009 (1967).
381 A. Opschoor, P. H. Pouwels, C. M. Knijnenburg, and J. B. T. Aten, *J. Mol. Biol.*, **37**, 13 (1968).
382 B. Hudson and J. Vinograd, *Nature*, **216**, 647 (1967).
383 D. A. Clayton and J. Vinograd, *Nature*, **216**, 652 (1967).
384 M. M. K. Nass, *Science*, **165**, 25 (1969).
385 M. M. K. Nass, *Nature*, **223**, 1124 (1969).

chapter 6
conformational changes

I INTRODUCTION

One of the ways to learn about the structure of macromolecules in solution is to observe structural changes. The ordered form of nucleic acids is only marginally stable against temperature increase, so that most samples show a drastic alteration in structure between the convenient limits of 0°–100°C.[1-4] Many physical properties are changed in the process, and the nature of these changes and characteristics of the transformation provide fertile ground for physical studies.

Environment-sensitive physical properties are not limited to the highly ordered double and triple helical species. Single-stranded molecules also show temperature-dependent properties, for example. However, there are certain sharp contrasts between this and the "melting" of the more ordered species. Probably the most marked is the relatively narrow temperature range of the "melting" transitions, which is why they are given that name. The physical basis for such cooperative transformations is to be sought in the structure of double and triple helixes. Furthermore, the way the physical properties change on denaturation frequently provides insight into the native structure, as does the way environment influences helix stability.

II EXPERIMENTAL METHODS

A. Ultraviolet Absorbance

The most common method of following the denaturation of DNA is the profile of ultraviolet absorbance against temperature, called the "melting curve." The advantage of this measurement is that it is easy and can be performed on a few micrograms of material. Its chief disadvantage is that it provides only modest amounts of information concerning the details of the underlying physical change. The observation is that the UV absorbance increases when the temperature is increased, as shown in Fig. 6-1 for DNA samples at several salt concentrations.[5] Similar observations apply to the temperature dependence of the UV absorbance of a single-stranded polynucleotide,[6–9] as illustrated for poly A in Fig. 6-2. The increased ultraviolet absorbance reflects a lessened interaction between the bases in the polymer; the theoretical basis for this effect is discussed in detail in Chapters 3 and 4. In the double helix, disruption of the ordered state, with its stacked base pairs, leads to less frequent contact between the bases and an increased absorbance. In the single-stranded polymer, temperature increase produces more disorder and consequently less stacking of the bases along the strand.

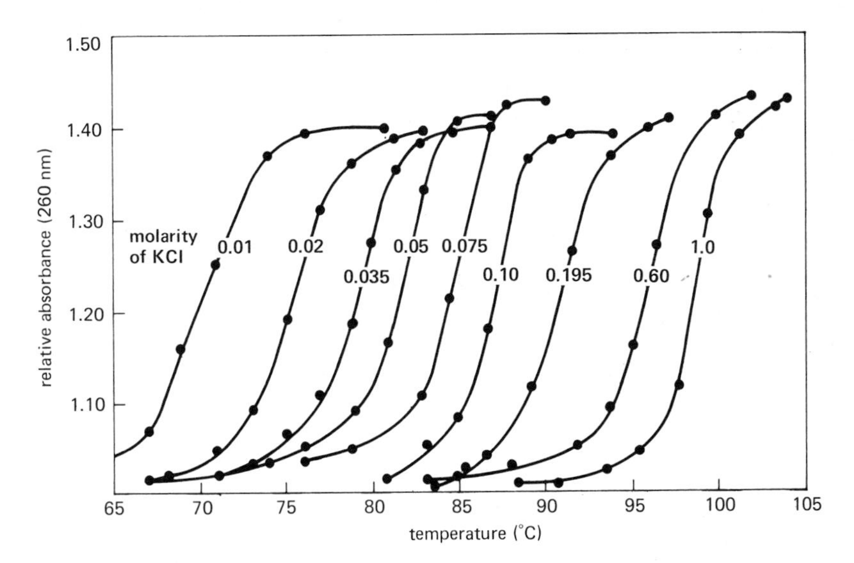

FIGURE 6-1 Dependence of thermal denaturation of *E. coli* K12 DNA on ionic strength. *E. coli* DNA, suspended in various concentrations of KCl in glass-stoppered quartz cuvettes, was heated in the Beckman model DU spectrophotometer chamber and the relative absorbance (corrected for thermal expansion) measured at the elevated temperatures. The temperature readings are uncorrected. [From J. Marmur and P. Doty, *J. Mol. Biol.*, **5**, 109 (1962). Reprinted with permission.]

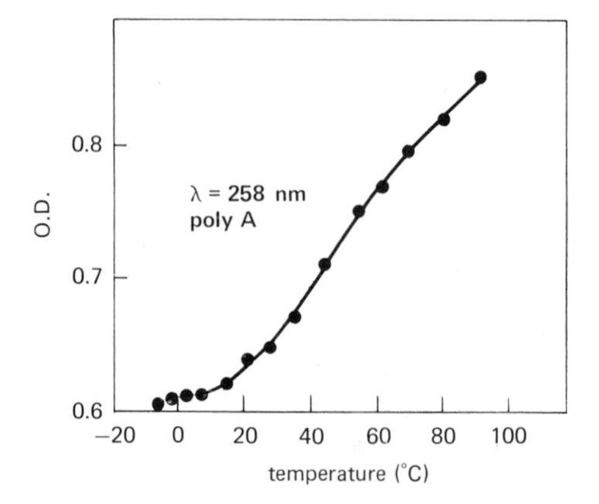

Figure 6-2 Absorbance at 258 mn of poly A as a function of temperature. Solvent is 0.1 M-LiCl-0.01 M-cacodylate buffer. [From M. Leng and G. Felsenfeld, *J. Mol. Biol.*, **15**, 455 (1966). Reprinted with permission.]

An important quantity is the characteristic transition temperature T_m, usually defined as the temperature at which the measured parameter has changed halfway from the value characteristic of helix to that for coil.

The cooperative nature of the melting transition is best seen by observing the degree of advancement of the transition as a function of pH. At high pH DNA is destabilized by the tendency of thymine and guanine to ionize. The titration of a polyelectrolyte would ordinarily cover several pH units, but the melting transition actually occurs within about 0.1 pH unit. This means that once the reaction is initiated, only small increases in pH are required to bring it to conclusion. The primary physical basis for this is the stacking interaction between adjacent base pairs. The first base pair that opens in the middle of the structure must do so at the expense of stacking interactions on both sides, whereas successive ionizations of opened base pairs can occur by breaking just one stacking interaction. This accounts for the difficulty of nucleating denaturation and the ease of bringing it to conclusion once it is begun.

The essentially noncooperative nature of the "melting" of single-stranded polymers is responsible for the wide temperature range of the transition shown in Fig. 6-2, although more detailed analysis is required for validation of the noncooperative character.[6-9] The physical basis is readily understood. The interactions that are broken are stacking forces between the adjacent bases. The first one broken in a set of stacked bases is not very different from subsequent steps, and consequently initiation of disorder presents no special problems.

The appearance of the UV melting curve depends very much on the nature of the nucleic acid sample. The sharpest transitions known are those for synthetic homopolynucleotides, for example that for the acid double helix of polyadenylic acid.[10] Viral DNA samples can also show sharp melting transitions, and sometimes exhibit biphasic behavior.[11] The origin of this is intramolecular variation in helix stability, arising from local fluctuations in G · C content. RNA samples usually show much broader melting transitions.[12] A general rule is that smaller polymers containing only limited helical regions show lower transition temperatures and broader melting curves.

Nucleic acids show an increase in UV absorbance on melting, except in the longer wavelength region. The isosbestic wavelength for melting poly A · poly U is around 280 nm; above this wavelength melting produces a decrease in absorbance. The wavelength dependence of the hyperchromic change depends on the base composition of the nucleic acid. This phenomenon can be used to obtain information about DNA base composition and the composition of regions that melt within any temperature interval.[13-17] The basic experiment is to measure melting curves at several wavelengths. If calibration experiments and a set of rational assumptions for interpreting them are available, the desired compositions can be calculated.

The first important observation required for treating such data is that the absorbance of DNA samples is not a strictly linear function of base composition.[16] Hence the extinction depends on neighbor contacts. The simplest assumption is that these result from pairwise interactions between the bases in the helix (not necessarily just betweeen nearest neighbors); hence the absorbance change $\Delta A(\phi, \lambda)$ at wavelength λ on complete melting of a DNA containing mole fraction ϕ of A · T pairs can be written[16]

$$\Delta A(\phi,\lambda) = C\{\phi^2 \Delta\epsilon_{AA}(\lambda) + 2\phi(1-\phi)\Delta\epsilon_{AG}(\lambda) + (1-\phi)^2\Delta\epsilon_{GG}(\lambda)\} \tag{6-1}$$

where $\Delta\epsilon_{AA}(\lambda)$, $\Delta\epsilon_{AG}(\lambda)$ and $\Delta\epsilon_{GG}(\lambda)$ are coefficients that represent the extinction change due to the sum of all A · T—A · T, A · T—G · C and G · C—G · C interactions, respectively. C is the molar concentration of base pairs. Equation (6-1) assumes that the base sequence is random, so that the mole fraction ϕ can be used to calculate the number of pairwise interactions.

Equation (6-1) can be rearranged to

$$\Delta\epsilon_{ap} = \phi\Delta\epsilon_{AA} + (1-\phi)\Delta\epsilon_{GG} + \phi(1-\phi)(1-2K)(\Delta\epsilon_{AA} + \Delta\epsilon_{GG}) \tag{6-2}$$

with

$$K = \frac{\Delta\epsilon_{AG}}{(\Delta\epsilon_{AA} + \Delta\epsilon_{GG})} \tag{6-3}$$

where

$$\Delta\epsilon_{ap} = \frac{\Delta A}{C} \tag{6-4}$$

is the apparent change in the DNA extinction coefficient on melting. Equation (6-2) is quadratic in ϕ and contains three parameters that are not directly known: $\Delta\epsilon_{AA}$, $\Delta\epsilon_{GG}$, and K. If ϵ_{ap} is measured for three different values of ϕ at any wavelength, three equations are produced, sufficient to determine the three unknown quantities. Use of more DNA samples overdetermines the parameters, and a least squares analysis can be used to find the set that fits the data best.

When K in Eq. (6-2) is 0.5, the quadratic term vanishes, and the extinction change on melting is a linear combination of extinction changes due to A · T and G · C base pairs. K is found to be close to 0.5 at most wavelengths.[16] Figure 6-3 shows the variation of ϵ_{AA} and ϵ_{GG} with λ, taken from the data of Felsenfeld and Hirschman.[16] There are large wavelength-dependent differences in the contributions of A · T and G · C pairs to the hyperchromism.

Once the parameters ϵ_{AA}, ϵ_{GG}, and K are established, measurement of the absorbance change on melting at several different wavelengths can be used to determine the base composition ϕ, either for the whole molecule, or for

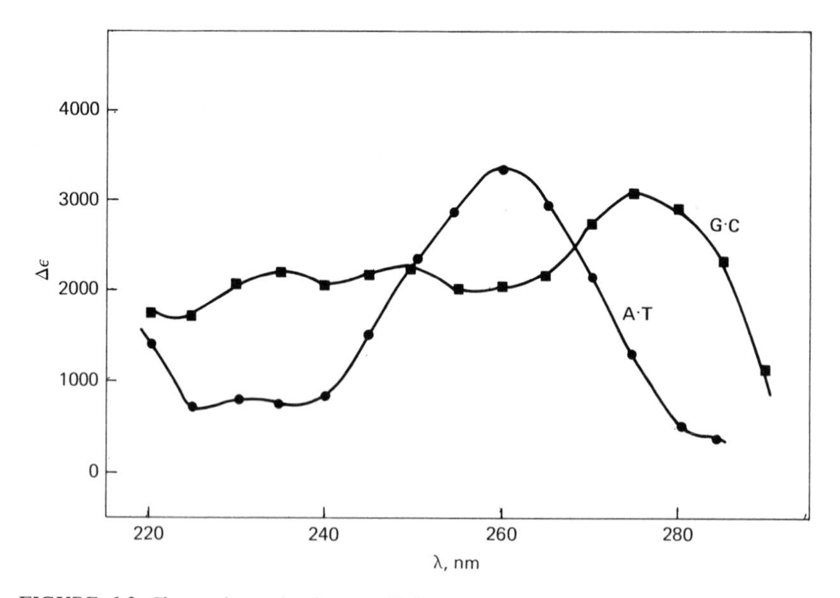

FIGURE 6-3 Change in extinction coefficient (Δε) for melting A · T and G · C pairs. [From data of G. Felsenfeld and S. Z. Hirschman, *J. Mol. Biol.*, **13**, 407 (1965). Reprinted with permission.]

the portion that melts within a given temperature interval. Since the concentration C is routinely known only from absorbance measurements, at least two wavelengths must be used. The original papers should be consulted for details.[13-17]

The hypochromism of a helical form also depends on the helix stoichiometry, the best-known example being the double and triple helical complexes of poly A and poly U. Observation of the UV absorbance at certain wavelengths will detect one kind of transition and not another[17]; this system is discussed in detail in Section III.

Another important problem is the chain length dependence of the absorbance change on melting a helix. This is a rather difficult experimental point to establish accurately. The theory of ultraviolet hypochromism was discussed in Chapter 0; calculations of the dependence on polymer size have been reported by Rich and Tinoco.[18] These calculations seem to be consistent with the measurements that have been made, although the accuracy of the latter does not provide a severe test of the theory.

Measurement of percentage hypochromism has been reported for oligomers that form double helixes, both dimers and hairpins, and for single-stranded oligomers that are hypochromic because of base stacking. We define the hypochromicity $H(n)$, which is a function of the length n of the oligomer, by

$$H(n) = \frac{A_c - A_h(n)}{A_c} \tag{6-5}$$

where A_c is the absorbance of the unstacked coil form and $A_h(n)$ the length-dependent absorbance of the helix or base-stacked single strand. The absorbance change on unstacking the bases in the single strands of oligoadenylic acid has been studied by several workers.[6-9] Applequist and Damle[9] fitted the experimental data with the empirical relation

$$\frac{H(n)}{H(\infty)} = 1 - \frac{2.32}{n+1} + \frac{3.82}{(n+1)^2} \tag{6-6}$$

Leng and Felsenfeld[6] observed that the hypochromism, when plotted against $1/n$, gave a straight line, suggesting the simpler form

$$\frac{H(n)}{H(\infty)} = 1 - \frac{1}{n} \tag{6-7}$$

which is not severely different from (6-6). In view of the evident difficulty of obtaining accurate extrapolated values of the absorbances of stacked and unstacked forms, Eq. (6-7) seems an adequate approximation of the data; its use has been suggested by Applequist.[19]

Equation (6-7) can be derived from simple considerations about the origin of hypochromism. If one assumes that each stacking interaction between bases or base pairs contributes equally to hypochromism, then the difference in absorbance between helix and coil can be written

$$A_c - A_h(n) = Cj\Delta\epsilon_0 \tag{6-8}$$

where j is the number of stacking interactions, $\Delta\epsilon_0$ is the change in extinction coefficient when one stacking interaction is broken and C the concentration. The absorbance of the unstacked coil is given by

$$A_c = Cn\epsilon_c \tag{6-9}$$

where n is the number of coil units and ϵ_c the extinction of the coil. Thus, using Eq. (6-5)

$$H(n) = \frac{j\Delta\epsilon_0}{n\epsilon_0} \tag{6-10}$$

When n approaches infinity the ratio of j to n approaches 1, so that

$$H(\infty) = \frac{\Delta\epsilon_0}{\epsilon_c} \tag{6-11}$$

and

$$H(n) = \frac{j}{n}H(\infty) \tag{6-12}$$

For stacking in a single-stranded oligomer or a dimer double helix, j is one less than the number of bases or base pairs respectively, or $j = n - 1$; this substitution yields Eq. (6-7).

The data on hypochromicity of hairpin helixes are also consistent with Eq. (6-12). Scheffler et al.[20] report measurement of $H(n)$ for oligomers of the form $(A - T)_n$ from $n = 8$ to 22. In these materials there are g bases unbonded in the loop at one end of the hairpin, and g must be even. Hence the number of stacking interactions between base pairs is $(n - 1 - g)/2$, which should be set equal to j in (6-12), giving for the predicted hypochromism

$$\frac{H(\text{n})}{H(\infty)} = 1 - \frac{1}{n} - \frac{g}{2n} \tag{6-13}$$

They found excellent fit for $g = 4$, which seems on other grounds as well to be a reasonable value for the minimum loop size.[20]

Absorbance measurements are often used to measure the degree of advancement of a melting transition. When high molecular weight polymers are studied, the dependence of hypochromism on helix length can probably be neglected because of the large average length of helical règions in the melting zone. Theoretical estimates place their size at several hundred base pairs. Consequently the degree of advancement of the transition is proportional to the fractional increase of the absorbance, or

$$\theta = \frac{A_c - A(T)}{A_c - A_h} \tag{6-14}$$

where θ is the fraction of bases hydrogen bonded.

In some cases the absorbances of both helix and coil forms are temperature dependent. For example, the coil may contain some stacked bases that melt out as the temperature is raised, or the helix may be protonated, as with poly A, and its extinction can change as further ionization occurs. Equation (6-14) may still be used, but A_h and A_c must refer to values at the measurement temperature. This requires extrapolating the absorbance of helix and coil forms into the transition zone. Figure 6-4 shows a geometrical construction by which θ may be calculated from absorbance data.

B. Infrared Spectroscopy

Infrared spectroscopy offers several advantages over measurements in the ultraviolet for detecting structural changes in nucleic acids. Foremost among these are the well-resolved absorption bands characteristic of the polymers in helix and coil states. For this reason the information content of an infrared spectrum is unquestionably higher than the ultraviolet analog. The main disadvantage of infrared measurements is that the extinction coefficients are about an order of magnitude lower, so that considerably more material is needed. In addition, all measurements are carried out in D_2O solution.

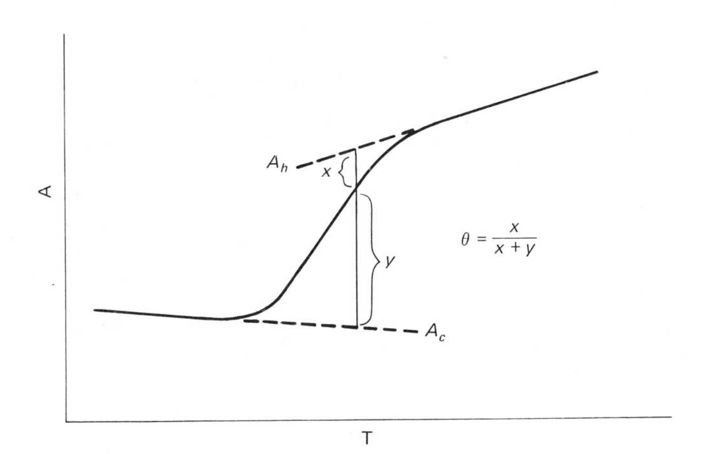

FIGURE 6-4 Graphical construction for calculating fraction helix (θ) when the absorbance of helix and coil forms depends linearly on temperature.

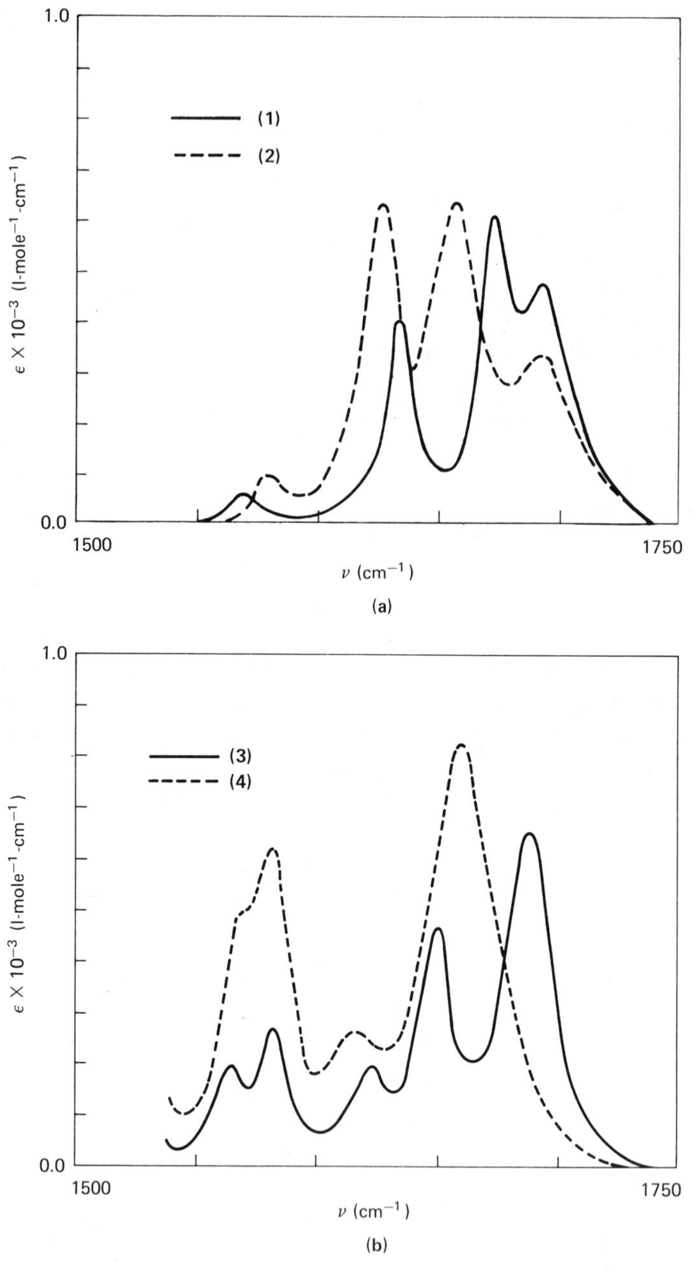

FIGURE 6-5 (a) Infrared spectra: (1) double-helical poly A + poly U complex in D_2O solution pD = 7. [The spectrum is unchanged over the temperature range $0 \leqslant T \leqslant 65°C$ then melts sharply to give a spectrum barely distinguishable from (2)]; (2) spectrum obtained by adding spectra of single-stranded poly A and single-stranded poly U, or spectra of the component nucleotides. (b) Infrared spectra: (3) double-helical poly G + poly C complex in D_2O solution, pD = 7 (the spectrum is unchanged over the temperature range $0 \leqslant T \leqslant 100°C$); (4) spectrum obtained by adding spectra of the component nucleotides. [From G. J. Thomas, *Biopolymers*, 7, 325 (1969). © 1969 by John Wiley & Sons, Inc. Reprinted with permission.]

The infrared absorption spectrum of a single-stranded polynucleotide is very similar to that of its component nucleotides, but drastic changes occur on formation of hydrogen-bonded helical structures. Figure 6-5[21] shows infrared spectra of double helical poly A · poly U and poly G · poly C compared with their single-stranded components (or nucleotides in the case of G · C). Whereas polymerization of nucleotides has little influence, double helix formation greatly alters the infrared spectrum. Infrared maxima of a number of polynucleotides are given in Table 6-1.

The bands observed in this region of the spectrum correspond to the vibrational modes of the purine and pyrimidine rings and their substituents. Miles and his collaborators[28] have been able, by isotopic substitution with ^{18}O, to show that certain normal modes contain a large contribution from a carbonyl stretching vibration. It is observed that on helix formation these carbonyl bands are shifted to higher frequencies, presumably because of hydrogen bonding to another base in the helical structure. Unlike ultraviolet spectroscopy, the infrared spectrum is little affected by stacking of the bases in the coil form, so it is an excellent tool for detection of hydrogen-bonded helical structure.

Infrared spectroscopy is also very useful for distinguishing one helical form from another. It is apparent from Table 6-1 that poly A · 2poly U and poly A · poly U have very different infrared spectra.[29] An important use of infrared spectroscopy is to follow the sometimes complicated equilibria between multistranded helical forms as temperature and other conditions are changed. Figure 6-6[30] shows the absorbance at 1657 cm^{-1} of a 1:1 mixture

TABLE 6-1 INFRARED ABSORPTION MAXIMA OF POLYNUCLEOTIDES

Polymer	Form	ν_{max} (cm^{-1})	Ref.[a]
Poly A	Coil	1628	22
Poly U	Coil	1692, 1657	22
Poly (A + U)	Helix	1691, 1672, 1631	22
Poly (A + 2U)	Helix	1696, 1677, 1657	22
Poly G	Helix	1682, 1587	23
5′-GMP	Monomer	1665, 1578, 1568	24
Poly I	Coil	1677	25
Poly C	Coil	1653, 1617	25
Poly (I + C)	Helix	1697, 1648, 1630	25
Poly (G + C)	Helix	1564, 1580, 1622, 1650, 1688	26
Poly (BC)	Coil	1485, 1527, 1590, 1617, 1644	27
$(I)_n \cdot (BC)_n$	Helix	1406, 1532, 1544, 1587, 1621, 1639, 1692	27

[a] [22] C. J. Thomas, *Biopolymers*, **7**, 325 (1969).
[23] H. T. Miles and J. Frazier, *Biochem. Biophys. Res. Comm.*, **14**, 21 (1964).
[24] H. T. Miles and J. Frazier, *Biochim. Biophys. Acta*, **79**, 216 (1964).
[25] F. B. Howard and H. T. Miles, *J. Biol. Chem.*, **240**, 801 (1965).
[26] H. T. Miles, *Proc. Nat. Acad. Sci., U.S.*, **47**, 791 (1961).
[27] F. B. Howard, J. Frazier, and H. T. Miles, *J. Biol. Chem.*, **244**, 1291 (1969).

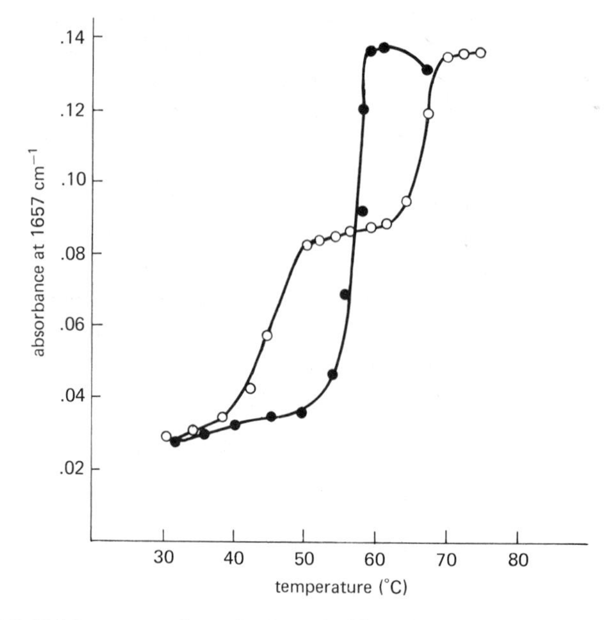

FIGURE 6-6 Melting curves for poly A · poly U in the presence of Mg^{2+} and in the absence of Mg^{2+}. In both cases the total polymer concentration (as phosphate) was 0.045 M, and the ordinate has been corrected for scale expansion to give true absorbance values. (Measured at 1657 cm^{-1}.) [From H. T. Miles and J. Frazier, *Biochem. Biophys. Res. Commun.*, **14**, 21 (1964). Copyright by Academic Press. Reprinted with permission.]

of poly A and poly U, in the presence and absence of Mg^{2+}. At this wave number, the triple helix has a strong absorption band, while the double helix absorbs more weakly. As the temperature is raised with Mg^{2+} present, the absorbance shifts from that characteristic of the double helix to the triple helix, corresponding to the disproportionation reaction

$$2[\text{poly A} \cdot \text{poly U}] \rightarrow \text{poly A} \cdot 2\ \text{poly U} + \text{poly U} \tag{6-15}$$

At a higher temperature the absorbance rises again, corresponding to melting of the triple helix to single strands. In the absence of Mg^{2+} the strand disproportionation does not occur under these salt conditions. Biphasic melting curves are also seen in the ultraviolet at appropriate wavelengths (see Section III for a discussion of the phase diagram of this system), but the IR observations, reported by Miles and Frazier,[30] played an important part in clarifying the complexities of the equilibria involving poly A and poly U.

Fritzsche[31] has shown that DNA base composition can be determined from IR studies, and Thomas[21] has calculated the helix content of ribosomal RNA from IR spectra.

C. Nuclear Magnetic Resonance

Nuclear magnetic resonance spectroscopy provides a tool of great potential power for investigating structural changes in nucleic acids. Its main disadvantage is the relatively high concentration needed for adequate detection. In general, one finds that the ordered double and triple helical forms of high polymers do not give a high resolution NMR spectrum because of their slow rotational diffusion. The result of the slow motion is that the magnetic environment is not rapidly averaged, and the resonance absorption lines are so broad that they are not detected. Upon melting, however, segmental motion of the polymer is sufficiently rapid to provide sharp NMR lines. Fast segmental motion of the multistranded helixes is prevented by their rigidity.

The NMR spectrum of single-stranded polymers is quite similar to that of their component nucleotides. However, as discussed in Chapter 3, stacking of nucleotide bases together results in a shift to higher field of the base protons. The same holds true in general of stacking of the bases in single-stranded polymers. There is a considerable effect of temperature on the NMR spectrum of poly A, but much less for poly U,[32,33] consistent with the general view that stacking is more pronounced in poly A than in poly U. As the temperature is raised, the adenine base proton resonances move to lower field, corresponding to reduction in the degree of stacking.

Probably the greatest potential use for NMR spectroscopy of nucleic acids is to detect fine details of structure which are inaccessible to other techniques in solution. An example is provided by the high resolution spectroscopy of single-stranded DNA reported by McDonald et al.[34] As seen in Fig. 6-7 the thymine methyl resonance is split into two lines located at 1.83 and 1.71 ppm. In mononucleotides and dinucleotides containing T, only one resonance line is seen, but its chemical shift depends on the material. ApT and GpT show lines at higher field than T or any other dinucleotide, including TpA and TpG. The interpretation is that when a purine is on the 5′ side of T, the thymine methyl group is in a position to be strongly influenced and shifted upfield by the purine ring current when the two bases are stacked together. The stereochemistry is quite different when the purine is on the 3′ side of T, and pyrimidines do not show the effect because of lesser stacking and smaller ring current. Thus, the resonance line at 1.71 ppm is due to thymines that have purines for 5′ neighbors, and the line at 1.83 ppm corresponds to thymines with pyrimidines for 5′ neighbors. The splitting of this resonance indicates that stacking effects are still present in single-stranded DNA at high temperatures.

Cross and Crothers[35] recently showed that proton magnetic resonance could detect double helix formation between oligonucleotides of complementary sequence. The small size of oligomers permits measurement of a high

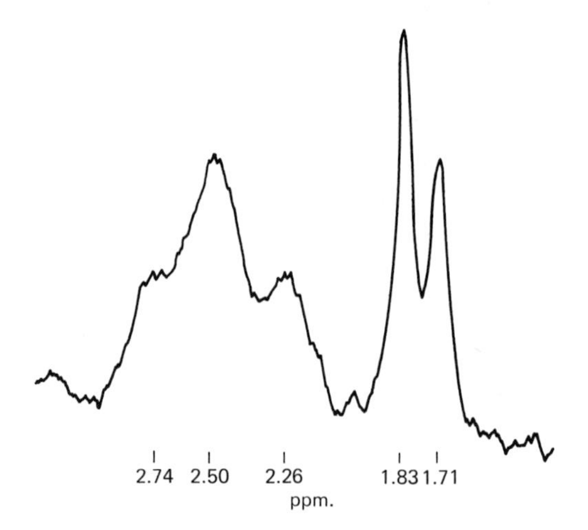

FIGURE 6-7 High-field regions of PMR spectrum of calf thymus DNA at 220 MHz: concentration 20 mg/ml in D_2O, pD 7.0, 93°, 20 spectra averaged in computer of average transients. [From C. C. McDonald, W. D. Phillips, and J. Lazar, *J. Amer. Chem. Soc.*, 89, 4166 (1967). Copyright 1967 by the American Chemical Society. Reprinted by permission of the American Chemical Society.]

resolution for the helix, in contrast to the very broad resonance of the double helical high polymer.

An especially promising recent application of NMR to nucleic acids is the demonstration by Shulman and co-workers[36] that the hydrogen bonding protons in double helical regions of tRNA are sufficiently slowly exchanging and lie far enough down field that they can be observed in H_2O solutions. Characteristic chemical shifts are found for A · U and G · C pairs. Further second-order effects, possibly due to ring current magnetic anisotropy, shift the resonance position of individual base pairs enough so that in good cases virtually all the base pair

$$\gt NH \ldots N \lt$$

resonances in a purified tRNA can be resolved. This clearly provides a powerful tool for investigation of conformational changes in small nucleic acids.

The potential information content of NMR spectra concerning denaturation kinetics has not yet been fully exploited. For synthetic polymers, mainly of short chain length, the resonances of the coil rise fully sharpened from the very broad resonance of the helix.[32,33] This indicates that the lifetime of the coil state is greater than 10^{-4} seconds, fully consistent with the view of an all-or-none transition between helical molecules and separated strands predicted for such small helixes. With DNA the observed resonances sharpen as the transition proceeds.[32]

D. Circular Dichroism and Optical Rotatory Dispersion

The CD and ORD properties of helical molecules are unusual among the spectroscopic methods for their sensitivity to the details of helix structure. As indicated before, UV absorbance is relatively nonspecific, while the IR spectrum is mainly responsive to whether certain groups are hydrogen bonded in the helix or not, that is, whether the helix is double or triple, and so on. The NMR method is not applicable to multistranded high polymers. CD and ORD, however, reflect the detailed mutual orientations of the bases in the helix.

The CD of a mononucleotide in the base absorbance region arises from the influence of the asymmetry of the sugar on the absorption bands of the base. This CD spectrum is very different from that of a dinucleotide. Current theories of the CD of di-, oligo-, and polynucleotides begin with the assumption that the effects are due primarily to interactions among the bases (see Chapters 2 and 4 for a full discussion). Hence, the crucial feature of the helix for determining CD and ORD properties is the mutual orientation of the transition moments of the bases. The measured properties reflect these details and will depend on the bases involved, because the electronic structure determines the direction of the transition moment within each base. These features are illustrated in Fig. 4-15 (page 134), which shows the measured circular dichroism spectra of double helical DNA and RNA, each compared with the sum of the spectra of the component mononucleosides. It is evident that helix formation produces large circular dichroism effects and that the structural difference between RNA and DNA results in marked differences in the CD spectrum.

The DNA CD spectrum is roughly conservative, meaning that the positive and negative bands arranged antisymmetrically about the absorbance maximum have equal areas. The RNA spectrum is nonconservative.[37-39] Another example of a conservative spectrum is the CD of double helical poly I · poly BC, observed by Howard et al.[27] Substitution of Br on the cytosine ring shifts the absorbance maximum of poly BC to 289 nm, where it can be observed with less interference from the other absorption bands. The conservative pair of CD bands is fairly well separated from the other UV bands at lower wavelength. The ORD spectrum shows approximately the character required for the pair of CD bands: a trough centered at the UV absorbance maximum, and two peaks, which should be of equal amplitudes, flanking the trough.

The CD spectrum of single-stranded polynucleotides can be roughly conservative, like poly A,[7,37] or not, like poly C.[40] The amplitude of the CD bands of these materials is relatively large, usually greater than most double helical forms. The CD and ORD of strongly stacked single-stranded polynucleotides are very dependent on temperature. As the temperature is raised, the rotational strengths diminish, corresponding to melting out of the stacked structure. In the high temperature limit the bases would have a random orientation relative to each other and the rotational strength would come only from the environmental asymmetry produced by the backbone. Polymers with lesser degrees of stacking, like poly U,[41-44] show a smaller temperature dependence of the CD and ORD.

CD and ORD are in general useful for detecting small structural changes of helical polymers. In the temperature region below the melting zone, the UV absorbance of a DNA sample is essentially constant, but the same is not true of the CD[45] or ORD.[46] The change in the optical rotation indicates structural changes that are not detected by the UV absorbance. Possible sources of the change are alterations in the degree of winding per base pair or the tilt of the bases relative to the helix axis. Another structural difference that can be detected by CD and ORD is that between ribo and deoxyribo single-stranded polynucleotides and also the corresponding oligomers.[47,48] For example, the CD of poly rA and poly dA show marked differences in the lower wavelength region. The structural origin of these differences is still debated.

Thus in summary, CD and ORD hold great potential for resolving structural features of polynucleotides, but a successful and reliable theory will be required to realize this potential fully. At present, structural changes and differences are readily detected by this technique, and one can hope that with further theoretical advances it will be possible to state more definitively the nature of the changes.

E. Hydrodynamic and Light-Scattering Properties

The hydrodynamic and light-scattering properties of native and denatured DNA have been discussed at some length in Chapter 5. It should be apparent that these methods are more sensitive to longer-range structural changes than the spectroscopic techniques so far discussed. The sedimentation coefficient and intrinsic viscosity of a polymer reflect the dimensions and configuration in solution, and one would expect these to change when the secondary structure is altered. In general, a drastic change occurs when a double helical molecule undergoes a cooperative melting transition, but a much more gradual change for melting out of structure in a single-stranded polymer. Light scattering detects similar kinds of changes in polymer structure, since it is suited for determining chain dimensions and molecular weight.

There is still work to be done on the variation of dimensions of single-stranded polymers with environment. One can safely predict that all, because they are polyelectrolytes, will show an increase in dimensions when the salt concentration is lowered.[49-51] A more difficult question is the influence of base stacking on the dimensions. A careful study of this question has been reported for poly A by Eisenberg and Felsenfeld,[50] who concluded that base stacking stiffens the polymer and increases its average dimensions at low temperatures. Figure 6-8 shows the radius of gyration (from light scattering) of poly A at θ-conditions as a function of temperature. The decrease of S_z at low temperatures is interpreted as melting out of base-stacked structure. The conformation of poly U is less sensitive to temperature[51] because of the lesser degree of stacking.

In a DNA melting transition the change in hydrodynamic properties generally parallels closely the change in spectroscopic properties.[52] The direction of change will depend on conditions. At low salt concentration, for example, the single-stranded form can be more extended than the helix because of electrostatic repulsions, so that the intrinsic viscosity can rise and the sedimentation coefficient decrease on denaturation. Figure 6-9 shows the variation with salt concentration of the intrinsic viscosity and sedimentation coefficient of native and denatured T7 DNA.[53] The salt concentration of crossover of intrinsic viscosity or sedimentation coefficient of native and

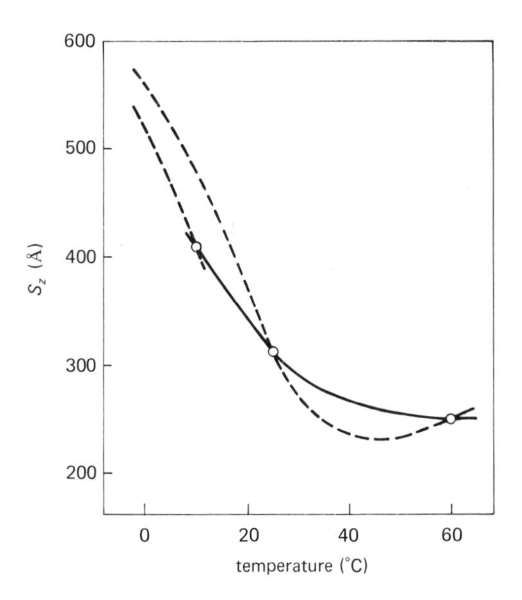

FIGURE 6-8 Unperturbed radii of gyration S_z^0 (solid line) from light scattering measurements at "ideal" θ temperatures, as a function of temperature. Upper and lower dotted curves correspond to values of S_z at 1.0 and 1.3 M NaCl, respectively. [From H. Eisenberg and G. Felsenfeld, *J. Mol. Biol.*, **30**, 17 (1967). Reprinted with permission.]

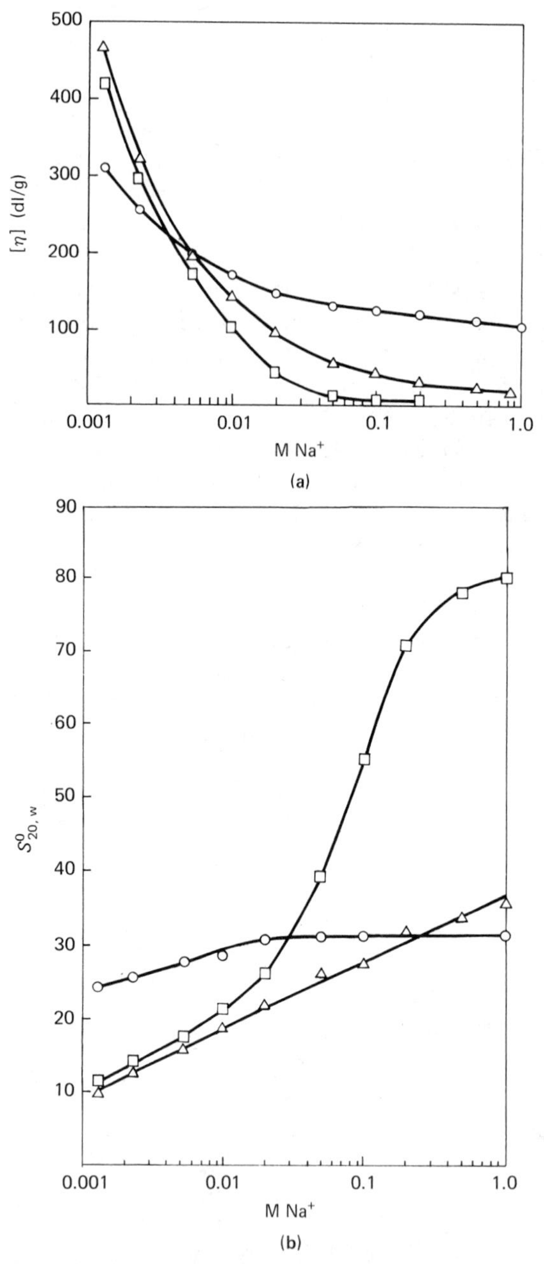

FIGURE 6-9 (a) Intrinsic viscosity of T7 DNA as a function of ionic strength: (○) native DNA at neutral pH; (Δ) alkaline single strands; (□) neutral single strands. Neutral solvents contained NaCl; alkaline solvents contained NaOH or 0.1 M NaOH plus NaCl. The native molecules were intact; 30–40% of the single strands were broken. (b) Sedimentation coefficient of T7 DNA as a function of ionic strength. Symbols and solvents the same as for part (a) except that solvents 0.2 M or less in Na^+ also contained 50% D_2O. $S^\circ_{20,w}$ refers to intact molecules. [From A. H. Rosenberg and F. W. Studier, *Biopolymers*, 7, 765 (1969). © 1969 by John Wiley & Sons, Inc. Reprinted with permission.]

alkaline-denatured forms will depend on molecular weight. The sedimentation coefficient of DNA denatured and returned to neutral pH is much greater than that of the strands in alkali, and the intrinsic viscosity is much smaller, presumably because of some local helix formation or folding in the metastable neutral single strands.[49,53]

One interesting example of a structural change that could be detected by viscosity measurements but not spectroscopically is to be found in the work of Inman and Baldwin[54] on melting of poly d(A–T). This polymer is self-complementary and can fold back on itself to make "hairpin" helixes. As the temperature is raised, the viscosity drops before there is a noticeable change in the absorbance, corresponding to the conversion of a long helix to a series of hairpins. The viscosity rises again when these hairpins melt out into the coil. On cooling, the viscosity does not return to the level of the long double helix, presumably because it is trapped in a state with interrupted hairpin helixes.

F. Calorimetry and Thermodynamics

An important factor in understanding structural changes of nucleic acids is the difference in energy between the initial and final states. Modern calorimetric techniques permit direct measurement of this quantity using only a few milligrams of material. Since measurement is usually made at constant pressure, the quantity actually determined is the enthalpy change ΔH, but because the volume change is small, this is essentially equal to the energy change.

Two different techniques have been used to measure heats of nucleic acid helix-coil transitions. In some cases, a transition can be brought about by mixing two solutions together, and this permits determination of the heat by mixing calorimetry. Examples are the heat of mixing poly A with poly U,[55,56] neutral poly A with acid to form the acid double helix,[55] and DNA with acid to determine the heat of acid denaturation.[57] The second technique is measurement of the heat capacity of a solution or the difference in heat capacity between solution and solvent.[58–63] Any transition that is brought about by increasing temperature will absorb heat, causing the heat capacity of the solution to be greater than the solvent. The extra energy that has to be put into the solution to raise it to the same temperature as the solvent is the enthalpy change of the reaction.

Each of these techniques has its own advantages. Measurement of heat capacity requires that one work at the transition temperature, necessitating a simultaneous variation of external parameters. On this basis it would be impossible to separate the variation of the measured heat with salt concentration from the temperature variation, since the two variables are not independent at T_m. They can be varied separately in mixing experiments. On the other hand, the heat of thermally induced transitions such as the melting of DNA is not readily accessible by mixing techniques. The indirect approach required is, for example, to study the acid denaturation and then correct for the ionization heats.[57] For such problems the heat capacity method is simpler.

For both methods the form of the output data is a plot of heat absorbed or released as a function of some external parameter. In Figure 6-10 the heat absorbed in a thermally induced transition is shown. The first step corresponds to conversion from double to triple helix, and the second step to melting of the triple helix. The difference between initial and plateau values is the measured enthalpy change, and this is divided by the number of moles of material present to get the molar enthalpy change, which is expressed in terms of base pairs if the helix formed is double, and in terms of base triplets for triple helix formation. By convention we write the heats in the direction of helix formation, causing the sign to be negative in most cases.

Table 6-2 shows a collection of calorimetric data for the melting

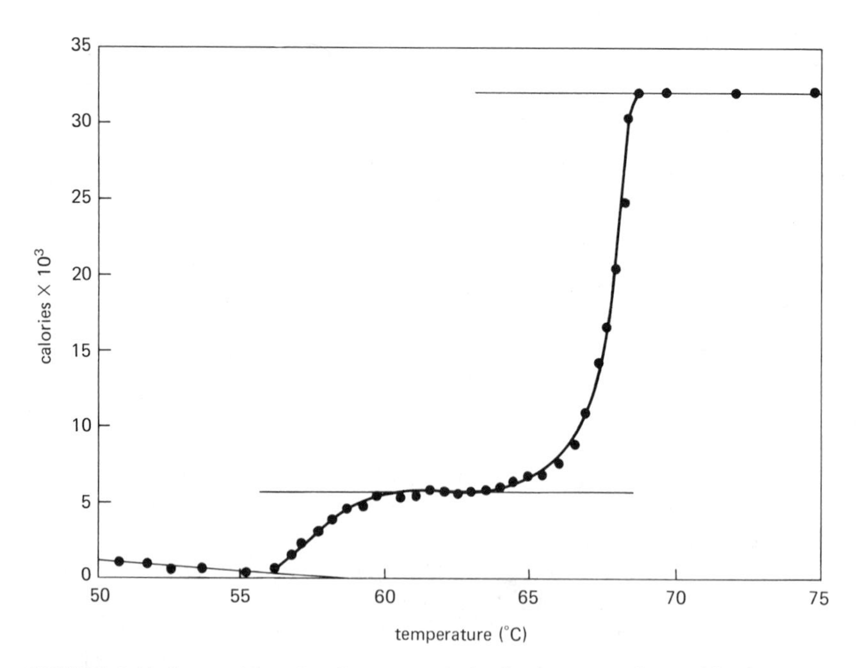

FIGURE 6-10 Input of heat by the automatic feedback system observed in the course of the reactions poly A · poly U → $\frac{1}{2}$ [poly A · 2 poly U] + $\frac{1}{2}$ poly A (step 1) and $\frac{1}{2}$ [poly A · 2 poly U] + $\frac{1}{2}$ poly A → poly A + poly U (step 2) at neutral pH, $[Na^+]$ = 0.263 M, concentration of nucleotides = 5.00 mM. The calorimeter cell contained 4.25 μmole nucleotide pairs (A + U). [From H. Krakauer and J. M. Sturtevant, *Biopolymers*, 6, 491 (1968). © 1968 by John Wiley & Sons, Inc. Reprinted with permission.]

TABLE 6-2 HEATS OF HELIX FORMATION

Reaction	ΔH (kcal/mole)[a]	Ref.[b]
$A_n + U_n \rightarrow A_n \cdot U_n$	−5.2 to −8.5	55–59
$(dA{-}dT)_{2n} \rightarrow (dA{-}dT)_n \cdot (dA{-}dT)_n$	−7.9	60
$A_n \cdot U_n + U_n \rightarrow A_n \cdot 2U_n$	−2.7 to −4.3	56, 58, 59
$2\{A_n \cdot U_n\} \rightarrow A_n \cdot 2U_n + A_n$	1.4 to 3.0	56, 58, 59
$A_n + 2U_n \rightarrow A_n \cdot 2U_n$	−9.1 to −12.7	58, 59
DNA (coil) → DNA (helix) (corrected for heat of protonation)	−8	57
DNA (coil) → DNA (helix)	−9.2	61
poly A(coil) → poly A(helix)	−3.4 to −7	56, 62, 63
$I_n + C_n \rightarrow I_n \cdot C_n$	−5.6	64
adenosine + $2U_n \rightarrow$ adenosine $\cdot\ 2U_n$	−12.8	64
2-aminoadenosine + $2U_n \rightarrow$ 2-aminoadenosine $\cdot\ 2U_n$	−15.8	64
adenine + $2U_n \rightarrow$ adenine $\cdot\ 2U_n$	−12.8	64
2,6-diaminopurine + $2U_n \rightarrow$ 2,6-diaminopurine $\cdot\ 2U_n$	−15.9	64
deoxyadenosine + $2U_n \rightarrow$ deoxyadenosine $\cdot\ 2U_n$	−12.8	64
2-methylaminoadenosine + $2U_n \rightarrow$ 2-methylaminoadenosine $\cdot\ 2U_n$	−14.5	64

[a] The mole in a base pair or base triplet.
[b] 55 M. A. Rawitscher et al., *J. Amer. Chem. Soc.*, **85**, 1915 (1963).
56 P. D. Ross and R. L. Scruggs, *Biopolymers*, **3**, 491 (1965).
57 L. C. Bunville et al., *Biopolymers*, **3**, 213 (1965).
58 E. Neumann and T. Ackermann, *J. Phys. Chem.*, **71**, 2377 (1967).
59 H. Krabauer and J. M. Sturdevant, *Biopolymers*, **6**, 491 (1968).
60 I. Scheffler and J. M. Sturtevant, *J. Mol. Biol.*, **42**, 577 (1969).
61 P. L. Privalov et al., *Biopolymers*, **8**, 559 (1969).
62 H. Klump et al., *Biopolymers*, **7**, 423 (1969).
63 D. W. Hennage, Thesis Yale University.
64 R. L. Scruggs and P. D. Ross, *J. Mol. Biol.*, **47**, 29 (1970).

transitions of double and triple helical nucleic acids. The most thoroughly investigated system is that involving the equilibria between poly A and poly U. Concerning the heat of forming the double helix, one can compare the results of mixing experiments (which are somewhat suspect because of possible transient formation of the triple helix[65]) with heat capacity measurements at various salt concentrations and comparable temperatures to conclude that there is no important variation of the heat with salt concentration at constant *T*. There is indication from the work of Krakauer and Sturtevant[59] that the heat depends on the nature of the counter-ion, Na^+ or K^+, although this conclusion is lost in the scatter when the data from several laboratories are compared.

One observation that seems secure is that the heat of melting poly A · poly U depends on temperature. Figure 6-11 shows the heat of forming $A_n \cdot U_n$ from A_n and U_n, and also the heat of triple helix formation from double helix and a poly U strand at various temperatures. The temperature dependence of the first reaction is much more marked. Stevens and Felsenfeld[17] suggest that this is because of the secondary structure in poly A.

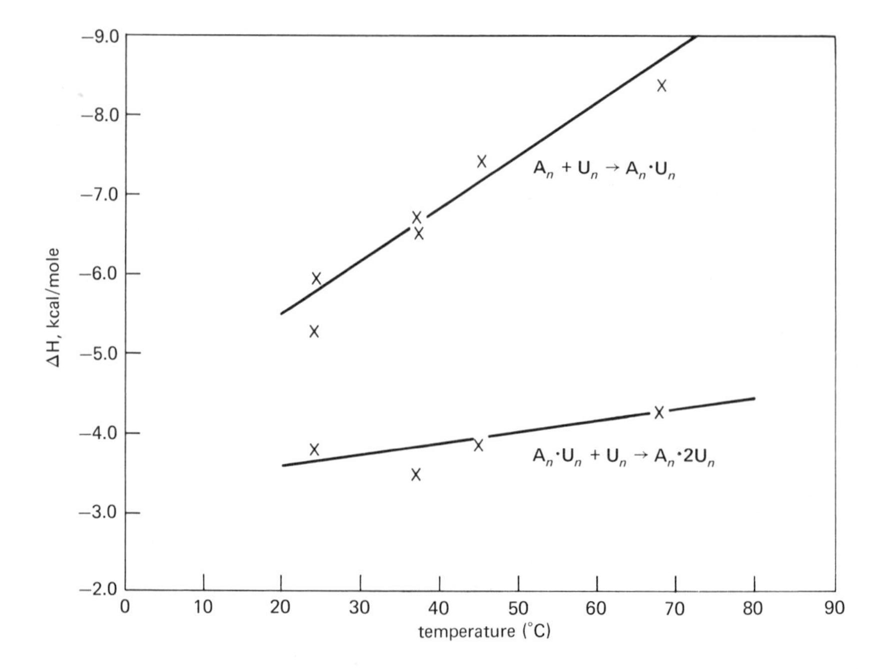

FIGURE 6-11 Heats of helix formation as functions of temperature. Data from references in Table 6-2.

As the temperature is raised, some of this structure is melted out, resulting in a heat of helix formation larger in magnitude at higher temperatures. This conclusion is supported by the smaller apparent dependence of the heat of triple helix formation on temperature. The free strand in this case is poly U, which has less base stacking by all indications. Krakauer and Sturtevant[59] argue that the situation is not that simple, since they do find an appreciable temperature dependence of the heat of triple helix formation. One should use with caution the assumption that unstacking of the bases in the single strand is the only source of difference in heat capacity between helix and coil.

Until recently, the studies of Bunville et al.[57] were the only measure of the heat of DNA denaturation. They denatured the polymer with acid and then corrected for the heat of protonation to estimate the heat of denaturing at neutral pH. They concluded that the heat of helix formation at 25° is about −8 kcal. This is in reasonable agreement with the heat capacity measurements of Privalov et al.[61] in the temperature range from 65° to 85°. They found that the magnitude of the heat increases slightly from about −9 kcal to −9.5 kcal per base pair in that range.

Heat capacity measurements on the melting of protonated helixes like the acid double helix of poly A are a little more complicated.[62,63] The reaction actually measured involves the disruption of helix structure and the transfer of bound protons to the buffer:

$$\alpha B^- + A_n \cdot A_n(\alpha H^+) \rightarrow 2A_n + \alpha HB \tag{6-16}$$

In order to compare with heats determined by the mixing method, where no buffer is used, one must correct for the heat of buffer neutralization, obtaining the heat of the reaction:

$$A_n \cdot A_n(\alpha H^+) \rightarrow 2A_n + \alpha H^+ \tag{6-17}$$

The heat of this process can still be expected to be pH dependent, because the degree of proton binding by the helix depends on pH.[57,66] The heat would also be expected to be temperature dependent because of the stacking in the single strand. Furthermore, there will be a substantial contribution to the heat capacity of the helix from the temperature-dependent proton binding. In partial recompense for all this complexity, there is a relation between the heat of melting, the degree of proton binding, and the pH dependence of the transition temperature,[66] so that heats can actually be determined without calorimetry.[67]

The entropy of helix formation may be calculated from the heat by setting the free energy of helix formation equal to zero at the transition temperature. The entropy is then $\Delta H/T_m$. The value of ΔH inserted in this relation must be either measured at the transition point from heat capacity measurements or extrapolated to T_m if mixing experiments are performed. The entropy of double helix formation is found to be about −20 to −25 cal/mol deg.

If we assume ΔH is approximately independent of temperature, ΔG can be easily calculated at any temperature

$$\begin{aligned} \Delta G &= \Delta H - T\,\Delta S \\ &= \Delta H\left(1 - \frac{T}{T_m}\right) \end{aligned} \tag{6-18}$$

where the identity of $\Delta S = \Delta H/T_m$ has been used. (In all the thermodynamic equations T_m in degrees Kelvin must be used.) This gives a value of ΔG of about −0.7 kcal for forming poly A · poly U at 30°C in good agreement with direct measurement.[68] If we further assume that ΔH is the same for A · U and G · C base pairs in RNA or A · T and G · C in DNA (see Table 6-1), then:

$$\Delta G_{GC} - \Delta G_{AU} = T\,\Delta H\left[\frac{1}{T_{m,AU}} - \frac{1}{T_{m,GC}}\right] \tag{6-19}$$

$T_{m,AU}$, $T_{m,GC}$ is the extrapolated absolute temperature for melting an RNA double strand which has 100% A · U (or G · C) base pairs.

From the data of Marmur and Doty[5] for DNA and Kallenbach[69] for RNA, plus ΔH of –8 kcal/mole base pair, we calculate at 25°C:

$$\text{RNA: } \Delta G_{\mathrm{GC}} - \Delta G_{\mathrm{AU}} = -1.25 \text{ kcal/mole base pair}$$
$$\text{DNA: } \Delta G_{\mathrm{GC}} - \Delta G_{\mathrm{AT}} = -0.75 \text{ kcal/mole base pair}$$

The goal of the thermodynamic studies is to be able to predict the effect of temperature and solvent on the stability of DNA, RNA, and hybrid multistranded molecules. We also want to know how base composition and sequence affect the stability. Implicit in the answers to these questions is a knowledge of the structures of the polynucleotides and the forces which govern them. This knowledge should be particularly helpful in the understanding of such biological processes as transcription and replication of the DNA. It should also be useful in interpreting experiments involving complementarity between denatured DNA and messenger RNA.

The most definitive thermodynamic information we could have would be H, S, and G as a function of temperature for the individual single strands and corresponding double strands. Near 100°C there is evidence for nearly complete loss of specific structure for single-strand polyribonucleotides.[51] That is, apparently both poly A, which has highly ordered, stacked bases at room temperature, and poly U, which is disordered, have the same local structure at high temperature. Measurements of heat capacity of single strands between 0° and 100°C would give their relative thermodynamic properties. We would know the stability of a single strand of poly (rA–rC) vs. that of poly (rA–rG), for example. If the stabilities depended only on nearest-neighbor interactions, it would take measurements on 13 independent single strands to be able to calculate the stability of any base sequence.[70] This would have to be done for both ribopolymers and deoxyribopolymers, of course. Measurement of heats of mixing of the single strands would then give the enthalpies of double strands relative to each other, as well as relative to the single strands. The advantage of this is that it would give insight into the differences in structures of the double strands. To obtain the entropies and free energies of the double strands, either heat capacity measurements up to T_m should be made or the free energy differences between single and double strands can be determined directly.[68,71] The purpose of these measurements is to find out how T_m and stability depend on the single strands and how they depend on the double-strand structures. This separation seems necessary to the understanding of the thermodynamics.

At present we can say little about the meaning of the ΔH and ΔS measurements which have been made (see Table 6-3). The enthalpy increase of 6–8 kcal/mole base pair on melting double strands can be compared with enthalpies of hydrogen bonds and hydrophobic interactions in water. To break a urea hydrogen bond in water requires a ΔH of about 1.5 kcal/mole.[72] For hydrophobic interaction measured as the transfer of a benzene molecule from water to liquid benzene, the ΔH is zero.[72] This emphasizes the fact that to understand the results, all the interactions discussed in Chapter 2 must be considered not only for bases, but also for the sugars and phosphates. The reasons for the decrease of entropy by about 22 units are also unclear. An approximately maximum configurational entropy decrease would correspond to having all the bonds completely rigid in the double strands and having three orientations in the single strands. This would give for the six bonds per nucleotide (assuming the sugar ring is rigid) in each strand:

$$\Delta S = -2(6\,R \ln 3) = -26 \text{ cal/deg mole base pair} \tag{6-20}$$

This estimate is undoubtedly too large in absolute value. A rough lower limit for the magnitude of the decrease in configurational entropy would be to assume that half the bonds remain fixed in the single strand and the other half acquire only two orientations.[73] This gives:

$$\Delta S = -2(3R \ln 2) = -8.3 \text{ cal/deg mole base pair} \tag{6-21}$$

It is also possible that there is some restriction on torsional oscillations about the equilibrium bond angle when a double helix is formed. This would give an additional entropy loss on helix formation.

The solvent entropy change must make up the difference between these figures and the measured ΔS of about -22 cal/deg mole base pair. Hydrophobic interactions which tend to order water molecules about the bases would lead to an entropy increase on helix formation and are therefore unlikely. If, however, the large electrostatic field of the ordered phosphate groups bound more water than the more disordered single strands, then a negative entropy contribution on helix formation would occur, consistent with experiment.

A direct measure of the free energy of a single strand to double strand conversion can be obtained by titration of the polynucleotides. The general theory is discussed by Ptitsyn and Birshtein[71]; Litan[68] has developed both the theory and experiments for acid titration. The method is based on the use of thermodynamic Maxwell relations[74] at constant temperature and pressure.

$$\left(\frac{\partial \bar{G}_j}{\partial \bar{G}_i}\right)_{n \neq n_i} = -\left(\frac{\partial n_i}{\partial n_j}\right)_{n \neq n_i\, n_j;\, \bar{G}_i} \tag{6-22}$$

$\bar{G}_i$, $\bar{G}_j$ are the partial molal free energies
n_i, n_j is number of moles of i, j

The subscripts on the partial derivatives specify that the number of moles of all components in the solution except n_i are held constant on the left-hand side of the equation. On the right-hand side the partial molal free energy of i plus all components except n_i and n_j are held constant. The equation relates the free energy of one component, such as a polynucleotide (j), to the free energy of another, such as $H^+(i)$. The dependence is directly proportional to the binding of H^+ by the polynucleotides. The free energy of H^+ is measured by the pH

$$dG_{H^+} = -2.303\, RT\, d\,\text{pH} \tag{6-23}$$

The binding of H^+ is measured by the degree of ionization (α) per mole of subunit in the polynucleotide.

$$-\left(\frac{\partial n_{H^+}}{\partial n_{\text{subunit}}}\right)_{\text{pH}} = \alpha \tag{6-24}$$

By integration one can measure the difference in free energy of the polynucleotide between two pH values.

$$\bar{G}(\text{pH}_2) - \bar{G}(\text{pH}_1) = -2.303\, RT \int_{\text{pH}_1}^{\text{pH}_2} \alpha d\,\text{pH} \tag{6-25}$$

Therefore, to measure the free energy change for the reaction

$$\text{poly A} + \text{poly U} \longrightarrow \text{poly A} \cdot \text{poly U}$$

one needs to titrate poly A and poly U separately and together. The concentration of subunit should be the same in each solution (moles of base in poly A and poly U, moles of base pair in poly A · poly U). At a high enough pH (pH*) the poly A · poly U will be completely dissociated so there will be no difference in free energy between poly A and poly U in separate containers or in the same solution. Therefore the free energy difference between single strands and double strands at any other pH can be obtained by titrating from this pH.

$$\bar{G}(\text{poly A} \cdot \text{poly U}) - \bar{G}(\text{poly A}) - \bar{G}(\text{poly U})$$

$$= -2.303\, RT \int_{\text{pH}^*}^{\text{pH}} [\alpha(\text{poly A} \cdot \text{poly U}) - \alpha(\text{poly A}) - \alpha(\text{poly U})]\, d\,\text{pH} \tag{6-26}$$

Litan[68] found a free energy change of −680 cal/mole of base pair for forming poly A · poly U at 30°C in 0.2 M KCl at pH 7.6. Titrations at different temperature would provide $\Delta\bar{H}$ and $\Delta\bar{S}$. This seems to be a very useful method for measuring thermodynamics of intermolecular double strands. Unfortunately it can not be applied directly to intramolecular double strands, without assumptions about the titration properties of the bases in the potentially double strand regions.

G. Buoyant Density and Specific Volume Changes

The increase in buoyant density of DNA upon its denaturation[75,76] has been a powerful tool for characterizing and separating DNA preparations.

The physical basis for this density change has received less attention.[77] Measurement of the specific volume of native and denatured DNA reveals no detectable difference between helix and coil,[78] a fact that might on the face of it seem to contradict the findings from banding density determinations. Careful attention to the detailed thermodynamics of these solutions[77,80-82] reveals that the two measurements are not equivalent, however.

According to the equations in Chapter 7 the banding density ρ_B is

$$\frac{1}{\rho_B} = \frac{\bar{v}_3 + \Gamma_1'\bar{v}_1}{1 + \Gamma_1'} \tag{6-27}$$

where $\bar{v}_3$ is the specific volume of the polymer, $\bar{v}_1$ is the same quantity for the solvent (water), and Γ_1' is a distribution coefficient equal to the weight of solvent that must be added to the solution for each gram of dry polymer in order to keep constant the chemical potential of solvent and electrolyte. A quantity closely related to $\bar{v}_3$ was measured through the DNA denaturation transition by Chapman and Sturtevant,[78] revealing only a gradual change as temperature increased, and no anomaly associated with the melting transition. Since the banding density increases on melting, and $\bar{v}_3$ is smaller than $\bar{v}_1$, Eq. (6-27) implies that Γ_1 (sometimes called the "hydration") decreases on melting.[78]

In alkali, where denaturation occurs concomitantly with ionization of guanine and thymine residues, there is a large increase in banding density, and also a change in the specific volume.[83] The difference between this and melting at neutral pH is that protons are replaced by Cs^+ in the course of the reaction, predictably leading to density effects. The volume changes found by Chapman and Sturtevant[83] seem to be correlated with the volume change expected for neutralization of the buffer by the protons released.

H. Unwinding of Closed Circular DNA

Very small structural changes of DNA can be detected by measuring the number of superhelical turns in artificially closed DNA circles.[84] The principle is to use the joining enzyme to form circles under various conditions, and to measure the number of superhelical turns under standard conditions. It is assumed that there are no net turns at the moment of joining. Following circle formation, the topological restrictions require that any changes in the average double helix rotation per base pair be compensated by changes in superhelical winding. Thus, differences at standard conditions in the number of superhelical turns in molecules joined at different temperatures reflect the temperature dependence of the double helix rotation. Wang[84] found a slight unwinding of the double helix as the temperature is raised, amounting to about 5×10^{-3} degrees of angle per degree centigrade per base pair. When the salt concentration is lowered, there is generally also a slight unwinding, for example 0.14° per base pair when the salt concentration is lowered from 3 M Cs^+ to 0.1 M Cs^+.

I. Electron Microscopy

Electron microscopy is particularly helpful in obtaining a realistic physical picture of the denaturation process. It is possible to observe partially denatured molecules in the microscope,[85-88] although it has been necessary to stabilize the denatured state by chemical reaction. Formaldehyde has generally been used for this purpose. It reacts with the base amino groups[89-94] and prevents helix formation, but at an appreciable rate only at moderately elevated temperatures. The electron micrographs show that

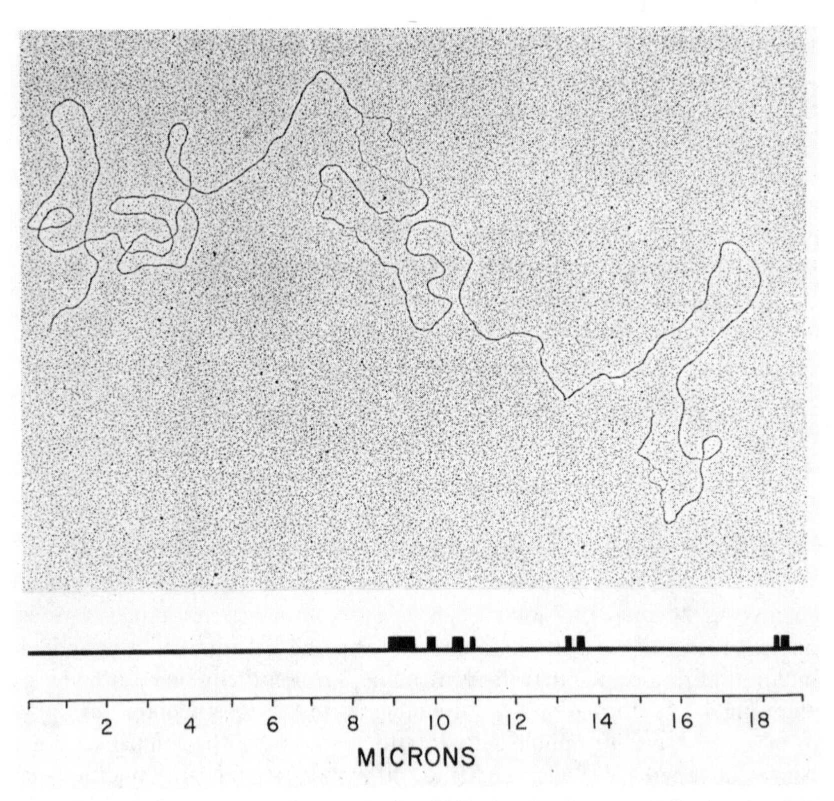

FIGURE 6-12 An electron micrograph of λ DNA that has been partially denatured by high pH (11 min at pH 11.29), stabilized by formaldehyde. The denaturation map for this particular molecule is shown below; the end of the molecule corresponding to the left end of the map is at the left of the photograph. The position and length of the denatured sites are indicated by black rectangles. [From R. B. Inman and M. Schnös, *J. Mol. Biol.*, **49**, 93 (1970).]

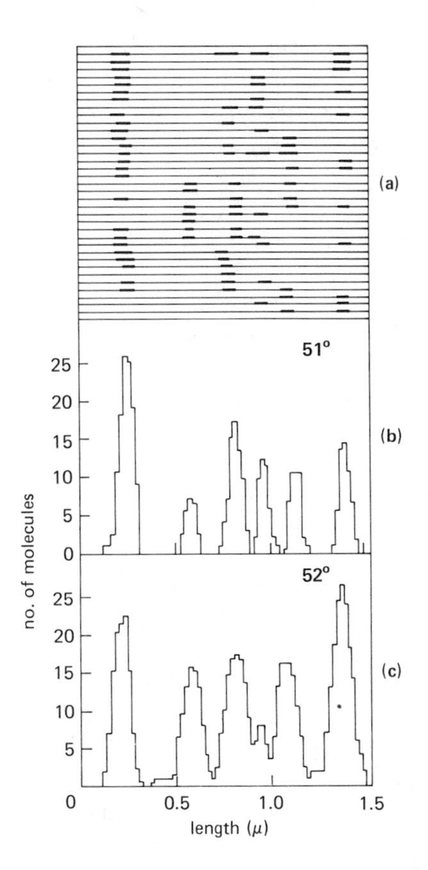

FIGURE 6-13 (a) Two-dimensional presentation of a set of 35 polyoma virus DNA molecules partially denatured by heating to 51°C with 12% formaldehyde. The circular molecules have been aligned to give maximum overlap of the regions of strand separation (denoted in position and extent by the thick lines). The circles are represented as though they had been opened at an arbitrarily chosen point. (b) Histogram constructed from (a) showing the frequency of occurrence of regions of strand separation at intervals of 200 Å around the molecule. (c) Histogram constructed from 400 molecules partially denatured at 52°C. The point at which the rings have been opened was chosen to illustrate the similarity between the overall pattern of distribution of denatured regions around the molecule at 52°C and the pattern obtained at 51°C. [From E. A. C. Follett and L. V. Crawford, *J. Mol. Biol.*, **34**, 565 (1968). Reprinted with permission.]

regions of denaturation open up along the molecule. The regions attacked have a definite location, as seen in Figs. 6-12 and 6-13 which show the distribution of denaturation in λ and polyoma virus DNA. The length scale of 1.0 μ corresponds to roughly 3000 base pairs. According to the diagram, the partially denatured molecules contain alternating helix and coil regions located at particular points in the molecule, and the successive regions are hundreds of base pairs long. Denaturation presumably begins at A · T-rich loci. This view is very similar to that given by the theory of DNA melting, discussed in Section IX. Formaldehyde denaturation is a complicated process, since kinetic factors may be involved in determining which regions are attacked. Nevertheless, there seems to be a definite similarity with purely thermal denaturation.

III STOICHIOMETRY OF HELIX FORMATION

One of the main problems that must be faced in studying polynucleotides is whether the ordered structure formed is a double or triple helix. If the helix contains two different components, like poly A and poly U, this question can be answered by solution studies designed primarily to determine the stoichiometry of the helix, for example whether it is $A_n \cdot U_n$ or $A_n \cdot 2U_n$. The experimental techniques of mixing curves and melting curves are both useful in this regard, and the results can be summarized in a phase diagram in which the stable forms are shown as functions of temperature and salt concentration.

The mixing curve is based on a study of the "continuous variation"[95] of the ultraviolet absorbance of variable-composition mixtures of the two polynucleotides that make up the complex. This technique has been applied to a number of polynucleotide systems.[17,95-102] It is based on the observation that helix formation leads to a change in UV absorbance, as discussed in Section II. Consider, for example, the double and triple helical complexes formed by poly A and poly U. Let ϵ_A, ϵ_U, $\epsilon_{A \cdot U}$ and $\epsilon_{A \cdot 2U}$ be the extinction coefficients of the polymeric species A_n, U_n, $A_n \cdot U_n$ and $A_n \cdot 2U_n$, respectively, based on the absorbance per mole of phosphate. Let the concentrations of these species be represented by C with corresponding subscripts, again in terms of phosphate concentration. The absorbance A is given by

$$A = \epsilon_A C_A + \epsilon_U C_U + \epsilon_{A \cdot U} C_{A \cdot U} + \epsilon_{A \cdot 2U} C_{A \cdot 2U} \tag{6-28}$$

Three separate regions of a typical mixing curve can be recognized. Consider first a region in which A_n is in excess, and the added U_n forms $A_n \cdot U_n$. Therefore the concentration of $A_n \cdot U_n$ phosphate is twice the total concentration of added U_n, which we call $C_U{}^\circ$, or $C_{A \cdot U} = 2\,C_U{}^\circ$. Similarly, $C_A = C_A{}^\circ - C_U{}^\circ$, where $C_A{}^\circ$ is the total concentration of added A_n. We define the apparent extinction coefficient ϵ_{ap} by

$$\epsilon_{ap} = \frac{A}{(C_A{}^0 + C_U{}^0)} \tag{6-29}$$

and the mole fraction of A_n, called X_A, by

$$X_A = \frac{C_A{}^0}{(C_A{}^0 + C_U{}^0)} \tag{6-30}$$

with the additional relation

$$X_U = 1 - X_A \tag{6-31}$$

Substitution of these last three equations in (6-28) and rearrangement yields

$$\epsilon_{ap} = 2\epsilon_{A \cdot U} - \epsilon_A + 2X_A(\epsilon_A - \epsilon_{A \cdot U});\ 1 \geq X_A \geq \frac{1}{2} \tag{6-32}$$

which says that the apparent extinction coefficient should be a linear function of X_A, with slope $2(\epsilon_A - \epsilon_{A\,\cdot\,U})$, for X_A greater than or equal to ½.

A different relation is derived for the region of the mixing curve in which enough U has been added to convert all the A to the double helix, and some to the triple helix. The two species present are $A_n \cdot U_n$ and $A_n \cdot 2U_n$, with $C_{A\,\cdot\,U} = 2\{C_A{}^\circ - (C_U{}^\circ - C_A{}^\circ)\}$ and $C_{A\,\cdot\,2U} = 3(C_U{}^\circ - C_A{}^\circ)$, giving

$$\epsilon_{ap} = 3\epsilon_{A \cdot 2U} - 2\epsilon_{A \cdot U} + 6X_A(\epsilon_{A \cdot U} - \epsilon_{A \cdot 2U});\ \frac{1}{2} \geq X_A \geq \frac{1}{3} \tag{6-33}$$

Again a linear variation of ϵ_{ap} with X_A is predicted, but with slope $6(\epsilon_{A\,\cdot\,U} - \epsilon_{A\,\cdot\,2U})$.

A third possible region is that in which all the A_n has been converted to the triple helical complex, and there is some free U_n present. In this case, $C_U = C_U{}^\circ - 2C_A{}^\circ$, and $C_{A\,\cdot\,2U} = 3C_A{}^\circ$, giving

$$\epsilon_{ap} = \epsilon_U + 3X_A(\epsilon_{A \cdot 2U} - \epsilon_U);\ \frac{1}{3} \geq X_A \geq 0 \tag{6-34}$$

which also predicts a linear variation of ϵ_{ap} with X_A, but with slope $3(\epsilon_{A\,\cdot\,2U} - \epsilon_U)$. Thus, if ϵ_{ap} is plotted against X_A, straight lines are expected with a discontinuous change of slope at $X_A = \frac{1}{2}$ and $X_A = \frac{1}{3}$. It should be pointed out, however, that accidental choice of an unfortunate wavelength can cause the slopes of Eqs. (6-32) and (6-33) to be the same, so that the discontinuity at $X_A = \frac{1}{2}$ can be overlooked.

A different case is that in which only the triplex helix is stable, and added U_n makes only this form, leaving some free A_n if that is in excess. Equation (6-34) still applies when X_A is between $\frac{1}{3}$ and 0, but for the region of A_n in excess one finds

$$\epsilon_{ap} = \frac{3}{2}\epsilon_{A \cdot 2U} - \frac{1}{2}\epsilon_A + \frac{3}{2}X_A(\epsilon_A - \epsilon_{A \cdot 2U}),\ 1 \geq X_A \geq \frac{1}{3} \tag{6-35}$$

According to this equation, a plot of ϵ_{ap} against X_A will yield only a single discontinuous change of slope, at the equivalence point $X_A = \frac{1}{3}$.

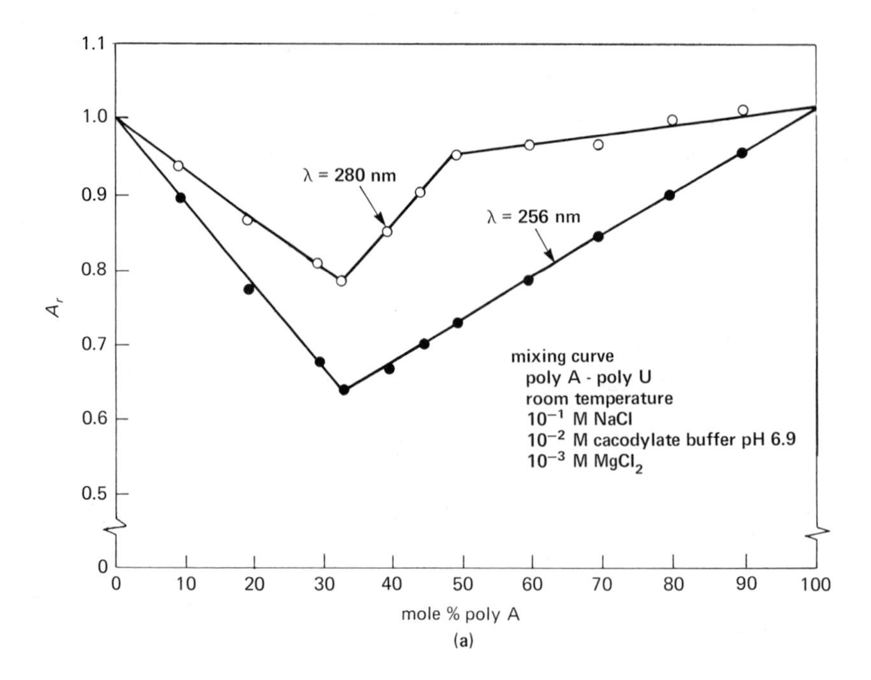

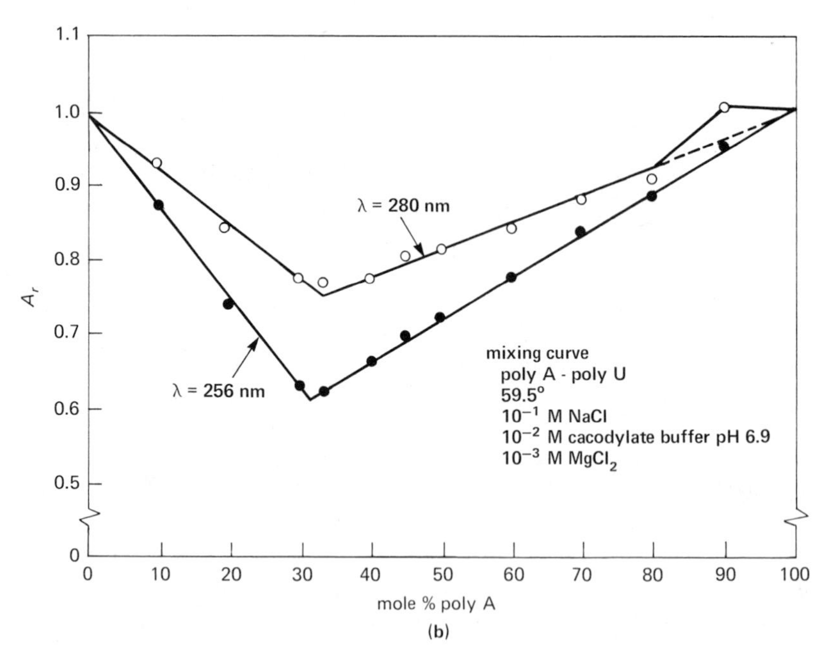

FIGURE 6-14 Mixing curves for poly A and poly U. The filled and open symbols represent relative absorbancy at 256 and 280 nm, respectively. The experiments were carried out at room temperature (a) and at 59.5° C. (b) The solvent for both experiments was 0.1 M in NaCl, 0.01 M in sodium cacodylate (adjusted to pH 6.95), and 0.001 M in $MgCl_2$. [From C. L. Stevens and G. Felsenfeld, *Biopolymers,* 2, 293 (1964). © 1964 by John Wiley & Sons, Inc. Reprinted with permission.]

To distinguish triple helix and double helix formation, one should measure ϵ_{ap} as a function of X_A (or X_U, which is equivalent), and look for changes of slope at the equivalence points $X_A = \frac{1}{3}$ and $X_A = \frac{1}{2}$. Figure 6-14, taken from the paper of Stevens and Felsenfeld,[17] illustrates such measurements, and brings out a very important point concerning the wavelength dependence of mixing curves. (In Fig. 6-14, the variable plotted is $\epsilon_{ap}/(\epsilon_U X_U + \epsilon_A X_A)$, a quantity that is strictly linear in X_A only if $\epsilon_U = \epsilon_A$. These two extinction coefficients are close enough that the curvature introduced is not detectable.) It can be seen in Fig. 6-14(a) that change in slope is seen only at 280 nm. The reason is accidental equality of the slopes of equations (6-32) and (6-33) at 256 nm, a coincidence that was responsible for much of the early controversy concerning the favored helix form in mixtures of poly A and poly U. Thus the failure to find, at one wavelength, a slope change at an expected equivalence point should not be taken as conclusive evidence against the existence of a particular form. It is essential that mixing curves be measured at several wavelengths when an unknown system is being examined. For the conditions of Fig. 6-14, the double helix $A_n \cdot U_n$ is formed when X_A is greater than 1/2, and triple helix when more U_n is added. At higher temperatures, shown in Fig. 6-14(b), only the triple helix is formed at all values of X_A.

A possible difficulty that can be encountered in mixing curves is the transient formation of metastable states. For example, Blake and Fresco[65] showed that mixing A_n and U_n under conditions where the double helix is the preferred form leads to transient production of $A_n \cdot 2U_n$. Several days may be necessary for establishing equilibrium. Also, slow helix interconversion reactions[99] may cause confusion.

The melting curve of a helical complex provides important diagnostic information concerning its composition, and is the basis for establishing the phase diagram that displays the stabilities of the various forms. As with mixing curves, it is important that melting curves be measured at several wavelengths in order to avoid overlooking a particular transition. Considering again the example of the well-studied complexes of poly A and poly U, there are four possible melting reactions. One can have either the double or the triple helix, and each of these can melt by two distinguishable paths. The double helix can melt directly to the single strands:

$$A_n \cdot U_n \xrightarrow{(1)} A_n + U_n \tag{6-36}$$

or it can melt to the triple helix, which then melts to the single strands:

$$2A_n \cdot U_n \xrightarrow{(2)} A_n \cdot 2U_n + A_n \xrightarrow{(3)} 2U_n + 2A_n \tag{6-37}$$

Which mechanism prevails depends on conditions, as discussed below. The triple helix can either melt directly to the single strands

$$A_n \cdot 2U_n \xrightarrow{(3)} A_n + 2U_n \tag{6-38}$$

which is just process (3) over again, or it can melt first to the double helix and then to the single strands:

$$A_n \cdot 2U_n \xrightarrow{(4)} A_n \cdot U_n + U_n \xrightarrow{(1)} A_n + 2U_n \tag{6-39}$$

These reactions differ in absorbance change, and respond differently at different wavelengths. Table 6-3 shows the direction of absorbance change that accompanies each of the four reactions. With this information the mechanism of a melting process and the composition of the starting material can be quickly diagnosed. Figure 6-15 shows an example of a melting curve measured at two wavelengths. The finding of a two-step melting curve at 260 nm means that one of the two-step denaturation processes must be involved. The second transition is not seen at 280 nm and consequently must be the melting of the double helix, process (1). The first step can then only be the formation of the double helix from the triple helix, process (4), and the starting material must be a triple helix. The stoichiometry is, in fact, two parts U to one part A, and the experiment was performed at low salt, conditions under which the triple helix melts first to the double helix. Examples of melting curves for all of the four possible melting mechanisms can be found in the paper by Stevens and Felsenfeld.[17]

The results of such melting experiments can be summarized in a phase diagram, which indicates the transition temperatures for each of the four numbered reactions at various salt concentrations. Figure 6-16 is a recent example taken from the work of Krakauer and Sturtevant.[59] The lines show the transition temperature for the corresponding reaction. In the region I the double helix is stable with respect to both reactions (1) and (2). Therefore all

TABLE 6-3 ABSORBANCE CHANGES ON MELTING DOUBLE AND TRIPLE HELIXES

	260 nm	280 nm
(1) Poly A · poly U → poly A + poly U	+	0
(2) Poly A · poly U → $\frac{1}{2}$ {poly A · 2 poly U} + $\frac{1}{2}$ poly A	0	−
(3) Poly A · 2 poly U → poly A + 2 poly U	+	+
(4) Poly A · 2 poly U → poly A · poly U + poly U	+	+

[a] A list of reactions of poly A and poly U is given on the left; the observability of the reaction by the hyperchromic effect and 260 and 280 nm is indicated at the right. The sign indicates the direction of change in absorbance accompanying the reaction. [From C. L. Stevens and G. Felsenfeld, *Biopolymers*, 2, 293 (1964).]

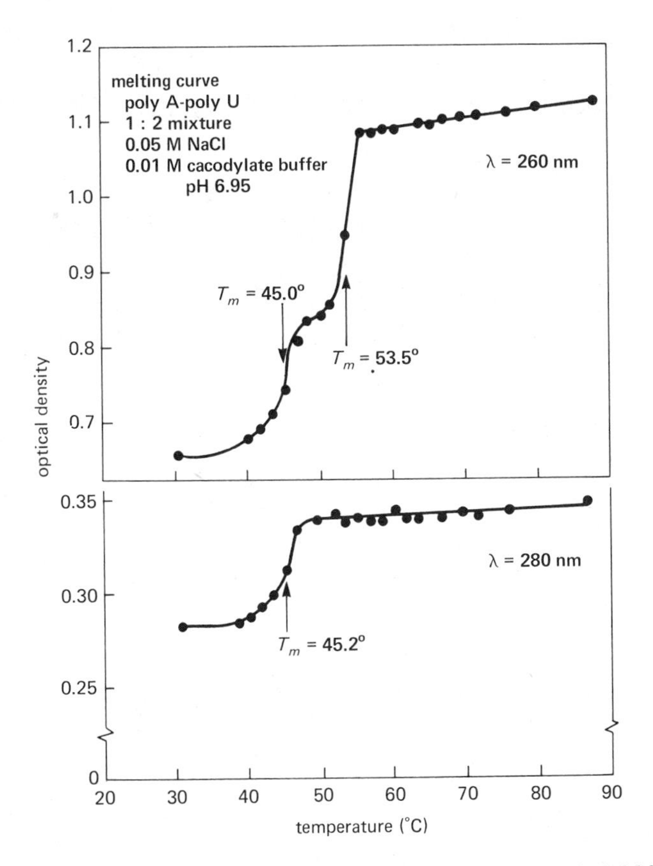

FIGURE 6-15 The melting of a 1 : 2 mixture of poly A and poly U in 0.05 M NaCl. [From C. L. Stevens and G. Felsenfeld, *Biopolymers*, 2, 293 (1964). © 1964 by John Wiley & Sons, Inc.. Reprinted with permission.]

the U added, up to equivalence with A, should form a double helix. The triple helix is also stable in this region, and so all the U added from equivalence with A to twice equivalence with A should form a triple helix. Consequently in region I the helix form present depends on the stoichiometry of mixing. In region II the triple helix is unstable with respect to reaction (4), which forms a double helix, so in this zone the only stable helical form is the double helix. If the triple helix is present in region I, a transition will be seen when the temperature crosses line (4), but not if the double helix was the starting material. Line (4) can be crossed only when the salt concentration is below about 0.1 M Na^+. In region III of the phase diagram, the double helix is unstable with respect to the disproportionation reaction (2) which forms a triple helix. Thus the only ordered form found in region III is the triple helix.

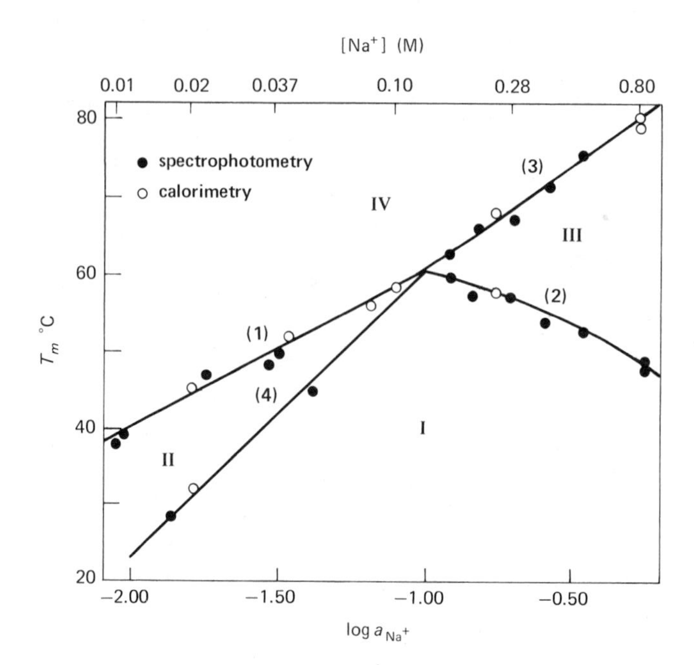

FIGURE 6-16 Dependence of T_m on the concentration of Na^+ at neutral pH for the reactions (1)–(4) described in the text. [From H. Krakauer and J. M. Sturtevant, *Biopolymers,* 6, 491 (1968). © 1968 by John Wiley & Sons, Inc. Reprinted with permission.]

If the double helix is the starting material at high salt, a transition will be seen when the temperature crosses line (2), but not when the triple helix is the starting material. No helical forms are present in region IV of the phase diagram. Line (1) is the melting of the double helix, and line (3) the melting of the triple helix, forming single strands in both cases.

It should be emphasized that such phase diagrams are very sensitive to conditions and the nature of the polymer constituents. For example, the phase diagram in K^+ is considerably different from that in Na^+.[59] Furthermore, substituting poly dA for poly rA seems to destabilize drastically the double helix, and only the triple helix $(dA)_n \cdot 2(rU)_n$ is found.[102] There are no obvious rules on such matters, and each system must be treated individually.

IV COMPOSITIONAL EFFECTS ON HELIX STABILITY

The stability of helical structures varies with the nature of the bases involved and the composition of the backbone chain. The best-known example of this is the linear variation of melting temperature with fractional

G · C composition. Marmur and Doty[5] found that

$$T_m(^\circ\mathrm{C}) = 69.3 + 41\, f_{GC} \qquad \text{(for DNA)} \tag{6-40}$$

for about 40 DNA molecules in a solution of 0.15 M NaCl plus 0.015 M sodium citrate (0.2 M Na^+). The $T_m(^\circ C)$ is the midpoint of the absorbance rise with temperature and f_{GC} is the mole fraction of G · C base pairs. Studies of RNA, DNA, and hybrid molecules show that RNA double strands generally have a higher T_m than the others.[69,103] For 11 double-stranded RNA molecules in 0.2 M Na^+ ion, the data of Kallenbach[69] can be approximated bv

$$T_m(^\circ\mathrm{C}) = 62 + 78\, f_{GC} \qquad \text{(for RNA)} \tag{6-41}$$

Linear variation of transition temperature with composition is sometimes found in other systems. For example, Howard et al.[27] reported a straight-line relation between T_m and the fraction of 5-bromocytidine in double helical I_n · $(C, BC)_n$. The pure BC-containing polymer is about 25° more stable than the pure C-containing material. On the other hand, Swierkowski et al.[104] reported decided nonlinearity between the T_m and fraction of rT in A_n · $(U, rT)_n$ double helixes. The T_m for melting of the $(U, rT)_n$ helix also depended nonlinearly on the fraction of rT, with the T-containing helixes more stable in both cases.

It is instructive to consider briefly the energetic aspects of such stability variations. An important point to recall is that even fairly large variations of T_m reflect quite small changes in the free energy of helix stabilization. At the transition temperature of a given polymer, the stabilization free energy ΔG is zero:

$$\Delta G = \Delta H - T_m\, \Delta S = 0 \tag{6-42}$$

The variation in ΔG with temperature is given by

$$\left(\frac{\partial \Delta G}{\partial T}\right)_P = -\Delta S \tag{6-43}$$

Using a heat of melting of 9 kcal per mole of base pairs in (6-42) gives an entropy change ΔS of 24 cal deg^{-1} $mole^{-1}$ for a polymer that melts at 100°C. Assuming ΔS is constant, the free energy ΔG that stabilizes this helix 40° lower in temperature (roughly the spread between G · C and A · T) is 1 kcal per mole of base pairs. This is a very small energy relative to all the interactions, attractive and repulsive, that differentiate the helix and coil states. The remarkable thing is that the transition temperatures of the known helixes are so similar.

A linear relation between T_m and base composition arises if the energetic contributions from the various base pairs are simply additive and do not depend on the details of sequence. In this case the heat and entropy per base pair of helix formation depend linearly on f_{GC}:

$$\Delta H = \Delta H_{AT}(1 + a f_{GC})$$
$$\Delta S = \Delta S_{AT}(1 + b f_{GC}) \qquad (6\text{-}44)$$

where ΔH_{AT} and ΔS_{AT} refer to the values for an A · T pair, and a and b are proportionality constants. (Experiment shows that these latter are considerably smaller than 1.) Since

$$T_m = \frac{\Delta H}{\Delta S} \qquad (6\text{-}45)$$

we obtain

$$T_m = T_{m,AT} \frac{(1 + a f_{GC})}{(1 + b f_{GC})}$$
$$\cong T_{m,AT}\,(1 + (a - b) f_{GC}) \qquad (6\text{-}46)$$

predicting a linear variation of T_m with f_{GC}. $T_{m,AT}$ is the transition temperature of the hypothetical pure A · T polymer.

The energetic contributions to ΔH from base pair formation come both from in-plane interactions such as hydrogen bonding, and from the stacking interaction with adjacent base pairs. It is easy to imagine that the total in-plane interactions should depend only on the number of base pairs of a given kind, but the stacking interactions along the helix axis could easily depend on sequence. In fact, most theoretical calculations of this interaction energy (see Chapter 2) indicate a much larger dependence on sequence than is found experimentally. Since T_m is linear with G · C composition, but relatively independent of base sequence (see below), the simplest interpretation is that each base pair contributes a fixed amount of stacking energy, roughly independent of what its neighbors are. In special cases, a favorable interaction between bases might lead to an energy that depends on pairwise contacts, and therefore to a nonlinear relation between T_m and composition. Under these circumstances, Eq. (6-44) is not valid.

Variations of helix stability with base sequence do exist, but the general rule seems to be that these effects are small compared to the difference in stability between A · T and G · C pairs. For example, the T_m of $(dA)_n \cdot (dT)_n$ in 0.1 M salt is 68°,[102] while $(dA{-}dT)_n \cdot (dA{-}dT)_n$ melts at 60°[105] under the same conditions. Similarly the polymers of identical composition but differing sequence $(dT{-}dG)_n \cdot (dA{-}dC)_n$ and $(dT{-}dC)_n \cdot (dA{-}dG)_n$ melt at 92° and 94°, respectively, in 0.15 M salt.[106] More data of this kind have recently been reported for DNA polymers of regular sequence.[107] It appears difficult to generalize about such stability variations. In the two examples quoted, the polymer with all purines on one strand is more stable than the alternating analog in one case and less stable in the other. A further complication is the possible difference in the single-strand stacking in the sequence isomers, which will also influence the transition temperature. Present theory is quite inadequate to predict or even rationalize such small energy differences.

Another important question that must be answered by experiment is the relative stability of ribo, deoxyribo, and hybrid helixes. A considerable amount of information on this topic has been accumulated, particularly by Chamberlin and his collaborators.[101-103] In the case of G · C pairs, for example,[103] $(rG)_n \cdot (rC)_n$ is the most stable, followed by $(rG)_n \cdot (dC)_n$, $(dG)_n \cdot (rC)_n$ and $(dG)_n \cdot (dC)_n$ as the least stable. There is about a 35° difference between the highest and lowest T_m's. For A · T or A · U pairs, the order of decreasing stability is $(rA)_n \cdot (rT)_n$, $(dA)_n \cdot (dT)_n$, $(rA)_n \cdot (dT)_n$. It is reported[102] that the double helix $(dA)_n \cdot (rU)_n$ does not form at all under the conditions examined. From these experiments the general observation is[103] that the ribohomopolymer pairs are more stable than the deoxy, but that the hybrid helixes may be more or less stable than the deoxy.

Only a few generalizations can yet be made about when triple helix formation will occur. All the cases known have all pyrimidines or all purines in a given strand. Examples can be given with three purine strands such as $I_n \cdot I_n \cdot I_n$[108] and $A_n \cdot 2I_n$,[109] with two purine strands and one pyrimidine, such as $(dC)_n \cdot 2(dI)_n$[99] and $C_n \cdot 2G_n$,[96,110] and with two pyrimidine strands and one purine, such as the well-known $A_n \cdot 2U_n$ or $(dT)_n \cdot (dA)_n \cdot (rU)_n$,[111] and $(dT{-}dC)_n \cdot (dG{-}dA)_n \cdot (rU{-}rC)_n$.[111] The rC in the third strand in this latter example is protonated, with a structure presumably analogous to the triple helix formed between oligo G and two strands of poly C.[112,113] Morgan and Wells[111] report that of the DNA-like structures with alternating sequence, only those with all purines in one strand and all pyrimidines in the other will add a third strand to make a triple helix.

The tendency toward base stacking in single-stranded helixes also depends on compositional effects, both on the nature of the base and on the backbone chain. Most conclusions about structure in this field are reached on the basis of spectroscopic evidence, and the theory is not yet secure enough to provide unambiguous answers. The general tendency is to interpret temperature-dependent optical properties as evidence for base stacking, and there is general agreement that this occurs extensively in poly and oligoadenylic acids,[6-9] and also in the case of other bases.[41-44,114,115] The greatest difficulties occur when one optical property shows temperature dependence, like the circular dichroism of U-containing dinucleotides,[43,44] while another, like the UV absorbance, changes very little in the same temperature range.[42] It seems safe to say at this stage that poly U shows less stacking than poly A or other ribopolynucleotides, but whether it stacks to an appreciable extent is still disputed. Techniques other than optical measurements will probably be required to answer this question.

Unusual internucleotidic links, like 2′—5′, definitely change the extent of stacking in dinucleotides,[43] and the optical properties of deoxyoligo and polynucleotides are demonstrably different from the ribo analogs.[47,48] Thus the 2′—OR definitely plays a role in establishing the conformation of single-stranded polynucleotides, but the exact structural influence is still debated.[116]

One method for studying the detailed nature of the conformational changes that occur in unstacking of single-stranded polymers is analysis of the NMR spectra of dinucleotides, as reported, for example, by Ts'o et al.[117] They suggest that the main effect of elevated temperature is increased rotation about the P—O bond in the ester linkage.

V ENVIRONMENTAL EFFECTS ON HELIX STABILITY

A. Salt Concentration

Multistranded helical forms of the nucleic acid are usually stabilized by increased concentrations of counterions[118]; this subject is included in the review by Felsenfeld and Miles.[116] The most studied systems are those involving monovalent cations, and the strikingly universal observation is that the transition temperature is linearly dependent on the logarithm of the activity of the counterion.[119] Examples may be seen in the phase diagram shown in Fig. 6-16 and data are available for a number of other systems.[17,96,98,102,105,119–125] Deviations from linearity occur mainly at very high salt concentrations. The theory of electrostatic effects around DNA has received considerable attention,[122–124,126,127] and is discussed in Chapter 7. We avoid here any consideration of the molecular theory, and take instead a thermodynamic and empirical approach to rationalizing the observed behavior.

When DNA is denatured, fewer counterions are closely associated with it than in the native state. (We circumvent the issue of describing the nature of the "association," whether by ion atmosphere or site binding.) Thus the denaturation reaction may be written

$$\text{DNA(helix}, r_h\text{M}^+) \rightleftharpoons \text{DNA(coil}, r_c\text{M}^+) + (r_h - r_c)\,\text{M}^+ \qquad (6\text{-}47)$$

where r_h and r_c are the number of ions "bound" per pair of bases to helix and coil, respectively. In general r_h is greater than r_c because the charge density on the helix is larger than for the coil and therefore creates a larger electrostatic potential, attracting the counterions more effectively.

Since counterions are released in reaction (6-47), the position of the equilibrium can be influenced by changing the activity of the counterion, just as would be the case with any chemical equilibrium. The equilibrium can also be shifted by changing the temperature. The effectiveness of the two intensive variables, temperature and ion activity, in shifting the equilibrium (6-47) is dependent on the extensive variables ΔH and $(r_h - r_c)$, respectively. At the transition temperature, the equilibrium is just in balance. If the salt concentration is raised, the equilibrium is shifted back toward helix, and the temperature must be raised further to return to the transition point (the reaction absorbs heat in the forward direction). This relation may be expressed quantitatively as[67]

$$\frac{d\,T_m}{d \ln a_m} = \frac{(r_h - r_c)\,RT_m^{\,2}}{(\Delta H_a{}')_m} \tag{6-48}$$

where a_m is the activity of the counterion at the transition point, $(r_h - r_c)$ the difference between binding to hypothetical pure helix and coil forms at the transition point, and $(\Delta H_a{}')_m$ the heat of helix formation, including the heat of binding the extra $(r_h - r_c)_m$ counterions, at constant ion activity a and at the transition temperature. Since there is a general tendency for the heat of helix melting to increase with temperature, this roughly compensates the increase in the term $RT_m^{\,2}$ and a nearly constant slope of the plot of T_m against counterion activity is predicted. It is beyond the scope of this simplified approach to rationalize the remarkable experimental constancy of this slope for a particular polymer. The slope does vary from one polymer to another, both for reasons of differences in ΔH and counterion binding.

Variations of transition temperature with ion activity are discussed in terms of Eq. (6-48) and measured heats by Krakauer and Sturtevant[59] for the poly A and poly U system. They calculated $r_h - r_c = 0.31$ per base pair for melting of $A_n \cdot U_n$ in Na^+, 0.29 for transformation of $A_n \cdot 2U_n$ to $A_n \cdot U_n + U_n$ (r_c in this case is the binding per base triplet in the helix plus coil product) and 0.62 per base triplet for melting of $A_n \cdot 2U_n$ to single strands. Interpretation of these phenomenological coefficients in terms of the electrostatic properties of polynucleotides requires a molecular theory.

Quantitative investigation of the effect of Mg^{2+} and similar divalent ions on the stability of helical polynucleotides has been less extensive. In low salt concentration solutions Mg^{2+} binds very tightly to DNA,[128] as evidenced by an activity coefficient of roughly 0.07.[129] There is some disagreement about whether it binds more tightly to native or denatured DNA,[128,129] but since Mg^{2+} markedly stabilizes the helical form,[119] Eq. (6-51) clearly indicates that it should have greater affinity for the helix at the transition point. Measurements of the binding affinity of "denatured" DNA are probably complicated by recovery of some secondary structure and nonspecific aggregation in quenched samples. Plots of T_m against the logarithm of the added concentration of Mg^{2+} are decidedly nonlinear.[119] A more revealing measurement would be the stability or T_m as a function of Mg^{2+} activity, which, because of the tight binding to DNA, is quite different from the added concentration.

Lowered salt concentration has a definite broadening effect on the melting transition of DNA. Dove and Davidson[119] found that the relative transition breadth was increased from 2.2°C in 0.01 M salt to 5.3°C in 1×10^{-4} M. This effect is not due to reduction in the differential stabilities of A · T and G · C pairs, since this is essentially unaltered in this salt concentration range.[125] From the observation of Scheffler et al.[130] that short helixes are relatively more stable than long helixes in low salt, one would expect that the effect is a long-range electrostatic one which tends to reduce the average helix or correlation length in the transition and hence make it broader.

B. pH

The stability of DNA is relatively insensitive to pH between 5 and 9 but decreases drastically on both sides of that range.[121,124,131] The reason is that at low and high pH the coil form is ionized more readily than the helix, and the helix-coil equilibrium is therefore shifted toward the coil. Both titration[55] and spectrophotometric[57,132-134] studies show that extensive protonation of the DNA helix precedes the actual cooperative melting transition at low pH. However, the melting transition can be clearly recognized from the sudden concomitant increase in proton binding.

Examination of the spectral changes prior to actual melting[132-134] reveals that they can be interpreted as due to protonation of cytosine. Furthermore, the number of protons taken up prior to melting is roughly proportional to the G · C content of the DNA sample. Thus it seems that the G · C pair is the main target for protonation prior to melting. The usual site of cytosine protonation is the ring nitrogen (see Chapter 2 for the acid-base properties of the bases), but it is difficult to imagine that it could be protonated there without breaking up the Watson-Crick structure. A possible hypothesis is that actual proton attachment occurs to guanine, which, by a tautomeric shift, transfers a proton to cytosine, still maintaining a hydrogen-bonded structure, as shown in Fig. 6-17. Further examination of this question will be needed for a definitive answer.

According to titration studies, the coil binds more protons than the helix, doubtless because of the increased accessibility of the protonation sites. Thus the denaturation reaction at low pH can be written

$$(r_c - r_h)\mathrm{H}^+ + \mathrm{DNA}(\text{helix}, r_h\,\mathrm{H}^+) \rightleftharpoons \mathrm{DNA}\,(\text{coil}, r_c\,\mathrm{H}^+) \qquad (6\text{-}49)$$

In this case increase of H^+ concentration shifts the equilibrium to the right, and the temperature must be decreased to maintain the same degree of transition; thus the T_m is lowered at low pH. By analogy with Eq. (6-48) the

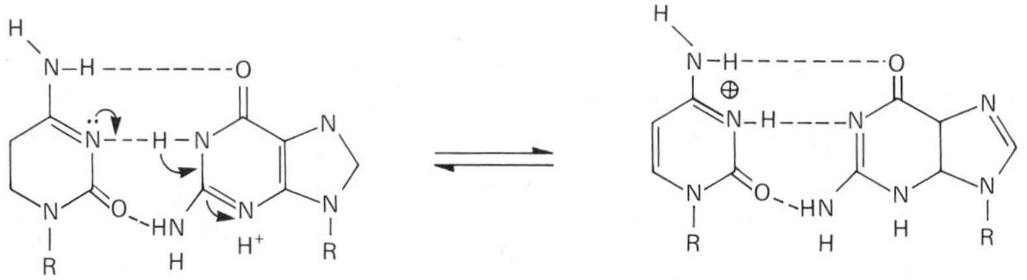

FIGURE 6-17 Possible structure of a protonated G · C pair.

relation between pH and T_m involves the heat of helix formation at constant H^+ activity at the T_m and $r_h - r_c$:

$$\frac{d\,T_m}{d\,\mathrm{pH}} = \frac{2.3(r_h - r_c)\,\mathrm{R}\,T_m^{\,2}}{(\Delta H_a')_m} \qquad (6\text{-}50)$$

Since all of the variables in this equation are experimentally accessible, it can be tested. From the titration measurements of Bunville et al.[57] the degree of binding to helix and coil can be extrapolated to the transition point. They also measured the heat of melting and the variation of T_m with pH. Using their data, one calculates a transition enthalpy from Eq. (6-50) of a little over 6 kcal/mole of base pairs, compared to a calorimetrically determined value of about 5 kcal (heat of melting in both cases). In view of the inaccuracies in titration data and the differences in concentration of the two determinations, this agreement should be considered satisfactory.

Hennage[63] tested Eq. (6-50) for the transition of the acid double helix of poly A, using the titration data of Holcomb and Timasheff.[66] The calculated value for the heat of helix formation was 4.6 kcal, compared with calorimetric values of –4.6[55] and –4.2[63] kcal per mole of base pairs.

DNA is destabilized in alkali because in the coil form the bases guanine and thymine are more readily deprotonated. Chapman and Sturtevant[83] have reported titration measurements on DNA in the alkaline melting transition that permit evaluation of the differential deprotonation of helix and coil. In contrast to the rather large premelting protonation in acid, there is little premelting deprotonation in alkaline medium. In a typical measurement at pH 11 and around 45°, r_h, the degree of proton removal form the helix, was roughly 0.1 per base pair. The site of this partial ionization is uncertain. Following the transition the guanine and thymine bases become extensively ionized. At pH 11 and 45° the difference between r_c and r_h is about 0.6 per base pair and is even larger at high pH. With the titration data this permits an estimate of the heat of melting in alkali from Eq. (6-50). The result is about 9.0 kcal per base pair at pH 11 in 0.15 M Na^+.

The pH of melting has a strong influence on the dependence of T_m on salt concentration. At all pH values one finds a linear variation with the log of counterion activity, but the slope of this line depends strongly on pH.[124] In acid, the coil binds more protons than the helix, and thus its charge density relative to the helix is still further reduced. Therefore the difference in counterion binding by the two will be larger than at neutral pH, and Eq. (6-49) predicts an increased dependence on salt concentration. In alkali, the coil is ionized more extensively than the helix, so its negative charge density approaches that of the helix. At some pH the degree of counterion binding by the two forms becomes equal, and there is no dependence of T_m on salt concentration. According to Record[124] this occurs at about pH 11. Above this pH the coil binds more counterions than the helix, and higher salt concentration destabilizes the helix.

C. Solvent

One of the main techniques for investigating the forces that stabilize polynucleotide helixes has been to change the nature of the solvent, using either organic solvents[135-138] or high concentrations of various salts.[125,139,140] An extensive survey of the relative denaturing power of various organic solvents was reported by Levine et al.[138] who measured the concentration of organic solvent necessary to bring about stated conditions of denaturation at a fixed temperature. Their results are summarized in Table 6-4. There is a clear tendency for more "organic" molecules to be more effective denaturing agents, indicated by a smaller concentration required for denaturation. For example, extension of the hydrocarbon chain on simple alcohols from methyl to *n*-butyl causes a continuous increase in denaturing power, and similar trends may be seen in other classes of compounds. Addition of a more hydrophilic group like —OH to cyclohexyl alcohol to make inositol markedly reduces denaturing power.

Attempts have been made to correlate the destabilizing effects of these solvents with other properties. There is no clear relation to solvent dielectric constant or to hydrogen bonding ability. There is, however, a definite correlation with the ability of the solvents to solubilize adenine.[143] The solvents that are most effective in increasing the solubility of adenine are the most effective denaturing agents. The implication is that the organic solvents interact preferentially in solution with the bases, and they are able to do this more effectively with the coil form. Consequently, they shift the helix-coil equilibrium toward the coil. Since this effect is brought about more effectively by agents generally considered "hydrophobic," what is called "hydrophobic bonding" is implicated in stabilizing the double helix. This subject is discussed in Chapter 2.

High concentrations of a number of salts have a dramatic destabilizing effect on DNA.[125,139,140] The effect is definitely related to the nature of the anion, with little effect of changing the cation. Again, it is found that the

TABLE 6-4 REAGENT CONCENTRATION (M) GIVING 50% DENATURATION OF T4 BACTERIOPHAGE DNA IN AQUEOUS SOLUTION AT 73°, IONIC STRENGTH 0.043[a]

	M		M
	Aliphatic alcohols		
Methyl alcohol	3.5	Isobutyl alcohol	0.45
Ethyl alcohol	1.2	*n*-Butyl alcohol	0.33
Isopropyl alcohol	0.90	*tert*-Amyl alcohol	0.39
n-Propyl alcohol	0.54		
Allyl alcohol	0.50	Ethylene glycol	2.2
sec-Butyl alcohol	0.62	Glycerol	1.8
tert-Butyl alcohol	0.60		
	Thio alcohol		
Dithioglycol	2.2		
	Cyclic alcohols		
Cyclohexyl alcohol	0.22	Phenol	0.08
Benzyl alcohol	0.09[b]	*p*-Methoxyphenol	0.09
Inositol	1.5[b]		
	Other cyclic compounds		
Aniline	0.08	1,4-Dioxane	0.64
Pyridine	0.09	γ-Butyrolactone	0.55
Purine	0.13	3-Aminotriazole	0.42
	Amides		
Formamide	1.9	Butyramide	0.46
N-Ethylformamide	1.0	Hexanamide	0.17
N,N-Dimethylform-amide	0.60		
Acetamide	1.1	Glycolamide	1.1
N-Ethylacetamide	0.88	Thioacetamide	0.32
N,N-Dimethylacet-amide	0.60	δ-Valarolactam	0.34
Propionamide	0.62		
	Ureas		
Urea	1.0	Ethyleneurea	0.53
Carbohydrazide	1.0		
1,3-Dimethylurea	1.0	Thiourea	0.41
Ethylurea	0.60	Allylthiourea	0.28
tert-Butylurea	0.22	Ethylenethiourea	0.32
	Carbamates		
Urethane	0.50	*N*-Propylurethane	0.24
N-Methylurethane	0.38		
	Other compounds		
Cyanoguanidine	0.21	Acetonitrile	1.2
Sulfamide	1.1	Tween 40	> 20%
Glycine	2.2	Triton X-100	> 10%

[a] From Levine et al., *Biochemistry*, **2**, 168 (1973).
[b] Based on a single experimental point.

TABLE 6-5 ORDER OF EFFECTS OF IONS ON ACTIVITY COEFFICIENTS OF PURINE AND PYRIMIDINE BASES[a] AND ON DENATURATION OF DNA[b]

Base	Cations[c]	Anions[c]
Thymine[a]	Ca^{2+}, Na^+, K^+, NH_4^+, Cs^+ Li^+, $(CH_3)_4N^+$	Tosylate$^-$ > Cl_3CCOO^- > SCN^-, ClO_4^- > I^- > Br^- > Cl, BrO_3^-, Ch_3COO^- > HSO_3^- > F^- > SO_2^{2-}
Cytosine[a]	Na^+, $(CH_3)_4N^+$	Cl_3CCOO^- > ClO_4^- > I^- > $(CH_3)_3CCOO^-$ > Cl^- > SO_4^{2-} > HPO_4^{2-}
Adenine[a]	$(CH_3)_4N^+$, Li^+ > Na^+, K^+, Cs^+	Cl_3CCOO^- > ClO_4^-, SCN^- > I^- > $(CH_3)_3CCOO^-$ > Br^- > Cl^-, CH_3COO^- > SO_4^{2-}
Deoxyadenosine[a]	$(CH_3)_4N^+$ > Na^+, K^+	Cl_3CCOO^- > ClO_4^- > I^- > Cl^- > SO_4^{2-}
Adenosine[a]		ClO_4^- > I^- > Cl^-
DNA[b]	Li^+, Na^+, K^+, $(CH_3)_4N^+$	CCl_3COO^- > SCN^- > ClO_4^- > I^- > CH_3COO^- > Br^-, Cl^-

[a] From Robinson and Grant.[140]
[b] From Hamaguchi and Geiduschek.[139]
[c] Ions are listed in order of decreasing ability to decrease activity coefficients of the bases or in decreasing ability to denature DNA.

same anions that destabilize DNA serve to solubilize the bases. Table 6-5 summarizes these comparative trends. Robinson and Grant[140] argue that there is no clear correlation of solubilizing power with whether the ion is structure forming or structure breaking in water solution, with effects on water activity, with electrostatic effects, or with surface tension or cavity formation energy. Their explanation in molecular terms for the observed thermodynamic effect of solubilization is that the anions form complexes with the polar bases. Theoretical examination of possible structures of such complexes, and explanations of the great differences between such ions as SO_4^{2-} and ClO_4^-, would provide support for this point of view.

VI MOLECULAR WEIGHT DEPENDENCE OF HELIX STABILITY

The stability of both DNA[141] and synthetic polynucleotides[142] decreased when the molecular size is reduced. (The melting behavior of even smaller materials, in the size range below roughly 30 nucleotides, is considered under the heading of oligonucleotides.) Formation of double and triple helical complexes involves a nucleation step, and this is responsible for the loss in stability as size is reduced. If two or three separate strands must come together to nucleate a helix, the first step can be thought of as formation of a single base pair or base triplet. This step is unfavorable because the bases must find each other in solution, resulting in a loss of entropy. Furthermore, the base pair that is formed is not stabilized by stacking on an adjacent pair. In the theory of melting processes, the equilibrium constant for forming an isolated base pair is symbolized by βs, and has a magnitude of roughly 10^{-3} M^{-1} [142,143] for A · U pairs. This means that very high concentrations, over 1 M, of mononucleotides would be required for substantial association by formation of isolated base pairs. Mononucleotides associate much more strongly by a stacking mechanism.

Since helix formation involves a nucleation step, processes that are favorable in thermodynamic terms must be provided if the final complex is to be stable. Successive steps of base pair formation or "helix growth" between two complementary strands serve this purpose. If the molecule is very long, the nucleation process is an inconsequential part of the free energy of helix formation, and the helix is in equilibrium with coil when "helix growth" has zero free energy change. Thus for very high molecular weights one expects the transition temperature to be independent of molecular weight. As the size is reduced there is a gradually more prominent drop in the T_m, since fewer growth steps are available to compensate for nucleation. Each growth step becomes more favorable as the temperature is lowered.

Figure 6-18 shows the reduction of T_m with size for DNA. There is practically no shift of the transition curve between molecular weights of 1.3×10^8 and 2.5×10^6, but below this the effect of reducing size is increasingly detectable. With a synthetic polynucleotide like the acid double helix of poly A, the effect of changing size can be detected at even higher molecular weights, since a material with a double strand molecular weight of 5×10^6 has a T_m different by about 0.06°C from the extrapolated value for infinite chain length.[142] Smaller variations can be detected with this material than with DNA because the transition is considerably sharper.

The failure to find a size dependence of the transition curve for DNA between molecular weights of a few million and 100 million has implications for the mechanism of denaturation. Specifically this indicates that the ends are not of great importance in determining the melting behavior, because, if they were, their contribution would be dependent on molecular size, and length-dependent melting curves would result. Therefore the bulk of the melting in high molecular weight samples must occur at internal points, forming alternating regions of helix and coil.

The effect of reducing size is to broaden the melting transition, as may be seen in Fig. 6-18 for DNA, and as is also found for acid poly A.[142]

VII OLIGOMER-POLYMER INTERACTIONS

The study of complementary interaction of oligonucleotides with polymers is a rapidly growing field. The advantage of such a system for physical studies is that structural, thermodynamic, and kinetic properties can

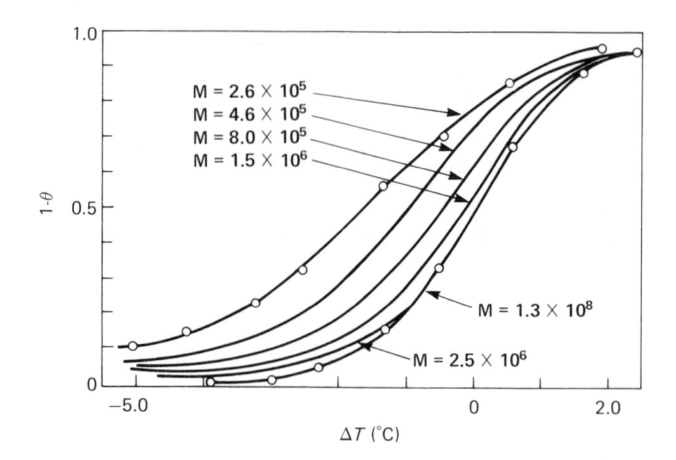

FIGURE 6-18 Experimental transition curves for T2 DNA after varying amounts of shear degradation, showing the fraction of bases unbonded as a function of temperature. The molecular weight M is indicated for each curve. [From Crothers, Kallenbach and Zimm, *J. Mol. Biol.*, **11**, 802 (1965). Reprinted with permission.]

be examined by techniques and with a precision not often possible for polymeric materials. Furthermore, several unexpected features of structure formation in nucleic acids have been revealed by such studies. This general topic was reviewed by Felsenfeld and Miles.[116]

It has been known for some years that polynucleotides will form complexes with oligonucleotides[144–152] and even with monomers.[153–158] In each of these systems it is found that combination occurs with a preferred stoichiometry, and that the interaction is specific with respect to base pairing. In general the base pairing rules mimic those found with the polymers, although there are some definite changes, for example, in the relative stabilities of double and triple helical forms. The experimental methods used are similar to those used with polymers, including UV and IR spectroscopy, mixing curves to establish stoichiometry, melting curves, rotatory dispersion and circular dichroism, and NMR. In addition, because the reaction is influenced by the concentration of oligomer or monomer, the thermodynamics of "adsorption" can be investigated in a manner not possible for polymers.

The tendency for base stacking among the purines is an important driving force for forming these complexes. The pyrimidines seem definitely weaker in this regard. For example, adenosine[154,155] and guanosine[153] monomers will stack on poly U and poly C, respectively, to form helixes, but the corresponding reactions where the monomers are pyrimidines are unknown.

The importance of base stacking in the mechanism of helix formation is

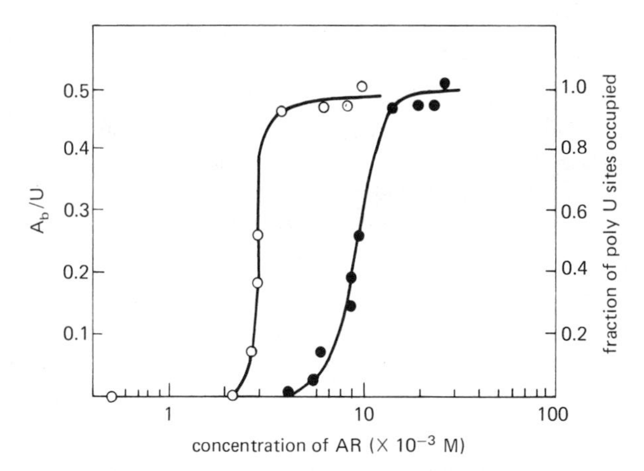

FIGURE 6-19 Adenosine bound per UMP of the poly U (1.5×10^{-2} M) versus adenosine input concentration at 5°C, –●–●–, 0.4 M NaCl, HMP. The fraction of poly U sites occupied versus free adenosine concentration under the same condition is also shown, –○–○–. [From W. M. Huang and P. O. P. Ts'o, *J. Mol. Biol.*, **16**, 523 (1966). Reprinted with permission.]

brought out by the cooperativity of the adsorption reaction. Figure 6-19, taken from the paper of Huang and Ts'o,[155] shows the progress of the binding of adenosine to poly U as a function of the free adenosine concentration. Up to a critical concentration of A, no binding occurs, but over a very narrow range of free concentrations the reaction proceeds to saturation. The molecular interpretation is that binding of the first A, which involves only hydrogen bonding or in-plane interactions, is not favored, but subsequent steps, which include stacking of the bound A on adjacent bases, are very favorable. The result is a cooperative binding reaction in which the state of the molecule changes drastically in response to small changes in external variables – for example, in adenosine concentration.

The stoichiometry of the binding reaction indicated in Fig. 6-19 is one A per two U, indicating formation of a triple helix,[155] in agreement with IR spectra.[154] In fact, in contrast with the phase diagram for the polymers, the only stable helix yet convincingly demonstrated in the adenosine + poly U system has a 1:2 stoichiometry. As higher oligomers of A (or dA) are used for the interaction with poly U (or poly dT), the stability of the double relative to the triple helix gradually improves, approaching the behavior of the polymers. On the other hand, when oligomers of T are used, the double helix is the preferred product.[150] This behavior can be rationalized in a simple manner.

Consider first the complex with oligo(A), compared with the corresponding complex with poly(A):

$$2\ \text{poly(A)} + 2\ \text{poly(U)} \rightleftharpoons 2\ [\text{poly(A)} \cdot \text{poly(U)}]$$
$$\rightleftharpoons \text{poly(A)} \cdot 2\text{poly(U)} + \text{poly(A)} \qquad (6\text{-}51)$$

$$\underset{\text{(single strands)}}{2m\ \text{oligo(A)} + 2\ \text{poly(U)}} \rightleftharpoons \underset{\text{(double helix)}}{2[m\ \text{oligo(A)} \cdot \text{poly(U)}]}$$
$$\rightleftharpoons \underset{\text{(triple helix)}}{m\ \text{oligo(A)} \cdot 2\text{poly(U)} + m\ \text{oligo(A)}} \qquad (6\text{-}52)$$

In reaction (6-52) the disproportionation of double helix to triple helix (the second step) releases *m* molecules of oligo(A), whereas in (6-51) only one molecule of poly(A) is released. Hence disproportionation is much more favorable in (6-52) because of the larger entropy gain, and the triple helix is more stable relative to the double helix when oligo(A) replaces poly(A).

These arguments do not exclude the possibility of the double helix as a transient intermediate, or that it could exist when the disproportionation reaction is blocked. An example of the latter is the observation by Howard et al.[154] that 9-methyl-6-aminopurine, in which the amino group can donate only one hydrogen bond, forms a double helix with poly(U). Similarly in the interaction of GMP with poly(C) a double helix is formed at high pH, whereas at lower pH a triple helix is formed with an extra (protonated) strand of poly(C).[153]

When the reaction involves oligo(U) and poly(A),

$$\underset{\text{(single strands)}}{2\ \text{poly(A)} + 2m\ \text{oligo(U)}} = \underset{\text{(double helix)}}{2[\text{poly(A)} \cdot m\ \text{oligo(U)}]}$$
$$= \underset{\text{(triple helix)}}{\text{poly(A)} \cdot 2m\ \text{oligo(U)} + \text{poly(A)}} \qquad (6\text{-}53)$$

and there is no special entropy gain that favors the disproportionation reaction in the oligomer system. When, as found by Cassani and Bollum[150] for poly(dA) and oligo(dT), the double helix is preferred, this reflects the preference of an oligo(dT) for binding to a single strand of poly(dA) rather than to the double helix poly(dA) · oligo(dT).

Several thermodynamic variables can be determined by studying the variation of transition temperature with concentration and chain length in the oligomer-polymer system. Illustrations are given by Tazawa et al.[159]; the theory for such processes has been developed by Damle[160] and by Crothers.[67]

Less extensive work than with polymer-polymer interactions has been done on the environment sensitivity of the oligomer-polymer helixes, although it has been shown, for example, that the T_m of the triple helical complex of adenosine with poly U varies linearly with the log of counterion concentration.[154]

VIII OLIGOMERS AND NATURAL RNA

A. Conformation and Free Energy

Natural RNA molecules such as tRNA and mRNA have base sequences that imply only partial double helix formation. It is generally agreed that the secondary structure in such molecules consists of regions of double helix alternating with single-stranded sections. In the case of transfer RNA, this is the well-known "cloverleaf" structure (Chapter 8). For other RNAs it is not generally clear which should be the most stable of the several possible patterns of hydrogen bonding of potential base pairs.

One approach to this problem is to analyze the contribution of particular structural features to the free energy of the ordered structure. The sum of all such terms is the free energy of folding the single strand into a particular state; the favored form is that with minimum free energy. Thus the problem is to calculate the conformational free energy of a particular state.[161,162]

The factors to be considered are the base pairs, including dependence on composition and sequence, and the single-stranded regions. The latter have been divided into three classes (see Fig. 6-20).[161]

The process of helix formation may be considered as two steps: nucleation followed by helix growth. Consider, for example, formation of a hairpin helix from a single strand (see Fig. 6-21). The first base pair, formed in the nucleation process, is not nearly so favorable as those subsequently formed in the growth steps. There are two reasons for this: (1) The first base pair is not stacked on an existing base pair and therefore lacks that favorable free-energy contribution; and (2) the bases that bond to close a loop and nucleate a helix are relatively far apart in the single-stranded state. Closure of the loop restricts the degrees of freedom of the intervening chain and provides an unfavorable free energy term. This term generally becomes more unfavorable as loop size increases, although there is an optimum size for hairpin loops.[162–164]

The conformational free energy ΔG_t of forming the hairpin helix in Fig. 6-24 (see below) can also be divided into two parts, corresponding to the nucleation and growth processes:

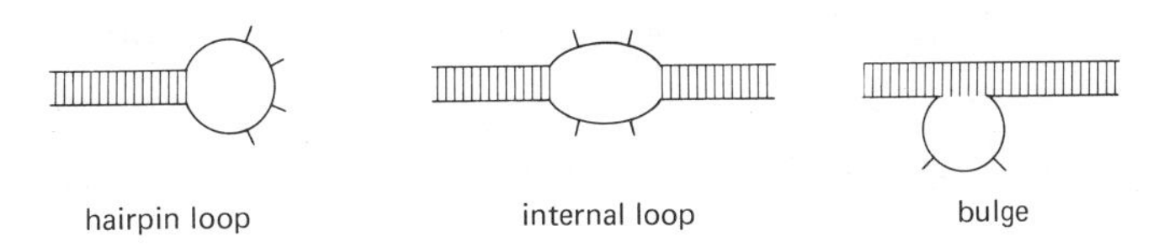

FIGURE 6-20 Single-stranded loops.

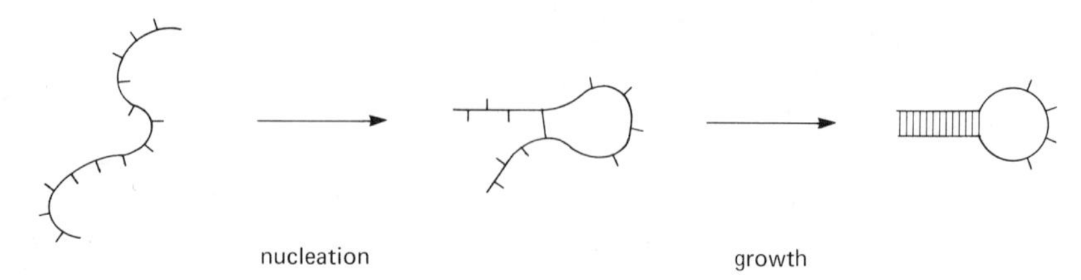

FIGURE 6-21 Formation of a hairpin helix from a single strand.

$$\Delta G_t = \Delta G_{\text{nuc}} + \Delta G_{\text{growth}} \tag{6-54}$$

ΔG_t can usually be measured from the melting transition of the hairpin helix; it is related to the equilibrium constant K for the whole reaction by the equation

$$\Delta G_t = -RT \ln K \tag{6-55}$$

The problem is to break this free energy down into the separate nucleation and growth contributions. This is usually done by making an independent estimate (by methods discussed below) of the growth free energy contributed by adding base pairs to an existing helix. Then the nucleation free energy is obtained by subtracting:

$$\Delta G_{\text{nuc}} = \Delta G_t - \Delta G_{\text{growth}} \tag{6-56}$$

When this analysis is complete for the possible nucleation and growth reactions, the conformational free energy of a more complicated structure can be estimated by adding the free energies of the reaction steps necessary to form the ordered state:

$$\Delta G_t = \sum_i \Delta G_{\text{nuc}}(i) + \sum_j \Delta G_{\text{growth}}(j) \tag{6-57}$$

B. Model Oligonucleotides

Oligonucleotides whose thermal stability has been studied fall into two classes, those that form helixes from two (or three) strands and those that form intramolecular bonds within one strand. An example of the first category is the series A_nU_n,[165] which forms dimers because the strands in a double helix are antiparallel:

$$\begin{array}{c} \overrightarrow{A_n U_n} \\ \bullet \quad \bullet \\ \overleftarrow{U_n A_n} \end{array}$$

Other two-stranded model oligomers have been studied by Applequist and Damle,[143] Uhlenbeck et al.,[166] Pörschke and Eigen,[167] and Craig et al.[168] Dimers with mismatching bases have been reported by Uhlenbeck et al.[166] and by Gralla and Crothers.[164] These are useful for determining the free energy of internal loops; an example is the series $A_4GC_xU_4$, which forms dimers when $x \leq 4$[164]:

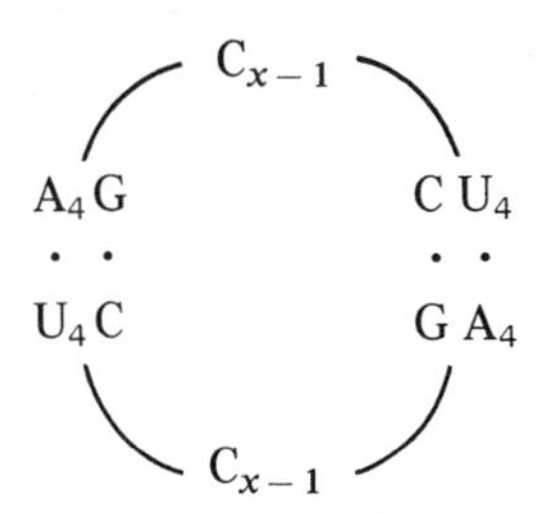

Several model oligomers that form hairpin helixes have also been studied. For example, Baldwin, Elson, and Scheffler[169-171] studied extensively the helixes formed by linear and circular $d(T \cdot A)_n$ oligomers, and model RNA hairpin helixes were synthesized and studied by Uhlenbeck et al.[163] and by Gralla and Crothers.[164]

C. Statistical Thermodynamics of Oligonucleotide Transitions

There is a considerable body of theory concerning the statistical mechanics of helix-coil transitions (see reference 172 for a brief review). A good introduction to this subject is the statistical thermodynamics of oligonucleotide transitions, where the mathematics is simpler but the physical principles are the same.

Our purpose is to describe the equilibrium over the possible states in melting a double helical oligomer, for example A_nU_n (see Fig. 6-22). In order to do this, we need to write equilibrium constants for the various steps in the reaction. By convention the bimolecular nucleation step that forms the first base pair is given the symbol βs,[143]

$$K_{\text{nuc}}\ (\text{bimolecular}) = \beta s \tag{6-58}$$

where βs is an unknown quantity that must be evaluated from experiment. The equilibrium constant for the growth reaction (see Fig. 6-23) in which a

FIGURE 6-22 Intermediates in melting an oligonucleotide.

FIGURE 6-23 Growth reaction; K (growth) = s.

new base pair is formed at the end of a helix, is given the symbol s. Again, s is unknown and must be evaluated from experiment.

It is now possible to write the equilibrium constant for forming any bonded intermediate from the separated strands, because the total equilibrium constant is the product of equilibrium constants for the individual steps. For example, in Fig. 6-24 the intermediate i with four bonded base pairs has formation equilibrium constant K_i. The equilibrium constant for forming all possible bonded states is the sum of K_i over all i:

$$q_B = \Sigma K_i \tag{6-59}$$

The quantity q_B is the conformational part of the *molecular partition function.* The partition function q_B for high polymer molecules has the same meaning as it does here, but the evaluation of the sum ΣK_i is much more difficult, requiring mathematical techniques that will not be developed here.

For the example of A_nU_n, the expression for q_B is simple. We count only those bonded states in which the two strands are aligned perfectly. (Other states are very unlikely since they would either contain an internal loop or have only a short helix.) A state with j base pairs can be formed in $2n - j + 1$ ways, so

$$q_B = \beta s \sum_{j=1}^{2n} (2n - j + 1)\, s^{j-1} \tag{6-60}$$

a series that can be summed analytically.

The utility of q_B is in calculating average quantities. Consider first the computation of the fraction of strands that are bonded together. If the strands are identical, the reaction is

$$2A \rightleftharpoons B$$

where B represents all the bonded states, and

$$q_B = \frac{[B]}{[A]^2} \tag{6-61}$$

Letting C_T be the total strand concentration

$$C_T = [A] + 2[B] \tag{6-62}$$

FIGURE 6-24 A specific intermediate; $K_i = (\beta s)\, s^3$.

and ζ the fraction of strands that are bonded,

$$\zeta = \frac{2[B]}{C_T} \tag{6-63}$$

combination of Eqs. (6-64), (6-65), and (6-66) yields

$$\zeta = \frac{4q_B C_T + 1 - (8q_B C_T + 1)^{1/2}}{4\,q_B C_T} \tag{6-64}$$

Half the strands are bonded, $\zeta = \frac{1}{2}$, when $q_B C_T = 1$.

Calculation of the average fraction of bases hydrogen bonded also involves q_B. The probability of a particular bonded state i depends on its relative contribution to q_B,

$$P_i^{(B)} = \frac{K_i}{q_B} \tag{6-65}$$

where $P_i^{(B)}$ is the probability that a bonded molecule will be in state i. Continuing with the example of dimers formed from A_nU_n, the probability that a bonded molecule has j pairs bonded is $(2n - j + 1)\,\beta s^j/q_B$. Therefore, the average number of bonds formed, averaged only over bonded states, is $\langle j\rangle^{(B)}$

$$\langle j\rangle^{(B)} = \beta \sum_{j=1}^{2n} j(2n - j + 1)s^j/q_B$$

$$= \frac{\partial \ln q_B}{\partial \ln s} \tag{6-66}$$

The fraction of bases bonded, θ, averaged over all states is thus

$$\theta = \frac{\zeta}{2n}\,\frac{\partial \ln q_B}{\partial \ln s} \tag{6-67}$$

a general result for dimer oligonucleotides.

The formal equations are somewhat simpler when helix formation is intramolecular. In that case

$$\zeta = \frac{q_B}{1 + q_B} \tag{6-68}$$

is the fraction of molecules with at least one bond. By analogy with (6-67),

$$\theta = \frac{q_B}{N(1+q_B)}\,\frac{\partial \ln q_B}{\partial \ln s}$$

$$= \frac{1}{N}\,\frac{\partial \ln q}{\partial \ln s} \tag{6-69}$$

where

$$q = 1 + q_B \tag{6-70}$$

and N is half the number of nucleotides. As an example, consider the hairpin helix formed from $A_4GC_6U_4$. The G · C pair is known to contribute heavily to the stability of this structure[163,164]; therefore a good approximation counts only those bonded states with the G · C pair intact. In this case

$$q_B = \gamma(5)\,(1 + s_{GA} \sum_{j=0}^{3} s_{AA}{}^{j}) \tag{6-71}$$

where two values of s are used, depending on whether the A · U pair to which s corresponds is stacked on a G · C(s_{GA}) or an A · U(s_{AA}) pair. The quantity $\gamma(5)$ is the nucleation constant for closing a hairpin loop of five bases.

D. The All-or-None Approximation

In the melting range of oligomers, s is generally considerably greater than 1. This is so because $s \simeq 1$ at the $T_m{}^{(\infty)}$ of a high polymer, and s increases as temperature decreases. The magnitude of s can be estimated from the van't Hoff equation

$$\frac{\partial \ln s}{\partial T} = \frac{\Delta H}{RT^2} \tag{6-72}$$

or, integrated,

$$\ln s = \int_{T_m(\infty)}^{T} -\frac{\Delta H}{R}\, d(1/T) \tag{6-73}$$

At room temperature and high salt concentration, s for an A · U pair is roughly 5–50, and approximately 30–500 for a G · C pair.[164] It is therefore likely that a bonded state has all base pairs formed, and the melting transition can be approximated as conversion of the coil form to the fully bonded helix. This is called an *all-or-none* transition; kinetic studies[167,168] indicate that it is a good approximation for short oligonucleotides.

The mathematical approximation introduced for the all-or-none model is that q_B can be replaced by the maximum term K_i in the series (6-59). This usually corresponds to the form with the maximum number of bases bonded.

For example, with the A_nU_n dimers, we set

$$q_B \approx \beta s^{2n} \tag{6-74}$$

Examination of Eq. (6-60) shows that q_B is actually

$$q_B = \beta s^{2n} (1 + 2s^{-1} + 3s^{-2} + \cdots) \tag{6-75}$$

so that the error factor in approximation (6-74) is $(1 + 2s^{-1} + \cdots)$. If $s = 10$, the error is 20%; it is difficult to measure experimental association constants to this accuracy, so that the all-or-none transition is realistic in the appropriate temperature range.

E. Evaluation of Free Energies

The free energy of the growth reaction is related to s by

$$\Delta G_{\text{growth}} = -RT \ln s \tag{6-76}$$

One way to estimate ΔG_{growth} is to use the van't Hoff equation, integrating from the T_m of a high polymer to the temperature of interest. This introduces considerable uncertainty, because ΔH is not always well enough known over the entire temperature range. Another approach is to compare the stability of oligonucleotides of varying size; for this we need a few additional equations.

According to Eq. (6-64), at the T_m of a dimer helix, $q_B C_T = 1$. Consequently,

$$d \ln q_B = -d \ln C_T \text{ at } T = T_m \tag{6-77}$$

The differential of q_B is, because q_B is an equilibrium constant,

$$d \ln q_B = -\frac{\Delta H}{R} d(1/T) \tag{6-78}$$

where ΔH is the heat of forming the dimer helix. Thus at T_m

$$\frac{d \ln C_T}{d(1/T_m)} = \frac{\Delta H}{R} \tag{6-79}$$

According to Eq. (6-79), a plot of $1/T_m$ vs. $\ln C_T$ should be linear, with slope $\Delta H/R$.

Consider now the comparison of two dimer-forming oligomers of different size, A_6U_6 and A_5U_5. As shown in Fig. 6-25, each of these shows a linear variation of $1/T_m$ with $\log C_T$. At a given temperature T_m

$$\frac{q_B(A_6U_6)}{q_B(A_5U_5)} = \frac{C_T(A_5U_5)}{C_T(A_6U_6)} \tag{6-80}$$

The ratio of q_B values is (see Eq. 6-78)

$$\frac{\beta s^{12}(1 + 2s^{-1} + 3s^{-2} + \cdots)}{\beta s^{10}(1 + 2s^{-1} + 3s^{-2} + \cdots)} = s^2 \qquad (6\text{-}81)$$

to a very good approximation. Therefore

$$s^2 = \frac{C_T(A_5U_5)}{C_T(A_6U_6)} \qquad (6\text{-}82)$$

and the growth free energy is

$$\Delta G_{\text{growth}} = -RT \ln s$$

$$= \frac{RT}{2} \Delta \ln C_T \qquad (6\text{-}83)$$

The principle here is simple: RT times the difference in the logarithm of total concentration required to produce the same T_m for two oligomers ($\Delta \ln C_T$) gives the difference in their free energy of formation. Since the two oligomers in this example differ by two growth reactions, the growth free energy is the difference in free energy of formation divided by 2.

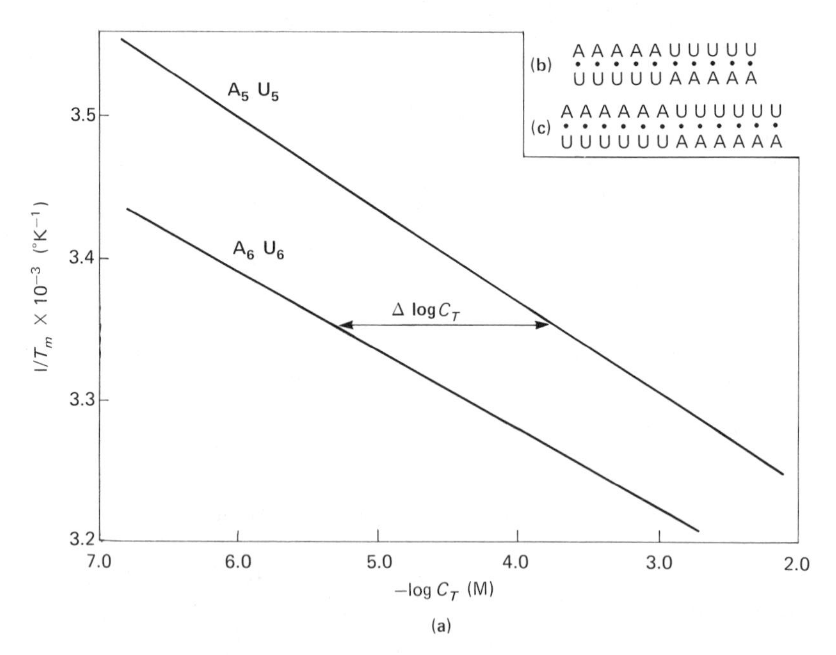

FIGURE 6-25 (a) Reported variation of melting temperature with strand concentration for A_5U_5 (b) and A_6U_6 (c) at 1 M Na^+. [From Martin et al. *J. Mol. Biol.*, **57**, 201 (1971) and Gralla and Crothers, *J. Mol. Biol.*, (in press). Reprinted with permission.]

TABLE 6-6 FREE ENERGY OF BASE STACKING AND PAIRING (1 M Na^+, 25°C)

adjacent base pairs				Δ G, kcal/mole
–A–A– · · –U–U–				−1.2
→ –A–U–, · · –U–A–	→ –U–A– · · –A–U–			−1.8
→ –A–C–, · · –U–G–	→ –C–A–, · · –G–U–	→ –G–A–, · · –C–U–	→ –A–G– · · –U–C–	−2.2
	→ –C–G– · · –G–C–			−3.2
	→ –G–C–, · · –C–G–	→ –C–C– · · –G–G–		−5.0
	–G– · –U–			0
	→ –G–U– · · –U–G–			−0.3

The same principle can be applied to determine s for other kinds of base pairs.[164] For example, comparison of A_4CGU_4 with A_5U_5 gives the free energy change for substituting two G · C pairs for two A · U pairs. More specifically, the free energies of the base pair are given in terms of the neighbor on which they are stacked. Thus A_4GCU_4 contains six A · U–A · U, two G · C–A · U and one G · C–C · G stacking interactions, while A_5U_5 contains eight A · U–A · U and one A · U–U · A stacking interactions. The free energy difference $RT\ \Delta \ln C_T$ is therefore that due to replacing two A · U–A · U and one A · U–U · A with two G · C–A · U and one G · C–C · G base pair stacking interactions. Table 6-6 summarizes the known free energies for adding a base pair on to an existing helix, obtained from comparison of a wide variety of oligonucleotides.

Loop free energies are determined in an analogous way. For example, when the hairpin helix $A_4GC_6U_4$ is half melted, the free energy of forming the fully bonded state is zero (all-or-none approximation). Therefore

$$\Delta G_{\text{growth}} + \Delta G_{\text{nucleation}} = 0 \text{ at } T = T_m \tag{6-84}$$

Considering the nucleation process to be formation of the G · C pair, the growth reactions involve one A · U–G · C stacking interaction and three

A · U–A · U interactions. Their free energies can be determined at $T = T_m$ from the information in Table 6-6, and $\Delta G_{\text{nucleation}}$ can be calculated from Eq. (6-84). A similar approach can be used to determine the free energy contributed by internal loops. Thus far, the free energy of bulge loops has only been calculated from the transition of randomly modified polymers.[173] Table 6-7 summarizes the known free energies of loops.

Tables 6-5 and 6-6 may be combined to calculate the conformational free energy of an RNA structure at 25°C. The simplest way to do this is to imagine building the structure up from the single-stranded form by a series of nucleation and growth reactions. (For simplicity, one should always nucleate a loop at its final size.) To each of these steps a free energy can be assigned from the tables; the sum of such terms is the conformational free energy.

TABLE 6-7 LOOP FREE ENERGIES (1 M Na^+, 75°C)

Bases unbonded (n)	
internal loops	
2–6 except next item	+2
⁄N⁄ —G C— · · —C G— ⁄N⁄	0
7	3
$n \geqslant 8$	$3 + 0.9 \ln(n/7)$
hairpin loops	
(a) closed by A · U (add 1 kcal for no U in loop	
3	+7
4–5	6
6–7	5
8–9	6
$n \geqslant 10$	$7 + 0.9 \ln(n/10)$
(b) closed by G · C	
3	8
4–5	5
6–7	4
8–9	5
$n \geqslant 10$	$6 + 0.9 \ln(n/10)$
bulge loop	
1	+3
2–3	4
4–7	5
$n \geqslant 8$	$6 + 0.9 \ln(n/8)$

F. Loop Free Energies in Oligomers and High Polymers

The free energy of forming the first base pair in a double helix is a quantity that must be determined by experiment. However, the variation of this nucleation free energy with the size of the loop that is closed is a quantity susceptible to theoretical estimation. One must calculate the probability that the second base is in the proper position and orientation around the first so that base pairing can occur. The free energy of nucleation is proportional to the logarithm of this probability. DeLisi and Crothers[174] and DeLisi[175] have calculated the detailed dependence of nucleation free energy on loop size. A model for the backbone chain was chosen that was consistent with crystallographic data[176] and gave the correct dimensions to the unperturbed (theta solvent) polymer chain.[51] For small loops the behavior is that of a stiff polymer chain[175] that does not follow a simple model. For large loops ($\gtrsim$35 nucleotides) the probability of loop closure varies with the 3/2 power of loop size, as predicted by the Jacobson-Stockmayer result.[177]

In real polynucleotide chains away from theta-solvent conditions, a polymer chain of any length is nongaussian. This changes the value of the exponent in the Jacobson-Stockmayer equation so that the probability of loop closure P_j is

$$P_j = j^{-\alpha} \tag{6-85}$$

where j is the number of nucleotides in the loop. Hennage[63] found that $\alpha = 1.8$ by fitting transition curves of the acid form of poly (A). This agrees well with the value 1.75 predicted by Fisher.[178]

This strong increase of loop free energy with loop size has a profound effect on the melting behavior of DNA. Consider Fig. 6-26 in which two partly melted states of a DNA molecule, each containing j unbonded nucleotides. The contribution of loop closure to conformation (a) is a factor $(j/2)^{-2\alpha}$ and to (b) it is $j^{-\alpha}$. The ratio of factor (a) to (b) is approximately $j^{-\alpha}$. Since j can be hundreds of nucleotides, this gives a strong preference to conformation (b), and is a major reason the melting of DNA is so highly cooperative.

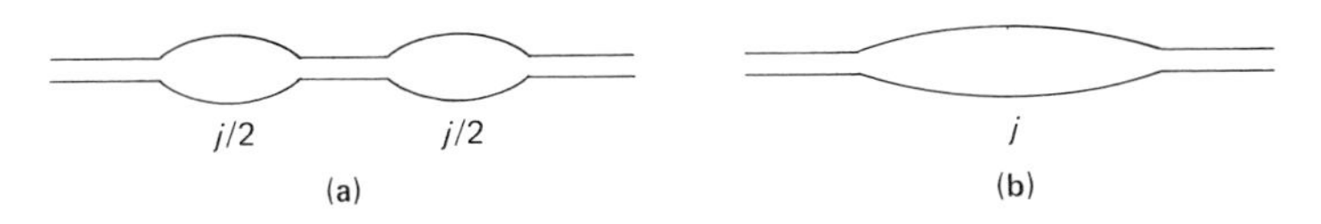

FIGURE 6-26 Two partly melted states of a DNA molecule.

IX MECHANISM OF DNA MELTING

Statistical mechanical theories for the melting of DNA are highly complex because of the necessity for including the effects of loops and the different kinds of base pairs. However, the results of a calculation[179] on a "typical" molecule are worth considering; they provide insight into the mechanism of melting. Figure 6-27 shows sequence fluctuations in a stretch of 5000 nucleotides of random sequence within a DNA molecule. (The actual molecule is, for reasons of computational simplicity, formed by periodic repetition of the 5000 nucleotide sequence.) The 5000 nucleotides have been broken up into 100 groups of 50 each, and the figure shows the excess of G · C pairs in each block of 50 above the random expectation value of 25 (50% G · C). In the calculation each block of 50 bases is assumed to act as a unit. This assumption, called "coarse graining," is reasonably good, because the calculation shows that the average length of helix and coil regions is considerably longer than 50 bases in the transition region.

Figure 6-28 shows the melting curve calculated for the molecule, with an experimental melting curve for T2 DNA bacteriophage DNA shown for comparison. (In the calculation, the loop closure free energy is assumed to vary with the $-3/2$ power of loop size, and the proportionality constant σ is adjusted to 10^{-4} to give a transition breadth that corresponds to experiment. The same values of these parameters are consistent with the melting transition of synthetic homogeneous polynucleotides.[180]) The theoretical transition

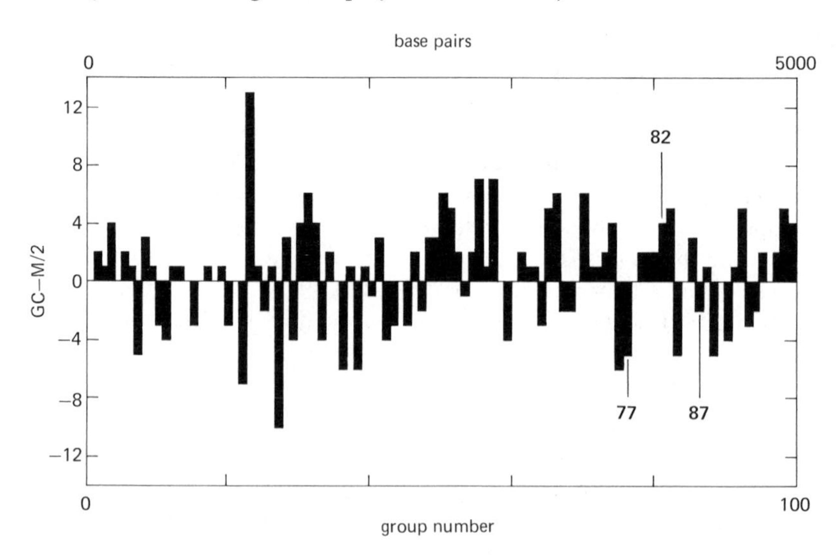

FIGURE 6-27 Composition fluctuations in a hypothetical DNA molecule, showing the excess of G · C pairs in a group of 50 over the random expectation value of 25. [From Crothers, *Accts. Chem. Res.*, **2**, 225 (1969).]

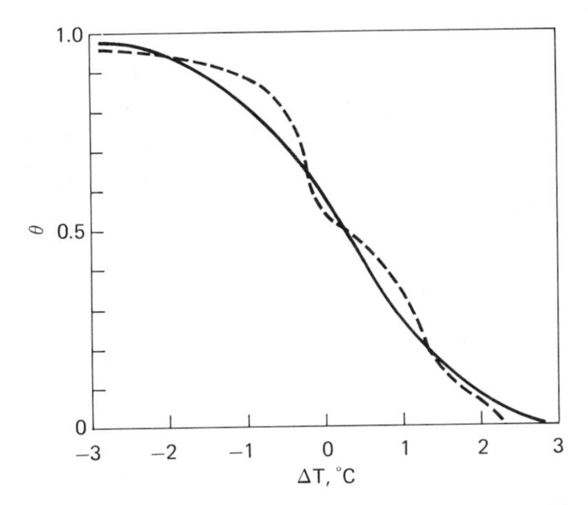

FIGURE 6-28 Calculated melting curve (– – – –) for the molecule diagrammed in Fig. 6-27 (with periodic repetition of the indicated sequence); an experimental melting curve for T2 DNA (————) is shown for comparison. [From Crothers, *Accts. Chem. Res.*, **2**, 225 (1969).]

curve in Fig. 6-28 is multiphasic, a result of the finite size of the sequence period and the tendency of specific sections to "melt out" in narrow temperature regions, as documented below.

The breadth of DNA melting curves is due primarily to the heterogeneity of base pair stability, as shown by the different characteristic transition temperatures of particular regions of the sequence. Examples are given in Fig. 6-29 for particular groups of base pairs. These differ widely depending on whether the sequence is locally G · C- or A · T-rich. The transition for residue 77 in an A · T-rich region occurs at the lowest temperature. This is expected; denaturation should begin in the less stable positions.

A further important difference between the melting of A · T- and G · C-rich regions is reflected in the breadth of the individual melting curves; that for the A · T-rich segment is considerably broader than that for the G · C-rich section. Inspection of the G · C-rich region around residue 82 reveals that this contains about 250 base pairs. The transition heat calculated by the van't Hoff equation from the transition slope is about 250 times the heat per base pair. Because the van't Hoff heat is about equal to the total heat available, the transition can be reasonably approximated by a two-state model.

Thus there results a view of the denaturation process as depicted in Fig. 6-30. Process (a), the nucleation and growth of denatured zones in A · T-rich regions occurs over several degrees, with a gradual increase in the size of the denatured portion. (The average length of helix or coil regions at the

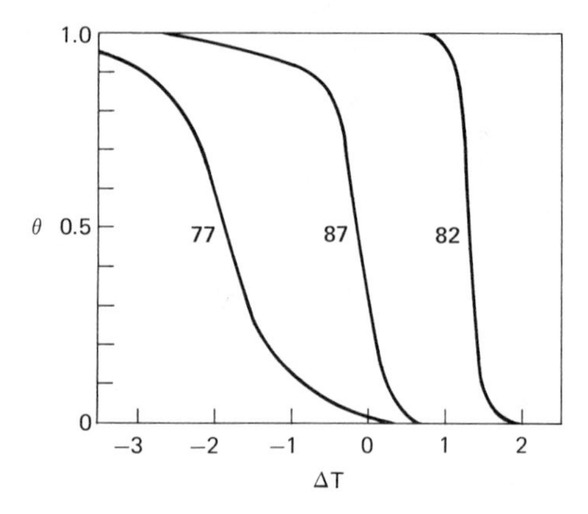

FIGURE 6-29 "Melting curve" for individual groups numbers in the sequence diagram in Fig. 6-27. θ is the probability that the base pairs in the particular group are in the helical state. [From Crothers, *Accts. Chem. Res.*, **2**, 225 (1969).]

transition midpoint is about 300 base pairs[179].) Then a critical temperature is reached at which process (b) occurs, with merging of two loops to form a single coil at the expense of the intervening G · C-rich helix. This transition is very sharp and is essentially an all-or-none change between the two states shown. The critical temperature for process (b) depends on the size and G-C composition of the helical section and on the sizes of the adjacent rings.

The calculation also predicts,[179] in agreement with experiment,[181] that the ends of the molecule do not contribute appreciably to the melting process when the DNA molecular weight is above a few million. There is a slight bias for melting at the ends, because no loops need be formed there. This factor, however, is much less important than the effect of composition fluctuations. If the ends are G · C-rich, they will remain intact while the interior of the molecule melts. These theoretical conclusions are consistent with the view of DNA melting presented in the electronmicrographs in Figs. 6-12 and 6-13.

X STRAND SEPARATION AND RENATURATION

A. Equilibrium Measurements

So far in this account of the melting transition of nucleic acids, attention has been focused on the loss of structure as the temperature is raised, called the melting curve. Another important stage of the melting process can be distinguished experimentally: the point at which the strands

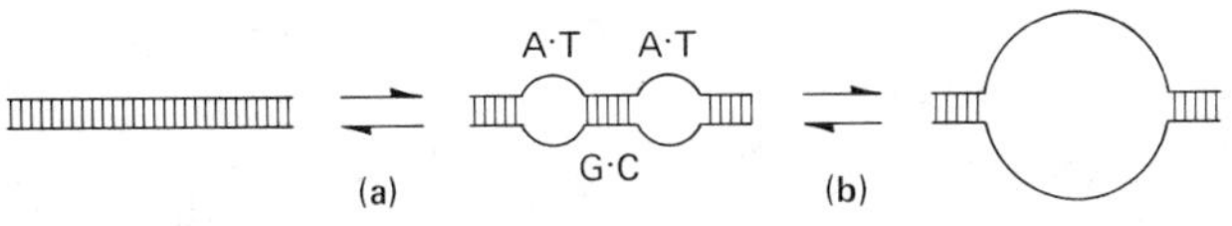

A·T A·T
(a) G·C (b)

FIGURE 6-30 Diagram of two distinguishable effects in the melting of DNA: (a) nucleation and growth of denatured regions at A · T-rich loci, and (b) disappearance of the helix section separating the two coils, to produce one large loop. [From Crothers, *Accts. Chem. Res.*, **2**, 225 (1969).]

become separated from one another. Consider the diagram of nucleic acid molecules undergoing melting shown in Fig. 6-31. When the molecular weight is large (greater than a few million), most of the melting occurs at interior regions of the molecule. Theory predicts (see Section IX) that the average length of the alternating helix and coil sections should be hundreds of base pairs. Physical quantities such as the UV absorbance are closely proportional to the unbonding of the base pairs and reflect this internal melting. Finally a temperature is reached at which the last regions holding the molecules together are melted, and the strands separate. This will be accompanied by only a very small change in absorbance. On the other hand, when the molecular weight is small, most of the early melting occurs at the ends of the molecule, and the melting of the helix is accompanied by separations of the strands. In this case the UV absorbance closely parallels strand separation. Dimer-forming oligonucleotides can be adequately treated by two-state models involving only the complete helix and the separated strands (see Section VIII).

Thus one predicts for high molecular weight DNA a strand separation transition that is not detected by a simple UV absorbance melting curve. The evidence that strand separation actually does occur has been carefully reviewed by Marmur et al.[182]

It has been known for some years that DNA heated beyond the

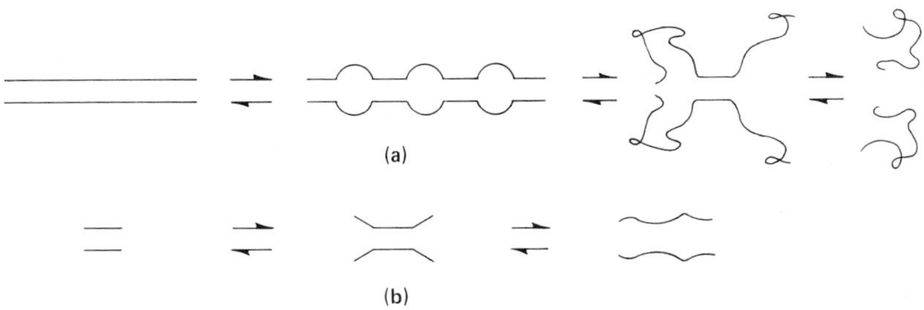

(a)

(b)

FIGURE 6-31 Nucleic acid molecules undergoing melting. (a) High molecular weight; (b) low molecular weight.

temperature at which strand separation occurs will slowly "renature," regaining its native physical properties and even its biological activity.[182-184] The renaturation process is initially a second-order reaction, and its rate depends on such factors as the type of DNA, molecular weight, temperature, and salt concentration.

The physical methods of assaying for strand separation depend on the slow rate of recombination of the strands to re-form the double helix when renaturing conditions are restored. If the molecule is in an intermediate state of denaturation and the temperature is suddenly lowered below the beginning of the melting curve, it rapidly zips up to re-form a helix. But if the strands have separated, the recovery of the double helical structure is much slower. This kinetic separation depends on the very low concentration of complementary regions in the DNA strands, and does not work for synthetic polynucleotides. These latter recombine much more rapidly, and the assay for strand separation is more difficult.

Two different kinds of measurement of strand separation provide complementary pictures of the nature of the process. One is to examine the properties of individual molecules, which can be carried out by assaying the biological activity. DNA that carries transforming activity, or the ability to

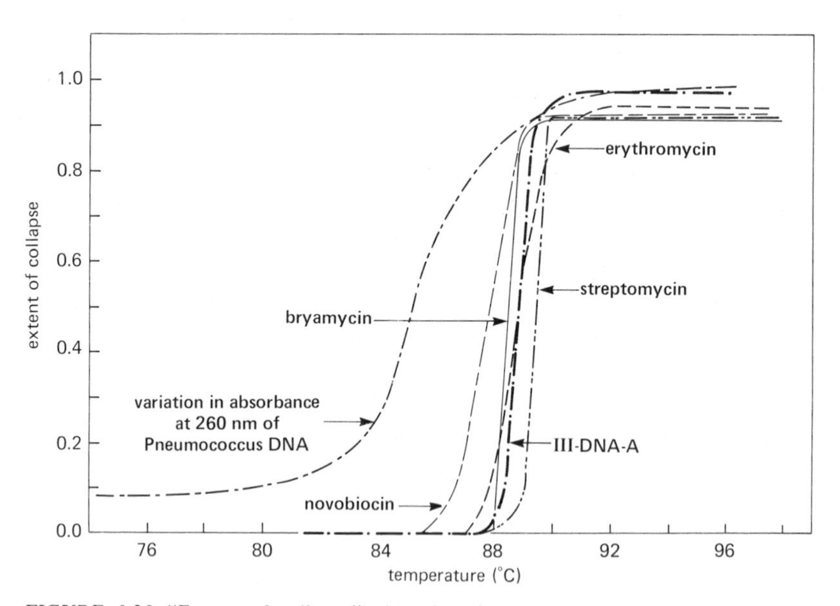

FIGURE 6-32 "Extent of collapse" plotted against temperature. The straight lines of Figs. 6-1 and 6-2 were extrapolated to the ordinate; the loss of activity of these extrapolated points is taken as the extent of collapse. Also included is the optical density curve of Marmus and Doty.[5] [From W. Ginoza and B. H. Zimm, *Proc. Nat. Acad. Sci. U.S.*, 47, 639 (1961). Reprinted with permission.]

change the genetic characteristics of an organism, is greatly reduced in activity when it loses the double helical configuration.[183,186-189] Therefore by heating to temperature T, cooling rapidly, and measuring the biological activity, one can assay for separation of the strands that carry the particular genetic marker. Since most bacterial DNA preparations contain many fragments of the original chromosome, most markers will be located on different molecules and can be expected to separate at different characteristic temperatures.

Figure 6-32 shows examples of this first type of experiment in which *Pneumococcus* DNA containing several markers was assayed after heat treatment. Strand separation is thought to parallel loss of transforming activity and occurs at temperatures considerably higher than the T_m measured from UV absorbance. This is in agreement with the mechanism shown in Fig. 6-31, because strand separation occurs only at the end of the reaction. Of further interest is the extreme sharpness of the transition curves for individual markers. Assuming the oversimplified model of an all-or-none transition for the last step in strand separation, that is, that a single helical section with a fixed number of base pairs holds the strands together, the length of that section can be estimated from the transition breadth and the heat of melting. The result is that the section is of the order of a hundred

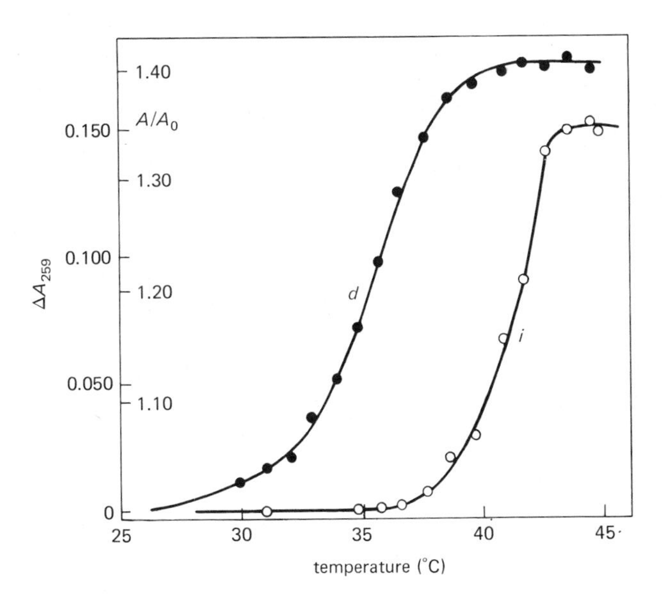

FIGURE 6-33 The *d*-assay and *i*-assay denaturation curves of bacteriophage T2 DNA, in 7.2 M $NaClO_4$. *i*-assay absorbances extrapolated to zero time at 25°C, to eliminate absorbance changes caused by type II reversibility. [From E. P. Geiduschek, *J. Mol. Biol.*, 4, 467 (1962). Reprinted with permission.]

base pairs in length. This is presumably a section locally rich in G · C pairs.

The other assay for strand separation is accomplished by measuring the UV absorbance of a quenched sample. Figure 6-33, taken from the work of Geiduschek,[190] shows an example of this kind of assay. The curve labeled d is the absorbance measured at temperature T and corresponds to the dissociation of base pairs. When the sample is quickly cooled from T and the absorbance measured, the result is the curve labeled i, for "irreversible." (Strand separation is "irreversible" only in a relative kinetic sense. In this solvent and with T2 DNA there is actually a quite appreciable rate of strand recombination, which has been corrected for in the i curve.)

The temperature of strand separation is strongly influenced by molecular weight,[181] even for high molecular weights where the other equilibrium melting properties are little influenced by size. Strand separation behavior is determined by G · C composition fluctuations, and increase of the molecular size always increases the probability of finding an extreme fluctuation that will hold the molecule together at still higher temperatures. The strand separation transitions usually measured for "undegraded" virus DNA actually refer to samples that undoubtedly contain single-strand breaks. For this reason there is some separation before the UV absorbance changes are completed. If the strands were absolutely intact essentially no strand separation should occur until all detectable absorbance changes were completed, because the final helical section that holds the strands together is such a small fraction of the whole molecule.

When the molecular weight is reduced, the strand separation transition begins to drop below the T_m of the UV melting curve for the high polymer. Thus the transition temperatures of DNA samples begin to decrease when the molecular weight is reduced below 1 million. As the molecular weight is lowered still further, an all-or-none (two-state) model for the melting transition becomes increasingly valid, approaching the behavior of oligonucleotides. Therefore as the polymer size is reduced, strand separation changes from an event at the very end of the transition to one that dominates and is synonomous with the melting process.

DNA that has been cross-linked by chemical action does not undergo strand separation. For such materials renaturation is always rapid when the sample is quenched, no matter how far beyond the melting temperature it has been heated.[190,191]

Increase of the average G · C composition of DNA sample increases the difference between the UV melting curve and strand separation.[190] This doubtless arises from increase in the probability of extreme composition fluctuations, although it is not obvious from simple arguments why this should be the case with high G · C samples.

B. Kinetics

When two polynucleotide strands with complementary sequence are mixed in solution, they can react to form double helical structures. The initial rate of this reaction is limited by the bimolecular process of strand combination. One therefore expects that the initial rate will be second order in the concentration of complementary strands. This expectation has been verified both for reaction of complementary DNA strands[192–197] and combination of poly A with poly U.[198–201] Relaxation measurements on the association of oligomer strands also reveal a second-order reaction step.[167–168] Furthermore, the magnitudes of the observed rates are consistent from one system to another, once corrections are made for the physical differences between them.

Consider a reaction between two complementary strands A and B to form the helical complex H:

$$A + B \longrightarrow H \tag{6-86}$$

We assume that the rate is limited by formation of the first bonds, after which the molecule quickly zips up to form the maximum number of bonded pairs. The rate of this second-order reaction is defined in terms of the rate of production of the helical molecules H,

$$\frac{dC_{HM}}{dt} = k_S C_{SA} C_{SB} \tag{6-87}$$

Where C_{HM} is the molar concentration of helix molecules, and C_{SA} and C_{SB} are the molar concentrations of strands A and B, respectively. The empirical constant k_S is the observed second-order rate constant.

Rates of strand combination reactions can also be expressed in terms of the concentration of nucleotides. Suppose that an amount of DNA containing T total base pairs (the DNA content of a single cell) is broken into pieces of length L. It is often found in higher organisms that there are unequal number of copies of different genes. Therefore let there be, in the total double helix length T, n_1 copies of sequence 1, n_2 copies of sequence 2, and n_i copies of sequence i. One strand of sequence i can renature only with the complementary strand. The concentration $C_{SA}{}^{(i)}$ (or $C_{SB}{}^{(i)}$) of strands containing sequence i is

$$C_{SA}{}^{(i)} = \left(\frac{n_i}{2T}\right) C_N \tag{6-88}$$

where C_N is the total single strand nucleotide concentration. Since each double helical molecule produced contains $2L$ nucleotides, the rate of production of nucleotides in helical form in molecules of type i,

$$\frac{dC_{HN}{}^{(i)}}{dt} = 2L \left(\frac{dC_{HM}{}^{(i)}}{dt}\right) \tag{6-89}$$

where C_{HN} expresses the helix concentration in nucleotides.
Therefore, assuming that k_S is the same for all species (at constant L),

$$\frac{dC_{HM}{}^{(i)}}{dt} = k_S C_{SA}{}^{(i)} C_{SB}{}^{(i)} \tag{6-90}$$

and, furthermore, that $C_{SA}{}^{(i)} = C_{SB}{}^{(i)}$ (complementary strands present at equal concentrations), we can substitute (6-88) and 6-89) into (6-90) to obtain

$$\frac{dC_{HN}{}^{(i)}}{dt} = \frac{1}{2} L k_S \left(\frac{n_i}{T}\right)^2 C_N{}^2 \tag{6-91}$$

Several simplifications of Eq. (6-91) are possible. In the simplest case, represented by the combination of oligonucleotides or of intact viral DNA strands, there is one copy of the sequence for each length L, so $n_i/T = 1/L$. A more common, but still simple, occurrence with DNA renaturation is that all genes i are present in equal numbers of copies. They will thus renature at equal rates, and each will be present at a certain fraction of the total helix nucleotide concentration C_{HN}; specifically, since one molecule is a fraction L/T of the total genetic content,

$$C_{HN}{}^{(i)} = \left(\frac{L n_i}{T}\right) C_{HN} \tag{6-92}$$

and hence

$$\frac{dC_{HN}{}^{(i)}}{dt} = \left(\frac{L n_i}{T}\right) \left(\frac{dC_{HN}}{dt}\right) \tag{6-93}$$

Combining (6-93) with (6-91) yields, for a DNA sample containing equal numbers of copies of all molecules,

$$\frac{dC_{HN}}{dt} = \frac{1}{2} k_S \left(\frac{n_i}{T}\right) C_N{}^2 \tag{6-94}$$

This can be rewritten in terms of a new second-order rate constant k_N

$$\frac{dC_{HN}}{dt} = \frac{k_N C_N{}^2}{2} \tag{6-95}$$

where

$$k_N = \frac{k_S n_i}{T} \tag{6-96}$$

Eq. (6-95) is the definition of renaturation rate given by Wetmur and Davidson[196]; the factor of 2 it contains is arbitrary in a sense but affects only the magnitude of the rate constant.

From Eq. (6-96) it is clear that when one expresses renaturation rates in terms of the rate constant k_N, the magnitude will be a function of the number of gene copies n_i per DNA length T. Furthermore, k_S and k_N are found to be functions of length L.[196]

One way to measure k_S for association of two nucleic acid strands is to study the relaxation kinetics of oligonucleotide melting. Suppose two oligonucleotide strands are in equilibrium with the helix H:

$$A + B \underset{k_{-1}}{\overset{k_1}{\rightleftharpoons}} H \tag{6-97}$$

Examples that have been examined are oligomers of adenylic acid forming the protonated double helix or interacting with oligo U to form double and triple helixes, both systems studied by Pörschke and Eigen,[167] and double helix formation by block oligomers of the form A_nU_n, studied by Craig et al.[168] In such cases, as discussed in Section VIII, it is a good approximation to consider an all-or-none or two-state model in which a given strand is either fully bonded in a helix or completely in the coil form. Thus the equilibrium constant K for helix formation is

$$K = \frac{k_1}{k_{-1}} \tag{6-98}$$

and k_1 is usually expressed in terms of strand concentration, making it equivalent to k_S.

When the equilibrium (6-97) is perturbed, usually by a temperature jump, the amount of helix and coil will shift to new equilibrium values, accompanied by a change in absorbance. The characteristic time for this process is called the relaxation time τ of the system. The time τ observed for a bimolecular reaction like (6-97) is given by

$$\tau = \frac{1}{k_{-1} + k_1(C_A + C_B)} \tag{6-99}$$

where C_A and C_B are the concentrations of free strands A and B, respectively. If the equilibrium constant K is known from measurement of the equilibrium transition curve, C_A and C_B can be calculated from knowledge of the total concentration of each strand. A plot of $1/\tau$ against $(C_A + C_B)$ yields a line with slope k_1 and intercept k_{-1}. Alternatively, if K is not known, the kinetic data can be used to measure it. Let

$$K = \frac{C_H}{C_A C_B} \tag{6-100}$$

where C_H is the concentration of the helical form. Furthermore, let the two strands be present in equal concentration, $C_A = C_B$, and let the total strand concentration C_T be

$$C_T = C_A + C_B + 2C_H \tag{6-101}$$

Squaring Eq. (6-99) and eliminating individual concentrations and K with Eqs. (6-98), (6-100), and (6-101) yields

$$1/\tau^2 = 2k_1 k_{-1} C_T + k_{-1}^2 \tag{6-102}$$

A plot of $1/\tau^2$ against C_T allows one to determine k_1 and k_{-1} from the slope and intercept.

It is found that $k_S \approx 4 \times 10^6$ M^{-1} sec^{-1}, varying little with temperature.[167-168] There is a slight tendency for the rate constant to decrease when the temperature is increased, indicating a small negative apparent activation energy. The value of 4×10^6 M^{-1} sec^{-1} is roughly two to three orders of magnitude smaller than would be expected for a diffusion-limited reaction rate, or one in which the strands react every time they collide.

The two main differences between reaction of oligonucleotides and the combination of separate DNA strands to form a double helix are: (1) the obvious point of a large size difference, which one might expect to influence the rate; and (2) the possibility in DNA for local secondary structure formation that can make complementary regions less accessible and therefore slow down strand combination. Experimental data bear on both of these points.

It has been recognized for a number of years that renaturation of DNA strands involves a second-order step as one of the rate-limiting features.[192,193] The reaction is more complicated than with oligonucleotides, however, since in many cases the strands in the mixture do not match exactly, especially if they are random breakage products of a very large DNA molecule or if the parent genome is circularly permuted. In such cases strand combination and zipping-up to the maximum extent still leaves many single-stranded regions, which can slowly anneal out to form more perfectly matched molecules. Also, secondary structure formation in one strand can block and therefore complicate the zipping-up process. As a result of both these features the reaction generally follows second-order kinetics over only about the first 75%,[196] deviating toward slower rates thereafter. The initial rate, however, is proportional to the square of the strand concentration, as required by Eq. (6-87).

The effect of chain length and genome size on phage DNA renaturation rates was studied quantitatively by Wetmur and Davidson.[196] They found, at 25° below the melting temperature T_m, and in 1 M salt, that k_N is given by

$$k_N = 3 \times 10^5 \frac{L^{1/2}}{X} \text{ M}^{-1} \text{ sec}^{-1} \tag{6-103}$$

where X is the length of DNA that contains just one copy of each sequence. Comparison with (6-96), setting $n_i/T = 1/X$, gives

$$k_S = 3 \times 10^5 L^{1/2}\ \mathrm{M}^{-1}\ \mathrm{sec}^{-1} \qquad (6\text{-}104)$$

which is, as it should be, independent of X. k_S increases with L because more sites are available for reaction; the increase is less than linear because of polymer excluded volume.[196]

This result can be compared with the value of k_S found for oligomers by setting L approximately equal to 10, yielding $k_S \simeq 10^6$. This is smaller by only a factor of 4 than the measured value for oligonucleotides, highly satisfactory agreement in view of the extrapolation from much higher molecular weights. Moreover, one would expect DNA to react somewhat more slowly than oligomers because of secondary structure formation in the DNA chain ("folding") which serves to make some combination sites inaccessible.

The effect of "folding" on DNA renaturation rates has been examined in detail by Studier.[197] The main observations are that the rate is very sensitive to salt concentration and that it goes through a maximum some degrees below the melting temperature, in agreement with earlier findings.[192,193] The strong increase with temperature of the renaturation rate at low temperatures is quite unexpected on the basis of the experiments with oligonucleotides. The interpretation is that at low temperatures the DNA single strands "fold" to make a secondary structure, which melts out as the temperature is increased. Since secondary structure formation makes many sites unavailable for renaturation, it greatly slows down strand combination, and a large rate increase with increasing temperature is expected. The decrease in rate as the melting temperature is approached is expected in general, and found, for example, with synthetic polymers like poly A reacting with poly U.[198,199] Finally, reaction of two strands to form a helix requires bringing two polyelectrolytes together, and it is not surprising that the rate of this increases with ionic strength.

Probably the most important application to date of the kinetics of DNA renaturation has been to detect redundant DNA sequences.[200,201] As shown in the development of Eq. (6-96) and as indicated experimentally in Eq. (6-103), the rate of renaturation depends on how many copies of the sequence are present. A full discussion of this problem is given by Britten and Kohne.[202] The rate of formation of helical molecules is given by Eq. (6-95); this can be integrated to give

$$\frac{1}{C_N} - \frac{1}{C_0} = k_N t/2 \qquad (6\text{-}105)$$

where C_0 is the total nucleotide concentration. Letting the fraction of nucleotides in helical form be f,

$$f = \frac{C_N}{C_0} \qquad (6\text{-}106)$$

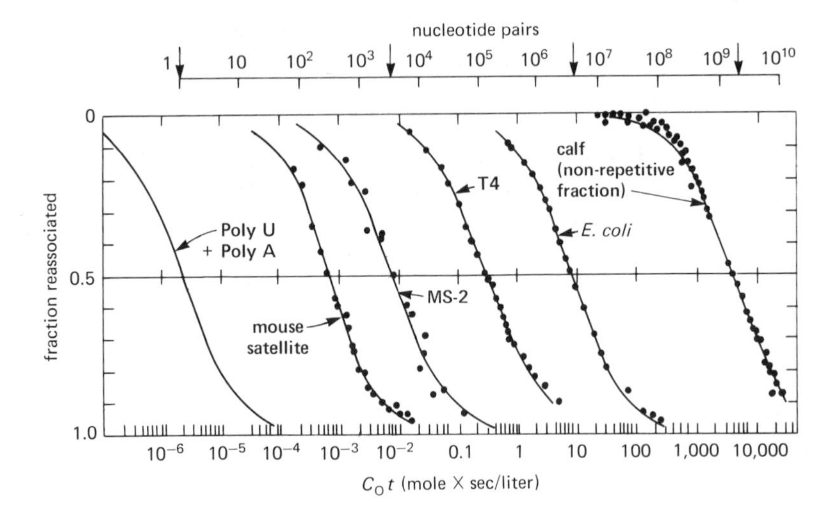

FIGURE 6-34 Reassociation of double-stranded nucleic acids from various sources. The genome size is indicated by the arrows near the upper nomographic scale. Over a factor of 10^9, this value is proportional to the C_0t required for half reaction. The DNA was sheared and the other nucleic acids are reported to have approximately the same fragment size (about 400 nucleotides, single-stranded). Correction has been made to give the rate that would be observed at 0.18 M sodium-ion concentration. No correction for temperature has been applied as it was approximately optimum in all cases. [From Britten and Kohne, *Science,* **161**, 529 (1968).]

Eq. (6-105) can be rearranged to give

$$f = \frac{1}{1 + k_N C_0 t/2} \tag{6-107}$$

This equation says that the fraction of a sample renatured depends only on the product of C_0 and t (assuming L constant). Furthermore, at half renaturation,

$$C_0 t = \frac{2}{k_N}, \qquad \text{when } f = 0.5 \tag{6-108}$$

Since k_N is proportional to n_i/T [see Eq. (6-96)] one concludes that the value of C_0t when $f = 0.5$ should be proportional to T/n_i, or the inverse frequency of occurrence of the sequence i within the DNA content T of the organism. T/n_i is equal to X, the average number of double helix base pairs that must be taken to obtain one copy of the sequence in question. Figure 6-34 shows "C_0t curves"[202] for several kinds of nucleic acids, ranging in complexity. The lower scale shows C_0t, and the vertical axis gives the fraction renatured. The upper scale gives X, a number linearly related to the lower C_0t scale. There are still some difficult and unanswered questions concerning sequence repetition, for example the extent to which two sequences must be identical in order to "renature" as defined in a particular experiment. But these results make it clear that there is enormous variation in the frequency of occurrence of certain classes of sequences in cellular DNA.

REFERENCES

1 J. M. Gulland, D. O. Jordan, and H. F. W. Tayler, *J. Chem. Soc.*, **1947**, 1131 (1947).
2 R. Thomas, *Biochim. Biophys. Acta,* **14**, 231 (1954).
3 J. Marmur, R. Round, and C. L. Schildkraut, *Prog. Nucleic Acid Res.*, **1**, 231 (1963).
4 B. H. Zimm and N. R. Kallenbach, *Ann. Rev. Phys. Chem.*, **13**, 171 (1962).
5 J. Marmur and P. Doty, *J. Mol. Biol.*, **5**, 109 (1962).
6 M. Leng and G. Felsenfeld, *J. Mol. Biol.*, **15**, 455 (1966).
7 J. Brahms, A. M. Michelson, and K. E. Van Holde, *J. Mol. Biol.*, **15**, 467 (1966).
8 D. Poland, J. N. Vournakis, and H. A. Scheraga, *Biopolymers,* **4**, 223 (1966).
9 J. Applequist and V. Damle, *J. Amer. Chem. Soc.*, **88**, 3895 (1966).
10 D. W. Hennage and D. M. Crothers, *Biochemistry,* **8**, 2298 (1969).
11 S. Yabuki, A. Wada, and K. Uemura, *J. Biochem.*, **65**, 443 (1969).
12 H. J. Gould and H. Simpkins, *Biopolymers,* 7, 223 (1969).
13 G. Felsenfeld and G. Sandeen, *J. Mol. Biol.*, **5**, 587 (1962).
14 J. Fresco, L. C. Klotz, and E. G. Richards, *Cold Spring Harbor Symp. Quant. Biol.*, **28**, 83 (1963).
15 H. R. Mahler, B. Kline, and B. D. Mehrotra, *J. Mol. Biol.*, **9**, 801 (1964).
16 G. Felsenfeld and S. Z. Hirschman, *J. Mol. Biol.*, **13**, 407 (1965).
17 C. L. Stevens and G. Felsenfeld, *Biopolymers,* **2**, 293 (1964).
18 A. Rich and I. Tinoco, Jr., *J. Amer. Chem. Soc.*, **82**, 6409 (1960).
19 J. Applequist, in *Conformation of Biopolymers*, G. N. Ramachandran, ed., Academic Press, London and New York (1967), p. 403.
20 I. Scheffler, E. Elson, and R. L. Baldwin, *J. Mol. Biol.*, **48**, 145 (1970).
21 G. J. Thomas, *Biopolymers,* 7, 325 (1969).
22 H. T. Miles and J. Frazier, *Biochem. Biophys. Res. Commun.*, **14**, 21 (1964).
23 H. T. Miles and J. Frazier, *Biochim. Biophys. Acta,* **79**, 216 (1964).
24 F. B. Howard and H. T. Miles, *J. Biol. Chem.*, **240**, 801 (1965).
25 H. T. Miles, *Proc. Nat. Acad. Sci., U.S.*, **47**, 791 (1961).
26 G. J. Thomas, Jr., *Biopolymers,* 7, 325 (1969).
27 F. B. Howard, J. Frazier, and H. T. Miles, *J. Biol. Chem.*, **244**, 1291 (1969).
28 H. T. Miles, *Proc. Nat. Acad. Sci., U.S.*, **51**, 1104 (1965).
29 F. B. Howard, J. Frazier, M. F. Singer, and H. T. Miles, *J. Mol. Biol.*, **16**, 415 (1966).
30 H. T. Miles and J. Frazier, *Biochem. Biophys. Res. Commun.*, **14**, 129 (1964).
31 H. Fritzsche, *Biopolymers,* **5**, 863 (1967).
32 C. C. McDonald, W. D. Phillips, and S. Penman, *Science,* **144**, 1234 (1964).
33 J. P. McTague, U. Ross, and J. H. Gibbs, *Biopolymers,* **2**, 163 (1964).
34 C. C. McDonald, W. D. Phillips, and J. Lazar, *J. Amer. Chem. Soc.*, **89**, 4166 (1967).

35 A. D. Cross and D. M. Crothers, *Biochemistry,* **10**, 4015 (1971).
36 D. R. Kearns, D. Patel, R. G. Shulman, and T. Yamane, *J. Mol. Biol.,* **61**, 265 (1971).
37 J. Brahms and W. F. H. M. Mommaerts, *J. Mol. Biol.,* **10**, 73 (1964).
38 J. Brahms, *J. Mol. Biol.,* **11**, 785 (1965).
39 T. Samejima, H. Hashizume, K. Imahori, I. Fuji, and K. Miura, *J. Mol. Biol.* (in press).
40 J. Brahms, J. C. Maurizot, and A. M. Michelson, *J. Mol. Biol.,* **25**, 465 (1967).
41 H. Simpkins and E. G. Richards, *J. Mol. Biol.,* **29**, 349 (1967).
42 E. G. Richards, C. P. Flessel, and J. R. Fresco, *Biopolymers,* **1**, 431 (1963).
43 J. Brahms, J. C. Maurizot, and A. M. Michelson, *J. Mol. Biol.,* **25**, 481 (1967).
44 A. M. Michelson and C. Monny, *Proc. Nat. Acad. Sci., U.S.,* **56**, 1528 (1966).
45 R. B. Gennis and C. R. Cantor, *J. Mol. Biol.,* **65**, 381 (1972).
46 T. Samejima and J. T. Yang, *J. Biol. Chem.,* **240**, 2094 (1965).
47 P. O. P. Ts'o, S. A. Rapaport, and F. J. Bollum, *Biochemistry,* **5**, 4153 (1966).
48 C. A. Bush and H. A. Scheraga, *Biopolymers,* **7**, 395 (1969).
49 J. Eigner and P. Doty, *J. Mol. Biol.,* **12**, 549 (1965).
50 H. Eisenberg and G. Felsenfeld, *J. Mol. Biol.,* **30**, 17 (1967).
51 L. D. Inners and G. Felsenfeld, *J. Mol. Biol.,* **50**, 373 (1970).
52 M. T. Record, Thesis, University of California.
53 A. H. Rosenberg and F. W. Studier, *Biopolymers,* **7**, 765 (1969).
54 R. B. Inman and R. L. Baldwin, *J. Mol. Biol.,* **5**, 172 (1962).
55 M. A. Rawitscher, P. D. Ross, and J. M. Sturtevant, *J. Amer. Chem. Soc.,* **85**, 1915 (1963).
56 P. D. Ross and R. L. Scruggs, *Biopolymers,* **3**, 491 (1965).
57 L. G. Bunville, E. P. Geiduschek, M. A. Rawitscher, and J. M. Sturtevant, *Biopolymers,* **3**, 213 (1965).
58 E. Neumann and T. Ackermann, *J. Phys. Chem.,* **71**, 2377 (1967).
59 H. Krakauer and J. M. Sturtevant, *Biopolymers,* **6**, 491 (1968).
60 I. Scheffler and J. M. Sturtevant, *J. Mol. Biol.,* **42**, 577 (1969).
61 P. L. Privalov, O. B. Ptitsyn, and T. M. Birshtein, *Biopolymers,* **8**, 559 (1969).
62 H. Klump, T. Ackermann, and E. Neumann, *Biopolymers,* **7**, 423 (1969).
63 D. W. Hennage, Thesis, Yale University (1969).
64 R. L. Scruggs and P. D. Ross, *J. Mol. Biol.,* **47**, 29 (1970).
65 R. D. Blake and J. R. Fresco, *J. Mol. Biol.,* **19**, 145 (1966).
66 D. N. Holcomb and S. N Timasheff, *Biopolymers,* **6**, 513 (1968).
67 D. M. Crothers, *Biopolymers,* **10**, 2147 (1971).
68 A. Litan, *J. Phys. Chem.,* **70**, 3107 (1966).
69 N. R. Kallenbach, *J. Mol. Biol.,* **37**, 445 (1968).
70 D. M. Gray and I. Tinoco, Jr., *Biopolymers,* **9**, 223 (1970).
71 O. B. Ptitsyn and T. M. Birshtein, *Biopolymers,* **7**, 435 (1969).
72 W. Kauzmann, *Adv. Prot. Chem.,* **14**, 1 (1959).
73 M. Sundaralingam, *Biopolymers,* **7**, 821 (1969).
74 R. A. Alberty, *J. Amer. Chem. Soc.,* **91**, 3899 (1969).
75 J. Meselson and E. Stahl, *Proc. Nat. Acad. Sci., U.S.,* **44**, 671 (1958).
76 N. Sueoka, J. Marmur, and P. Doty, *Nature,* **183**, 1429 (1959).
77 M. J. B. Tunis and J. E. Hearst, *Biopolymers,* **6**, 1325 (1968).
78 R. E. Chapman, Jr., and J. M. Sturtevant, *Biopolymers,* **7**, 527 (1969).
79 G. Cohen and H. Eisenberg, *Biopolymers,* **6**, 1077 (1968).
80 J. E. Hearst, *Biopolymers,* **3**, 57 (1965).
81 E. F. Casassa and H. Eisenberg, *Adv. Prot. Chem.,* **19**, 287 (1964).

82 J. Vinograd, J. Morris, N. Davidson, and W. F. Dove, Jr., *Proc. Nat. Acad. Sci., U.S.*, **49**, 12 (1963).
83 R. E. Chapman, Jr., and J. M. Sturtevant, *Biopolymers,* **9**, 445 (1970).
84 J. C. Wang, *J. Mol. Biol.,* **43**, 25 (1969).
85 R. B. Inman, *J. Mol. Biol.,* **18**, 464 (1966).
86 R. B. Inman and M. Schnös, *J. Mol. Biol.,* **49**, 93 (1970).
87 E. A. C. Follett and L. V. Crawford, *J. Mol. Biol.,* **28**, 461 (1967).
88 E. A. C. Follett and L. V. Crawford, *J. Mol. Biol.,* **34**, 565 (1968).
89 H. Fraenkel-Conrat, *Biochim. Biophys. Acta,* **15**, 307 (1954).
90 M. Staehlin, *Biochim. Biophys. Acta,* **29**, 410 (1958).
91 R. Haselkorn and P. Doty, *J. Biol. Chem.,* **236**, 2738 (1961).
92 D. Stollar and L. Grossman, *J. Mol. Biol.,* **4**, 31 (1962).
93 C. A. Thomas, Jr., and K. I. Berns, *J. Mol. Biol.,* **4**, 309 (1962).
94 K. I. Berns and C. A. Thomas, Jr., *J. Mol. Biol.,* **3**, 289 (1962).
95 G. Felsenfeld and A. Rich, *Biochim. Biophys. Acta,* **26**, 457 (1957).
96 J. R. Fresco, in *Informational Macromolecules*, Academic Press, New York (1963), p. 121.
97 A. M. Michelson, M. Dondon, and M. Gunberg-Manago, *Biochim. Biophys. Acta,* **55**, 529 (1962).
98 R. B. Inman and R. L. Baldwin, *J. Mol. Biol.,* **8**, 452 (1964).
99 R. B. Inman, *J. Mol. Biol.,* **10**, 137 (1964).
100 P. B. Sigler, D. R. Davies, and H. T. Miles, *J. Mol. Biol.,* **5**, 709 (1962).
101 M. J. Chamberlin and D. Patterson, *J. Mol. Biol.,* **12**, 410 (1965).
102 M. Riley, B. Maling, and M. J. Chamberlin, *J. Mol. Biol.,* **20**, 359 (1966).
103 M. J. Chamberlin, *Fed. Proc.,* **24**, 1446 (1965).
104 M. Swierkowski, W. Szer, and D. Shuga, *Biochem. Z.,* **342**, 429 (1965).
105 R. B. Inman and R. L. Baldwin, *J. Mol. Biol.,* **5**, 172 (1962).
106 R. D. Wells, E. Ohtzuka, and H. G. Khorana, *J. Mol. Biol.,* **14**, 221 (1963).
107 R. D. Wells, J. E. Larson, R. C. Grant, B. E. Shortle, and C. R. Cantor, *J. Mol. Biol.,* **54**, 465 (1970).
108 A. Rich, *Biochim. Biophys. Acta,* **29**, 502 (1958).
109 A. Rich, *Nature,* **181**, 521 (1958).
110 M. Lipsett, *Biochem. Biophys. Res. Commun.,* **11**, 224 (1963).
111 A. R. Morgan and R. D. Wells, *J. Mol. Biol.,* **37**, 63 (1968).
112 M. Lipsett, *J. Biol. Chem.,* **239**, 1256 (1964).
113 F. B. Howard, J. Frazier, M. N. Lipsett, and H. T. Miles, *Biochem. Biophys. Res. Commun.,* **17**, 93 (1964).
114 M. M. Warshaw and I. Tinoco, Jr., *J. Mol. Biol.,* **13**, 54 (1965).
115 M. M. Warshaw and I. Tinoco, Jr., *J. Mol. Biol.,* **20**, 29 (1966).
116 G. Felsenfeld and H. T. Miles, *Ann. Rev. Biochem.,* **36**, 407 (1967).
117 P. O. P. Ts'o, N. S. Kondo, M. P. Schweizer, and D. P. Hollis, *Biochemistry,* **8**, 997 (1969).
118 R. Thomas, *Biochim. Biophys. Acta,* **14**, 231 (1964).
119 W. F. Dove and N. Davidson, *J. Mol. Biol.,* **5**, 467 (1962).
120 R. B. Inman, *J. Mol. Biol.,* **9**, 624 (1964).
121 C. Zimmer and H. Venner, Naturwissenschaftfen, **49**, 86 (1962).
122 C. Schildkraut and S. Lifson, *Biopolymers,* **3**, 195 (1965).
123 M. T. Record, Jr., *Biopolymers,* **5**, 975 (1967).
124 M. T. Record, Jr., *Biopolymers,* **5**, 993 (1967).
125 D. Gruenwedel and C. Hsu, *Biopolymers,* **7**, 557 (1969).
126 T. L. Hill, *Arch. Biochem. Biophys.,* **57**, 229 (1955).

127 L. Kotin, *J. Mol. Biol.*, **7**, 309 (1963).
128 J. Shack and B. S. Bynum, *Nature*, **184**, 635 (1959).
129 J. W. Lyons and L. Kotin, *J. Amer. Chem. Soc.*, **86**, 3634 (1964).
130 I. Scheffler, E. Elson, and R. L. Baldwin, *J. Mol. Biol.*, **36**, 291 (1968).
131 P. L. Privalov, K. A. Kafiani, and D. P. Monaselidzc, *Biophysics*, **10**, 433 (1965). [Translation of *Biofizika*, **10**, 343 (1965).]
132 W. F. Dove, A. Wallace, and N. Davidson, *Biochem. Biophys. Res. Commun.*, **1**, 312 (1959).
133 Ch. Zimmer and H. Venner, *Biopolymers*, **4**, 1073 (1966).
134 Ch. Zimmer, G. Luck, H. Venner, and J. Fric, *Biopolymers*, **6**, 563 (1968).
135 G. K. Helmkamp and P. O. P. Ts'o, *J. Amer. Chem. Soc.*, **83**, 138 (1961).
136 T. T. Herskovitz, S. J. Singer, and E. P. Geiduschek, *Arch. Biochem. Biophys.*, **94**, 99 (1961).
137 E. P. Geiduschek and T. T. Herskovits, *Arch. Biochem. Biophys.*, **95**, 114 (1961).
138 L. Levine, J. A. Gordon, and W. P. Jencks, *Biochemistry*, **2**, 168 (1963).
139 K. Hamaguchi and E. P. Geiduschek, *J. Amer. Chem. Soc.*, **84**, 1329 (1962).
140 D. R. Robinson and M. E. Grant, *J. Biol. Chem.*, **241**, 4030 (1966).
141 D. M. Crothers, N. R. Kallenbach, and B. H. Zimm, *J. Mol. Biol.*, **11**, 802 (1965).
142 D. W. Hennage, Thesis, Yale University (1969).
143 J. Applequist and V. Damle, *J. Amer. Chem. Soc.*, **87**, 1450 (1965).
144 M. N. Lipsett, L. A. Heppel, and D. F. Bradley, *J. Biol. Chem.*, **236**, 857 (1961).
145 M. N. Lipsett, *J. Biol. Chem.*, **239**, 1256 (1964).
146 M. N. Lipsett, L. A. Heppel, and D. F. Bradley, *Biochim. Biophys. Acta*, **41**, 175 (1960).
147 A. M. Michelson and C. Monny, *Biochim. Biophys. Acta*, **149**, 107 (1967).
148 G. Cassani and F. J. Bollum, *J. Amer. Chem. Soc.*, **89**, 4798 (1967).
149 P. M. Pitha and P. O. P. Ts'o, *Biochemistry*, **8**, 5206 (1969).
150 G. R. Cassani and F. J. Bollum, *Biochemistry*, **8**, 3928 (1969).
151 C. R. Cantor and N. W. Chin, *Biopolymers*, **6**, 1745 (1968).
152 O. Uhlenbeck, R. Harrison, and P. Doty, in *Molecular Association in Biology*, B. Pullman, ed., Academic Press, New York (1968), p. 107.
153 F. B. Howard, J. Frazier, M. N. Lipsett, and H. T. Miles, *Biochem. Biophys. Res. Commun.*, **17**, 93 (1964).
154 F. B. Howard, J. Frazier, M. F. Singer, and H. T. Miles, *J. Mol. Biol.*, **16**, 415 (1966).
155 W. M. Huang and P. O. P. Ts'o, *J. Mol. Biol.*, **16**, 523 (1966).
156 P. O. P. Ts'o and W. M. Huang, *Biochemistry*, **7**, 2954 (1968).
157 P. O. P. Ts'o and M. P. Schweizer, *Biochemistry*, **7**, 2963 (1968).
158 P. M. Pitha, W. M. Huang, and P. O. P. Ts'o, *Proc. Nat. Acad. Sci., U.S.*, **61**, 332 (1968).
159 I. Tazawa, S. Tazawa, and P. O. P. Ts'o, *J. Mol. Biol.*, **66**, 115 (1972).
160 V. N. Damle, *Biopolymers*, **9**, 353 (1970).
161 I. Tinoco, Jr., O. Uhlenbeck and M. D. Levine, *Nature*, **230**, 362 (1971).
162 C. DeLisi and D. M. Crothers, *Proc. Nat. Acad. Sci., U.S.*, **68**, 2682 (1971).
163 O. C. Uhlenbeck, p. N. Borer, B. Dengler, and I. Tinoco, Jr., *J. Mol. Biol.*, **73**, 483 (1973).
164 J. Gralla and D. M. Crothers, *J. Mol. Biol.*, **73**, 497 (1973).
165 F. H. Martin, O. C. Uhlenbeck, and P. Doty, *J. Mol. Biol.*, **57**, 201 (1971).
166 O. C. Uhlenbeck, F. H. Martin, and P. Doty, *J. Mol. Biol.*, **57**, 217 (1971).
167 D. Pörschke and M. Eigen, *J. Mol. Biol.*, **62**, 361 (1971).
168 M. E. Craig, D. M. Crothers, and P. Doty, *J. Mol. Biol.*, **62**, 383 (1971).

169 I. E. Scheffler, E. Elson, and R. L. Baldwin, *J. Mol. Biol.*, **36**, 291 (1968).
170 I. E. Scheffler, E. Elson, and R. L. Baldwin, *J. Mol. Biol.*, **48**, 145 (1970).
171 E. Elson, I. E. Scheffler, and R. L. Baldwin, *J. Mol. Biol.*, **54**, 401 (1970).
172 D. M. Crothers, *Acts. Chem. Res.*, **2**, 225 (1969).
173 T. R. Fink and D. M. Crothers, *J. Mol. Biol.*, **66**, 1 (1972).
174 C. DeLisi and D. M. Crothers, *Biopolymers,* **10**, 1809 (1971).
175 C. DeLisi, *Biopolymers* (in press).
176 M. Sundaralingam, *Biopolymers,* 7, 821 (1969).
177 H. Jacobson and W. H. Stockmayer, *J. Chem. Phys.*, **18**, 1600 (1950).
178 M. E. Fisher, *J. Chem. Phys.*, **45**, 1469 (1966).
179 D. M. Crothers, *Biopolymers,* **6**, 1391 (1968).
180 D. M. Crothers and B. H. Zimm, *J. Mol. Biol.*, **9**, 1 (1964).
181 D. M. Crothers, N. R. Kallenbach, and B. H. Zimm, *J. Mol. Biol.*, **11**, 802 (1965).
182 J. Marmur, R. Round, and C. L. Schildkraut, *Prog. Nuc. Acid. Res.*, **1**, 231 (1963).
183 J. Marmur and D. Lane, *Proc. Nat. Acad. Sci., U.S.*, **46**, 451 (1960).
184 P. Doty, J. Marmur, J. Eigner, and C. L. Schildkraut, *Proc. Nat. Acad. Sci., U.S.*, **46**, 461 (1960).
185 J. Marmur and P. Doty, *J. Mol. Biol.*, **3**, 585 (1961).
186 P. Doty, J. Marmur, and N. Sueoka, *Brookhaven Symposia,* **12**, 1 (1959).
187 W. Ginoza and B. H. Zimm, *Proc. Nat. Acad. Sci., U.S.*, **47**, 639 (1961).
188 R. M. Herriott, *Proc. Nat. Acad. Sci., U.S.*, **47**, 146 (1961).
189 R. M. Herriott, *J. Chim. Phys.*, **58**, 1101 (1961).
190 E. P. Geiduschek, *J. Mol. Biol.*, **4**, 467 (1962).
191 E. P. Geiduschek, *Proc. Nat. Acad. Sci., U.S.*, **47**, 950 (1961).
192 J. Marmur and P. Doty, *J. Mol. Biol.*, **3**, 585 (1961).
193 J. Marmur, R. Round, and C. L. Schildkraut, *Prog. Nucl. Acid. Res.*, **1**, 231 (1963).
194 J. A. Subirana, *Biopolymers,* **4**, 188 (1966).
195 J. A. Subirana and P. Doty, *Biopolymers,* **4**, 171 (1966).
196 J. G. Wetmur and N. Davidson, *J. Mol. Biol.*, **31**, 349 (1968).
197 F. W. Studier, *J. Mol. Biol.*, **41**, 194 (1969).
198 P. D. Ross and J. M. Sturtevant, *Proc. Nat. Acad. Sci., U.S.*, **46**, 1360 (1960).
199 P. D. Ross and J. M. Sturtevant, *J. Amer. Chem. Soc.*, **84**, 4503 (1962).
200 R. D. Blake and J. R. Fresco, *J. Mol. Biol.*, **19**, 145 (1966).
201 R. D. Blake, L. C. Klotz, and J. R. Fresco, *J. Amer. Chem. Soc.*, **90**, 3556 (1968).
202 R. J. Britten and D. E. Kohne, *Science,* **161**, 529 (1968).

chapter 7
binding of small molecules

Nucleic acids function through interaction with other substances. Transcription and translation, for example, require enzyme systems for readout of genetic information by recognition of complementary base pairing. Also, genetic control can be exercised by substances that bind to particular regions of a DNA molecule. Detailed physical information on these highly specific interactions is only beginning to become available. On the other hand, considerable work has been done on the binding of smaller, simpler molecules to nucleic acids, revealing many general features of nucleic acid interactions and forming an essential basis for future development of this important field.

The most pervasive, and least specific, of the nucleic acid interactions is that with the components of the solvent: water and neutral electrolytes. We begin our discussion with these, and then consider more specific binding of metal ions, drugs, and, very briefly, polyamino acids and proteins.

I HYDRATION OF NUCLEIC ACIDS

Foremost among the interactions of nucleic acids with other molecules is the interaction with the solvent. Since the main component is almost invariably water, we shall term such interaction *hydration*. The degree of hydration of DNA is extremely important in determining the stability of the Watson-Crick double helical structure. Franklin and Gosling[1] showed by X-ray diffraction in 1953 that the B configuration of NaDNA was stable only at relative humidities (RH) above 75–80%. Below 50% RH, ordered structure seems generally to disappear. Very similar conclusions were arrived at by use of polarized infrared and ultraviolet spectra of oriented films of NaDNA.[2] At RH above 75%, the IR and UV dichroic ratios ($A_{\parallel}/A_{\perp}$) and A_{260} are low and roughly constant, indicating that the bases are stacked perpendicular to the fiber (helix) axis, and that the full hypochromism characteristic of the native state is present. Between 75% and 55% RH, the dichroic ratios and the absorbance at 260 nm all increase congruently, reaching a plateau below 55% RH. These results indicate that lowering the RH below 75% leads to configurations in which the bases are unstacked and no longer perpendicular to the helix axis.

Because of the fundamental nature of the nucleic acid–water interaction, and the importance of water in stabilizing nucleic acid structure, hydration of nucleic acids has been the subject of many studies. These have been very well reviewed by Tunis and Hearst.[3] We shall draw heavily on this review throughout this section. Nearly all of the studies have been made on native DNA, while a few have been devoted to hydration changes attendant upon denaturation. Studies of RNA hydration have been very rare to date, despite the potential interest in the effect of 2′–OH group on hydration.

A great variety of experimental techniques has been used to elucidate hydration, including gravimetry, IR spectroscopy, NMR, calorimetry, self-diffusion of labeled water in DNA solutions, sedimentation velocity, sedimentation equilibrium in density gradients, isopiestic and other thermodynamic measurements, and low-angle X-ray scattering. Each of these techniques is sensitive to hydration in a different way, and therefore will indicate a different extent of hydration under similar conditions. However, as will be seen in the ensuing discussion, most of these techniques give reasonably concordant results which can be subsumed in a consistent picture of nucleic acid hydration. We shall discuss the experimental results first, and follow these with a discussion of the molecular mechanism of hydration.

A. Hydration in Fibers and Films

Some of the most detailed measurements of DNA hydration, over a wide range of relative humidities, have been made by gravimetric and IR studies of water absorption on DNA fibers and films.[4,5] The RH is maintained at a constant, known value by equilibrating the DNA with saturated aqueous salt solutions of predetermined RH.

In Fig. 7-1 is shown a plot of the hydration Γ (moles H_2O/moles nucleotide) of NaDNA fibers as a function of RH.[4] Similar data were

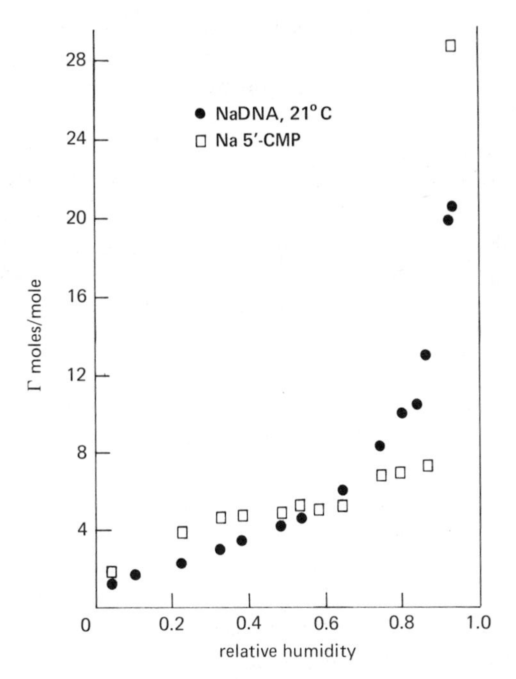

FIGURE 7-1 Hydration of NaDNA (•) and Na 5′–CMP (□) as a function of relative humidity. [From M. Falk, K. A. Hartman, Jr., and R. C. Lord, *J. Amer. Chem. Soc.*, **84**, 3843 (1962); M. Falk, *Canad. J. Chem.*, **43**, 314 (1965). Reprinted with permission.]

obtained for LiDNA. We have neglected small hysteresis effects – that is, slightly different values of Γ obtained during absorption and desorption – which are probably due to nonequilibrium packing effects of the individual chains of the fibers.

Falk and co-workers[4] found that the curve in Fig. 7-1 could fit well with the BET[6] adsorption isotherm,

$$\Gamma = \frac{Bcx}{(1 - x + cx)(1 - x)} \tag{7-1}$$

which is obtained assuming strong binding in a first adsorption layer, with identical, weaker binding in all subsequent layers. In Eq. (7-1), x is the relative humidity; B is the maximum number of moles of H_2O per mole nucleotide adsorbed on the first layer; and $c = \exp[E_1 - E_L]/RT$, where E_1 is the energy of adsorption of the first layer, E_L is the energy of adsorption of subsequent layers, and RT is the product of gas constant and absolute temperature. Analysis of the data led to $B = 2.2$ moles/mole for NaDNA and 2.0 moles/mole for LiDNA, while $E_1 - E_L = 1.7$ kcal/mole of H_2O adsorbed for NaDNA and 2.1 kcal/mole for LiDNA. The values of $E_1 - E_L$ are obtained assuming that the partition function for adsorbed water is the same in the first and subsequent layers. This assumption, which is unlikely to be true, generally leads to an underestimate of $E_1 - E_L$ by as much as 1 kcal/mole.

Deviations from the BET curve, in the direction $\Gamma_{\text{observed}} < \Gamma_{\text{calculated}}$, were noted at relative humidities above 80%. This is probably due to the fact that, with RH < 80%, there are void spaces between the DNA chains which are being filled with water; while above 80% RH, additional energy is needed to push the DNA chains further apart to accommodate more water. This explanation is rendered more plausible by the X-ray diffraction observations[7,8] that the intermolecular spacing in DNA fibers increases rapidly at RH = 80–85%. Of course, the rapid increase in Γ observed at RH > 90% mainly reflects addition of bulk H_2O to wet fibers, and has little significance in terms of DNA—H_2O interactions.

Similar gravimetric studies of water adsorption by nucleic acid components[9] show that purines, pyrimidines, nucleosides, and nucleotides generally remain unhydrated even at 93% RH. The most highly hydrated, guanosine and 5′-adenylic acid, form only dihydrates. Contrastingly, salts of the nucleotides, which contain ionic phosphate groups, become hydrated at low RH and may bind more than 20 moles H_2O/mole nucleotide at 93% RH. This is shown in Fig. 7-1. Thus the ionic phosphate group seems responsible for most water binding by nucleotides, and, by extension, by polynucleotides. One notes, as in Fig. 7-1, plateaus in water binding to individual nucleotide salts, corresponding to formation of particular hydrates. These plateaus are averaged out in DNA, because of the admixture of different nucleotides and of environmental effects such as base stacking and interchain interactions.

Infrared absorption spectral studies of hydration of DNA films[5] have tended to confirm the picture obtained gravimetrically. The asymmetric OH stretching band of pure liquid water is found at $3400 \pm 10\ \text{cm}^{-1}$. The absorption in this band increases with increasing RH of solid films of DNA, and the shape of the absorption vs. RH curve is very similar to the gravimetric Γ vs. RH curve in Fig. 7-1. Furthermore, it is observed that the frequency of maximum absorption increases from $3390 \pm 10\ \text{cm}^{-1}$ at low RH to $3420 \pm 10\ \text{cm}^{-1}$ at high RH. This shows that: (a) hydrogen bonding is stronger at lower RH, which agrees with the gravimetric results interpreted according to the BET theory; and (b) at high RH, water has IR absorption corresponding to normal liquid water, and not to some form of ice. This tends to make improbable the existence of icelike layers of H_2O surrounding the DNA chains.

Infrared study of DNA absorption bands belonging to the phosphate, deoxyribose, and base groups confirms the above quantitative results, and in addition gives information on the sequence of occupation of hydration sites on the DNA.[5] These experiments will therefore be reviewed below, in the section on molecular mechanisms of hydration.

B. Hydration in Aqueous Solutions

Measurements have also been made of hydration of nucleic acids in dilute or moderately concentrated solutions. In such solutions, of course, the hydration is not meaningfully defined simply as the amount of water present in equilibrium with the nucleic acid, for then Γ would approach infinity with increasing dilution of the macromolecular component. What is of interest, and what is measured by the techniques discussed in this section, is the amount of water that is immobilized or otherwise physically affected by the presence of the nucleic acid.

The most definitive results are those obtained by Wang[10,11] by studying the self-diffusion of ^{18}O-labeled water in aqueous DNA solutions. It is evident that the self-diffusion will be retarded, relative to that in pure water, by two effects. The first is the obstruction effect: the diffusion paths of the labeled water molecules are longer, since they are forced to go around the macromolecule. This effect will depend on the shape of the macromolecule and on its volume fraction ϕ in solution. The second effect, and the one of primary interest here, is the hydration effect: at any given time, a certain number of H_2O molecules will be bound to DNA, and thus will be stopped from diffusing since $D_{DNA} \ll D_{H_2O}$. This effect depends on ϕ and Γ, or, on a weight basis, on the weight fraction w of DNA and Γ'. Both of these effects will cause the observed self-diffusion coefficient, D_{H_2O}, to fall below the self-diffusion coefficient $D^{\mathrm{O}}_{H_2O}$ of pure water.

By considering the flux J of $H_2{}^{18}O$ in a capillary tube where the concentration gradient of labeled water was dc/dx, and neglecting at first the hydration effect, Wang[10] found

$$J = -D^{\circ}_{H_2O}[1 - \bar{\alpha}\phi]\,\frac{dc}{dx} \tag{7-2}$$

$\bar{\alpha}$ is a weakly shape-dependent obstruction factor which varies from 3/2 for spherical macromolecules to 5/3 for prolate ellipsoids of infinite axial ratio. The calculation leading to Eq. (7-2) assumed that the macromolecules were stationary and uniformly distributed in the tube. Since the apparent diffusion coefficient is

$$D'_{H_2O} = \frac{-J}{(dc/dx)} \tag{7-3}$$

one has

$$D'_{H_2O} = D^{\circ}_{H_2O}(1 - \bar{\alpha}\phi) \tag{7-4}$$

The hydration effect is taken into account as follows. Assuming that exchange between bound and free water is very fast compared to the diffusion rate, the isotopic fractions of bound and free water are equal to the isotopic fractions in the initial solution. Then, if c_h and c_0 are the concentrations of bound and total H_2O, in grams per gram of solution, the experimentally observed self-diffusion coefficient is

$$D_{H_2O} = D'_{H_2O}\left(\frac{1 - c_h}{c_0}\right) \tag{7-5}$$

This is true because only the fraction of free water, $(c_0 - c_h)/c_0$, is diffusible at any given time. Combination of Eqs. (7-4) and (7-5) yields

$$D_{H_2O} = D^{\circ}_{H_2O}\,(1 - \alpha\phi)\left(\frac{1 - c_h}{c_0}\right) \tag{7-6}$$

If c_p is the concentration of anhydrous macromolecule in grams per gram of solution,

$$\frac{c_h}{c_0} = \Gamma'\frac{c_p}{c_0} = \Gamma'\frac{w}{(1 - w)} \tag{7-7}$$

Also,

$$\phi = c_p\left(\tilde{v}_p + \frac{\Gamma'}{\rho_0}\right) \tag{7-8}$$

where $\tilde{v}_p$ is the apparent specific volume of macromolecule and ρ_0 is the density of pure water. Using the relations $c_p\tilde{v}_p + c_0/\rho_0 = 1$, and $c_p/c_0 = w/(1 - w)$ one finds

$$\phi = \frac{\tilde{v}_p + \Gamma'/\rho_0}{\tilde{v}_p + \dfrac{1}{\rho_0}\left(\dfrac{1 - w}{w}\right)} \tag{7-9}$$

Thus, one finally obtains

$$D_{H_2O} = D^{\circ}_{H_2O}\left\{1 - \alpha\left[\frac{\tilde{v}_p + \dfrac{\Gamma'}{\rho_0}}{\tilde{v}_p + \dfrac{1}{\rho_0}\left(\dfrac{1 - w}{w}\right)}\right]\right\}\left[1 - \left(\frac{w}{1 - w}\right)\Gamma'\right] \tag{7-10}$$

which enables determination of the mass hydration Γ'.

Wang[11] measured $D_{H_2O}/D^{\circ}_{H_2O}$ as a function of w for NaDNA over the range $0 < w \leqslant 0.124$, in solutions with and without added NaCl. There was no significant effect of added salt on the diffusion coefficient ratio. Using $\bar{\alpha} = 5/3$, which is appropriate for highly elongated molecules, $\tilde{v}_p =$

0.53 cm^3/g, and $\rho_0 = 1.0$ g/cm^3, he found $\Gamma' = 0.35$ g H_2O/g dry NaDNA. This corresponds to about 6.5 water molecules per nucleotide, which is roughly sufficient to fill the grooves of the double helix.[3] It is rather lower than, though in the same range as, the value of $\Gamma = 8.4-10.1$ moles/mole determined gravimetrically[4] (Fig. 7-1) at relative humidities of 75–80% where the intermolecular distance between DNA molecules begins to increase markedly. Part of this difference, which is probably statistically significant, may be due to the different environments of the DNA in the fiber and in solution. Another part of the difference is undoubtedly due to the fact that the H_2O self-diffusion method is sensitive only to relatively firmly bound water, which is translationally immobilized during a significant part of its lifetime near the DNA.

Several other experiments have been performed which should be sensitive to the same sort of hydration as is the H_2O self-diffusion technique, although exact numerical agreement on values of Γ is not to be expected. Analysis of the molecular weight dependence of the sedimentation coefficient of DNA,[12] discussed in Chapter 5, yields a value of 26–27 Å for the hydrodynamic diameter of the DNA double helix. This corresponds to about one extra layer of hydrodynamically immobilized water molecules surrounding the helix, and a Γ of 13 moles/mole. However, because of theoretical approximations in the hydrodynamic analysis, this may not be significantly different from the 22 Å diameter defined by the phosphate residues. Thus the hydration determined by s vs. M studies is in rough accord with that determined by Wang.[11]

Heat capacity studies of DNA in the presence of varying amounts of added water[13] show that 9–10 water molecules per nucleotide can be added without observing a phase transition as the temperature is varied around 0°C. This may be interpreted as a (probably partially irrotational) binding of water to the DNA in such a way as to prevent the normal $H_2O(l) \rightleftharpoons H_2O(s)$ transition. Twenty to thirty water molecules per nucleotide, an amount sufficient to partially surround the helix exterior with about two layers of water molecules, must be added before the influence of the DNA on the liquid water–ice equilibrium is no longer observed.

Early nuclear magnetic resonance studies indicated that the water-proton NMR line is broadened in the presence of DNA. This was attributed to a shortening of the spin-lattice relaxation time due to binding of H_2O by the DNA. It has been suggested, however, that this increased relaxation is due instead to paramagnetic ion impurities associated with the DNA.[14] Another type of experiment,[15] which dealt with DNA and ribosomal RNA, seems less open to this objection. At −35°C and concentration of 100 mg/ml, the nucleic acid solution has a water proton resonance much narrower than that of ice, though broader than that of supercooled water at the same temperature. The bound water signal is proportional to the nucleic acid concentration. These results suggest that water of hydration is not irrotationally bound, although its rotation is restricted relative to free H_2O. The amount bound in this fashion was estimated at 0.59 g/g (11 moles/mole) for DNA and 0.63 g/g (12 moles/mole) for ribosomal RNA.

In summary, all of the above methods demonstrate that in aqueous solutions of nucleic acids, at least sufficient water is translationally (and probably rotationally) immobilized to fill the grooves of the DNA double helix. There are also indications, based on the *s* vs. *M* analysis and the calorimetric and NMR experiments, that an additional layer or so of water is at least partially immobilized, translationally and rotationally, at the periphery of the double helix, perhaps bound to the phosphates. However, until more experiments are forthcoming, such indications should be viewed with caution.

C. Preferential Hydration in Binary Solvent Systems

In studies of sedimentation equilibrium in a density gradient, and in other thermodynamic studies on three-component systems, it is convenient to define a purely thermodynamic hydration parameter. As discussed in Chapter 5, this is

$$\Gamma_1 = \left(\frac{\partial m_1}{\partial m_3}\right)_{\mu_1} = -\frac{\left(\frac{\partial \mu_1}{\partial m_3}\right)_{m_1}}{\left(\frac{\partial \mu_1}{\partial m_1}\right)_{m_3}} \tag{7-11}$$

Thus Γ_1 represents the number of moles of H_2O (component 1) which must be added per mole of nucleotides (component 3) in order to keep the chemical potential of water constant. It should be noted that, in order to ensure positive values of Γ_1, Hearst and Vinograd[16] defined the molalities m_1 and m_3 per unit mass of salt (component 2). On a mass basis, one defines

$$\Gamma_1' = \frac{M_1}{M_3}\,\Gamma_1 \tag{7-12}$$

as the number of grams of H_2O which must be added per gram of DNA to maintain μ_1 constant. M_1 is the molecular weight of water, M_3 the mean nucleotide molecular weight.

It should be emphasized that Γ_1 is a measure of the thermodynamic nonideality of the solution as a whole, and, as will be seen later, can be only partly attributed to short-range H_2O–DNA interactions. This is in contrast to the hydration measures discussed in Sections A and B, in which only direct solvent-polymer interactions are detected.

The most extensive studies of thermodynamic hydration have been carried out using sedimentation equilibrium in a density gradient.[3,16-21] The operative equation is Eq. (5-6)

$$\frac{1}{\rho_0} = \frac{\bar{v}_3 + \Gamma_1{}'\bar{v}_1}{1 + \Gamma_1{}'} \qquad (7\text{-}13)$$

$\bar{v}_1$, the partial specific volume of H_2O, is taken to be 1 cm^3/g. $\bar{v}_3$, the partial specific volume of anhydrous DNA (as the appropriate salt), is determined as the reciprocal of the intercept of a plot of buoyant density, ρ_0, vs. water activity a_w at $a_w = 0$. For CsDNA, $1/\bar{v}_3 = 2.12$ g/cm^3, while for LiDNA, $1/\bar{v}_3 = 1.783$ g/cm^3.[18]

The hydration determined in this way might be expected to depend on both the cation and the anion of the salt used as component 2 to form the density gradient. This is because of possible differences in size, affinity for the DNA, and structure making or breaking effects on the water. However, it has been found that Γ_1 is simply a monotonic function of water activity alone. A plot of Γ_1 vs. a_w for various Cs and Li salts is shown in Fig. 7-2[18]; points for Na and K salts fall on the same curve. It will be noted that, despite the quite different method of measurement and rather different definition, the curve of Γ_1 vs. a_w in Fig. 7-2 strongly resembles that of Γ_1 vs. RH, in Fig. 7-1.

By studying hydration at density gradient equilibrium as a function of temperature, it is possible to determine an apparent ΔH of hydration. This was done for T7 DNA in aqueous potassium trifluoroacetate solutions, where $a_w = 0.7$ at 25°C.[3] It was found that $(\partial\Gamma/\partial T)_{a_w} = -0.024$ moles/mole-°C, and that $\Delta H = 397$ cal/mole H_2O for native DNA; for denatured T7 DNA, also at 25°C, the corresponding values are -0.015 moles/mole-°C and 341 cal/mole H_2O.

An investigation has also been made of the effect of base composition on the hydration of DNA.[21] After extrapolation of ρ_0 vs. a_w to $a_w = 0$ for DNA from *B. megatherium* (38% G · C), *E. coli* (50% G · C), and *M. leisodeikticus* (72% G · C), no difference in the anhydrous partial specific volumes of the Li salts of these DNAs was found, and only a very slight difference in $\bar{v}_{DNA}$ of the Cs salts. Thus, according to Eq. (7-13), the substantial difference in buoyant densities of these DNAs as a function of their G·C content (see Chapter 5) is almost entirely attributable to differential hydration. Quantitatively, it was found that at $a_w = 0.7$, A · T base pairs have about two more molecules of water associated with them than G · C base pairs, thus accounting for the increase in ρ_0 with percent G · C.

Thermodynamic hydration may also be determined by isopiestic measurements.[18,22] In these experiments, equal amounts of water and salt are placed in weighed containers, and a known amount of DNA is added to one of the containers. After equilibrium has been reached, it is found that a certain amount of water has distilled over from the vessel without the DNA to that with the DNA, so that a_w is equalized between the containers. The

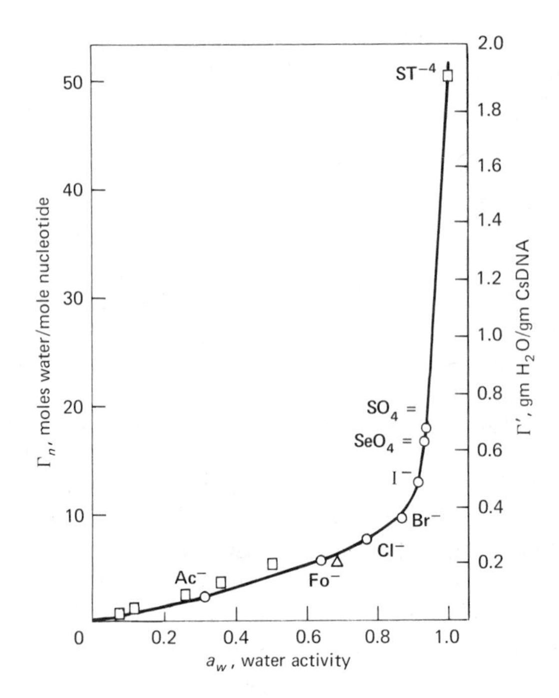

FIGURE 7-2 The net hydration of DNA from bacteriophage T4 as a function of water activity, for various Cs and Li salts. [From J. E. Hearst and J. Vinograd, *Proc. Nat. Acad. Sci., U.S.*, **47**, 1005 (1961). Reprinted with permission.]

weight of water transferred, divided by the weight of DNA added, is equal to the mass hydration Γ_1'. Studies done with sonicated salmon sperm DNA and with NaCl, $NaClO_4$, and Na_2SO_4 as salts give a Γ_1' vs. a_w curve very similar to that in Fig. 7-2, and confirm that Γ_1' depends only on a_w. It is of particular interest that the perchlorate anion, which is a very strong denaturing agent, has no special effect on the hydration of native DNA.

Similar results have also been obtained from measurements of density increments.[23] In these experiments, water and salt are placed on one side of a dialysis membrane, and water, salt, and DNA on the other. After dialysis equilibrium is achieved, the density change in the solution which contains DNA is measured as a function of its volume concentration C_3. Since the process occurs at constant chemical potentials of water and salt, the density increment is defined as $(\partial\rho/\partial C_3)_{\mu_1\,\mu_2}$. For incompressible solutions which are dilute in DNA,

$$\left(\frac{\partial\rho}{\partial C_3}\right)^{\circ}_{\mu_1\,\mu_2} = 1 + \Gamma_1' - \rho^{\circ}(\bar{v}_3{}^{\circ} + \Gamma_1'\bar{v}_1) \tag{7-14}$$

where superscript zero denotes vanishing polymer concentration, and ρ^0 is

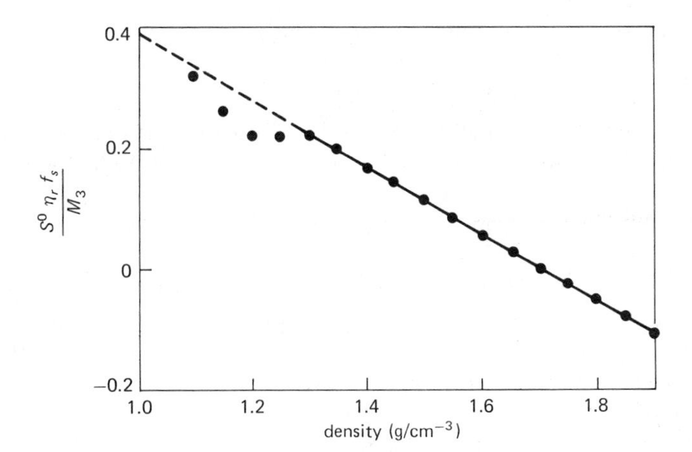

FIGURE 7-3 Corrected sedimentation coefficient of CsDNA as a function of solvent density in concentrated CsCl solutions. [From R. Bruner and J. Vinograd, *Biochim. Biophys. Acta*, **108**, 18 (1965). Reprinted with permission.]

the density of the solvent mixture without polymer. Measurements of Γ_1' vs. a_w for NaDNA in NaCl and CsDNA in CsCl solutions fall nicely on the curve in Fig. 7-2.

It is interesting that the values of Γ_1' obtained in *equilibrium* experiments can fairly satisfactorily explain the dependence of the sedimentation velocity of CsDNA in concentrated CsCl solutions.[24] The sedimentation coefficient of the solvated DNA is given in Chapter 5:

$$S^\circ = \frac{M_s(1 - \bar{v}_s\rho)}{N_A f} \tag{7-15}$$

Using Eqs. (5-62) and (5-63) for the hydrolated molecular weight M_s and hydrated partial specific volume $\bar{v}_s$, substituting into Eq. (7-15), and rearranging, one finds

$$S^\circ\eta_r = \frac{M_3}{f_s}(1 + \Gamma_1') - \frac{M_3}{f_s}(\bar{v}_3 + \Gamma_1'\bar{v}_1) \tag{7-16}$$

η_r is the relative viscosity of the solvent, and f_s is the frictional coefficient of the DNA divided by η_r. In CsCl (but not in other salts such as $NaClO_4$ or Na_2SO_4) the dependence of $S^0\eta_r$ on ρ is found experimentally to be linear, as is the dependence of Γ_1' on ρ. It is also plausible to expect that f_s remains constant, at least in the high salt concentration region. In these circumstances, Eq. (7-16) predicts that a plot of $S^0\eta_r f_s/M_3$ vs. ρ should be linear if the values of Γ_1' determined thermodynamically for CsCl[18] are used. This prediction is nicely borne out, as is shown in Fig. 7-3. The deviation from linearity at low density (high a_w) is most plausibly attributed to inaccuracies in $\bar{v}_1$ and $\bar{v}_3$,[3,24] although changes in frictional coefficient due to polyelectrolyte expansion may also play a part.

It has been pointed out[3] that data on low-angle X-ray scattering from DNA solutions[25] can be plotted against density in a way exactly analogous to the sedimentation coefficient, yielding a curve very similar to that in Fig. 7-3.

D. Molecular Mechanisms of Hydration

The experimental results cited, and summarized in Figs. 7-1 and 7-2, show that the hydration Γ_1 of DNA is a monotonic function of water activity or relative humidity. It goes to zero at $a_w = 0$, remains below about 10 moles H_2O/mole nucleotide until a_w rises above 0.8, and then rises rapidly with increasing a_w, becoming indeterminate as a_w approaches unity. In this section we shall attempt to account for this behavior.

It should first be noted that the enormous thermodynamic hydration observed at high a_w cannot be ascribed with any plausibility to direct interactions between the DNA and the solvent. Consider, for example, the value of $\Gamma_1 = 50$ moles/mole found at $a_w = 0.99$. Taking the radius of the DNA double helix as 11 Å, defined by the exterior of the phosphate groups, and assuming that 6.5 moles/mole of water are used to fill the large and small grooves of the DNA, this value of Γ_1 corresponds to a sheath of water 10 Å thick surrounding the double helix, or about three to four layers of water molecules. If salt were not entirely excluded from this layer, it would be even thicker. There is no evidence, from X-ray diffraction or spectroscopy, that any such ramified structure of icelike or otherwise extensively hydrogen-bonded water surrounds the DNA. Further, no electrostatic forces could produce "binding" over such a range.

Thus, it is clear that most of the very large preferential hydration which is observed at high a_w must be attributed to general features which influence the thermodynamic behavior of ternary, polyelectrolyte solutions, and not to specific short-range interactions. In fact, the general features of the curve in Fig. 7-2 will be observed in any ternary system, and do not depend on the polymeric or charged character of solute component 3, or even on solution nonideality, whether it arises from electrostatic or other causes.

It is well known from Raoult's law that the chemical potential of the solvent in an ideal solution is

$$\mu_1 = \mu_1{}^\circ + RT \ln a_1 = \mu_1{}^\circ + RT \ln X_1 \tag{7-17}$$

where the mole fraction of solvent is

$$X_1 = \frac{m_1}{m_1 + m_2 + m_3} \tag{7-18}$$

Substitution of these equations into Eq. (7-11) then gives

$$\Gamma_1 = \frac{X_1}{1 - X_1} = \frac{a_w}{1 - a_w} \tag{7-19}$$

Therefore, in accord with Fig. 7-2,

$$\lim_{a_w = 0} \Gamma = 0, \qquad \lim_{a_w = 1} \Gamma = \infty \tag{7-20}$$

The behavior of Γ at the two extremes $a_w = 0$ and $a_w = 1$ is thus essentially a consequence of the definition of preferential hydration. Physically, Eqs. (7-20) are rather obvious. At very low water activity, the amount of H_2O available to hydrate the nucleotide is very small, and goes to zero at $a_w = 0$. Therefore, addition of more nucleotides will require the addition only of vanishing amounts of water to keep a_w, or Γ_1, constant. Conversely, at very high water activity the amount of H_2O present per nucleotide is very large. Then the addition of a small amount of nucleotide will require the compensating addition of a large amount of water to bring Γ_1 up to its former value.

The foregoing very simple derivation has neglected all sources of nonideality in the solution. These include the polyelectrolyte character of the solute, the ionic nature of the solution, and the specific short-range interactions. Hearst[22] has developed a theory which takes these factors into account and explicitly separates the electrostatic factor from the other sources of nonideality. He obtains the following equation for the preferential hydration Γ^* (moles H_2O/mole of macroion).

$$\Gamma^* = -\frac{n + m_2 \left(\frac{\partial \ln \gamma_{el}}{\partial m_2}\right)_{m_3}}{\ln a_w \left[1 + \frac{M_2}{\Phi}\left(\frac{d\Phi}{dm_2}\right)\right]} + a_w \sum_{n=0}^{\infty} \alpha_n a_w{}^n \tag{7-21}$$

The number of charges per macroion is n, and Φ is the osmotic coefficient of the salt solution[26]

$$\Phi = -\left(\frac{55.5}{\nu m_2}\right) \ln a_w \tag{7-22}$$

where ν is the number of particles into which the salt dissociates (e.g., $\nu = 2$ for NaCl). The activity coefficient of the polyelectrolyte has been separated into two parts: an electrostatic contribution γ_{el}, and a part due to short-range interactions, given by the infinite series in a_w on the right-hand side of Eq. (7-21).

The effective charge on the DNA is

$$n_{\text{eff}} = n + m_2 \left(\frac{\partial \ln \gamma_{el}}{\partial m_2}\right)_{m_3} \tag{7-23}$$

This can have a maximum value of n, while a perhaps more plausible value is[27] $n_{eff} = 0.4\,n$ (see Section II). Using a presumed value for n_{eff}, and known values of Φ and a_w at particular salt molalities m_2, Hearst[22] calculated the fraction of the preferential hydration which could be attributed to long-range electrostatic interactions. For example, in 1.5 M NaCl, $a_w = 0.95$, and the observed hydration is $\Gamma_1' = 1.20$ g/g. With $n_{eff} = n$, 75% of this is attributable to electrostatic effects; with $n_{eff} = 0.4\,n$, 30%.

Taking the latter value as most reasonable, this leaves 0.84 g H_2O/g DNA or 15.6 moles H_2O moles H_2O/mole nucleotide, to be accounted for by short-range interactions. This is an amount reasonably consistent with those determined by other means as discussed above.

The nature of the short-range interactions responsible for DNA hydration has been most clearly elucidated by the infrared spectroscopic studies of Falk et al.[5] The IR absorption at 1220–1240 cm^{-1} has been assigned to the asymmetric PO_2^- stretching mode.[28-30] This absorption band decreases in frequency as the relative humidity rises from 0 to 65%. At this point about six water molecules have been taken up per nucleotide by the DNA fibers, and most of these may therefore be assumed to occupy the hydration sphere of the $PO_2^- \cdot Na^+$ group. The P—O stretching frequency[31,32] at 962–970 cm^{-1} also shifts over the range of RH = 0–60%, while the deoxyribose bond C—O stretch[31,32] at 1066–1052 cm^{-1} shifts at RH about 50%. Thus it appears that the P—O—C and C—O—C oxygen atoms also become hydrated at RH below 65%. It may not be coincidental that the gravimetrically determined hydration[4] at this RH, 6.1 ± 0.5 moles/mole is very close to the value of 6.5 moles/mole found by Wang[11] from H_2O self-diffusion studies. In both cases, it would appear that only the most tightly bound water is being observed.

Above 65% RH, changes are noted in the 1550–1720 cm^{-1} region of the spectrum. In deuterated DNA, this region includes the C=O and ring stretching vibrational modes of the bases; in nondeuterated DNA, there are also contributions from NH and NH_2 bending motions.[28,29,31] Thus it appears that C=O groups and ring nitrogen atoms of the bases only become hydrated above 65% RH. It should be remembered that this is the humidity region in which significant changes in the X-ray diffraction pattern and dichroism of DNA films and fibers are taking place. These changes appear to be largely due to alterations in base orientations, which in turn depend on hydration of the bases.

Lewin[33] has pointed out that, whereas the NH···C=O interbase hydrogen bonds are located in both the wide and narrow grooves of the DNA double helix, the free thymine C_2=O groups, the H-bonded amino groups of guanine and deoxyribose ring oxygens are found only in the narrow groove. Thus it seems likely that the narrow groove is fully hydrated before the wide groove.

A wide variety of forces can be invoked to account for the hydration.[3] Dipole-dipole forces between the $M^+{-}PO_2^-$ group and the H_2O molecule are doubtless very important; and, particularly in denatured DNA, the base dipoles may also play a part. Hydrogen bonding of H_2O to ring nitrogens and to C=O and C—O—C oxygens in the base and sugar rings must also be considered. Hydrogen-bonded water bridges between proton-donating and -accepting groups on the DNA have been proposed.[33] As noted above, there is no evidence that extensively hydrogen-bonded icelike sheaths surround the periphery of the double helix; but the existence of at least one layer of partially ordered water surrounding the DNA seems consistent with much of the evidence cited above. The extent to which hydrophobic interactions[34] and surface tension forces[35,36] may be involved in structuring water around the DNA, particularly in the denatured state, is unclear. The influence of salt on these interactions, which could substantially influence the thermodynamic hydration (since exclusion of salt = "binding" of water) is particularly obscure. For more extensive consideration of these questions, the reader is referred to the reviews by Tunis and Hearst[3] and by Lewin.[33]

II POLYELECTROLYTE BEHAVIOR

The polyelectrolyte behavior of nucleic acids is of interest for several reasons. The ionic environment of the polynucleotide chain will in many cases influence its chemical behavior. Thus, thermodynamic and electrophoresis studies, which give information about ion binding, may help to define the environment in the neighborhood of the phosphate groups. Long-range electrostatic repulsions will expand the polymer beyond the average dimensions expected for an uncharged macromolecule; this effect will be the more prominent the lower the ionic strength. This expansion will in turn influence observable properties of the polymer in solution, such as sedimentation and viscosity. Studies of the response of polymer configuration to ionic strength can also give information about chain stiffness. Finally, the midpoint and breadth of the helix-coil transition in DNA is markedly dependent upon both ionic strength and specific ion effects. In the present chapter we shall examine ion binding and chain expansion effects. Ionic influence on the helix-coil transition is considered in Chapter 6. The reader is referred to several excellent reviews[37,38] for a more detailed account of some of the material discussed here.

We shall first consider the interaction of small ions with the polynucleotide macroion, neglecting polymer chain expansion effects. This is not a rigorous procedure, as can be seen from the following considerations. Neglecting specific chemical effects between small ions and macroion, the interaction of the small ions with the polynucleotide chain will depend on the electrostatic potential in the polyelectrolyte domain. Since coulombic interactions are quite long-ranged, decaying only as $1/r$, the potential experienced by a small ion in the polymer domain would be expected to depend significantly upon the positions of all the charges on the polymer chain (as well as upon the positions of all the other small ions). However, the expansion of the polyelectrolyte will be markedly affected by the interactions of the small ions with the charged polymer sites. The higher the degree of screening or neutralization of the negatively charged phosphate groups by positively charged counterions, the less the expansion. Thus, small ion–macroion interactions and polyelectrolyte chain expansion are interdependent phenomena. It is this fact which has made the development of polyelectrolyte theory so difficult.

However, substantial success in the interpretation of the thermodynamic properties of polyelectrolyte solutions has been achieved by neglecting the chain expansion problem, and treating the polyelectrolyte chain as if it were a rigid rod with a specified charge distribution.[39–41] This procedure is made physically plausible by remarking that ionic interactions in solution are screened by other ions, and hence of shorter range than would be predicted from Coulomb's law. Moreover, because close approach between negatively charged phosphates far from each other along the chain contour will be energetically unfavorable, it will occur relatively rarely, and a small ion will interact mainly with charged polymer sites near each other along the chain. These arguments are buttressed by the fact that thermodynamic quantities, such as ionic activity coefficients, which depend mainly on polyion–small ion interactions, are found to be independent of the polyelectrolyte molecular weight.[37] The further approximation, of treating the polyelectrolyte chain as a rigid rod, seems particularly appropriate for native DNA; and even single-stranded polynucleotides are probably rather stiff over several adjacent nucleotides. More precisely, the validity of this approximation depends on the average length of a local rodlike segment being larger than the radius of the ion atmosphere due to the small ions.[41]

Insofar as the small ion–macroion interaction aspects are concerned, the polyelectrolyte behavior of the nucleic acids is very similar to that of other polyelectrolytes. In all such systems, it is found that the osmotic second virial coefficient, as calculated below on the basis of Donnan equilibrium theory, is higher by one or two orders of magnitude than that observed experimentally. Equally interesting, the activity coefficient of the byion – the small ion of same charge as the polyion, thus the anion in the case of polynucleotides – is near unity; while the activity coefficient of the counterion is very low, often just a few tenths. Both of these observations may be explained by postulating that the counterions are strongly attracted to the polyelectrolyte domain, thereby reducing the polymer charge (and thus its second virial coefficient) and becoming less "available" themselves. The byions, on the other hand, are strongly repelled from the polyelectrolyte domain, and thus their thermodynamic activity is close to their concentration. In the following paragraphs we shall put this discussion on a more quantitative basis.

Consider an osmotic pressure experiment, in which a compartment containing a charged polynucleotide, water, and added salt is separated by a membrane from a compartment containing only water and salt. The membrane is permeable to the small molecules, but not to the polynucleotide. Donnan[42,43] pointed out that the requirements of electroneutrality would lead to an equilibrium condition in which there was an excess of counterions, and a deficiency of byions, on the polymer side of the membrane. We let a, m, and γ be the activity, molality, and activity coefficient, respectively; denote salt by 2, cation by +, and anion by −, and polymer by 3; and let primed quantities be those on the side of the membrane without the polymer, while unprimed quantities are those on the polymer side. Each polynucleotide molecule has Z (negative) charges. Then electroneutrality requires

$$Zm_3 + m_- = m_+, \tag{7-24}$$

$$m_+' = m_-', \tag{7-25}$$

while the equality of salt activities across the membrane demands

$$a_2 = a_2' \tag{7-26}$$

or

$$m_+ m_- \gamma_\pm^2 = m_+' m_-' \gamma_\pm'^2 \tag{7-27}$$

where $\gamma_\pm$ is the mean ionic activity coefficient. Solution of these simultaneous equations gives

$$\begin{aligned} (m_+') &= (\gamma_\pm/\gamma_\pm')[m_+(m_+ + Zm_3)]^{1/2} \\ (m_-') &= (\gamma_\pm/\gamma_\pm')[m_-(m_- - Zm_3)]^{1/2} \end{aligned} \tag{7-28}$$

from which it is seen that, with $Z<0$, $m_+' < m_+$ and $m_-' > m_-$, so that the positive counterion has concentrated on the polymer side of the membrane, and the negative byion has been depleted on that side.

This redistribution of ions can be conveniently expressed in terms of a "Donnan salt-exclusion factor" Γ_2:

$$\Gamma_2 = \lim_{Zm_3 \to 0} \left(\frac{m_2' - m_2}{Zm_3} \right) = - \left[\frac{\partial m_2}{\partial (Zm_3)} \right]_{\mu_2} . \tag{7-29}$$

Zm_3 is the equivalent molality of polyelectrolyte. The salt-exclusion factor

Γ_2 and the thermodynamic hydration Γ_1 are simply related, in solution infinitely dilute with respect to polymer, by[44]

$$\Gamma_1 = \left(\frac{55.5}{m_2}\right) \Gamma_2 \tag{7-30}$$

It is readily shown, by substituting Eq. (7-28) into (7-29) and expanding the square root, that the ideal value of Γ_2 (when the activity coefficients are equal) is 1/2. Measured values of Γ_2[45] are less than this ideal value by severalfold, particularly at low ionic strengths. This discrepancy arises because we have assumed the ionic activity coefficients identical on both sides of the membrane. Equivalently, one might say that the charge on the polymer is not Z, but $(1 - \alpha)Z$, where α represents the fraction of phosphate charge which is neutralized by counterion. Since the polymer is effectively less highly charged, it will exclude byions less strongly, and the exclusion factor Γ_2 will be decreased.

The redistribution of ions affects the osmotic pressure π. Equality of solvent chemical potential across the membrane requires $\mu_1 = \mu_1{}'$, or since in ideal solutions $\mu_1 = \mu_1{}^0 + RT \ln X_1$,

$$RT \ln X_1{}' = RT \ln X_1 + \overline{V}_1 \pi \tag{7-31}$$

where $\overline{V}_1$ is the partial molar volume of solvent. One has

$$X_1{}' = \frac{1000}{1000 + M_1(m_+{}' + m_-{}')}$$

$$X_1 = \frac{1000}{1000 + M_1(m_+ + m_- + m_p)} \tag{7-32}$$

Combination of these equations, and expansion in powers of polymer concentration $c_3 = m_3 M_3 M_1 / 1000\, \overline{V}_1$ g/ml, gives

$$\frac{\pi}{c_3} = RT\left[\frac{1}{M_3} + \frac{1000\, Z^2 v_1}{4\, m_2 M_3{}^2}\, c_3 + 0(c_3{}^2) + \cdots\right] \tag{7-33}$$

m_2 is the equilibrium salt concentration on the polymer side of the membrane, and v_1 is the solvent specific volume. The coefficient of c_3 in Eq. (7-33) is B, the second virial coefficient.

For polynucleotides, $|Z|/M_3 = 1/330$, so with $v_1 = 1$ g/ml and $m_2 =$ 0.1 M, B = 0.023 mole-cm^3/g^2. Even at this relatively high ionic strength, 0.1 M, B is much higher than the 3×10^{-4} mole-cm^3/g^2 observed[46] from light scattering by T7 DNA. Here again, the discrepancy may be attributed to the reduction of the polyion charge by a fraction α by tightly bound counterions. α will depend on electrostatic interactions between counterion and polyion, as well as more specific chemical interactions.

It is easiest to deal with the nonspecific electrostatic effects first. A limiting law theory has been constructed[41] which serves the same purpose for dilute polyelectrolyte solutions as does the Debye-Hückel theory for simple electrolyte solutions. This theory treats the polyelectrolyte chain as a rigid, infinite line charge with continuous charge density

$$\beta = Z_p{}^{e/b} \tag{7-34}$$

Z_p is the monomer valence (−1 for polynucleotides)
e is the electron charge
b is the charge spacing (1.7 Å for native DNA, corresponding to two charges every 3.4 Å)

Interactions between polyions are neglected, regardless of the concentration of added salt, and the dielectric constant ϵ is taken as that of pure, bulk solvent.

The crucial physical point of this theory arises from the realization that, for all values of charge density β above a critical value, the partition function (and hence all thermodynamic quantities) for an infinite line charge model diverges. This divergence arises because of the logarithmic dependence of electrostatic potential energy on distance away from the line charge. The critical value of β, for univalent charged groups and mobile ions, arises when

$$\xi = \frac{e^2}{\epsilon kTb} = 1 \tag{7-35}$$

Divergence occurs when $\xi \geq 1$.

Since such divergence is physically impossible, the theory requires that, if $\xi \geq 1$, sufficient counterions will "condense" on the polyion to lower the charge-density parameter ξ to unity. The remaining uncondensed mobile ions may then be treated in the Debye-Hückel approximation.

At 25°C in H_2O, $e^2/\epsilon kT = 7.135$ Å, so for native DNA, with $b = 1.7$ Å, $\xi = 4.19$. This is much greater than one, so a fraction $\alpha = 1 - 1/\xi = 0.76$ of the negatively charged phosphate groups will be neutralized by condensation of univalent counterions such as Na^+ or K^+.

The observed thermodynamic properties of the solution may then be calculated by employing a Debye-Hückel (screened Coulomb) interaction between the remaining uncondensed small ions and the line charge with an effective ξ of one. In this way one obtains

$$\Gamma_2 = (4\xi)^{-1}, \xi > 1 \tag{7-36}$$

which for native DNA equals 0.06. It should be remembered that this is a limiting theory, becoming valid only for dilute polyelectrolyte solutions as the salt concentration tends to zero. Experimental results[45] for the system KDNA–KBr are in fairly good agreement, running from $\Gamma_2 = 0.09$ at the lowest KBr concentration measured, 0.002 M, to $\Gamma_2 = 0.26$ at [KBr] = 0.234 M. Similar values were obtained for the Li, Na, and tetramethylammonium salts of DNA in solutions with added bromide salts.

Other thermodynamic parameters related to polyelectrolyte interactions with small ions are the osmotic coefficient ϕ (roughly, the fraction of ions that are osmotically active), the activity coefficients of counterion and byion, γ_+ and γ_-, and the mean ionic activity coefficient, $\gamma_\pm$. The limiting law theory[41] yields for these quantities, all for $\xi > 1$.

$$\phi = \frac{\left(\frac{1}{2}\xi^{-1}X + 2\right)}{(X+2)}$$

$$\gamma_+ = (\xi^{-1}X + 1)(X+1)^{-1} \exp\left[\frac{-\frac{1}{2}\xi^{-1}X}{(\xi^{-1}X+2)}\right]$$

$$\gamma_- = \exp\left[\frac{-\frac{1}{2}\xi^{-1}X}{(\xi^{-1}X+2)}\right]$$

$$\gamma_\pm^{\,2} = \gamma_+\gamma_- = (\xi^{-1}X + 1)(X+1)^{-1} \exp\left[\frac{-\xi^{-1}X}{(\xi^{-1}X+2)}\right] \tag{7-37}$$

where X is the ratio of equivalent concentrations of polyion to added salt, Zm_3/m_2. Note that, in the absence of added salt, $X = \infty$, so $\phi = (2\xi)^{-1} = \Gamma_2$, $\gamma_+ = \xi^{-1} = 1 - \alpha$, the fraction of free counterions, and $\gamma_- = 1$. In the presence of added salt, γ_+ cannot be identified simply with the fraction of unneutralized phosphate groups, because Debye-Hückel interactions modify the thermodynamic behavior of the ions.

In Fig. 7-4 is shown the Na^+ ion activity in DNA solutions of varying concentration with no added salt.[47] Activity coefficients were measured with a cation-exchange membrane electrode. It is observed that γ_{Na^+} is lower in solutions of native DNA than in solutions of denatured DNA, consistent with the higher density of negative phosphate charge in native DNA. As the polymer is diluted, γ_{Na^+} rises somewhat, but still remains low. Some of this rise may be due to inadvertant denaturation of native DNA at the lowest concentrations.

In addition to the general electrostatic interactions between polyion and counterion discussed above, more specific interactions can be envisioned. These may be divided into two classes: ion pairing and complex formation. Ion pairs were first proposed by Bjerrum[48] to account for deviations from Debye-Hückel theory in concentrated solutions of strong electrolytes. If there is no screening by intervening ions, the probability of finding oppositely charged ions, of charge z_+ and z_-, at a distance r from one another, is, from the

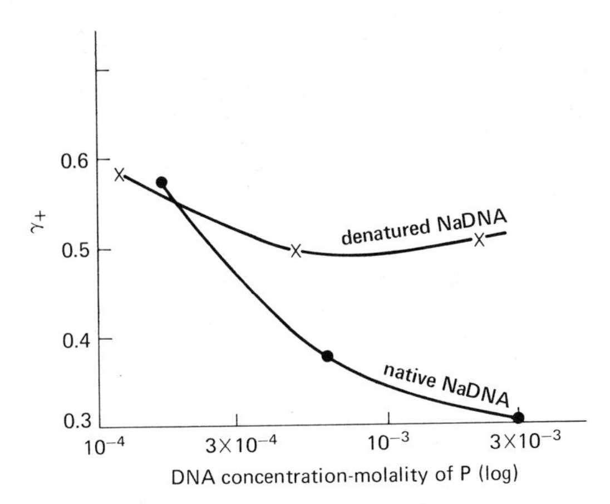

FIGURE 7-4 Concentration dependence of the counterion activity coefficient in solutions of native and denatured NaDNA, without added simple salt. [From J. W. Lyons and L. Kotin, *J. Amer. Chem. Soc.*, 87, 1670 (1965). Reprinted with permission.]

Boltzmann equation

$$P(r)dr = \frac{N_A c}{1000} [\exp(-z_+ z_- e^2 / \epsilon r k T)] \; 4 r^2 dr \tag{7-38}$$

This probability distribution has a maximum when $dP(r)/dr = 0$, or when the ions are within a distance $r_{\min}$ of one another:

$$r_{\min} = \frac{-z_+ z_- e^2}{2\epsilon kT} \tag{7-39}$$

For 1:1 electrolytes at 18°C, $r_{\min} = 3.52$ Å. Thus, ions at least this close to each other are to be regarded as paired. Evidently, the degree of association, α, is given by

$$\alpha = \int_a^{r_{\min}} P(r)dr \tag{7-40}$$

where a is the distance of closest approach. The ions are imagined to remain hydrated in the ion-pairing process, so α should correlate inversely with the hydrated ionic radii.

Covalent complex formation is a more specific type of interaction than ion pairing. Coordination complexes of substantial strength are formed when more than one ligand can bind to the metal ion. It is likely that chelation of this type occurs when Ag^+, Hg^{2+} or other transition metals interact with the heterocyclic bases in the nucleic acids. These binding reactions are considered in detail in Section V. Cations with filled electron shells interact mainly with the phosphorus atoms. Since coordination complexes involve displacement of

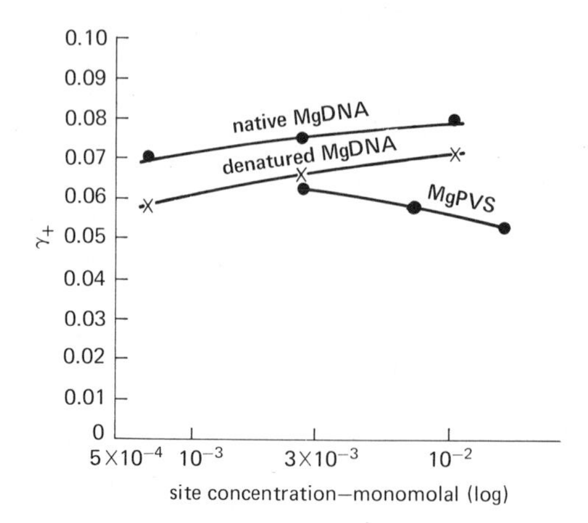

FIGURE 7-5 Concentration dependence of the counterion activity coefficient in solutions of MgDNA and Mg polyvinylsulfate, without added simple salt. [From J. W. Lyons and L. Kotin, *J. Amer. Chem. Soc.*, 87, 1670 (1965). Reprinted with permission.]

solvent by ligand from the coordination sphere of the ion, a rough correlation between α and unhydrated ionic radii of the cations might be expected.

Figure 7-5 shows the dependence of the activity coefficient of Mg^{2+} on the DNA concentration in solutions of MgDNA without added salt. Comparison with Fig. 7-4 indicates that $\gamma_{Mg^{2+}}$ is 3–10 times lower than γ_{Na^+}. Simple electrostatic theory would predict a lowering by a factor of 2, because of the +2 charge on Mg^{2+}. The additional lowering may reflect some specific interaction, whether ion pairing or complexation, with the phosphates. The relative strength of binding of Mg^{2+} to the double-helical and denatured forms of DNA is discussed in Section VI,A.

Electrophoresis, the study of the translational motion of a polyelectrolyte in an applied electric field, is one of the best ways to study ion binding phenomena in polynucleotides. The theory of electrophoresis is very difficult and by no means satisfactorily completed.[37] However, a simple theory due to Hermans[49] leads to qualitatively adequate, and surprising, results for flexible polyelectrolytes.

Hermans uses a simplified model of a flexible polymer as a porous sphere of uniform density.[50,51] The polymer contains N segments in a radius R, so the segment density is $\nu = N/(4\pi R^3/3)$. The translational frictional coefficient f of this sphere is

$$f = 6\pi\eta_0 R\psi(\sigma) \tag{7-41}$$

where $\psi(\sigma)$ is a function of the shielding depth σ, which is the distance below the polymer surface at which the fluid velocity has decayed to e^{-1} of its exterior value:

$$\sigma = (\eta_0 v_e / N\zeta)^{1/2} \tag{7-42}$$

v_e is the monomer volume. When σ is large, the polymer is free-draining, and Eq. (7-41) becomes $f = N\zeta$, where ζ is the monomer frictional coefficient. When σ is small, the polymer is nondraining, $\psi(\sigma) = 1$, and the polymer behaves like a Stokes law sphere of radius R. Clearly $R\psi(\sigma)$ corresponds to an effective hydrodynamic radius.

When the total charge on the polymer is Q, Hermans found that the electrophoretic mobility u – that is, the velocity produced by unit electric field – was given by

$$u = \frac{V}{300\zeta} = \frac{1}{300}\frac{Q}{N\zeta}\left[1 + \frac{1 + 2\eta\sigma}{3\eta^2\sigma^2(1 + \eta\sigma)}\right] \tag{7-43}$$

The factor of 300 is required when the charge is measured in esu and the potential, ϵ, in volts per centimeter. It is seen from Eq. (7-43) that the mobility depends on the ratio of the hydrodynamic shielding depth σ to the thickness of the ion atmosphere, $1/\eta$. If the polymer is very short (small N), $\sigma \gg 1$, and $u = (1/300)(Q/N\zeta)$. This is the free-draining limit, since Q/N is the charge per monomer, q; and the monomer mobility would be $(1/300)(q/\zeta)$.

For high-molecular-weight polymers, the mobility will depend on the ionic strength. At very low ionic strength the polyion will be highly expanded and will thus be free draining. At moderate or high ionic strength the thickness of the ion atmosphere will be much less than the size of the polymer domain, $R\eta \gg 1$; and the polymer will be hydrodynamically nondraining, $R/\sigma \gg 1$. In the limit of $\eta\sigma\ (= R\eta/(R/\sigma)) \gg 1$, Eq. (7-43) states that the polymer will again appear electrophoretically free-draining. At intermediate values of $\eta\sigma$, higher values of u will be found.

The prediction that high-molecular-weight polyelectrolytes will appear electrophoretically free draining in sufficiently high salt can be translated to mean that u will be independent of molecular weight, and depend only on linear charge density $q = Q/N$. This prediction has been strikingly borne out for DNA in 0.01 M NaCl.[52] Over a range of molecular weights from 0.26×10^6 to 130×10^6, native DNA has a mobility of $2.15 \pm 0.03 \times 10^{-4}$ cm^2/V; while denatured DNA has $u = 1.87 \pm 0.03 \times 10^{-4}$ cm^2/V. The apparent contradiction between hydrodynamic nondraining and electrophoretic free draining has been explained[53] by noting that at high ionic strength the net charge density vanishes at all points within the polymer domain. The net force acting on each point is therefore vanishingly small; and hydrodynamic interaction effects, which are proportional to the forces (Chapter 5, Section VI, A) will therefore vanish.

It is possible to use these results to calculate roughly the average charge per phosphate. In the free-draining limit, Eq. (7-43) is

$$u = \frac{(1-\alpha)en}{300\zeta} \tag{7-44}$$

where n is the number of phosphate sites per frictional element. The frictional element may be taken to be a sphere of diameter b, so $\zeta = 3\pi\eta_0 b$. In native DNA, $n = 2b/3.4$, since there are two phosphates every 3.4 Å along the helix axis. Substitution of these values into Eq. (7-44), with the experimental mobility, gives for the fraction of free charge, $1 - \alpha = 0.22$ in NaDNA. This value may be low, however, because the ion atmosphere will partly shield the macroion from the field, thus reducing the mobility.

Figure 7-6 shows the mobility of several DNA salts over a wide range of concentration.[54] The mobility decreases with increasing salt because of the increasing shielding of the polyion by the ion atmosphere. However, the fractional charge per phosphate, $1 - \alpha$, was found to be essentially the same for a given cation regardless of salt concentration. Values of $1 - \alpha$ are 0.33 for Li^+, 0.36 for Na^+, 0.43 for tetramethyl ammonium ion, and 0.51 for K^+. The effects of these ions in depressing the electrophoretic mobility of DNA

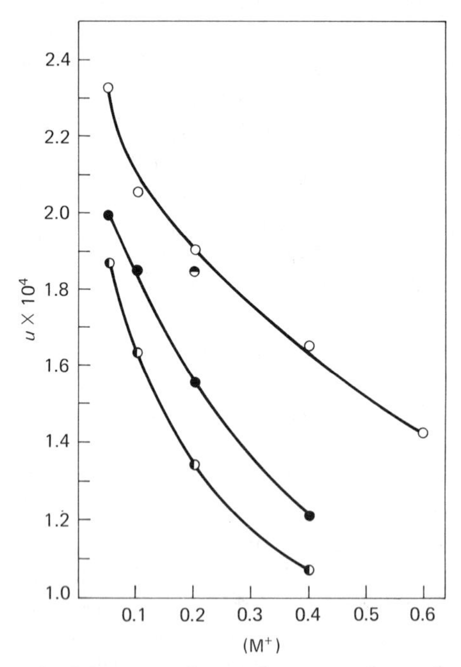

FIGURE 7-6 Effect of salt concentration on the average electrophoretic mobility of the DNA anion. ○, tetramethylammonium chloride; ○; KCl; •, NaCl; ○, LiCl. [From P. D. Ross and R. L. Scruggs, *Biopolymers,* **2,** 231 (1964). Reprinted with permission.]

are in the same order as they are for interaction with polyphosphate,[55] indicating that the cations are indeed interacting with the phosphates. The order of effectiveness of the ions in neutralizing phosphate charge also correlates with the crystal radii, rather than the hydrated radii, of the ions.[54] This suggests that some water of hydration is lost from the counterions when they interact closely with the phosphates.

The decrease in mobility of denatured DNA relative to native DNA[52,56] is also explicable in terms of Eq. (7-44). As noted in Fig. 7-4, γ_{Na^+} (which is approximately $1 - \alpha$ when no salt is added) is greater for denatured DNA than for native. However, the charge density, or n/ζ, is greatly reduced in the denatured form since there is only one phosphate every 7 Å rather than two every 3.4 Å. It is interesting to note that decreasing the pH lowers the mobility of the native and denatured forms equally.[52] Upon changing the pH from 7.50 to 4.10, the "titration charge" (−1 per phosphate + the number of protons bound) goes from −1.0 to −0.75 for native DNA, and from −1.0 to −0.52 for denatured DNA. The greater change in the denatured form is due to the fact that the cytosine and adenine bases are protonated more readily in the denatured state. However, the ratio of mobilities at pH 4.10 and 7.50 is 0.90 for both forms, because the true charge is substantially less negative than the titration charge due to ion binding, and ion binding is greater to the native form.

The second aspect of polyelectrolyte effects which is important in discussion of nucleic acid properties is chain expansion. This occurs under the influence of coulombic repulsions among the backbone charges. Ion binding, which reduces the effective charge per phosphate, will somewhat reduce the expansion; and the chain stiffness of native DNA will make it somewhat less susceptible to polyelectrolyte expansion than single-stranded polynucleotides, but substantial effects are noted in all cases.

The theory of polyelectrolyte chain expansion is, if anything, more difficult than that of excluded volume effects, since the perturbing forces are transmitted over a longer range. A number of theories have been devised, all of which have deficiencies as well as virtues in conception and efficacy. We shall briefly consider three of these here. The reader is referred to other sources for a more comprehensive and detailed account.[37,38,57]

The first theory is that of Katchalsky and Lifson.[58] It has the virtue of maintaining the connectivity of the chain and of recognizing the long-range nature of the electrostatic forces, but it severely oversimplifies the statistical problem. This theory starts from the quite general equation for the distribution function $W(L)$ for the end-to-end distance L

$$W(L)\,dL = W_0(L)\exp\left[\frac{-G_{el}(L)}{kT}\right]dL \tag{7-45}$$

in terms of the distribution function $W_0(L)$ for the unperturbed chain, which is taken to be the gaussian Eq. (5-6). $G_{el}(L)$ is the electrostatic free energy, and is a function of L. From $W(L)$ one may calculate the mean-square end-to-end distance

$$\langle L^2 \rangle = \frac{\int_0^\infty L^2 W(L)\, dL}{\int_0^\infty W(L)\, dL} \tag{7-46}$$

or the most probable end-to-end distance, L^*, for which $(dW/dL)_{L^*} = 0$. The electrostatic energy is approximated by the interaction between charges screened by a Debye-Hückel ion atmosphere, summed over all pairs of backbone charges whose distances are consistent with the end-to-end length L. $G_{el}(L)$ is further approximated by averaging these distances using the unperturbed distribution $W_0(L)$. The result, for high-molecular-weight chains in sufficient salt (a few thousands molar) that $\eta L^* > 1$, is that the chain expansion coefficient is

$$\alpha_L{}^2 \equiv \frac{\langle L^2 \rangle}{\langle L^2 \rangle_0} \approx 1 + \frac{9\sqrt{2}}{\sqrt{3}\ \eta^2 DkT} \left(\frac{Q}{\langle L^2 \rangle_0} \right)^2 \langle L^2 \rangle^{1/2} \tag{7-47}$$

η is the reciprocal thickness of the ion atmosphere, $(8\pi e^2 I/1000\, N_A \epsilon kT)^{1/2}$. But $Q\langle L^2 \rangle_0^{-1}$ is proportional to the charge per residue in the polymer, which will be constant at constant salt concentration and pH, so that if $\alpha_L{}^2 \gg 1$, which is easily the case for large polyelectrolytes, $\langle L^2 \rangle^{1/2}$ is proportional to $\langle L^2 \rangle_0$, or to N. Thus, the theory predicts that the end-to-end distance becomes a constant fraction of the chain length.

This conclusion greatly overestimates the chain expansion, at least in reasonable salt concentrations. The fault seems largely to lie in the incorrect averaging of G_{el}, particularly in using the unperturbed distribution function, for this will overweight configurations in which negatively charged groups are close to each other and have high repulsive potential energy. This in turn will yield too high a value for G_{el}, and thus too large an expansive force.

A second theory, that of Harris and Rice,[37,39] maintains chain connectivity but treats polyelectrolyte repulsions as a relatively short-ranged phenomenon. These workers consider that because of ion atmosphere screening, the major effect of the repulsions is to change the average angle θ between neighboring segments. By analogy with Eq. (5-14), one has

$$\frac{\langle L^2 \rangle}{\langle L^2 \rangle_0} = \frac{1 - \langle \cos\theta \rangle}{1 + \langle \cos\theta \rangle} \tag{7-48}$$

where θ is now the complement of the bond angle, $\langle \cos\theta \rangle$, which is zero for uncharged segments, is given by

$$\langle \cos\theta \rangle = \frac{\int_0^\pi \cos\theta \, e^{-Gel(\theta)/kT} \sin\theta \, d\theta}{\int_0^\pi e^{-Gel(\theta)/kT} \sin\theta \, d\theta} \tag{7-49}$$

If the charges are localized at the midpoints of the segments of length b, and interact only with charges on nearest-neighbor segments by a screened Coulomb potential,

$$G_{el}(\theta) = \frac{q^2 e^{-\eta b} \sin(\theta/2)}{\epsilon_e b \sin(\theta/2)} \tag{7-50}$$

where ϵ_e is an effective dielectric constant. With judicious choice of ϵ_e (taken to be 5.5, intermediate between the dielectric constant of water and that of organic material through which most of the electrical lines of force will flow), good agreement of calculated and experimental $\alpha_L{}^2$ is obtained for stiff polymers. However, Eq. (7-48) demands that, because of the restriction to nearest-neighbor interactions, $\langle L^2 \rangle$ varies simply as N, a slower variation that is observed experimentally.

Finally, we may consider the theory of Flory,[59] in which the connected character of the chain is neglected. The polymer is considered instead as a (microscopic) gel, within which the Donnan electroneutrality condition must apply. The swelling of the molecule may then be attributed to osmotic forces, owing to an excess of small ions in the polymer domain, instead of explicitly to charge repulsions. Flory considers in addition the nonelectrostatic swelling characteristic of a polymer in a good solvent. These swelling forces are opposed by retractive elastic forces due to the entropy decrease of the elongated polymer chain, and the equilibrium degree of expansion is obtained when the free energy is a minimum. The result is

$$\alpha_L{}^5 - \alpha_L{}^3 = 2c_m \psi_1 (1 - \theta/T) M^{1/2} + \frac{2c_I q^2 M^{1/2}}{I} + \cdots \tag{7-51}$$

where

$$C_I = \frac{27{,}000}{16\sqrt{2}\, \pi^{3/2} (\langle L^2 \rangle_0 / M)^{3/2} N_A M_0{}^2} \tag{7-52}$$

M_0 is the segment molecular weight. The first term in Eq. (7-51) is that which will occur in the absence of polyelectrolyte effects.

At large polyelectrolyte expansions, Eq. (7-51) predicts a limiting dependence of $\langle L^2 \rangle^{1/2}$ on $M^{0.8}$, which is in better agreement with experiment than either the $M^{1.0}$ or the $M^{0.5}$ dependencies of the theories discussed above. It also predicts a dependence of $\langle L^2 \rangle^{1/2}$ on $I^{0.2}$, compared with the I^{-1} dependence of the Katchalsky-Lifson[58] theory. This, too, appears to be in better agreement with experiment. The major drawback of the Flory theory would appear to be its smearing out of the polyion charge distribution. This leads to a lower charge density than actually exists in the neighborhood of polymer sites adjacent to one another along the backbone. The extent of ion pairing will thus be underestimated, and the charge q per site overestimated. This deficiency can be remedied by introduction of an association parameter α, but only at the cost of introducing an additional variable.

The above discussion has concentrated on long-range polyelectrolyte expansion effects. It should be pointed out that interactions between neighboring charges along the chain can cause redistribution of populations of bond rotational isomeric states, and thus lead to changes in the segment length b. This effect has already been noted in our discussion of b of single-stranded polynucleotides as determined from sedimentation and viscosity measurements in solvents of varying ionic strength (Chapter 5).

Polyelectrolyte expansion manifests itself most commonly, in studies of nucleic acids, in changes of hydrodynamic properties. (See also the light scattering study on ϕX 174 DNA, Fig. 5-8). Figure 7-7 shows the concentration dependence of the reciprocal sedimentation coefficient of native DNA from *Proteus mirabilis* at various ionic strengths.[60] It is observed that as I decreases, $S^0_{20,w}$ decreases, save at the highest ionic strengths, where it remains

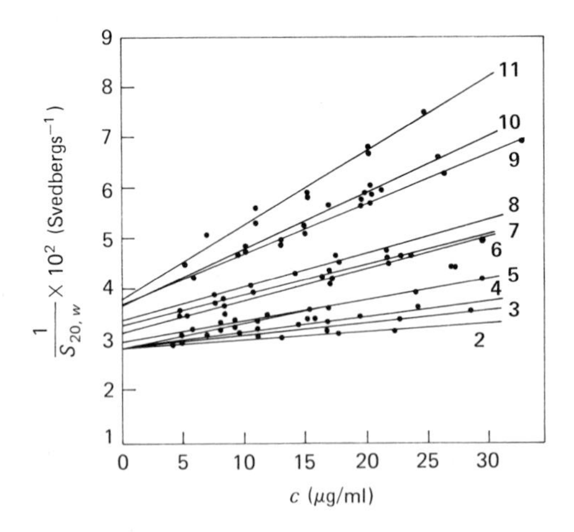

FIGURE 7-7 Dependence of the reciprocal sedimentation coefficient on the DNA concentration at various salt concentrations. The Na^+ concentration for the numbered curves are 1, 2.0 M; 2, 0.4 M; 3, 0.2 M; 4, 0.05 M; 5, 0.02 M; 6, 0.012 M; 7, 0.008 M; 8, 0.004 M; 9, 0.002 M; 10, 0.0015 M; 11, 0.0008 M. [From H. Treibel and K. E. Reinert, *Studia Biophys.*, **10**, 57 (1968).]

constant. The decrease in $S^0_{20,w}$ with I is due to an increase in frictional coefficient as the molecule expands. Figure 7-7 also shows that the concentration dependence of s increases with decreasing I. This again reflects the expanding hydrodynamic volume of the chain as I decreases.

Figure 7-8 shows the increase of the intrinsic viscosity of T4 DNA as the salt concentration drops below 0.6 M.[61] At ionic strengths above 0.6 M, $[\eta]$ levels off at a value which is independent of the nature of the monovalent cation.

The ionic strength dependence of the intrinsic viscosity can be predicted from the Flory-Fox intrinsic viscosity equation

$$[\eta] = \Phi \frac{\langle L^2 \rangle^{3/2}}{M} \tag{7-53}$$

and Eq. (7-51), which, neglecting nonpolyelectrolyte expansion effects and assuming $\alpha_L{}^5 \gg \alpha_L{}^3$, can be rearranged to give

$$\langle L^2 \rangle = \alpha_L{}^2 \langle L^2 \rangle_0 = (2C_I q^2 M^{1/2} \langle L^2 \rangle_0)^{2/5} I^{2/5} \tag{7-54}$$

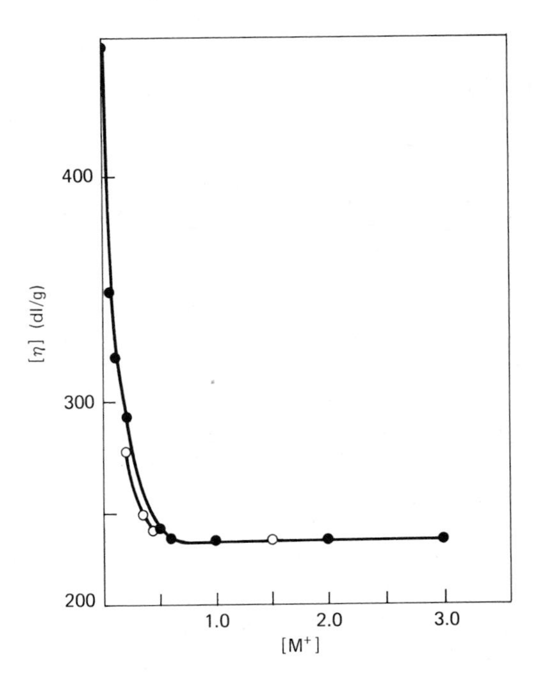

FIGURE 7-8 Intrinsic viscosity of T4 bacteriophage DNA as a function of the molarity of added NaCl, Kcl, or LiCl at 25°C (•) and 61°C (○) [From P. D. Ross and R. L. Scruggs, *Biopolymers*, 6, 1005 (1968). Reprinted with permission.]

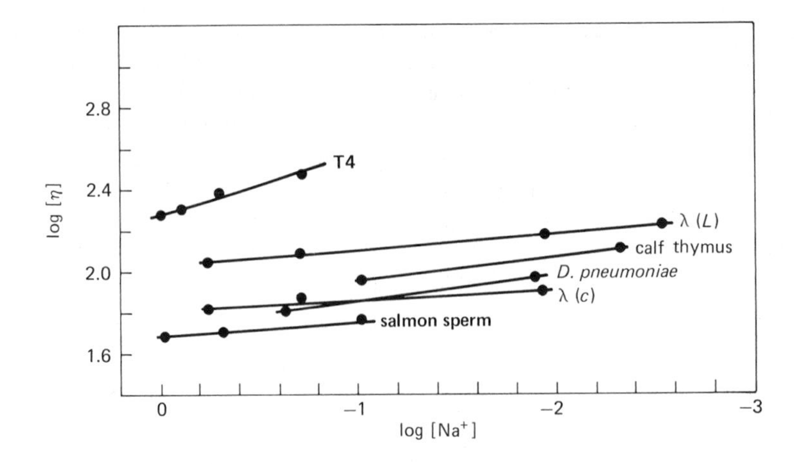

FIGURE 7-9 Log-log plot of intrinsic viscosity as a function of sodium ion molarity for linear, λ(L), and circular, λ(C), DNA and for other linear DNAs. [From R. J. Douthart and V. A. Bloomfield, *Biopolymers,* **6**, 1297 (1968). Reprinted with permission.]

Thus, $[\eta]$ is predicted to vary as $I^{-0.6}$ over the range in which the polynucleotide dimensions are influenced by ionic strength. This reasoning is due to Cox,[62] and is a special case of the relation

$$\frac{[\eta]_I}{[\eta]_{I'}} = \left(\frac{I}{I'}\right)^{-m} \tag{7-55}$$

with $m = 0.6$. Chain stiffness or free-draining effects might lead to a slightly lower value of m.[62] In fact, the variation of $[\eta]$ with ionic strength is substantially less than this for native DNA, as shown in Fig. 7-9.[63] Here m varies from 0.06 to 0.28. The relatively slight dependence of $[\eta]$ on I may be attributed to the great stiffness of the double helix. Figure 7-10 shows a similar plot for RNA, mainly ribosomal, isolated from *E. coli.*[62] Here also there are substantial deviations from the simple behavior of Eq. (7-55), due this time, undoubtedly, to base stacking interactions or local double helix formation at the higher ionic strength. Only below about 10^{-2} M salt are the electrostatic repulsions strong enough to overcome these, and the molecule expands very rapidly between $I = 10^{-2}$ and $I = 10^{-3}$. Similar observations on $[\eta]$ and $S^0_{20,w}$ for native and denatured T7 DNA have been made by Rosenberg and Studier.[64]

We have previously expressed deviations from gaussian random walk statistics by the parameter ϵ:

$$\langle L^2 \rangle = b^2 N^{1+\epsilon} \tag{7-56}$$

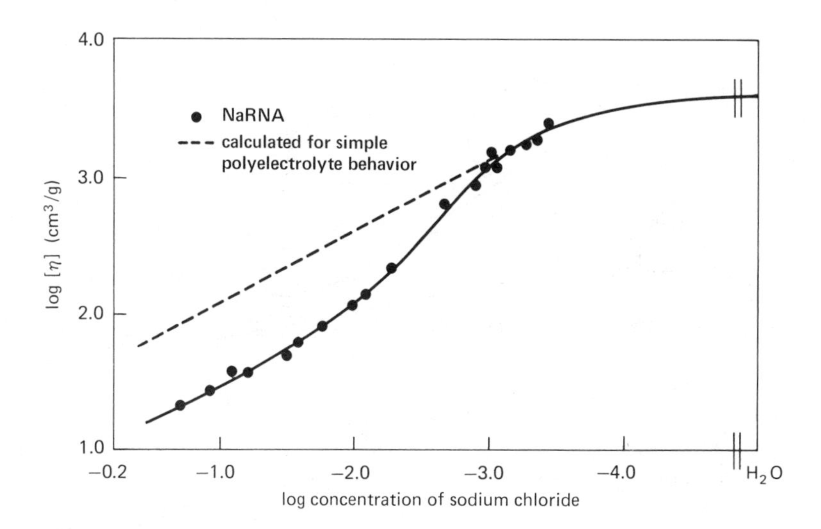

FIGURE 7-10 The dependence of the intrinsic viscosity of RNA on the concentration of added NaCl at 25°C. (•), NaRNA in neutral solution; (−−−), values calculated if Eq. (7-55) were followed with $m = 0.6$. [From R. A. Cox, *J. Polymer Sci.*, **47**, 441 (1960). Reprinted with permission.]

Clearly, ϵ should increase with decreasing ionic strength, and be reflected in the viscosity-molecular weight and sedimentation-molecular weight exponents a_η and a_s. The relations are

$$\epsilon = \frac{1}{3}(2a_\eta - 1) = 1 - 2a_s \tag{7-57}$$

In Fig. 7-11 we see plots of ϵ vs. log $[Na^+]$ determined from sedimentation[60] and intrinsic viscosity[63] studies of native NaDNA of a variety of molecular weights. The equation for ϵ vs. $[Na^+]$ determined from viscosity, which is probably more reliable due to the absence of complicating electrophoretic and relaxation effects,[65] is[61,63]

$$\epsilon = 0.05 - 0.063 \log[Na^+] \tag{7-58}$$

The flow dichroic properties of polyelectrolytes change markedly with ionic strength. Figure 7-12 shows the dichroism of T4 DNA as a function of shear rate, at various Na ion concentrations.[66] Not surprisingly, both the magnitude of the dichroism, and its rate of increase with shear, increase as the ionic strength is lowered. This is attributable to the greater expansion of the DNA coil, and thus its greater orientability in shear.

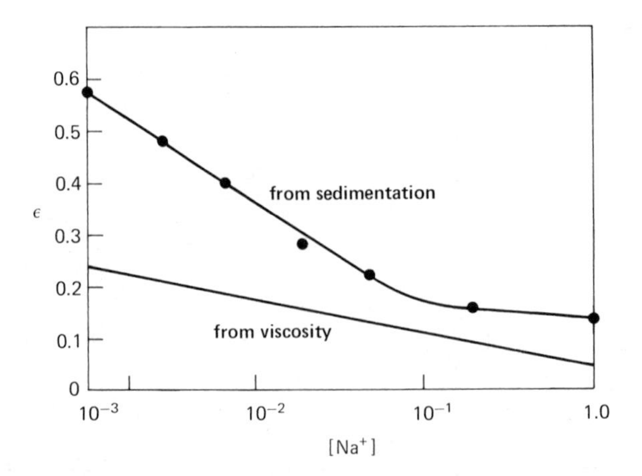

FIGURE 7-11 Expansion coefficient ϵ as a function of $[Na^+]$ for native NaDNA. Upper curve, from sedimentation data of H. Treibel and K. E. Reinert [*Studia Biophys.*, **10**, 57 (1968)]; lower curve, from viscosity data of R. J. Douthart and V. A. Bloomfield [*Biopolymers*, **6**, 1297 (1968) Reprinted with permission.]

III THERMODYNAMICS OF BINDING REACTIONS

A number of substances form reversible complexes with nucleic acids; thorough understanding of such an interaction requires characterization of the binding equilibrium. This problem can be approached on several levels. In the simplest case one assumes that the polymer contains a certain number of binding sites which are independent of each other, and the methods of equilibrium thermodynamics are applied by assuming a binding constant for each kind of site. Relatively simple analysis leads to expressions for the degree of binding as a function of the concentrations of the various species and the binding constants as demonstrated in this section.

If the binding sites are not independent, statistical methods must be used to take account of their interaction. The usual approach is to take a model in which molecules bound within a certain distance of each other have a given free energy of interaction. Analysis by the techniques of statistical mechanics leads to expressions for the degree of binding in terms of concentrations, the binding constants, and free energies of interaction. Comparison with experiment permits evaluation of these latter.

In some cases the concept of binding at a definite site may not be adequate to account for the interaction. The best example is the association of counterions with nucleic acids, or with any polyelectrolyte. Problems in this category were discussed in Section II.

One of the most common observations in studying binding equilibria of small molecules with DNA is that the polymer can accept only a restricted,

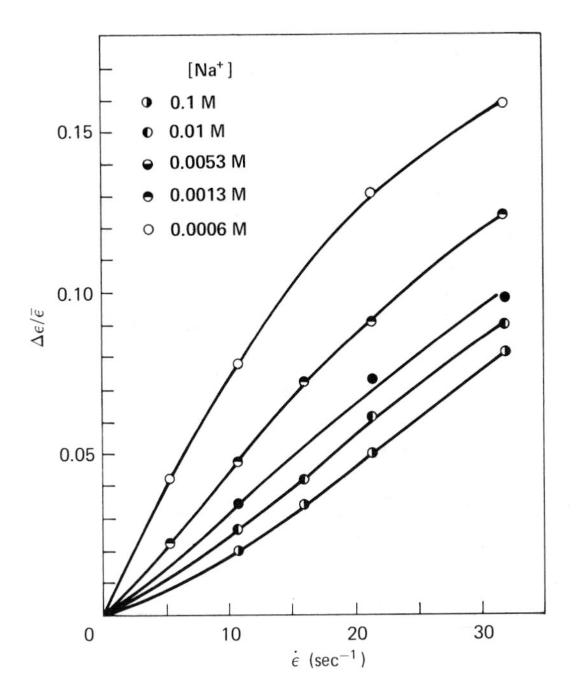

FIGURE 7-12 Flow dichroism of T4 DNA as a function of shear gradient at various ionic strengths. [From C. S. Lee and N. Davidson, *Biopolymers,* **6**, 531 (1968). Reprinted with permission.]

and definite, number of bound molecules. For example, there is evidence from several systems that the intercalation mode of binding saturates at one dye for every two base pairs. There are two simple ways to explain this. On one hand, one can assume that some base pairs can serve as binding sites while others cannot, so that the number of binding sites available is some fraction B_{ap} of the total number of base pairs. This is the assumption inherent in classical Scatchard[67] binding isotherms, and B_{ap} can be determined from these plots, as described below. Alternatively, one can adopt what we call the "neighbor exclusion" model, in which all base pairs are potential binding sites for intercalation, for example, but insertion at one site prevents binding to the immediately adjacent sites. The result would be to restrict the saturation binding to half the total number of base pairs. An advantage of the neighbor exclusion model is that it does not require assumption of an arbitrary and unknown difference between the base pairs in order to explain the restricted binding. The disadvantage of this approach is that more elaborate calculations are necessary to compute binding equilibria. The Scatchard equation, in contrast, is mathematically simple and provides a very convenient means for plotting data, which accounts for its popularity.

In general, the methods of statistical mechanics are needed to compute predicted binding equilibria for the neighbor exclusion model. However, there is one limiting condition under which simple statements can be made about this model: when the degree of binding approaches zero, restriction of the binding sites by neighbor exclusion vanishes, and the number of sites available to a small molecule is just the number of potential sites. Equations in this section for the neighbor exclusion model will therefore be given in terms of the limiting behavior at zero binding.

Detection of a complex between a small molecule and DNA often relies on finding some physical property of the complex that differs from the sum of the properties of the separate species. Alternatively, one can detect physical association by dialysis equilibrium, by solubility tests, or by membrane filter techniques. These latter three methods lead directly to the concentrations of free and bound species, but measurement of the differential physical properties requires an extra step of analysis. The most common example is a system in which a small molecule has different optical properties when bound to DNA than when free in solution. In order to calculate the degree of binding from observed optical properties, one must know the optical characteristics of the bound form, and this can only be obtained from the observed properties by extrapolating to high concentrations where all the added small molecules are bound.

To be more specific, consider a small molecule like a dye or a drug that has a visible absorption band, with a color change when bound to DNA. Absorbance in the visible range is contributed only by the dye. We define the following quantities, using the Scatchard model of a restricted number of binding sites.

$C_N{}^0$ is total concentration of nucleic acid base pairs
B_{ap} is apparent number of binding sites per base pair (from Scatchard plots; see below)
C_B is concentration of bound dye
C_F is concentration of free dye in solution (7-59)
C_T is total concentration of added dye $= C_F + C_B$
r is ratio of bound dye to base pairs $= C_B/C_N{}^0$
ϵ_F is the molar extinction coefficient of free dye
ϵ_B is the molar extinction coefficient of bound dye
$\Delta\epsilon$ is $\epsilon_B - \epsilon_F$

The absorbance A of a solution of dye in the presence of nucleic acid is given by

$$A = \epsilon_F C_F + \epsilon_B C_B \qquad (7\text{-}60)$$

The difference ΔA between A and the absorbance of a solution of pure dye at the same total concentration is

$$\Delta A = A - \epsilon_F C_T \qquad (7\text{-}61)$$

which can readily be rearranged to

$$C_B = \frac{\Delta A}{\Delta\epsilon} \tag{7-62}$$

from which C_B can be calculated if $\Delta\epsilon$ is known.

To see how $\Delta\epsilon$ can be determined, consider the equilibrium between free and bound dye. The binding constant is K_{ap}

$$K_{ap} = \frac{\text{(complex)}}{\text{(free dye)} \cdot \text{(binding sites)}} \tag{7-63}$$

Assuming unit activity coefficients, this is

$$K_{ap} = \frac{C_B}{C_F(B_{ap}C_N{}^0 - C_B)} \tag{7-64}$$

Experimental conditions can usually be arranged so that the nucleic acid is in large excess over the dye, in which case one can set

$$B_{ap}C_N{}^0 - C_B \approx B_{ap}C_N{}^0 \tag{7-65}$$

Inserting (7-62), (7-65), and the definition of C_T into (7-64) and rearranging yields

$$\frac{C_T C_N{}^0}{\Delta A} = \frac{1}{K_{ap}B_{ap}\Delta\epsilon} + \frac{C_N{}^0}{\Delta\epsilon} \tag{7-66}$$

Expression (7-66) is usually referred to as the Benesi-Hildebrand[68] equation. The left-hand side is an experimentally accessible quantity, since it involves only the total concentrations of dye and nucleic acid and the absorbance change. This can be plotted against $C_N{}^0$, and Eq. (7-66) predicts a linear variation with slope $1/\Delta\epsilon$ and intercept $(K_{ap}B_{ap}\Delta\epsilon)^{-1}$. In this way $\Delta\epsilon$ and the product $B_{ap}K_{ap}$ can be determined from the region of the binding equilibrium in which r is small.

Equation (7-66) suffers from the disadvantage that the extrapolation to infinite DNA concentration is not clear cut, since the equation is linear in $C_N{}^0$. For the purpose of determining $\Delta\epsilon$, a reciprocal plot is sometimes a better choice. Let ϵ_{ap} be the apparent extinction coefficient of the dye in the presence of nucleic acid,

$$\epsilon_{ap} = \frac{A}{C_T} \tag{7-67}$$

Combination of (7-60) and (7-67) gives

$$C_B = C_T \frac{(\epsilon_{ap} - \epsilon_F)}{\Delta\epsilon} \tag{7-68}$$

Insertion of (7-68) and (7-65) in (7-64) yields[69]

$$\frac{1}{(\epsilon_{ap} - \epsilon_F)} = \frac{1}{\Delta\epsilon \, C_N{}^0 B_{ap} K_{ap}} + \frac{1}{\Delta\epsilon} \tag{7-69}$$

The left-hand side, which is an experimentally measurable quantity, should be plotted against $1/C_N{}^0$. The intercept is $1/\Delta\epsilon$, extrapolated to complete binding at infinite nucleic acid concentration. This procedure is, in principle, more accurate than the more common practice of simply measuring ϵ_{ap} at high DNA concentration and equating to ϵ_B, assuming that all the dye is bound. Care should be taken in its application, however, since ϵ_B sometimes depends on r, or on the concentration of complex. In particular, values of K_{ap} determined from (7-69) may be unreliable as a consequence.

Equations (7-66) and (7-69) are adequate representations of binding data only when r is small, so that approximation (7-65) is valid. For larger r, one can substitute r for $C_B/C_N{}^0$ and rearrange to obtain

$$\frac{r}{C_F} = K_{ap}(B_{ap} - r) \tag{7-70}$$

If r/C_F is shown as a function of r, the result should be a straight line with slope $-K_{ap}$ and intercept B_{ap} on the r axis; this is commonly called a Scatchard[67] plot. Its utility is that it gives both K_{ap} and B_{ap} directly.

Experimentally determined Scatchard plots for interaction of DNA with small molecules often deviate from linearity, especially for large r. There are several circumstances that can give rise to this, and it is not easy to distinguish them in practice. It is possible, for example, that the polymer contains more than a single class of binding sites. If there are two such sites, two equations analogous to (7-70) can be written

$$\frac{r_1}{C_F} = K_1(B_1 - r_1); \qquad \frac{r_2}{C_F} = K_2(B_2 - r_2) \tag{7-71}$$

Many experimental methods determine just the total binding, or r_T,

$$r_T = r_1 + r_2 \tag{7-72}$$

Addition of Eqs. (7-71) and use of (7-72) gives for the measured Scatchard plot

$$\frac{r_T}{C_F} = K_1 B_1 + K_2 B_2 - K_1 r_T + (K_1 - K_2) r_2 \tag{7-73}$$

By elimination of C_F from Eq. (7-71) one can show that r_2 is a quadratic function of r_T. Therefore (7-73) is not, in general, linear in r_T, although it may be nearly so in some regions. The qualitative behavior expected is sketched in Fig. 7-13. If the binding sites have quite different affinities, and if the weaker site is present in larger numbers, then the slope and intercept extrapolated from the initial portion of the curve give K_1 and B_1, and those

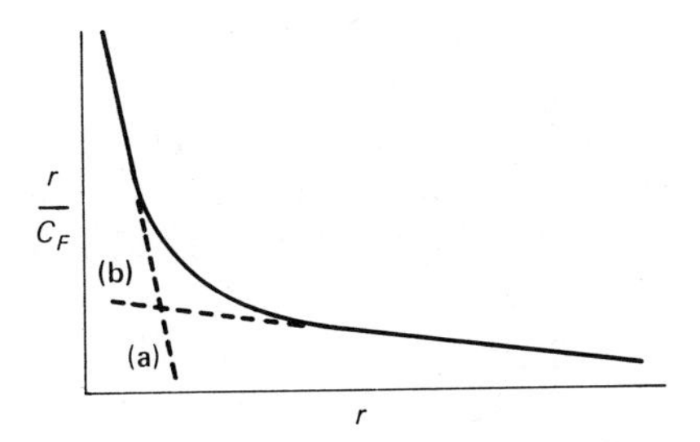

FIGURE 7-13 Scatchard plot for two types of binding sites. Line (a) slope $-K_1$, intercept B_1. Line (b) slope $-K_2$, intercept $B_1 + B_2$.

determined from the final portion of the curve give K_2 and $B_1 + B_2$. Curve fitting from the exact equations may also be used to calculate these quantities.

Unfortunately, there are other circumstances that can give rise to Scatchard plots that are very similar in appearance to Fig. 7-13. One source occurs when the binding sites are identical, but there is interaction among them. In this case Eq. (7-64) is no longer valid, and one is obliged to use the methods of statistical mechanics to calculate the binding equilibrium. We sketch the results qualitatively here.

The binding constant for a dye molecule at a particular site can be influenced by the presence of a bound molecule at an adjacent site. If the affinity is increased by the neighboring molecule, the binding is said to be cooperative, which means that the binding process is facilitated by neighbor interactions. When the affinity at an adjoining site is decreased by the presence of bound molecule, the binding is anticooperative, or neighbor excluded. The effect of these two cases on Scatchard plots is illustrated in Fig. 7-14. Curvature in the opposite sense arises in the two instances.

It is important to note that the Scatchard plot for neighbor exclusion

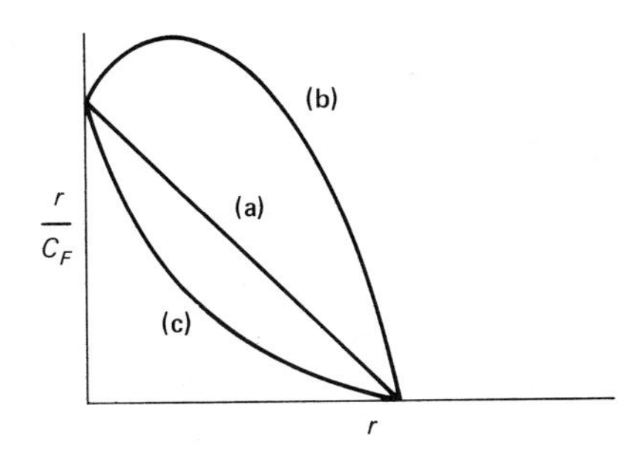

FIGURE 7-14 (a) Linear Scatchard plot, slope $-K$, intercept B. (b) Cooperative binding. (c) Anticooperative binding.

(line c) is very similar in appearance to Fig. 7-13 for two distinct binding sites. For this reason, the presence of two classes of binding sites should not be automatically inferred from a Scatchard plot of this shape. Other evidence must be adduced for this conclusion, such as, for example, spectroscopic indication that the nature of the complex changes as r increases.

Quantitative information on the neighbor exclusion model is readily extracted from a Scatchard plot of experimental binding data. Let B_0 be the number of *potential* binding sites per base pair, that is, the number of sites available to a small molecule when $r = 0$. We define a binding constant $K(r)$ in terms of the unoccupied potential sites, whose concentration is $B_0 C_N{}^0 - C_B$,

$$K(r) = \frac{C_B}{(B_0 C_N{}^0 - C_B) C_F} = \frac{r}{(B_0 - r) C_F} \tag{7-74}$$

Furthermore, let $K(0)$ be the limiting value of $K(r)$ when r approaches zero,

$$K(0) = \lim_{r \to 0} \left(\frac{r}{B_0 C_F} \right) \tag{7-75}$$

where $K(0)$ is now the intrinsic binding constant for an isolated potential binding site. $K(r)$ and K_{ap} are related by

$$\frac{K(r)}{K_{ap}} = \frac{B_{ap} - r}{B_0 - r} \tag{7-76}$$

In a Scatchard plot, r/C_F is plotted against r; the limiting value at $r = 0$ is given by

$$\lim_{r \to 0} \left(\frac{r}{C_F} \right) = K(0) B_0 \tag{7-77}$$

for the neighbor exclusion model, and

$$\lim_{r \to 0} \left(\frac{r}{C_F} \right) = K_{ap} B_{ap} \tag{7-78}$$

for the Scatchard model. Clearly,

$$\frac{K(0)}{K_{ap}} = \frac{B_{ap}}{B_0} \tag{7-79}$$

The main uncertainty in the neighbor exclusion model is that one must assume a value for B_0, the number of potential binding sites, which is possible only if the nature of the complex is thoroughly understood. If B_0 is known, Eq. (7-77) permits calculation of $K(0)$.

Standard thermodynamic equations permit evaluation of the heat of a reaction from the temperature variation of the equilibrium constant. Serious questions arise concerning this approach when binding equilibria are studied, because of uncertainty about what actually restricts the number of binding sites. If K_{ap} is determined from a Scatchard plot, but the binding is actually limited by neighbor exclusion, then K_{ap} has no real physical significance. In particular, its value would be influenced by the strength of the exclusion interaction, which may vary with temperature. Since both the Scatchard and neighbor exclusion models give simple results in the limit when r approaches zero, heats can best be determined from the limiting value of r/C_F when $r = 0$. According to Eqs. (7-77) and (7-78), this quantity is proportional to a binding constant multiplied by the number of binding sites. In neither case does it seem dangerous to assume that the number of binding sites is independent of temperature, so that temperature variation of r/C_F results from variation of the binding constant. Thus,

$$\frac{\partial \ln}{\partial T}\left\{\lim_{r\to 0}\left(\frac{r}{C_F}\right)\right\} = \frac{\Delta H^0}{RT^2} \tag{7-80}$$

where ΔH^0 is the heat of binding to an isolated binding site.

These considerations are complicated when there is more than one kind of binding site, in which case (7-76) is modified to read

$$\lim_{r\to 0}\left(\frac{r_T}{C_F}\right) = \sum_i B_i K_i \tag{7-81}$$

and Eq. (7-80) for the heat of binding is no longer valid.

It is not appropriate to use a Scatchard plot for a highly cooperative binding reaction; the best choice is a plot of r against the logarithm of C_F. In some cases it is possible to calculate directly from the slope of such plots the parameters that characterize the degree of cooperativity of the reaction; Section IV should be consulted for details. The most familiar example of a simple reaction plotted in this manner is a simple acid-base titration, in which the percentage neutralization is determined as a function of pH. Such a reaction covers roughly two pH units, or two orders of magnitude change in the H^+ concentration. Cooperative binding reactions cover a smaller range of concentration change; the larger the slope on a plot of r vs. log C_F, the more cooperative is the reaction (see Eq. 7-95).

Finally, we consider a third source of curved Scatchard plots of the kind shown in Figs. 7-13 and 7-14. The analysis developed so far assumes that in the solutions of the dye the activity can be set equal to the total concentration C_F free in solution. In some cases, such as with dyes that aggregate severely, this is not a good assumption. If it is made erroneously, serious mistakes can result. As an example, let the monomeric dye M dimerize to M_2, with equilibrium constant K_D

$$M + M \rightleftharpoons M_2$$

$$K_D = \frac{C_{M2}}{(C_M)^2} \tag{7-82}$$

The total free concentration C_F is the sum of monomeric and dimeric species:

$$C_F = 2C_{M_2} + C_M \tag{7-83}$$

Assuming that only the monomeric species binds, the Scatchard equation still applies, but C_F is replaced by the concentration C_M of the species that reacts, or, from Eq. (7-70),

$$\frac{r}{C_M} = K_{ap}(B_{ap} - r) \tag{7-84}$$

The quantity previously plotted was r/C_F, which, with the help of (7-82) can be expressed as

$$\frac{r}{C_F} = \frac{\dfrac{r}{C_M}}{1 + 2C_M K_D} \tag{7-85}$$

The initial slope of the "binding isotherm" obtained by plotting r/C_F is

$$\lim_{r \to 0} \frac{\partial [r/C_F]}{\partial r} = -(K_{ap} + 2K_D) \tag{7-86}$$

instead of $-K_{ap}$ as in the simple case of Eq. (7-70). Extrapolating this initial

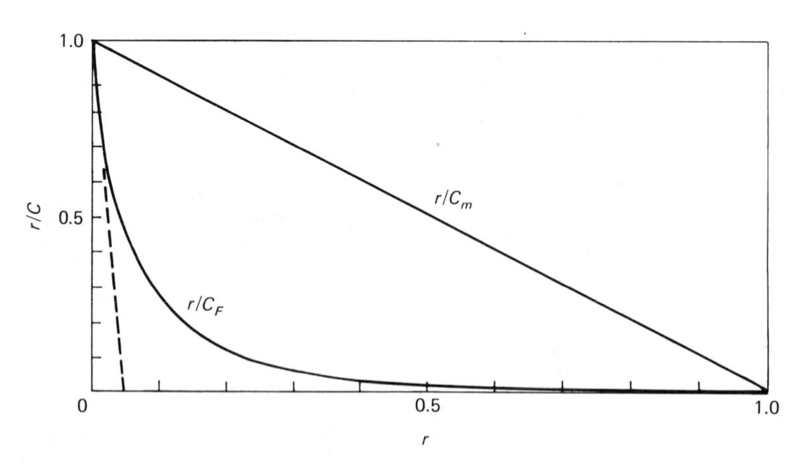

FIGURE 7-15 Hypothetical binding isotherm plotted as r/C or r/C_m, with $K_{ap} = 10^4\ M^{-1}$ and $K_D = 10^5\ M^{-1}$.

slope to the r axis gives what would be called the apparent number of binding sites B_{ap}', which is

$$B_{ap}' = \frac{B_{ap}}{1 + 2\dfrac{K_D}{K_{ap}}} \tag{7-87}$$

If K_D is much larger than K_{ap}, a binding isotherm in which C_F is incorrectly used instead of C_M will produce a much smaller estimate B_{ap}' of the number of binding sites than the actual value B_{ap}. An example of a hypothetical case is shown in Fig. 7-15, in which $K_D = 10^5\ M^{-1}$ and $K_{ap} = 10^4\ M^{-1}$. When r/C_F is plotted, a curved binding isotherm results, in contrast to the linear variation of r/C_M with r; in this case, use of C_F instead of C_M grossly misrepresents the actual binding equilibrium.

IV STATISTICAL MECHANICS OF BINDING REACTIONS

When molecules bound to a polymer interact with each other, the chemical equilibrium involved can become rather complicated. Such problems require more sophisticated mathematical analysis than discussed in the previous section. Interactions among bound molecules can take many forms, and it is always necessary to idealize these to some extent in order to make a model suitable for treatment by statistical mechanics. In general one assumes that the interactions are limited to closely adjoining neighbors, and tries to put in the features that seem physically significant.

We will consider first the class of problems in which the binding sites are all the same. Solutions to most such problems can readily be written in analytical form. A more difficult situation arises when different bases, base pairs, or base sequences have different affinities for the bound molecule. In this case resort to numerical methods is usually the simplest solution.

Most of the systems that have been examined in which small molecules bind to nucleic acids reveal two particular features that must be taken into account in devising models. Some systems show very strong cooperative interactions between the bound molecules. This would include the stacking of dyes on the outside of polynucleotides (as opposed to intercalation) and the binding of metals such as Hg(II) that destroy the native DNA structure. In other cases, there is evidence for strong anticooperative interactions between the small molecules. For example, when a dye molecule intercalates between the DNA base pairs, there is every indication that it can do so without regard to base composition or sequence. Nevertheless, this mode of binding seems to saturate at one dye per two base pairs. This can be explained by assuming that the binding forces, perhaps as expressed through changes in the bond angles in the double helix, prevent intercalation at adjacent potential sites.

The statistical mechanics of binding reactions may be approached in the language of chemical equilibrium, expressing the interactions in terms of binding constants and free energies of binding. The free energy of binding a small molecule to a particular site on a polymer depends on the activity a of the small molecule, and may be written

$$\Delta G = \Delta G^0 - RT \ln a \tag{7-88}$$

where ΔG^0 is the free energy change of the adsorption reaction under standard conditions of unit activity. (In many cases the activity can be replaced by the concentration.) ΔG^0 will depend on the presence of adjacent bound molecules with which there is interaction.

The total free energy of binding can be defined in terms of "type" reactions.[70-72] One may suppose that binding occurs by starting at one end of the polymer and placing the bound small molecules in the prescribed array. It is therefore necessary to describe reactions in which a small molecule is bound with a certain arrangement of bound neighbors on one side, but nothing on the other side. Restricting our attention to nearest neighbors, one can write $\Delta G^0(j)$ for the standard free energy of binding a small molecule when the closest neighbor on one side is j binding sites away ($j - 1$ empty binding sites in between) with no neighbors on the other side. A typical idealized view of $\Delta G^0(j)$ is shown in Fig. 7-16. When j is less than a number n that characterizes the range of the interaction, the free energy of binding is very large and unfavorable, corresponding to exclusion of neighbors from adjoining sites. When $j = n$ it is possible that the effect is reversed, and a bound molecule exerts a favorable influence on binding a neighbor. Thus the

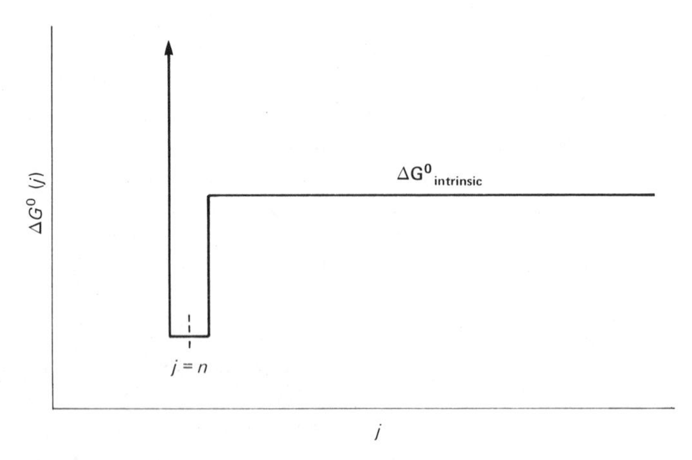

FIGURE 7-16 Idealized dependence of the free energy of binding on j, the separation between bound molecules.

low free energy at $j = n$ will produce cooperative binding effects, and the high free energy for $j < n$ will result in anticooperative effects. Although most systems probably show primarily one behavior or the other, mixed characteristics can be observed.[73]

When $j > n$, the idealized free energy $\Delta G^0(j)$ is independent of j, and is given by $\Delta G^0_{\text{intrinsic}}$. This means that molecules bound that far from neighbors are not influenced by them. We call the binding constant for these isolated molecules the intrinsic binding constant K,

$$K = e^{-\Delta G^0_{\text{intrinsic}}/RT} \tag{7-89}$$

If there are cooperative interaction effects, the binding constant will be greater when $j = n$, which we symbolize by a parameter $\tau > 1$:

$$\tau K = e^{-\Delta G^0(n)/RT} \tag{7-90}$$

where τK is the equilibrium constant for binding n sites away from a bound monomer on one side ($n - 1$ empty sites in between). The anticooperative effects are accounted for by simply forbidding binding closer than n sites away from an already bound monomer. More complicated models are possible, but are justified only for highly accurate binding data or well-understood systems.

Standard methods[72] yield two limiting cases of interest. When only the neighbor exclusion effects are important, $\tau = 1$, one obtains

$$Ka = \frac{1 - y}{y^n} \tag{7-91}$$

and

$$r = \frac{1 - y}{(1 - n)y + n} \tag{7-92}$$

where $y \leq 1$ is a parametric variable.

Bauer and Vinograd[74] showed that this model, with $n = 2$, predicts accurately the binding isotherm for ethidium bromide with DNA. Figure 7-17 shows a Scatchard plot of their data compared with the theoretical prediction. The curved binding plot is characteristic of anticooperative effects.

Neighbor exclusion of this sort is probably a general phenomenon in intercalating systems. With slightly more complicated dye molecules, n can be considerably larger.[73] The physical basis for these effects is still uncertain. They can be propagated to a distance larger than the bound molecule, so it cannot always be simple mutual steric exclusion. A possible general explanation is that the bond angles and structure of the double helix will not tolerate intercalation at adjacent sites. This conclusion is not obvious from a cursory examination of molecular models.

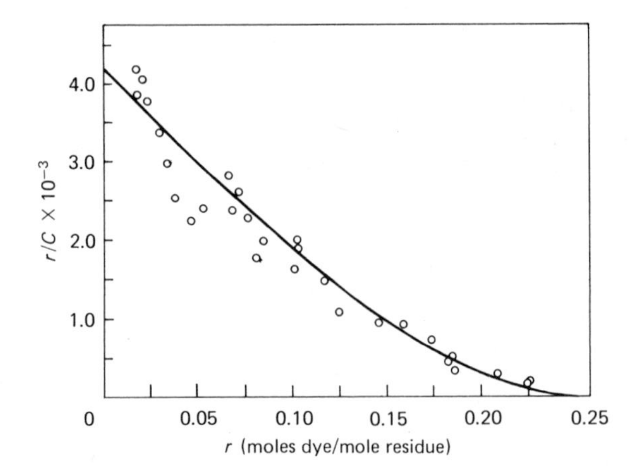

FIGURE 7-17 The quotient ν/C plotted as a function of the binding ratio, ν, in moles dye per mole nucleotide for the binding of ethidium bromide to nicked circular SV40 DNA in 5.8 M CsCl. The free dye concentration, C, is expressed in moles per liter. The experimental data (○) are compared with the theoretical prediction of the neighbor exclusion model with at least one empty site between adjacent bound dyes. [From W. Bauer and J. Vinograd, *J. Mol. Biol.*, **47**, 419 (1970). Reprinted with permission.]

The other limiting case of special interest is when only the cooperative effects are important, so that $n = 1$. Then

$$Ka = \frac{1 - y}{(1 - \tau)y^2 + \tau y} \tag{7-93}$$

and

$$r = \frac{(\tau - 1)y^2 + (1 - 2\tau)y + \tau}{(\tau - 1)y^2 + 2(1 - \tau)y + \tau} \tag{7-94}$$

The appropriate plot for highly cooperative reactions is r vs. log a. The slope of this at the midpoint is given by

$$\left(\frac{dr}{d \ln a}\right)_{r=1/2} = \frac{\sqrt{\tau}}{4} \tag{7-95}$$

Cooperative binding effects arise either when adjacent bound molecules interact physically, or when they induce a cooperative structural change in the nucleic acid. In the latter case it is the nucleic acid interactions, particularly the stacking effect, that are responsible for the cooperativity. The distinction between such reactions and cooperative melting transitions is not clear cut. For example, acid denaturation of DNA can be looked on as cooperative proton binding. A similar comment can be made about the binding of several metals that disrupt the native DNA structure; these are discussed in Section V.

A common example of a cooperative binding reaction that depends on physical interactions between the bound molecules is the stacking of charged dyes on a negative polyelectrolyte lattice such as DNA.[75,76] The favorable dye-dye interaction is demonstrated by the tendency of many dyes to aggregate in solution by a stacking mechanism analogous to that observed for nucleoside bases. Thus the binding of the first dye to the polymer does not benefit from a stacking interaction, but the second can interact with the first. The binding constant for the second process, $K\tau$, is much greater than that for the first, K. According to Eqs. (7-89) and (7-90), τ is directly related to the difference in standard free energies for these two processes:

$$\tau = \exp\left\{\frac{(\Delta G^0_{\text{intrinsic}} - \Delta G^0(n))}{RT}\right\} \tag{7-96}$$

Fitting of the theory to measured equilibrium binding curves allows one to evaluate the three parameters K, τ, and n.

Cooperative effects and neighbor exclusion can also occur in binding processes that depend on base composition or sequence. For example, Hg(II), which binds cooperatively to nucleic acids, has greater affinity for A · T pairs than for G · C.[77] Similarly, actinomycin, which strongly prefers G · C

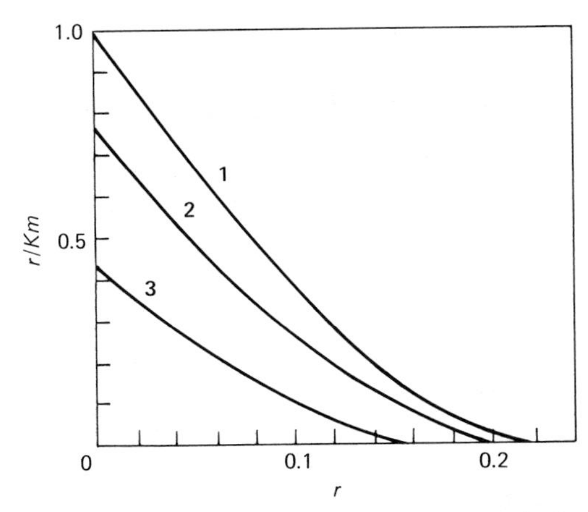

FIGURE 7-18 Calculated binding isotherms for a model in which a small molecule intercalates between two base pairs when at least one of them is G · C. In addition, there must be at least four base pairs between adjacent bound monomers. (r/Km vs. r) where K is the intrinsic binding constant, r the ratio of bound monomers to base pairs, and m the concentration of monomers free in the solution. The curves are for different base compositions: The fractional G · C contents are (1) 1.0; (2) 0.505; (3) 0.257. The base sequence is assumed to be random. [From D. M. Crothers, *Biopolymers,* **6**, 575 (1968). © 1968 by John Wiley & Sons, Inc. Reprinted with permission.]

pairs,[78–81] finds only a limited number of binding sites, an effect that can be accounted for by neighbor exclusion.[73] These problems are considerably more difficult from the theoretical standpoint, but can be treated by numerical methods.[72]

Figure 7-18 shows calculated binding isotherms for a model in which $n = 4$, and binding occurs by intercalation between base pairs only when at least one of them is G · C.

Figures 7-17 and 7-18 illustrate the danger of using as thermodynamic quantities the slope and intercept extrapolated from the linear region of Scatchard plots. In Fig. 7-17, for example, the number of *potential* binding sites is equal to the number of base pairs, although saturation occurs at one dye per two base pairs. The extrapolated intercept B_{ap} has no simple significance, nor does the slope K_{ap}.

The extrapolated intercept on the r/C_F axis is more meaningful. According to Eq. (7-77) this should be equal to K times B_0, where B_0 is the number of potential binding sites per base pair. In Fig. 7-18, where r/KC_F is plotted, the intercept gives B_0 directly. For that case, this is equal to the fraction of nearest neighbor pairs of which at least one is G · C, or

$$B_0 = 2\beta_{\mathrm{G \cdot C}} - \beta_{\mathrm{G \cdot C}}^2$$

for a random sequence, where $\beta_{\mathrm{G \cdot C}}$ is the fractional G · C content.

V COORDINATE BINDING OF METALS TO NUCLEIC ACIDS

In contrast to the electrostatic binding of metal ions like Na^+, K^+, Mg^{2+}, Ca^{2+}, etc., to the phosphate groups of nucleic acids, some metals can form a coordinate attachment to the bases. These bonds result from electron donation by either N or O atoms in the ring or appended to the ring, with unfilled metal orbitals acting as acceptors. Among the metals that behave this way are Ag(I), Hg(II), Cu(II), Cu(I), Cd(II), and Pb(II). The most studied are the first three in this list, and most of this section will concern them.

There is no single rule for recognizing ions that bind to the DNA bases rather than the phosphate groups. Those that do usually lower the T_m because they can interact better with denatured DNA, although this is not the case with Ag(I). Similarly, the metals that bind to the bases usually influence the UV absorption spectrum, although the effects of Cu(II) on native DNA are quite small. In some cases binding to the bases will be accompanied by proton release, but by no means always. Similarly, some metals that bind to the bases cause drastic changes in the hydrodynamic properties of the double helix, and some do not.

A. Binding to the Bases in Monomeric Form

An understanding of the nature of metal binding to DNA requires that one know the site of attachment of the metal to the bases. This question can best be approached by examining the bases in monomeric form. The results of extensive spectrophotometric, titration, and magnetic resonance studies with Ag(I), Hg(II), and Cu(II) are summarized in Table 7-1. The structures given there for metal complexes in solutions are mostly tentative: in some cases certain assumptions have been made. For example, for Hg(II), it is assumed[83] that the tautomeric form in the complex is the same as in the base itself. Thus the structure of the high pH complex with adenosine is given as I rather than II, even though this latter cannot be strictly ruled out. A similar comment applies to the cytidine complex.

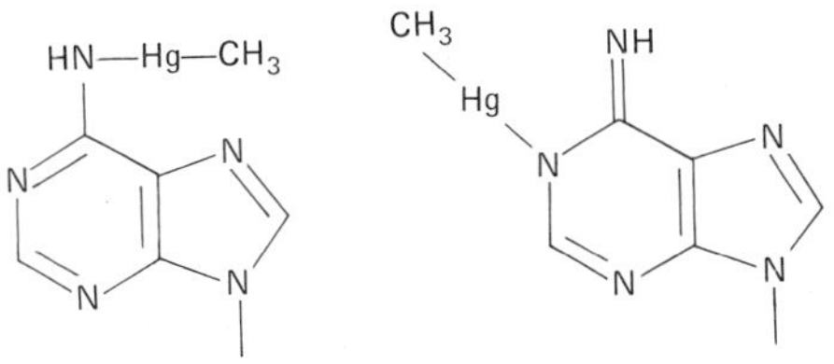

HN—Hg—CH_3 CH_3—Hg—N   NH

(I)   (II)

The chelation of Ag(I) by the N_7O^6 of guanosine seems relatively well established on the basis of IR evidence.[87] In the case of adenosine, the only information available is that a proton is displaced.[85,86] Therefore, a different tautomer could actually be present, as for I and II above, or the Ag^+ might actually be chelated to N_7, as in the analogous complex with guanosine. Thus most of the structures given in Table 7-1 should be considered only as probable, based on the best evidence presently available. It is also possible that different complexation sites are used in binding to polymers.

B. Binding to Polymers

The binding of Hg(II),[89-96] Ag(I),[86,97-99] and Cu(II)[88,100-107] to DNA and synthetic polynucleotides has been studied extensively. These show some sharp contrasts in complexation behavior. Of the three metals, Hg(II) makes the strongest complexes with the bases.[83] When mercuric ion is added to DNA in the absence of a competing ligand, binding is essentially quantitative up to a ratio of 1 Hg^{2+} per base pair, and the reaction occurs rapidly.[96] The spectrum of the nucleic acid is altered[91] in a manner analogous to the effect of Hg(II) on the spectra of the component nucleotides. There is an isosbestic point around 262 nm as long as the ratio of mercury to nucleotide phosphates is less than 0.5 (1 per base pair).

TABLE 7-1 PROBABLE SITES OF PRIMARY ATTACHMENT OF METALS TO NUCLEIC ACID BASES IN MONOMERIC FORM

	Hg(II) (CH_3Hg^+)	Ag(I)	Cu(II) (no H^+ removal)
A	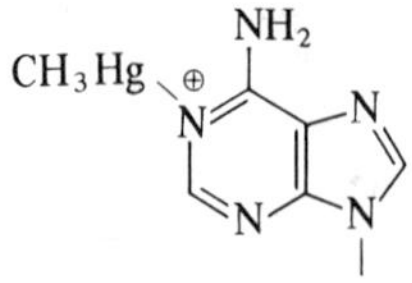 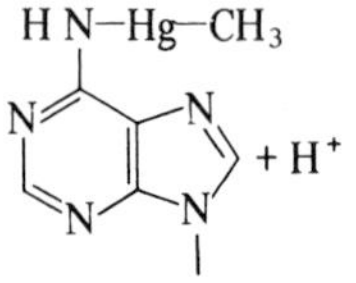 NH_2, $CH_3Hg^{\oplus}$–N (low pH); HN–Hg–CH_3 + H^+ (high pH) (83)	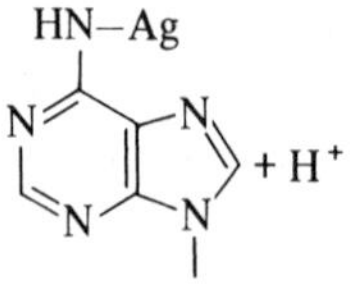 HN–Ag + H^+ (→ ppt) (above pH 7) (85, 86)	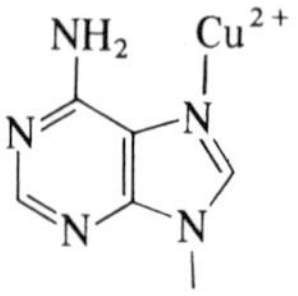 NH_2, Cu^{2+}–N (88)
U	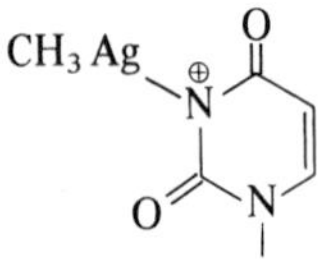 $CH_3Ag^{\oplus}$–N, O, O (83)	$N_1O^6 + H^+$ (above pH 7)	no complex (88)
G	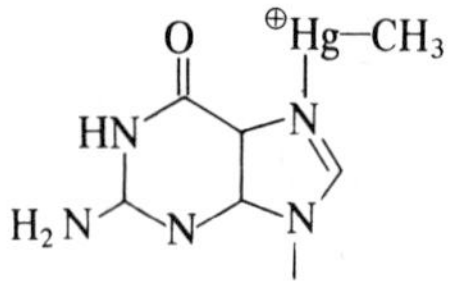 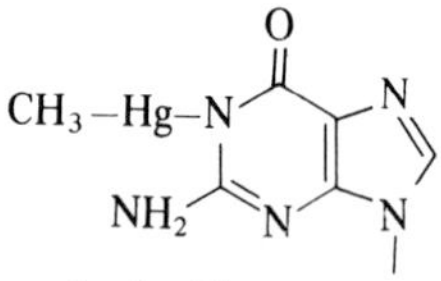 O, $^{\oplus}$Hg–CH_3, HN, H_2N (low pH); O, CH_3–Hg–N, NH_2 (high pH) (83) (also, possibly to –NH_2 84)	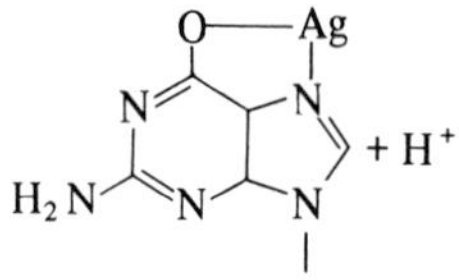 O—Ag, H_2N + H^+ (above pH 5) (86, 87)	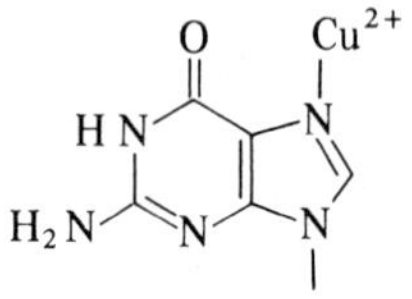 O, Cu^{2+}, H N, H_2N (88)
C	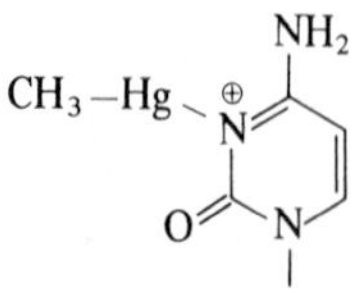 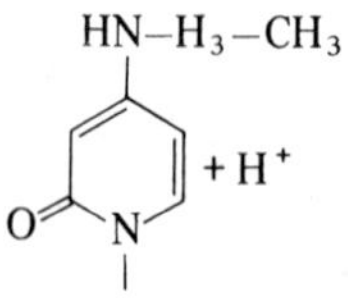 NH_2, CH_3–Hg$^{\oplus}$–N, O (low pH); HN–H_3–CH_3, O + H^+ (high pH) (83)	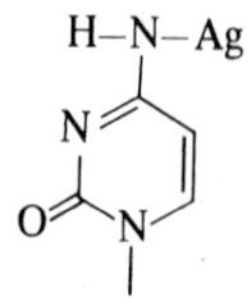 H–N–Ag, N, O (above pH 7) (86)	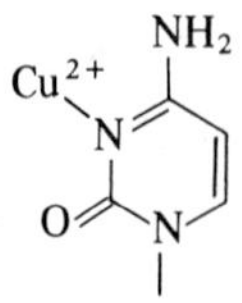 NH_2, Cu^{2+}–N, O (88)

The binding of Hg(II) to DNA is quite cooperative in some cases. Figure 7-19 shows the ratio of mercury bound per nucleotide phosphate as a function of the negative logarithm of the Hg^{2+} concentration.[94] One should note that poly d(A · T) has a higher affinity for mercury than does DNA, although the former binding saturates at lower values of r_b. DNA of high G · C content has even lower affinity for mercury, since higher Hg^{2+} concentrations are required for half-saturation. At high Hg^{2+} concentrations, poly d(A · T) begins to bind further Hg(II).

The dependence of mercury affinity on base composition can be understood from the affinities of the single bases for Hg(II), measured by Simpson[83] with methyl mercury, $CH_3{-}Hg^+$. His results are summarized in Fig. 7-20. At pH 9 the binding site of highest affinity is uracil (or thymine), followed by G, with A and C of lower and roughly equal affinity. Therefore, the reaction at low Hg^{2+} concentration with poly d(A · T) must involve complexation with T. There is one mercury per two base pairs because Hg(II) prefers linear complexes with coordination number 2, and must therefore be cross-linking two T's. Furthermore, Gruenwedel and Davidson[95] showed that the affinity of DNA for methyl mercury could be

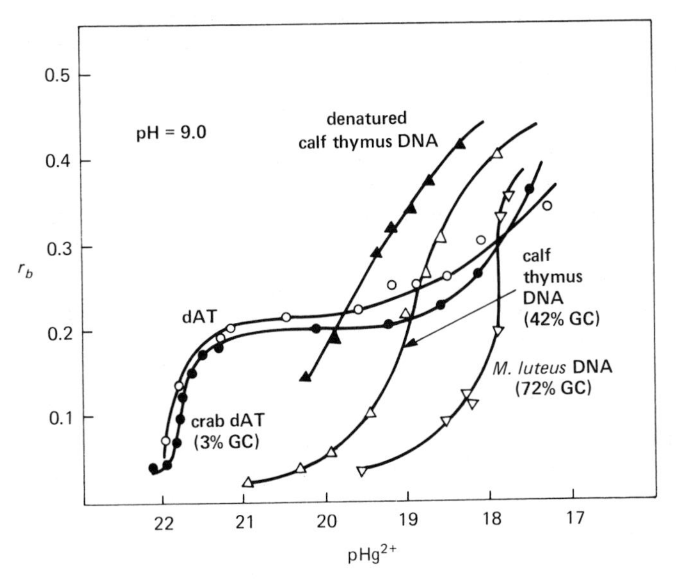

FIGURE 7-19 Binding curves for the complexing of DNA by Hg(II) at 27° in 0.5 M $NaClO_4$ and 0.01 M $Na_2B_4O_7$, pH 8.95. (O), synthetic dAT; (•), crab dAT; (Δ), calf thymus DNA; (▲), denatured calf thymus DNA; (▽), *M. lysodeikticus* DNA. [From W. S. Nandi, J. C. Wang, and N. Davidson, *Biochemistry*, 4, 1687 (1965). Copyright 1965 by the American Chemical Society. Reprinted by permission of the American Chemical Society.]

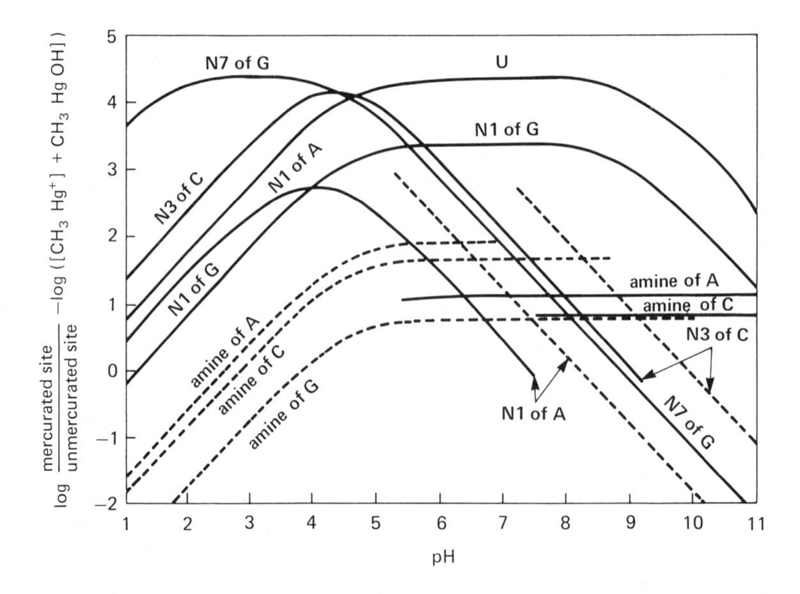

FIGURE 7-20 Effect of pH on mercuration with CH_2Hg. Solid lines refer to mercuration at the site indicated when no other site is mercurated. The dashed lines refer to mercuration at the site indicated when another site is mercurated. [From R. B. Simpson, *J. Amer. Chem. Soc.,* **86**, 2059 (1964). Copyright 1964 by the American Chemical Society. Reprinted by permission of the American Chemical Society.]

predicted with reasonable accuracy from the monomer binding constants measured by Simpson.

The binding reaction with poly d(A · T) is cooperative because the structure is drastically changed, for which evidence is shown below. The reaction with DNA is less cooperative because of the wide variation in binding affinity of the bases. The complex that forms up to one mercury per base pair probably involves linking of two bases by a single mercury.

Protons are released when Hg(II) binds to DNA,[91,94] and the number depends on the pH. This is as would be predicted from the monomer complex structures proposed by Simpson shown in Table 7-1. At low pH, binding to A, C, and G does not involve proton release, whereas it does at high pH. Hg^{2+} binding to DNA at pH 9 releases two protons per Hg^{2+}, corresponding to base-base cross-linking.[95]

The viscosity of DNA is greatly reduced by mercury binding. Figure 7-21 shows comparative results for Cu(II), Ag(I), and Hg(II). Only Hg(II) produces a large reduction in the viscosity. It is clear that the rigidity of the double helix structure has been lost, and the viscosity is roughly what would be expected of the coil form. In fact reaction with Hg(II) is probably similar structurally to DNA denaturation. With methyl mercury the analogy is very

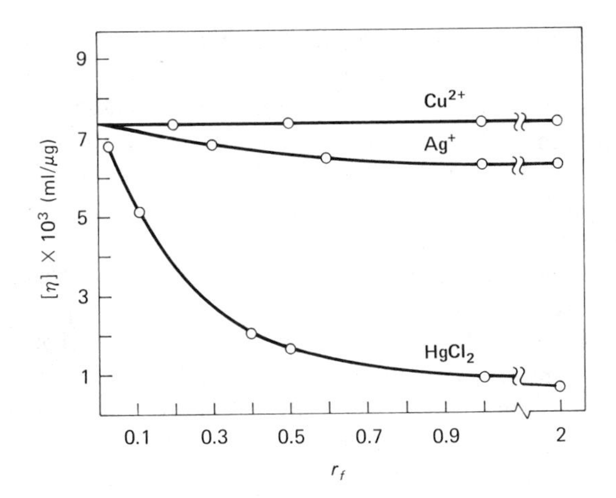

FIGURE 7-21 Effect of Ag(I) on the intrinsic viscosity of calf-thymus DNA solution in 0.10 M $NaClO_4$ at pH 5.7; temperature, 25.0°; r_f = moles of Ag(I) added per mole of nucleoside phosphate. The effects of Cu(II) and Hg(II) under identical conditions are shown for comparison. An intrinsic viscosity of 7×10^{-3} ml/μg is 70 dl/g. [From T. Yamane and N. Davidson, *Biochim. Biophys. Acta,* 55, 609 (1962). Reprinted with permission.]

close,[95] but the double coordination of Hg^{2+} produces one strong difference in that instance. Unlike the behavior of methyl mercury, disruption of DNA structure by Hg^{2+} can be completely reversed by removing the metal, for example by adding a competing ligand.[94] This has been interpreted to mean that Hg^{2+} cross-links the two strands and maintains their register sufficiently to permit rapid renaturation when conditions favoring the double helix are restored.

Kinetic measurements[96] confirm that mercury binds to the hydrogen bonding positions in poly A · poly U, since the double helix reacts orders of magnitude more slowly than does poly U. The reaction seems to occur by binding of mercury to uracil bases which are unbonded because of thermal fluctuations in the structure. With DNA the reaction is faster and much more complicated, and probably involves an initial attachment of Hg^{2+} to the outside of the double helix at the N_7 position on guanine (see Table 7-1).

An important practical use of metal ion binding, especially Hg^{2+}, has been to aid density gradient separation of nucleic acids on the basis of base composition.[94] Mercury binding increases the buoyant density of DNA because of the high density of the heavy metal. Since DNA rich in A · T binds more strongly, its density is increased more than G · C-rich DNA. The quantitative separation is larger than that resulting from the simple base composition dependence of the banding density.

Among the metals that bind to DNA, Ag(I) poses perhaps the most interesting structural problem. Figure 7-21 shows that the viscosity is only slightly decreased by Ag^+, indicating that the double helix structure remains essentially intact. Yet spectral data show large changes in the UV absorbance, and titration studies show instances of proton release accompanying highly cooperative binding reactions. The model-building problem is thus one of incorporating Ag^+ into the double helix, in some instances displacing protons and in others not.

Silver ion will bind to all of the nucleic acid bases at high pH,[85–87,98] displacing a proton in each case. Table 7-1 shows the probable sites of attachment of the metal. Binding is accompanied by large changes in the UV spectra of the bases. Guanosine has a considerably higher affinity for Ag^+ than any of the other bases,[85,98] and probably chelates the metal at the N_7O^6 positions, bringing the base into the enol tautomer.

DNA and polynucleotides have considerably higher affinity for Ag^+ than do any of the separate bases,[98,99] although silver binding produces similar changes in the UV absorbance of monomers and polymers. The observed spectral changes indicate clearly that there are three distinguishable binding processes. At low r (silver ions bound per nucleotide) the absorbance at 260 nm decreases, with a minimum that is pH dependent. A different complexing reaction at intermediate r then produces a rise in absorbance,

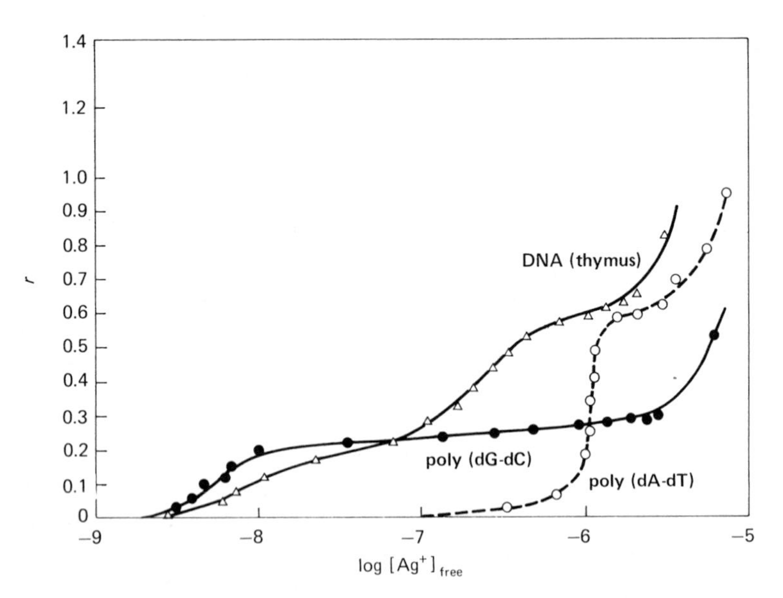

FIGURE 7-22 Binding of Ag^+ to double-stranded deoxyribonucleic acids. Plots of r vs. log $[Ag^+]$ for double-stranded deoxyribonucleic acids in 10^{-3} M phosphate. [From M. Daune, C. A. Dekker and H. K. Schachman, *Biopolymers,* 4, 51 (1966). © by John Wiley & Sons, Inc. Reprinted by permission.]

followed by a further decrease when there is more than one Ag^+ per base pair.

Binding curves showing the degree of binding as a function of the log of free Ag^+ concentration[54,55] are presented in Fig. 7-22. There are three separable binding reactions, all cooperative, and dependent on the G · C content of the sample. The reactions are well separated at lower pH values. The binding reaction of highest affinity (type I binding) involves G · C pairs, and saturates at no more than 1 Ag^+ per two base pairs, somewhat less for a DNA rich in A · T. Type II binding is weaker at low pH, and saturates at 1 Ag^+ per base pair. At high pH the two binding reactions tend to converge. A third kind of binding is evident at even higher silver ion concentrations.

The reason that the two binding reactions differ in their pH dependence is that they differ in proton release; no protons are released in type I binding, while somewhat less than 1 proton per Ag^+ is released in type II.[97,98] The lack of proton release in type I binding may seem surprising in view of the results with mononucleotides showing that binding is always accompanied by H^+ release.

Examination of a G · C pair reveals that Ag^+ binding of the type found for guanosine can occur, coupled with transfer of the proton to cytosine, as shown below:

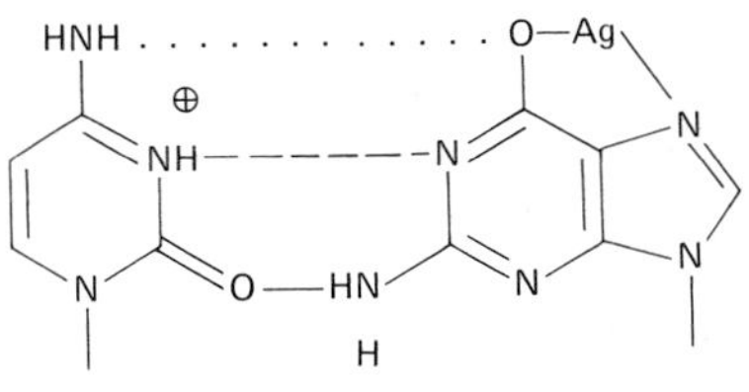

which is a model suggested for Cu(I) binding by Minchenkova and Ivanov.[100] Since cytosine serves as proton acceptor, no protons are released to the solution. Furthermore, the presence of a proton acceptor should increase the affinity for Ag^+ over that observed for guanine at neutral pH, since C takes the proton more readily than water or the highly dilute OH^-. Since the binding saturates at one metal per two base pairs, it would appear that the principle of neighbor exclusion is operating, preventing binding of Ag^+ at adjacent sites. In such cases, statistical calculations predict[72] that the saturation level of binding will not vary much with G · C composition as long as the fraction of G · C does not drop much below 50%, which agrees with the observations of Yamane and Davidson[97] on the number of type I binding sites. Why neighbor exclusion should be operative and why the binding reaction should be cooperative remain open questions.

The proton release data indicate that in type II binding even some of the G · C pairs release protons on silver binding.[98] This is a highly cooperative process that reaches saturation at one metal per base pair. Jensen and Davidson[98] suggest that Ag^+ replaces H^+ in a double helix hydrogen bonding position. Their choice is along the N—H—N hydrogen bond in A · T and G · C pairs, although one could equally well argue for insertion between an amino nitrogen and an enolized oxygen, by analogy with the binding to guanine and the known tendency of Ag^+ to bind to amino groups. The cooperativity of the reaction probably results from the DNA structural changes that are necessary to accommodate the Ag^+ ions, making it favorable to transform only large regions of the molecule at once. In the course of type II binding, some of the G · C pairs may be accepting silver by the type I mechanism, since there is less than one proton released per silver ion bound.

The highly cooperative binding of silver and the preference for G · C-rich DNA can be used to separate samples on the basis of composition.[98] For example, in a mixture of *E. coli* and T4 DNA, the *coli* DNA has higher G · C content, and therefore greater affinity for Ag^+, and binds nearly to saturation before T4 begins to take up silver. Thus a large density separation results at intermediate Ag^+ concentrations.

A number of unanswered questions remain concerning Ag(I) binding to polynucleotides. For example, Daune et al.[99] showed that specific complexes are formed with poly A (at neutral pH) and with poly C; that the reaction is cooperative; and that it does not occur with mononucleotides or with small oligomers. Thus some cooperative structure formation is involved, and one is tempted to postulate base pairing with Ag^+ replacing H^+, but there is not sufficient information for detailed models.

Cu(II) binds more weakly to DNA than do Hg(II) or Ag(I), in accordance with the behavior of the monomers. As shown in Table 7-1, Cu^{2+} binds to the N_7 of A and G, and to N_1 of C, but not in appreciable amounts to U or T.[88] The complex stabilities are in the order $G > C > A$.[107] In no case is a proton displaced by Cu^{2+} binding.[98]

When Cu(II) is mixed with DNA at low temperatures, there is essentially no change in the viscosity, as shown by Fig. 7-21. However, when the sample is heated, a melting transition occurs at considerably lowered temperature from that observed in the absence of the metal.[101,102] The viscosity drops drastically in the course of melting, indicating destruction of the native structure. The UV spectrum of DNA denatured in the presence of Cu(II) is different from that of thermally denatured DNA, indicating an interaction with Cu^{2+} in the former case. The spectral effects are smaller than with Ag(I) or Hg(II). The UV spectrum of native DNA at low temperature is only slightly altered by the presence of Cu(II).[88,108] Thus the picture that emerges is of some Cu(II) binding at low temperatures, and extensive binding to the bases at elevated temperatures, resulting in a decreased transition temperature.

The melting transition of DNA in the presence of Cu(II) cannot be reversed by lowering the temperature, but it can be by removing the copper.[101] One would conclude that Cu(II), like Hg(II) and Ag(I), acts to cross-link the two strands. This ability of the polymer to chelate Cu(II) probably accounts for its increased affinity over that of the monomers.

Since Cu(II) generally destabilizes the double helix, one would expect from the monomer results that DNAs rich in G · C would be more destabilized because of the higher affinity of Cu(II) for G and C.[107] This is the case, since there is an inversion in the relative stability of G · C- and A · T-rich DNA in the presence of Cu^{2+}. In fact, Cu(II) does not destabilize dAT under these conditions.

One of the most effective ways to screen metals for interaction with the bases is to examine the effect on the melting transition temperature. In addition to Cu^{2+}, both Cd^{2+} and Pb^{2+} reduce the T_m.[102,104] Furthermore, Zn^{2+} does not stabilize nearly as much as other divalent metals.[102] Shin and Eichhorn[109] showed that the melting transition is completely reversible in the presence of Zn^{2+}, indicating strongly that Zn^{2+} serves to cross-link the strands. Why the transition should be reversible under temperature change with Zn^{2+} but not with Cu^{2+} is not clear at the present time. The structural change induced by or in the presence of Ag^{+}, Hg^{2+}, Cu^{2+} and Zn^{2+} can be reversed by removing the metal ion. A lone exception is CH_3Hg^{+}, which, because of its single strong coordination position, cannot act as a cross-link.

The binding of metals to nucleic acids is a highly complex subject, and the current literature is occasionally contradictory. The reason is the large number of ways that metals can interact with polynucleotides, and the variation of the binding mechanism with conditions. Since much remains to be learned on this subject, it is unfortunately not possible at present to make many simplifying generalizations.

VI MODES OF BINDING DRUGS TO DNA

There are a number of substances such as dyes and various antibiotics that are able to bind to DNA. In so doing they usually exert some sort of pharmacological effect, and we therefore consider them under the heading of drugs. These materials will be considered individually in Section VIII, following a discussion of experimental methods in Section VII. In this section we recount some of the general modes of binding that have been proposed for these compounds. Reviews of this topic have been written by Waring[110] and by Hartmann et al.[111]

Perhaps the most studied mechanism of binding small molecules to DNA is intercalation. According to this model, proposed by Lerman,[112] flat polycyclic or heterocyclic aromatic substances like proflavine can be inserted between the base pairs. The result is that adjacent base pairs separated by a

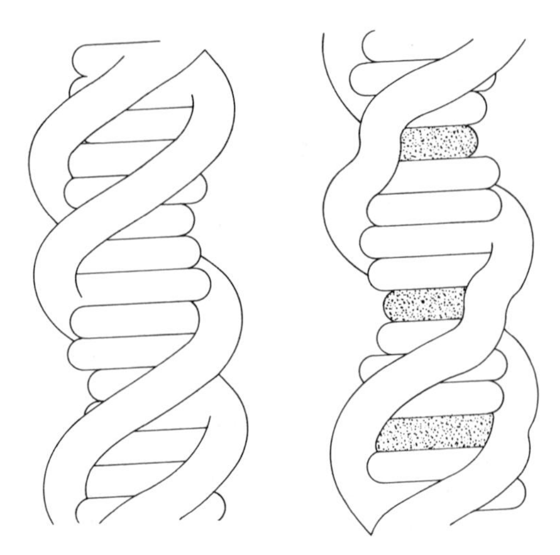

FIGURE 7-23 Sketches representing the secondary structure of normal DNA (left) and DNA containing intercalated proflavine molecules (right). The helix is drawn as viewed from a remote point, so that the base pairs and the intercalated proflavine appear only in edgewise projection, and the phosphate-deoxyribose backbone appears as a smooth coil. [From M. J. Waring, *Nature*, **219**, 1320 (1968). Reprinted with permission.]

proflavine are about twice their normal distance apart, leading to an increase in the molecular length. Figure 7-23 shows a sketch of the proposed structure. As indicated, the double helix must be unwound to some extent in order for intercalation to occur. The extent of unwinding has been estimated from 12°[113] to 36,°[114] but the lower of these figures, proposed for a model of ethidium intercalation, seems to be more generally accepted (see section on binding to closed circular DNA).

Another widely accepted mechanism of binding dyes to DNA is by stacking the planar molecules together along the polymer lattice.[75,76,115] Acridine dyes, for example, will bind in this manner to a number of polyanions. These dyes all have a tendency to aggregate in solution, presumably by a stacking mechanism, and the provision of a negative lattice to attract the positively charged dyes (they are generally protonated at neutral pH) serves to facilitate greatly the aggregation process. Figure 7-24 shows a diagram of binding by this mechanism.

Small molecules can also bind externally to nucleic acids in monomeric form. The middle case in Fig. 7-24, found when the concentration of dye (D) is very low relative to polymer (P), is an example. When acridines are bound to DNA (with large DNA excess), a small fraction is not intercalated, but instead bound to the outside. This minor mode of binding is best detected by kinetic methods.[64] Electrostatic interactions obviously provide an important

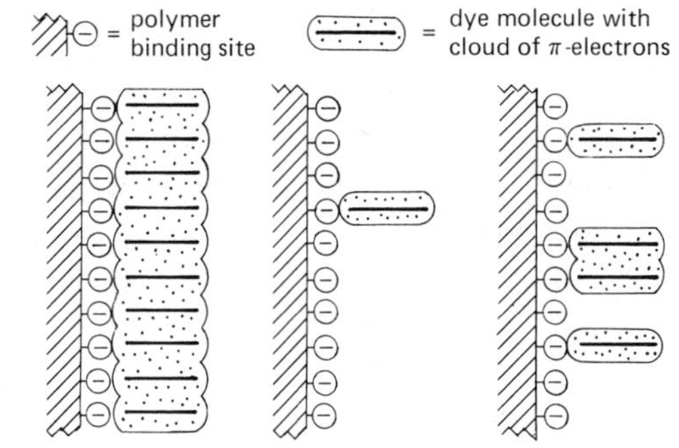

FIGURE 7-24 Schematic representation of the aggregation of dye molecules bound to the surface of the polyelectrolytes. [From D. F. Bradley and M. K. Wolf *Proc. Nat. Acad. Sci., U.S.*, **45**, 944 (1959). Reprinted with permission.]

contribution to the binding energy, but the similarity of the spectral shift to that observed on intercalation indicates that there may be considerable interaction by partial stacking with the exposed region of a base pair. Outside binding could also be favored by formation of specific hydrogen bonds, such as proposed for actinomycin binding by Hamilton et al.[116] The nature of specific outside binding of isolated small molecules is especially difficult to demonstrate. One reason is that changes in polymer shape are not expected to be nearly as dramatic as those produced by intercalation.

A fourth mechanism by which small molecules can bind to DNA is by covalent bond formation. An example is the antibiotic mitomycin, which acts to cross-link the DNA structure for forming such bonds, as is shown by reversibility of denaturation of treated DNA.[117] We will not give an extensive account of covalent modification of DNA structure, such as, for example, by alkylating agents, etc.

VII EXPERIMENTAL METHODS

In this section we consider the question of methodology, using several kinds of binding reactions as examples. In the following section the specific substances will be discussed, along with the evidence favoring a particular mode of binding.

A. Optical Absorbance

Easily the most convenient tool for detecting complex formation between nucleic acids and a wide range of small molecules is the change in absorbance. Many of the drugs that bind by intercalation have, by virtue of their heterocyclic structure, a visible absorption band. When these are mixed with DNA, a metachromatic complex results, whose change in color from the free dye is apparent to the eye. Figure 7-25 shows an example for ethidium bound to DNA. As increasing amounts of DNA are added, the absorption maximum shifts to the red, and there is a reduction in the peak absorbance. These measurements may be used to determine binding equilibria by the analysis described in Section III.

It should be noted that all the spectra in Fig. 7-25 pass through a common point, called the isosbestic point. The most common circumstance under which an isosbestic is obtained is when there are only two forms of the dye, in this case free in solution and intercalated. If these have the same absorbance at some wavelength, then shifts in the relative amounts of each will produce no change in absorbance (at constant total dye concentration) at

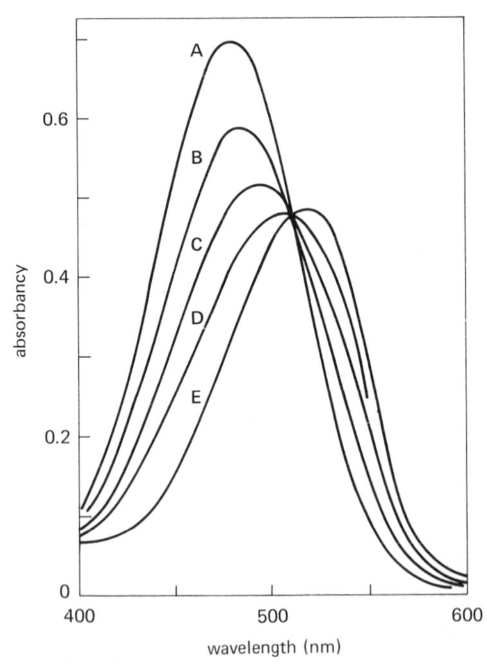

FIGURE 7-25 Shift of the absorption spectrum of ethidium bromide in the presence of DNA. (A) Spectrum of the drug alone in buffer. (B–E) With increasing concentrations of T2 DNA added. [From M. J. Waring, *Nature,* **219**, 1320 (1968). Reprinted with permission.]

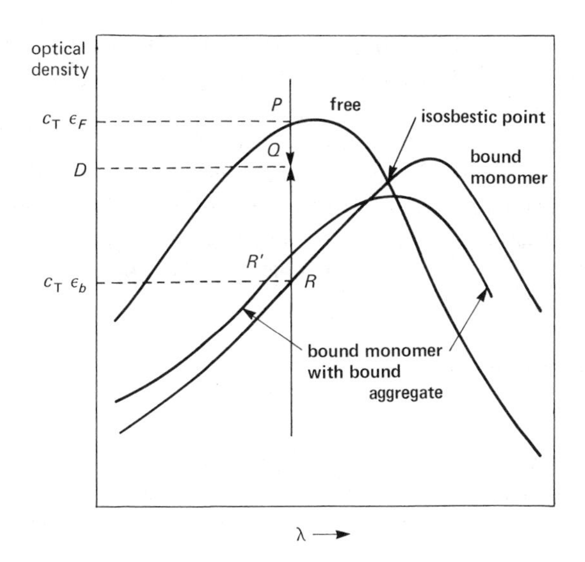

FIGURE 7-26 Spectra of proflavine, free and bound as monomer. The spectrum of proflavine bound both as monomer and as aggregate is schematic only, though based on experimental observations. [From A. Blake and A. R. Peacocke, *Biopolymers,* **6**, 1225 (1968).]

that wavelength. However, observation of an isosbestic wavelength is not proof that only two forms are present. For example, there could be a mixture of free dye with two forms of bound dye having different spectra, but if these latter are always present in constant ratio, there will still be an isosbestic wavelength.

The converse behavior permits a more definite statement. If there is not an isosbestic point even though the spectra of complex and free dye cross, then there must be more than two forms of the dye. An example is shown in Fig. 7-26, taken from a paper by Blake and Peacocke[118] on the binding of proflavine to DNA. When the DNA concentration is high, proflavine is intercalated and there is an isosbestic point as indicated. However, as the dye becomes more crowded on the polymer, the spectrum begins to shift in a different way and the isosbestic disappears. This is thought to be due to binding of the dye by a different mechanism, namely stacking on the outside of the double helix. The spectral behavior makes it clear that different attachment mechanisms apply in the limiting extremes of high and low binding.

The usual observation when a dye intercalates between the DNA base pairs seems to be that the visible absorption band is shifted to the red, with an accompanying reduction in absorbance. This presumably arises from interaction with the electron system of the base pairs. However, such spectral shifts should not be regarded as diagnostic for intercalation. Outside binding, for example, can probably produce similar spectral changes. An isolated proflavine bound to the outside of DNA has a difference spectrum closely parallel to that of the intercalated form.[69]

Large changes in absorbance are often observed when dyes stack on the surface of a polyelectrolyte. In addition there is often a blue shift in the absorbance maximum, in contrast to the red shift on intercalation. It should be emphasized again, however, that the nature of the absorbance change is not well enough understood to be taken as proof of a particular kind of complex structure.

B. Fluorescence and Phosphorescence

The energy absorbed from a light beam by a polymer-dye complex can be given up in a number of ways, of which fluorescence and phosphorescence are two. These both involve emission of a quantum of light at an equal or lower energy than the absorption process, and are distinguished experimentally by the time scale of the process. Fluorescence occurs when an excited state singlet returns to the ground state singlet, and is consequently much more rapid than the phosphorescence when a triplet state returns by a forbidden transition to the ground singlet. The usual experimental distinction is to count as phosphorescence the emission still present a few milliseconds after the excitation is removed.

Consider the schematic diagram of polymer and dye energy levels shown in Fig. 7-27, taken from a paper by Galley.[124] The polymer has an

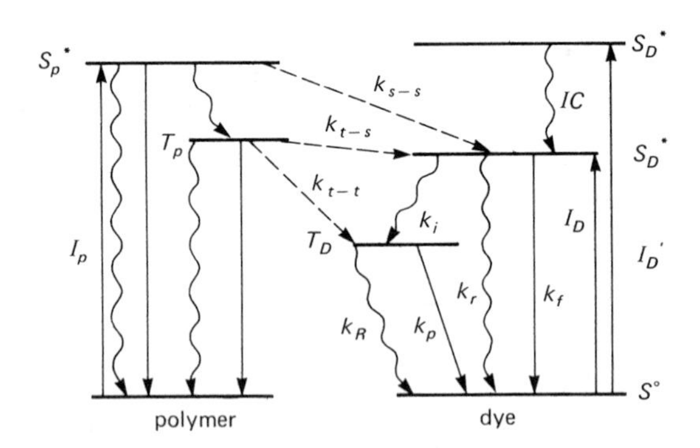

FIGURE 7-27 Excited state processes in the dye for a hypothetical polymer-dye complex; S°, ground state; S^*, lowest excited singlet; $S^{*\prime}$, higher excited singlet; T, lowest triplet (subscripts denote whether the state is primarily a polymer or dye state); I, rate of absorption; k_f, k_r, k_i, rate constants for dye fluorescence, internal conversion, and intersystem crossing; k_p, k_R, phosphorescence and internal conversion; k_{s-s}, k_{t-s}, k_{t-t}, singlet-singlet, triplet-singlet, and triplet-triplet, polymer-dye transfer; IC, $S_D^{*\prime} \rightarrow S_D^*$ internal conversion. [From W. C. Galley, *Biopolymers*, 6, 1279 (1968). © 1968 by John Wiley & Sons, Inc. Reprinted with permission.]

excited singlet S_p*, which corresponds to the state reached upon absorption of a UV quantum, and from which fluorescence can occur. In addition, there is a triplet, labeled T_p, from which phosphorescence[119,120] occurs around 458 nm. Thymine is thought to be responsible for this triplet emission of DNA.[121] Nonradiative transitions, indicated by wavy lines, can also occur from these excited states.

When the dye energy levels are also considered, several new transition possibilities arise. The dye itself is shown with two excited singlets, corresponding to the visible and UV absorption bands. The higher lying excited singlets presumably decay rapidly by nonradiative processes to the lowest excited singlet, from which fluorescence can occur. The singlet excitation can also be transferred to the dye triplet, T_D, via intersystem crossing in a nonradiative process. Transitions from T_D to ground can be accompanied by phosphorescence, or they can occur nonradiatively.

In addition, excitation in the polymer could conceivably be transferred to the dye. In general, the excited states of the polymer lie higher, and transfers are therefore considered only in the direction from polymer to dye. The three possibilities are shown, namely from polymer singlet to dye singlet, from triplet to singlet, and from triplet to triplet. There is evidence that all of these paths occur.

The evidence for singlet-singlet transfer is that there is a very considerable increase in dye fluorescence when the exciting wavelength corresponds to the DNA absorption maximum, compared to the fluorescence when excitation occurs via the visible dye absorption band.[112,122,123] The increase cannot be accounted for by increased absorbance in the UV band of the dye. Thus energy that does not come from absorbance by the dye alone is entering the dye singlet, and part of this energy is emitted as fluorescence. When the exciting wavelength is in the visible, there is no energy transferred from the polymer singlet, and the dye emission efficiency is much smaller.

An indication of triplet-singlet energy transfer from polymer to dye is found in the delayed fluorescence emission from the dye when the complex is excited in the UV.[120] Figure 7-28 shows the total emission (curve B) and the much smaller delayed emission (curve A) from a proflavine-DNA complex excited in the UV. The delayed emission would usually be classified as phosphorescence from a triplet state, but the wavelength of the emission band around 500 nm corresponds closely to the fluorescence emission peak, curve B. The same correspondence is seen for several dyes. It is therefore concluded[120,124] that the emission is actually from the dye singlet, and that it is slow because the excitation is transferred from the polymer triplet. This establishes the path t–s in Fig. 7-27. The delayed emission band at longer wavelength in Fig. 7-28 is the dye phosphorescence.

Galley[124] has provided evidence that triplet-triplet energy transfer between polymer and dye occurs. If all the energy transferred from polymer

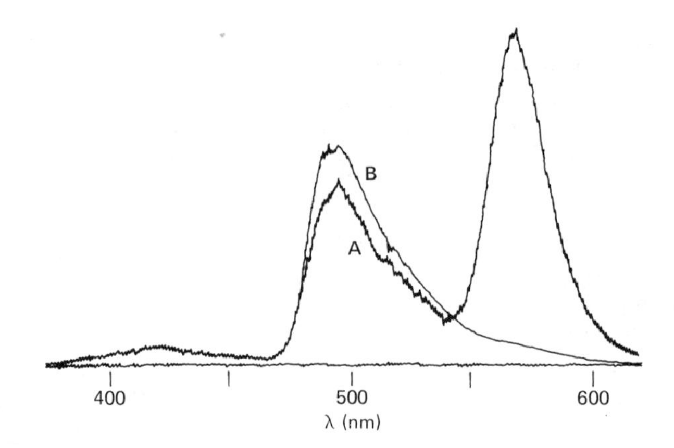

FIGURE 7-28 Emission of proflavine. Salmon sperm DNA complex 3.3×10^{-3} M in DNA phosphate, 1.1×10^{-4} M proflavine. (a) Emission with rotating shutter in place. (b) Total emission. [From Isenberg et al. *Proc. Nat. Acad. Sci., U.S.*, **52**, 379 (1964). Reprinted with permission.]

to dye went through the dye singlet, then there should be parallel increases in dye fluorescence and phosphorescence when the exciting wavelength is changed from visible to UV. This is not the case experimentally, since with 9-aminoacridine the ratio of dye phosphorescence to fluorescence is much larger for UV excitation than for visible excitation.[124] One concludes that energy is being transferred into the dye triplet without going through the singlet, thereby increasing phosphorescence more than fluorescence. Since the direct transfer from polymer singlet to dye triplet is considered sufficiently unlikely to neglect, the only way remaining is to transfer the energy from the polymer triplet.

Evidence from quenching of DNA phosphorescence by Mn^{2+} [119] and from dye studies[121] indicates that the DNA triplet excitation is delocalized. Triplet-triplet energy transfer requires close proximity of the chromophore systems, and probably occurs between adjacent base pairs. The observed triplet-triplet transfer between polymer and dye indicates close association of the dye and base π-electron systems, and is consistent with an intercalation mode of binding.

There remain some unresolved phenomena in the emission spectra of bound dye molecules. For example, the fluorescence of some dyes like proflavine[122] and polycyclic hydrocarbons[125] is quenched by binding to DNA, while that of others like ethidium[123] is enhanced. The underlying basis for this remains to be clarified. A further observation is that the ability to quench the fluorescence of acriflavine depends on the base composition of the DNA sample.[126]

C. Flow Orientation

The optical properties of dyes bound to DNA provide a means for determining the orientation of the chromophore relative to the helix axis. This technique is based on the optical anisotropy of the dye molecule. The observed singlet-singlet transitions are generally $\pi \rightarrow \pi^*$, so that the transition moment lies in the plane of the chromophore. A light beam of appropriate wavelength polarized parallel to this transition moment produces the excited state from the ground state at the expense of absorption of energy from the beam. The photon may be reemitted as fluorescence, with a polarization that depends on the orientation of the moment of the emitting transition. If the exciting beam is polarized perpendicular to the absorption moment, no absorption occurs and there is no fluorescence.

When molecular rotation is rapid, the observed fluorescence emission is not polarized, because by the time emission occurs the molecule has rotated so far that it has "forgotten" the orientation of polarization of the exciting radiation. However, when the chromophore is held rigid, as in a viscous solvent or by fixing to a macromolecule, emission polarization can be observed. If the emission intensity is larger with polarization parallel to the excitation polarization, then the absorption and emission transitions must be more parallel than perpendicular. On the other hand, if the emission intensity is greater when the detector is polarized perpendicular to the excitation, then the two transitions are more nearly perpendicular than parallel. For a wavelength of absorption close to that of emission, the two transitions are always parallel, since the same transition is responsible for both absorbance and emission. However, for absorbance at shorter wavelengths, the two transitions can be perpendicular.

Consider now the fluorescence properties of a dye when it is attached to a DNA molecule which has been oriented in a flowing liquid.[127,128] Because of the shear gradient, the polymer molecules tend to line up with the helix axis parallel to the flow direction. (Orientation is never perfect, but that does not affect the qualitative conclusions.) Suppose that the dye has an absorbance moment in its long axis, and an emission moment perpendicular to that, along the short axis of the chromophore. Three possible limiting alignments of the chromophore relative to the helix axis are shown in Fig. 7-29. In (1) and (2), the chromophore plane is parallel to the flow or helix axis and perpendicular to the base pairs, while in (3) the chromophore is perpendicular to the helix axis and parallel to the base pairs. Orientation (3) corresponds to intercalation. A number of possible combinations of excitation and detection polarization are possible, but we illustrate the argument with the case that is critical for distinguishing (3) from (1) and (2). Let both the exciting radiation and the detecting system be polarized perpendicular to

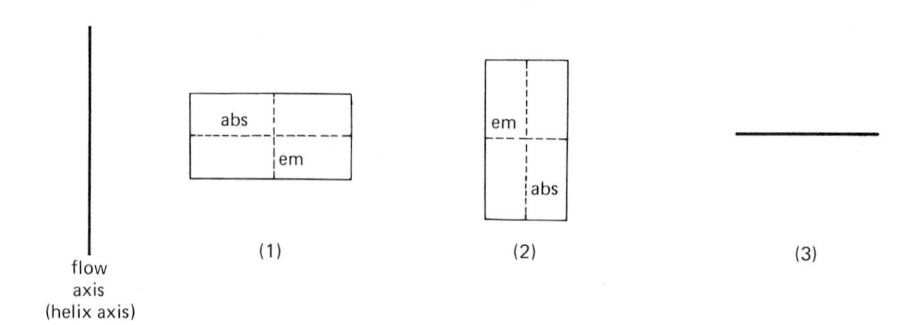

FIGURE 7-29 Three possible orientations of a dye molecule (with perpendicular absorbance and emission transitions) relative to the flow axis (helix axis).

the flow axis. The detection direction is perpendicular to the exciting beam. In the case (3) flow should increase the observed fluorescence, since both transitions become favorably oriented for absorbance and for emission with a polarization that will be detected. In cases (1) and (2) however, flow should decrease the observed fluorescence, since in case (1) the emission polarization will become perpendicular to the detector and will not be seen, and in case (2) the absorbance will diminish as that transition becomes perpendicular to the excitation. All the relevant cases have been considered by Lerman,[127] with the conclusion for acridines that case (3) best fits the observations. Intermediate cases are possible, but it seems clear that the chromophore is more nearly perpendicular to the helix axis than parallel to it.

The absorbance from a polarized beam can also be measured on DNA molecules oriented by flow, a technique called flow dichroism. For example, if the absorbance moment is perpendicular to the helix axis, the absorbance should increase when flow is initiated if the beam is polarized perpendicular to the flow direction, and decrease if the beam is parallel polarized.[127] It should be noted that this technique is intrinsically less powerful than the fluorescence measurements where the mutually perpendicular absorbance and emission transitions allow one to specify the orientation of the chromophore plane. With absorbance, one can specify only the orientation of a single vector, which, without further information, is not sufficient to determine the orientation of a plane. For example, cases (1) and (3) in Fig. 7-29 have the absorbance moment perpendicular to the helix axis, and cannot, in principle, be distinguished by flow dichroism. However, it is known that stretching of DNA fibers, presumably introducing tilt of the base pairs, reduces the dichroism of the double helix. Thus, if binding of dye molecules does not diminish the flow dichroism, one can reasonably conclude that there is no tilting of the base pairs.[127] This is found to be the case for binding of acridines. It is furthermore found that the extent of dichroism of the bases and dye chromophore is very similar, which is consistent with (although it does not prove) a structure in which the dye is parallel to the base pairs, as in intercalation. In appropriate cases, flow dichroism measurements can prove negative conclusions, for example, that a chromophore is *not* perpendicular to the helix axis. The ambiguity that then remains is which of the possible nonperpendicular structures actually occurs.

D. Circular Dichroism and Optical Rotatory Dispersion

Dye molecules that are themselves optically inactive become active when bound to DNA or other polymers.[125,128-133] The main problem in such systems is to separate the effects due to various factors. The largest induced circular dichroism bands arise from dye-dye interactions, since the observed amplitudes are larger when the dyes are crowded together on the polymer. Figure 7-30 shows circular dichroism spectra for the proflavine-DNA complex when the ratio of polymer to dye has three different values. For the curve marked 110, the dyes are widely separated, and the CD spectrum follows closely the absorption spectrum of the complex. When less

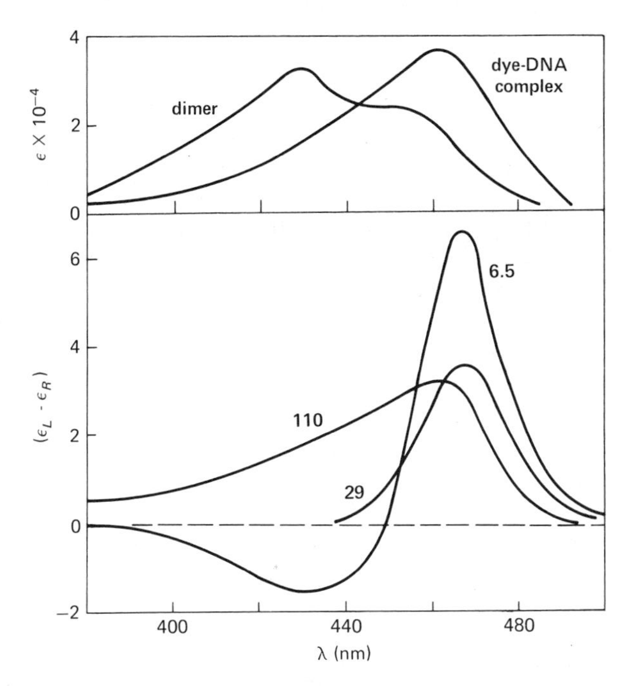

FIGURE 7-30 Comparison of absorption (upper) and circular dichroism spectra of proflavine in various forms of complex. The numbers in the lower graph indicate the ratio of nucleotide to dye concentration. [From H. J. Li and D. M. Crothers, *Biopolymers,* 8, 217 (1969). © 1969 by John Wiley & Sons, Inc. Reprinted with permission.]

polymer is present, the long wavelength band increases in intensity, and a new negative band appears at shorter wavelength.

The CD spectrum shown when the ratio of polymer to dye is 110 is nearly that obtained upon extrapolation to infinite separation of the dyes. This may be performed by plotting the amplitude of the band against r^2, the square of the concentration of bound dye. A measurable intercept is found at high salt concentrations, but this limiting CD is near zero when the salt concentration is low. Li and Crothers[133] found the change between the two limiting behaviors over a narrow range of salt concentrations. The origin of the CD band when the dyes are widely separated must be the asymmetry of the helix surrounding the binding site. The fact that this changes with salt concentration implies some kind of geometric change in the complex structure, but this cannot be further specified at this time.

The increased CD intensity at higher levels of saturation must result from dye-dye interactions. The problem of interpretation is compounded by the presence of both intercalated dyes and dyes stacked on the outside of the helix when the degree of binding is large. The linearity of the plot of amplitude against r^2 implicates dye-dye pairwise interactions as the source of these effects at relatively low degrees of binding, but the underlying physical basis remains an open question. There are probably some effects due to interaction of intercalated dyes. This could arise from electronic interactions, or from influence of neighboring dyes on the complex geometry. Furthermore, stacking effects are expected to give rise to induced circular dichroism. The tentative resolution of the spectra offered by Li and Crothers[133] indicates that at least part of the negative band is due to interaction of externally stacked dyes.

As with the study of double helix structure, circular dichroism offers good potential for investigation of the structure of complexes of dyes with helical molecules. However, because of the complexities noted above, the results with dyes are probably not presently adequate to permit sound conclusions concerning complex structure.

E. Resonance Techniques

ESR and NMR have not yet found wide application in the study of nucleic acid complexes. In the case of electron paramagnetic resonance, the reason is that few of the materials of interest are radicals, while the use of nuclear magnetic resonance is limited because fixation of the small molecule to DNA generally destroys the high resolution NMR spectrum due to the slow rotation time.

That ESR can have great utility for investigation of the structure of DNA complexes was shown by Ohnishi and McConnell[134] in a study of binding of the chlorpromazine ion radical. Figure 7-31 shows the para-

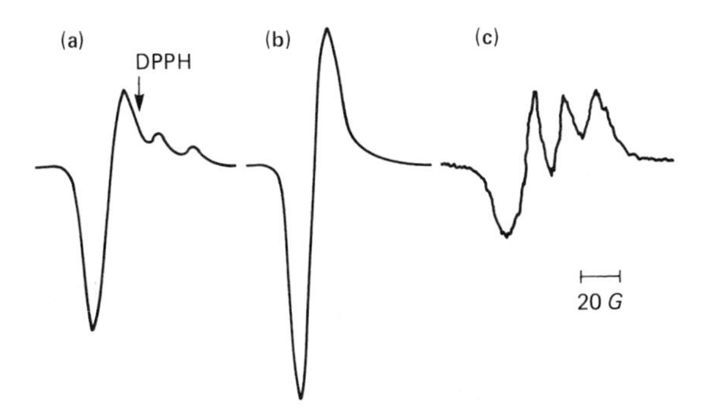

FIGURE 7-31 Paramagnetic resonance of the chlorpromazine cation bound to DNA: (a) "No flow" (see text). (b) Perpendicular flow. (c) Parallel flow. [From S. Ohnishi and H. M. McConnell, *J. Amer. Chem. Soc.*, 87, 2293 (1964). [Copyright 1964 by the American Chemical Society. Reprinted by permission of the American Chemical Society.]

magnetic resonance spectrum of the complex, with the typical broad resonances of a bound and immobilized small molecule. As shown in Fig. 7-31(b) and (c), the resonances observed when the DNA is flow-oriented perpendicular or parallel to the applied field are very different from one another. The explanation is to be found in the anisotropy of the ^{14}N hyperfine interaction, depending on whether the applied field is parallel or perpendicular to the aromatic molecular plane of the radical. The results shown in Fig. 7-31 are those predicted if this plane is perpendicular to the helix axis.

A few other radicals that bind to DNA have been studied by means of paramagnetic resonance. For example, the equilibrium characteristics of binding 5-methylphenazinium cation could be determined by ESR methods.[135] Again, the fine structure of the ESR spectrum of the radical disappeared on binding, leaving a characteristic broad resonance.

Blears and Danyluk[136] have reported an NMR study of acridine binding to solutions of DNA in D_2O–methanol mixtures. Under conditions of strong binding, presumably corresponding to intercalation, the high resolution resonances could not be detected, presumably because of the inability of the complex to rotate. When more dye was added, a weaker binding was found, producing a broadened but measurable NMR spectrum. This is probably due to outside binding, which kinetic measurements[69] show to be in rapid equilibrium with free dye. An interesting finding reported by Blears and Danyluk concerned the behavior of acridine bound to a quenched sample of denatured DNA. The acridine resonances were found to be sharp but shifted to higher field relative to the free dye. This was interpreted as due to the ring current interaction with adjacent bases, analogous to the NMR effects of stacking bases together. Whether the dye interacts with a single base in the single-stranded structure, or whether it is intercalated between adjacent bases could not be determined.

F. Hydrodynamic Properties

This and the next few sections are concerned with methods of detecting changes in macromolecular shape when small molecules bind to nucleic acids. One of the properties expected of intercalation is that the length of the helix should increase. Considerable attention has been focused on this problem, and there are several applicable techniques. One of the simplest and most widely available methods is study of the hydrodynamic properties.

Early observations by Lerman[112] indicated that acridine dyes produce an increase in the viscosity and a decrease in the sedimentation coefficient of DNA samples. This result could be produced by either an increase in molecular length, an increase in chain stiffness, or both. Either of these factors leads to an increase of the hydrodynamic volume occupied by the chain molecule, and thus to the observed sedimentation and viscosity changes. The qualitative effect of a small molecule on the observed viscosity of a DNA solution is an important and simple screening test for substances that might be intercalators.[114]

The quantitative aspects of viscosity determinations on nucleic complexes were improved by Drummond et al.[137] In order to determine the intrinsic viscosity of a polymer, it is necessary to extrapolate to zero concentration. If this is done with a complex by simply diluting with buffer, the complex will dissociate, and the nature of the complex at infinite dilution is not well defined. This can be avoided by working at constant chemical potential of the small molecule, since the degree of binding r is determined by this quantity. In practice, this means holding the free concentration C_F of dye constant in the dilution series. Examination of any of the equations for binding equilibria, such as (7-70) for a Scatchard plot, reveals that r depends on C_F but not on the total polymer concentration. Thus the intrinsic viscosity at any value of r is readily determined.

It is difficult to give a quantitative interpretation to the changes in viscosity and sedimentation coefficient of a sample of high molecular weight DNA. The reason is that essentially equivalent effects should be produced by stiffening or by lengthening the molecule. However, if a low molecular weight sample is used, this problem can be circumvented. Since the molecule is essentially a rod, an increase in stiffness should have negligible effect on the viscosity or sedimentation coefficient, and the hydrodynamic changes can be interpreted in terms of the dimensions of the rod. This principle was first applied by Müller and Crothers[73] to the actinomycin-DNA interaction and by Cohen and Eisenberg[138] to the proflavine-DNA complex. (Low molecular weight DNA samples were used for analogous reasons in light scattering measurements by Mauss et al.[139]) Sedimentation and viscosity measurements can be combined to calculate the change in both the major and minor axes of the equivalent hydrodynamic particle, the approach used by Müller and Crothers. Alternatively, if it can be assumed that the minor axis does not change greatly, which is likely to be the case for acridine intercalation, then length changes can be calculated independently from viscosity or sedimentation measurements, as done by Cohen and Eisenberg.

The rate of sedimentation of a macromolecule depends on the buoyancy of the particle. As written by Cohen and Eisenberg, the sedimentation coefficient s is

$$s = \frac{M_n}{fN}\left(\frac{\partial \rho}{\partial c_n}\right)_\mu \tag{7-98}$$

where M_n is the molecular weight of the nucleic acid, N Avogadro's number, f the particle frictional coefficient, and $(\partial\rho/\partial c_n)_\mu$ the density increment due to nucleic acid at constant chemical potential of solutes such as buffer and proflavine. $(\partial\rho/\partial c_n)_\mu$ is commonly approximated by $1 - \bar{v}\rho^s$, where $\bar{v}$ is the partial specific volume of the macromolecule and ρ^s the density of the solvent. Equation (7-98) applies in the presence and absence of bound molecule, but in all cases M_n and c_n refer to the molecular weight and weight concentration of only the DNA (actually, usually the Na^+ salt). If the density increment can be measured, the only unknown in Eq. (7-98) is the frictional coefficient, which can therefore be calculated from the measured sedimentation coefficient.

For a rodlike macromolecule, the ratio of the frictional coefficient f_r at degree of binding r to the frictional coefficient f of the DNA alone, is closely approximated by[138]

$$\frac{f_r}{f_0} = \left(\frac{L_r}{L_0}\right)\left(\frac{\ln 2p_0}{\ln 2p_r}\right) \tag{7-99}$$

where p is the axial ratio of the rod and L the length. Alternatively, one can use the model of an ellipsoid of revolution, with major axis L and minor axis b. The ratio of frictional coefficients is then approximately[73]

$$\frac{f_r}{f_0} = \left(\frac{L_r}{L_0}\right)^{0.803}\left(\frac{b_r}{b_0}\right)^{0.197} \tag{7-100}$$

these can be combined with (7-98) to yield either

$$\frac{L_r}{L_0} = \frac{s_0}{s_r}\,\frac{\left(\dfrac{\partial\rho}{\partial c_n}\right)_{\mu,r}}{\left(\dfrac{\partial\rho}{\partial c_n}\right)_{\mu,0}}\,\frac{\ln(2p_r)}{\ln(2p_0)} \tag{7-101}$$

for a rod, or

$$\frac{L_r}{L_0} = \left[\frac{s_0}{s_r} \frac{\left(\frac{\partial \rho}{\partial c_n} \right)_{\mu, r}}{\left(\frac{\partial \rho}{\partial c_n} \right)_{\mu, 0}} \right]^{1.24} \left(\frac{b_0}{b_r} \right)^{0.245} \tag{7-102}$$

for an ellipsoid of revolution. To a first approximation, the term involving p in Eq. (7-101) or b in Eq. (7-102) can be considered constant, so the change in length can be calculated from sedimentation data and the density increment. Schmechel and Crothers[140] found that the equation for an ellipsoid of revolution predicts length changes that agree better with the expected value of one base pair spacing per intercalated molecule than does the equation for a rod.

Occasions may arise where one does not have enough of the small molecule to measure the density increment. In this circumstance one can estimate the partial specific volume $\bar{v}_A$ of the small molecule A, and assume additivity of partial specific volumes in the complex.[73] The ratio of density increments appearing in (7-101) and (7-102) is then

$$\frac{\left(\frac{\partial \rho}{\partial c_n} \right)_{\mu,r}}{\left(\frac{\partial \rho}{\partial c_n} \right)_{\mu,0}} \cong \left[\frac{\left(1 - \bar{v}_D \rho^s + \frac{M_A}{M_D} (1 - \bar{v}_A \rho^s) r \right)}{(1 - \bar{v}_D \rho^s)} \right] \tag{7-103}$$

where M_A is the molecular weight of the bound molecule A, M_D the residue molecular weight of DNA, r the ratio of bound A per DNA residue, and $\bar{v}_D$ the partial specific volume of DNA.

The experimental observation is that the sedimentation coefficient of short DNA molecules decreases when either proflavine[40] or actinomycin[73] is bound. Consequently, the ratio s_0/s_r increases, as does the density increment ratio, when r increases. Thus, barring a very large change in b, Eqs. (7-101) and (7-102) predict an increase in length of complex over DNA alone. The sedimentation coefficient of the actinomycin complex decreases somewhat less than does that of the proflavine complex. The reason is the greater mass of actinomycin, which tends to make s increase. The actual calculated length changes are quite similar in the two cases, and are roughly equal to the expected value of one base pair spacing per intercalated chromophore.

Equations relating length changes to the viscosity of small DNA molecules can also be derived from the equations for rods or for ellipsoids of revolution. These are, respectively,

$$\frac{L_r}{L_0} = \left(\frac{[\eta]_r}{[\eta]_0} \frac{f(p)_0}{f(p)_r} \right)^{0.333} \tag{7-104}$$

for rods,[140] where $f(p)$ is

$$f(p) = \frac{3}{(\ln 2p - 1.5)} + \frac{1}{(\ln 2p - 2.5)} \tag{7-105}$$

while for an ellipsoid of revolution (obtained from the equations of Müller and Crothers by setting the particle volume proportional to Lb^2),[73]

$$\frac{L_r}{L_0} = \left(\frac{[\eta]_r}{[\eta]_0}\right)^{0.356} \left(\frac{b_0}{b_r}\right)^{0.0677} \tag{7-106}$$

Again barring a drastic increase in b, an increase in the viscosity of complex over DNA indicates an increase in molecular length. This is found to be the case for both proflavine-DNA and actinomycin-DNA complexes. Müller and Crothers[73] avoided uncertainty about variations in b by eliminating that parameter from Eqs. (7-102) and (7-106), resulting in a calculated length increase on actinomycin binding close to that expected for intercalation. Cohen and Eisenberg[138] assumed that the axial ratio p did not change greatly, and calculated length increases from both sedimentation and viscosity data, with good agreement between the two.

Müller and Crothers[73] demonstrated another difficulty that can arise in hydrodynamic measurements on high molecular weight DNA. It was found that intrinsic viscosity of large DNA molecules decreases when actinomycin is bound. The explanation given was that actinomycin serves to increase the intrachain interactions in the molecule, in a sense forming weak "cross links" that cause the molecule to contract. The result is to mask the effects of a length increase and give quite a false preliminary idea of the structural change on binding. Both for quantitative measurement of shape changes and for avoiding qualitative misinterpretation, low molecular weight samples of DNA are strongly to be preferred in hydrodynamic experiments.

G. Light and Low-Angle X-Ray Scattering

The scattering of electromagnetic radiation by solutions of DNA is altered by complexation with small molecules. Changes in shape can be quantitatively measured by these methods; in particular, the change in the mass per unit length of the particle can be determined.[139,141-147] If a dye molecule with a mass smaller than that of a base pair is intercalated into the double helix, thereby producing an increase of length by one base pair spacing, the mass per unit length should diminish. Such a decrease has been detected for acridine dyes both by light scattering[139] and by low-angle X-ray scattering.[141] Chapter 5 should be consulted for detailed discussion of these methods.

Light scattering determinations of the mass per unit length require correction for the optical anistropy of the particle. The polarizability of a base pair or a dye molecule depends on its orientation, and there is excess scattering due to fluctuations in orientation. This correction does not seem to be important for DNA alone, since the measured mass per unit length agrees well with expectation. However, Mauss et al.[139] found that the calculated length increase was not as large as would be expected for 1 base pair spacing per intercalated dye, a result they attributed to anisotropy of the scattering particles. This interpretation must be regarded as tentative, since it is possible that the real length increase might be somewhat smaller than the simple expectation value.

H. X-Ray Diffraction

X-ray diffraction studies can be made on fibers of DNA to which small molecules are bound.[113,116,148,149] The general observation in such instances is that as increasing amounts are bound, the quality of the diffraction pattern deteriorates. The intensity maxima become more diffuse and the information content of the photographs decreases. This is the result that is to be expected from the random attachment of molecules that distort the DNA structure and therefore diminish the regularity. The observations are further complicated by the finding of rather different patterns depending on the water content of the fiber.[149] As a result of these difficulties, X-ray studies have not provided an unambiguous answer to the question of the structure of small molecule complexes, although they have provided important evidence.

Neville and Davies[149] reported a systematic examination of the diffraction patterns observed for acridine dye-DNA complexes. Easiest to interpret were the photographs taken at high degrees of hydration of the fiber, under which conditions the patterns were fairly closely related to those given by the B form of DNA. Figure 7-32 shows a comparison of diffractions in the presence and absence of proflavine. The strong 3.4 Å meridional reflection (on the vertical axis of the photograph) is present in both cases, but the remainder of the pattern is altered by the dye. In particular, the layer line spacings (along the diagonal in the photograph) are more diffuse in the dye pattern, and are closer to the center of the photograph, indicating an increase in the corresponding dimension in the fiber. These spacings arise from the repeat distance of one full turn of the helix, 34 Å in the B form, and larger in the presence of the dye. The position of the equatorial spacing (along the horizontal axis of the photograph) is a measure of the distance between DNA molecules, and depends on the water content of the fiber. In the presence of dye, the layer line spacing varies with the degree of hydration, or the distance between DNA centers. At lower degrees of hydration the DNA molecules are closer together, and the layer line spacing in the presence of proflavine is essentially identical with that in the B form of DNA. When the water content

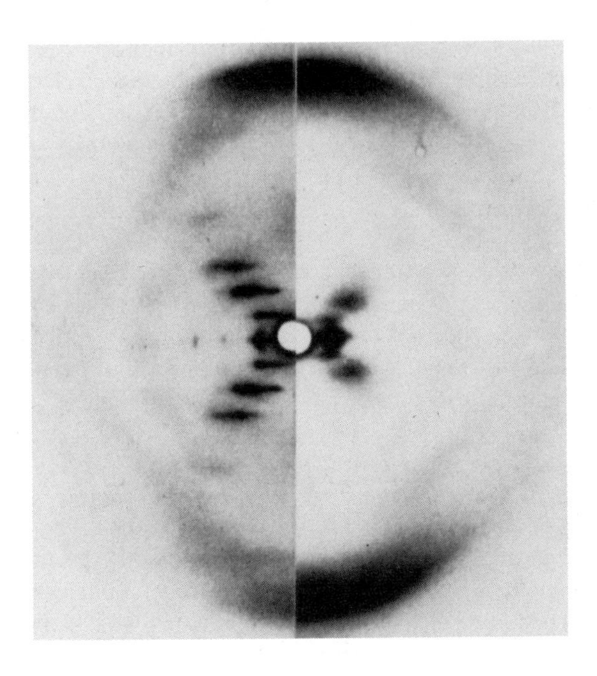

FIGURE 7-32 Comparison of diffraction patterns observed for the B form of sodium–DNA (left) and the DNA–proflavine complex (right). [From D. M. Neville Jr. and D. R. Davies, *J. Mol. Biol.*, **17**, 57 (1966). Reprinted with permission.]

is increased, the layer line spacing increases in the presence of proflavine or acridine orange, but not in the absence of dye or in the presence of *p*-rosaniline, a nonplanar, nonintercalating dye.

The increase in the layer line spacing implies an increase in the repeat distance for one turn of the helix. This can arise from an increase in the base pair spacing, from an unwinding of the helix, or both. Neville and Davies point out that a model in which the dye is bound externally and causes extensive unwinding of the helix, would fit the X-ray diffraction data. However, such a model would not predict the length increase found by a number of measurements on solutions of the complex. Therefore they prefer an intercalated model, in which the layer line spacing is increased by insertion of the dye and by some unwinding of the helix. The presently preferred model of a length increase of 3.4 Å and unwinding of 12° for each dye intercalated is in satisfactory agreement with the measured increase of 14 Å in the layer line spacing for proflavine at 0.17 dyes per phosphate and high water content.[149]

When the water content of the fiber is reduced, the layer line spacing returns to the value for the B form of DNA. It is clear that there cannot be extensive intercalation of the dye under these conditions, and a model of external binding is to be preferred. Relaxation kinetic measurements[69] show that even in solution, several percent of the bound proflavine is attached externally. Thus the balance between intercalation and external binding is a delicate one, readily influenced by environmental conditions.

I. Autoradiography

The technique of autoradiography can be used to detect length changes in DNA when small molecules are bound. Extensive use has not been made of this method in application to intercalation problems, but one of the first demonstrations of a length increase on binding acridine dyes came from the autoradiographic experiments of Cairns.[150]

J. Unwinding of Supercoiled DNA

Highly specific structural information on DNA complexes can be obtained from studying the effect of binding on the properties of closed circular double helical DNA. All structural models for intercalation require at least some unwinding of the double helix to accommodate the dye. If such a change in the average winding angle occurs in a closed circular molecule, there must be a corresponding change in the number of tertiary or superhelical turns. In all of the natural molecules yet discovered, the tertiary turns have a sense such that unwinding the double helix leads to initial unwinding of the superhelix, followed by formation of turns of the opposite sense if unwinding of the double helix is continued. Thus, one expects that the effect of dye intercalation will be first to remove superhelical turns, producing a circle without tertiary turns, followed by a rewinding in the opposite sense.

This technique has been applied to intercalating dyes, notably to ethidium by Crawford and Waring[151] and Bauer and Vinograd[152] and to a range of dyes by Waring.[153] The change in properties of the closed circular DNA can be conveniently observed by examining sedimentation velocity or banding density. A closed circular DNA with superhelical turns is distinctly more compact than the nicked circle with no net turns (see Chapter 5). Therefore one expects a decrease in sedimentation coefficient during binding of the first dye molecules, responding to the loss of tertiary turns. The sedimentation coefficient of the closed form rises again after a narrow minimum, responding to formation of tertiary turns of the opposite sense. From these experiments one can determine, by the r value at the minimum in the sedimentation coefficient, the number of dye molecules required to compensate exactly the superhelical turns in the closed circular molecule. Bauer and Vinograd[74,152] showed that the same information can be obtained from study of banding densities.

These experiments can be used in one of two ways. If one knows the degree of unwinding produced by intercalation of the dye, one can calculate the number of superhelical turns in a given closed circular DNA; most applications have taken this point of view. Alternatively, if one knows the number of tertiary turns in a closed circle, measurement of the number of dyes required to unwind these yields the angle of unwinding provided by insertion of the dye. It is not easy to obtain accurate measurement of the number of superhelical turns by independent means, so the general procedure has been to use the figure of 12° for insertion of ethidium as a standard for determining superhelix winding numbers. Since other techniques give reasonable agreement with these figures,[154] the 12° estimate cannot be far in error. Once the superhelix content of a particular DNA is well established, the degree of unwinding produced by any dye or other bound molecule can be accurately established, which is a great help in proposing specific structural models.

K. Model Building

The structure of DNA is relatively well known, and many of the intercalating dyes are small and simple molecules, so there is reason to be optimistic about the chances for reaching structural conclusions from examination of molecular models. However, there seems to be considerable latitude in constructing intercalated models. For example, Lerman proposed models in which the helix is unwound by 45°[112] or by 36°[114] to accommodate the inserted dye, while Neville and Davies[149] concluded that there was uncertainty in this figure from their examination of models. Probably the best estimate from models is that of Fuller and Waring,[113] who proposed the 12° minimum unwinding figure. In only a few cases other than the simple intercalators is there sufficient information to warrant formulation of detailed models. One example is the actinomycin-DNA complex, for which Hamilton et al.[116] proposed an outside complex stabilized by hydrogen bonds, and Müller and Crothers[73] suggested an intercalated structure, with a specific role for the peptide side chains. Sobell et al.[155] proposed a modified intercalation model based on the crystal structure of the deoxyguanosine-actinomycin complex.

L. Changes in Melting Properties

When another substance binds to a nucleic acid, the helix-coil transition is usually altered; in particular the midpoint (T_m) is often shifted, and the transition may become broader. Such measurements provide a convenient means for detecting complex formation, and have been widely applied.[156-165] In the case of a reversible binding equilibrium, one can express quantitatively the relation between binding properties and transition curves in the presence and absence of bound molecules. Here we consider the matter on a qualitative level. Suppose that substance M is bound to the helical form to the extent r_h and to the coil in amount r_c. This equilibrium is then superimposed on the helix-coil transition equilibrium:

$$\text{helix} \cdot r_h M \rightleftharpoons \text{coil} \cdot r_c M + (r_h - r_c)M \tag{7-107}$$

If the binding is tighter to the helix, then r_h is greater than r_c and $r_h - r_c$ is positive. In this circumstance, according to the law of mass action, addition of more M will shift the equilibrium to the left. The result will be to stabilize the helix at the expense of the coil, and consequently raise the T_m. A substance that binds more tightly to coil than to helix should produce the opposite effect, or a decrease in T_m. Dramatic effects on thermal transition curves are seen when polyamino acids or basic proteins are bound to nucleic acids[159-161]; examples are shown in Fig. 7-33 for the complex of poly-L-lysine with poly I · poly C (a) and protamine with DNA (b). In both instances the melting curve becomes biphasic when the polypeptide is present. In case (a), the first portion of the melting process occurs always at the transition temperature characteristic of the nucleic acid alone; as more polylysine is added, the fraction melting at this temperature decreases steadily. The transition of the complex occurs at higher temperature. In the case of protamine, there is a steady increase in the midpoint temperature of the lower melting transition as the amount of bound material increases. While it is possible, in principle, to produce a biphasic melting transition with a reversible, noncooperative binding process (see Chapter 6), it is not likely that this is the case here. At low salt concentrations, binding of polyamino acids is largely irreversible, and the thermodynamic considerations of Chapter 6 are not applicable. Irreversible binding which is cooperative could produce the behavior of Figure 7-33, since it would serve to segregate complexed molecules from the noncomplexed, yielding two separate transitions for the two species. If the irreversible binding were less cooperative, so that the bound molecules tended to be more uniformly distributed, the division into two separate transitions would be less sharp, and the two characteristic midpoints might be expected to shift as the binding ratio changes. This latter is closer to the behavior observed with protamine-DNA complexes, although in this case the higher transition temperature appears to be constant, independent of the degree of binding. Possibly the upper transition is characteristic of the protein which would account for its constancy.

M. Kinetic Measurements

The techniques discussed so far are primarily concerned with structural and equilibrium aspects of the complex, leaving unanswered many questions concerning the mechanism of complex formation. In some cases this kinetic

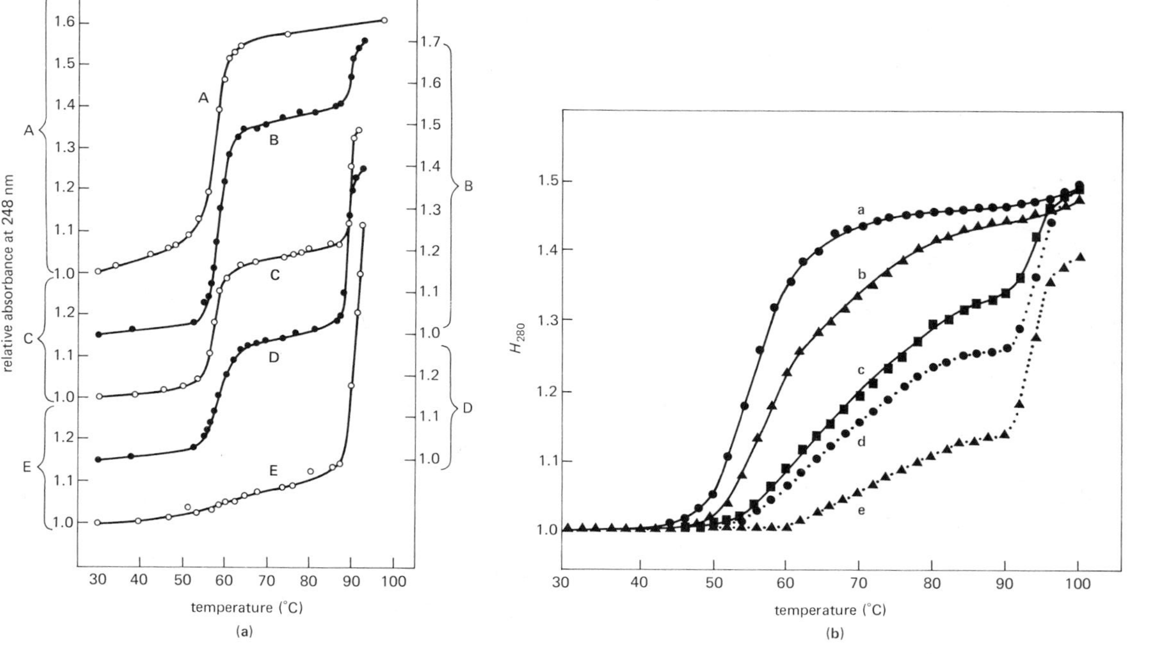

FIGURE 7-33 (a) Absorbance-temperature profile of poly-L-lysine polus poly (I + C) at 248 nm. Solvent: 0.05 M NaCl in 0.001 M sodium citrate buffer (pH 7.0).

Curve A, poly I 2.5×10^{-5} M plus poly C 2.5×10^{-5} M.

Curve B, poly I 2.5×10^{-5} M plus poly C 2.5×10^{-5} M plus poly-L-lysine $0.8_3 \times 10^{-5}$ M (NH_2 : P = 0.16_6 : 1).

Curve C, poly I 2.5×10^{-5} M plus poly C 2.5×10^{-5} M plus poly-L-lysine $1.2_5 \times 10^{-5}$ M (NH_2 : P = 0.25 : 1).

Curve D, poly I 2.5×10^{-5} M plus poly C 2.5×10^{-5} M plus poly-L-lysine $1.6_6 \times 10^{-5}$ M (NH_2 : P = 0.33 : 1).

Curve E, poly I 2.5×10^{-5} M plus poly C 2.5×10^{-5} M plus poly-L-lysine 2.5×10^{-5} M (NH_2 : P = 0.5 : 1). [From M. Tsuboi, K. Matsuo, and P. O. P. Ts'o, *J. Mol. Biol.*, **15**, 256 (1966). Reprinted with permission.] (b) Thermal denaturation profiles of DNA–protamine complexes. Buffer: 0.001 M-sodium cacodylate, pH 7.0. Peptide cation/DNA phosphate ratios: curve a, DNA; b, 0.26; c, 0.52; d, 0.78; e, 1.04, monitored at 280 nm. H_{280} : relative hyperchromicity at 280 nm. [From D. E. Olins, A. L. Olins, and P. H. von Hippel, *J. Mol. Biol.*, **33**, 265 (1968). Reprinted with permission.]

information may be crucial for understanding the biological function of the bound molecule, and it is therefore wise not to neglect it.

The same problems concerning the nature of the binding site that arose in formulation of binding equilibrium constants must also be faced in kinetic studies. For binding of intercalating dyes, antibiotics, and so on, the main decision is again a choice between the Scatchard and neighbor exclusion models (see Section III). If binding is strongly cooperative, solution of the kinetic model becomes highly complicated, and is analogous to the problem of helix-coil transition kinetics. We will not consider these problems here, but will restrict our attention to the Scatchard and neighbor exclusion models.

The rate of a binding reaction should be proportional to the concentration C_F of free dye and to the concentration C_S of unoccupied sites:

$$\text{rate} = k_1 C_F C_S \tag{7-108}$$

where k_1 is the rate constant for complex formation. The main problem is to know what to insert for the concentration of free sites, and the Scatchard and neighbor exclusion models (Section III) give different answers. According to the Scatchard model,

$$\begin{aligned} C_S &= B_{ap} C_N{}^0 - C_B \\ &= C_N{}^0 (B_{ap} - r) \end{aligned} \tag{7-109}$$

where B_{ap} is the number of binding sites per base pair, $C_N{}^\circ$ the total concentration of nucleic acid base pairs, C_B the concentration of bound small molecules, and r the ratio $C_B/C_N{}^\circ$. For this model we write the rate of complex formation as

$$\text{rate} = k_{1,ap} C_F C_N{}^0 (B_{ap} - r) \tag{7-110}$$

The dissociation rate constant $k_{-1,ap}$ is directly measurable and does not depend on the definition of the binding site.

For the neighbor exclusion model, a simple expression for the concentration of available binding sites can be given only when r approaches zero. In this case the number of binding sites is equal to the number of potential sites $B_0 C_N{}^\circ$, where B_0 is the number of potential sites per base pair. By custom we write the rate of the complex formation reaction in terms of the number of unoccupied potential sites, $B_0 C_N{}^\circ - C_B$

$$\begin{aligned} \text{rate} &= k_1(r) C_F (B_0 C_N{}^0 - C_B) \\ &= k_1(r) C_F C_N{}^0 (B_0 - r) \end{aligned} \tag{7-111}$$

where $k_1(r)$ is a reaction rate constant that will depend on the value of r. The limiting value of the reaction rate when r approaches zero is

$$\lim_{r \to 0} (\text{rate}) = k_1(0) C_F C_N{}^0 B_0 \tag{7-112}$$

where $k_1(0)$ is the intrinsic rate constant for complex formation at an isolated potential binding site. It is clear that

$$\lim_{r \to 0} k_1(r) = k_1(0) \tag{7-113}$$

The dissociation rate constant is again given by k_{-1}.

The rates of complex formation are written in terms of the appearance of bound molecules, which will be independent of the choice of a model for the binding site. Thus expressions (7-110) and (7-111) for the rate can be equated, yielding

$$\frac{k_1(r)}{k_{1,ap}} = \frac{B_{ap} - r}{B_0 - r} \tag{7-114}$$

and

$$\frac{k_1(0)}{k_{1,ap}} = \frac{B_{ap}}{B_0} \tag{7-115}$$

as the analogs to Eqs. (7-76) and (7-79), respectively. Since B_{ap} must be less than or equal to B_0, $k_1(r)$ must be less than or equal to $k_{1,ap}$. The rates of all first-order steps in complex formation or dissociation are independent of the definition of the binding site, and will be the same for the Scatchard or neighbor exclusion models.

Both mixing techniques and relaxation methods have been applied to study the kinetics of DNA complex formation. The general finding is that the reaction does not occur in a simple one-step reaction, but that there are intermediates formed on the path to the final complex. These intermediates are often present in very small amounts in the final complex, so that the only way they can be discovered and studied is by transient techniques. The general purpose of these experiments is to clarify the mechanism of complex formation.

As an example of the application of mixing techniques to the kinetics of a DNA binding reaction, we consider the reaction of actinomycin with DNA.[73] If the two components are mixed at low concentrations, the absorption spectrum of actinomycin changes to that characteristic of the bound form by a kinetic path that follows approximately the prediction of a simple second-order reaction, with a rate constant of about 10^4 M^{-1} sec^{-1}. However, if the reactants are mixed at higher concentrations, much more complicated kinetic curves are observed. Because of the high concentration, the second-order parts of the reaction become very fast, and concentration-independent first-order steps now limit the rate of formation of the final complex. Figure 7-34 shows the mechanism that was postulated to explain the kinetic data. The initial second-order reaction forms a transient intermediate, which is converted to the final form by reactions involving conformational changes of the peptide rings. The dissociation kinetics can be

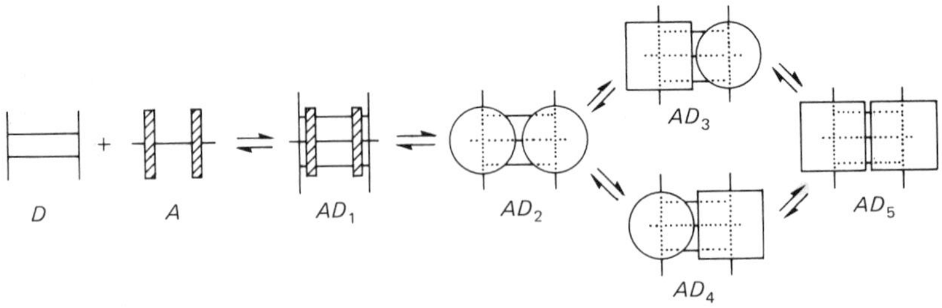

D A AD1 AD2 AD3 AD4 AD5

FIGURE 7-34 Schematic drawing of a proposed mechanism for the reaction of actinomycin with DNA. The reaction begins at the left with DNA (D) and actinomycin (A); the DNA pairs are represented by horizontal lines, as is the actinomycin chromophore. The actinomycin peptide rings are viewed from the edge; it is supposed that their faces interact with each other in solution. In complex AD_1, the chromophore has been inserted between the base pairs, and in complex AD_2 the peptide rings reorient themselves to interact more favorably but nonspecifically with the DNA backbones; the rings are now viewed from their faces. In complex forms AD_3 and AD_4, one or the other of the peptide rings undergoes a conformational change which adapts it to interact specifically with the DNA, and in form AD_5 both rings have been so rearranged. At equilibrium, only AD_2, AD_4 and AD_5 are present in significant amounts with about half the complex present as AD_5. [From W. Müller and D. M. Crothers, *J. Mol. Biol.*, **35**, 251 (1968). Reprinted with permission.]

observed by adding Duponol to the complex; several decay times are seen, indicating a mixture of the species AD_3, AD_4 and AD_5 in the equilibrium complex.

The important feature of this kind of experiment is to reveal the intermediate steps in the complexation reaction. When the concentration of reactants is low, the second-order step is rate-limiting for the total reaction, and nothing is learned about the intermediate steps. However, advantage can be taken of the concentration dependence of the second-order step to increase its rate until it is no longer limiting. If the intermediate complex forms are of sufficient stability to be formed in appreciable amounts when the reactants are mixed at high concentration, first-order steps corresponding to conversion of the transient intermediates to the final complex will be seen. Thus it is important to emphasize that observation of simple second-order reaction kinetics at low concentration does not mean that there are no first-order steps in the reaction, or that no transient intermediates are formed.

The other main kinetic technique used to study nucleic acid complexes is that of relaxation, especially temperature-jump methods.[166] In particular, the latter have been applied to the complex of acridine dyes with nucleic acids.[69,140] The results are fitted to a mechanism of the form

$$\mathrm{P} + \mathrm{D} \underset{k_{21}}{\overset{k_{12}(r)}{\rightleftharpoons}} (\mathrm{P}-\mathrm{D})_{\mathrm{out}} \underset{k_{32}}{\overset{k_{23}}{\rightleftharpoons}} (\mathrm{P}-\mathrm{D})_{\mathrm{in}} \tag{7-116}$$

where P is proflavine, D is DNA, and the complex forms $(\mathrm{P}-\mathrm{D})_{\mathrm{out}}$ and $(\mathrm{P}-\mathrm{D})_{\mathrm{in}}$ are the outside-bound and intercalated species, respectively. Using the neighbor exclusion model, the concentration of potential sites is $C_N{}^\circ\ (B_0 - r)$. Assuming $k_{12}(r)[C_N{}^\circ(B_0 - r) + C_F] + k_{21} \gg k_{23} + k_{32}$, the relaxation times are

$$\frac{1}{\tau_1} = k_{12}(r)\ [C_N{}^0(B_0 - r) + C_F] + k_{21}$$

$$\frac{1}{\tau_2} = k_{32} + \frac{k_{23}\,[C_N{}^0(B_0 - r) + C_F]}{\dfrac{1}{K_{12}(r)} + [C_N{}^0(B_0 - r) + C_F]} \qquad (7\text{-}117)$$

where $K_{12}(r) = k_{12}(r)/k_{21}$.

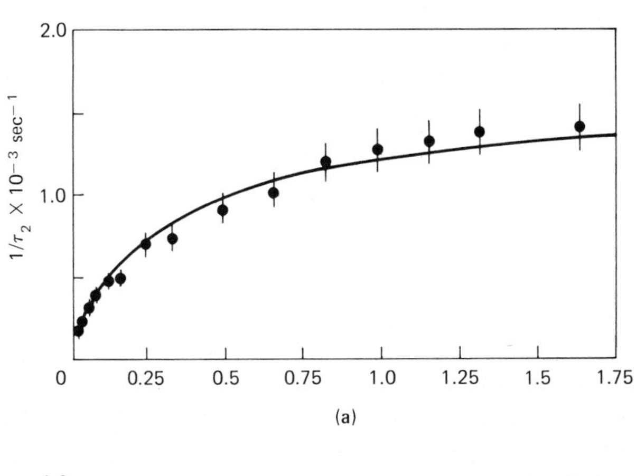

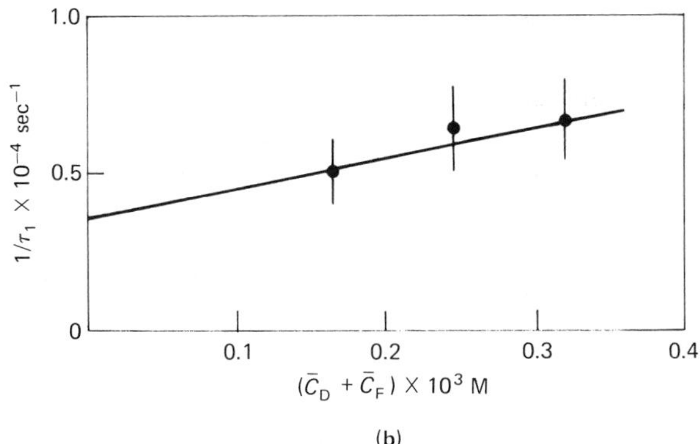

FIGURE 7-35 (a) Variation of the reciprocal of the slower relaxation time τ_2 with the concentration $\bar{C}_D$ of free DNA binding sites and free proflavine $\bar{C}_F$. $T = 10°C$. The solid curve was calculated from Eq. (7-117) with adjustment of parameters for best fit. Calf thymus DNA, $Na^+ = 0.2$ M, pH 6.9, $r = 0.046$. (b) Variation of $1/\tau_1$ with concentration for calf thymus DNA, $T = 10°C$, $Na^+ = 0.2$ M, pH 6.9, $r = 0.016$. The straight line corresponds to Eq. (7-117). [From H. J. Li and D. M. Crothers, *J. Mol. Biol.*, 39, 461 (1969). Reprinted with permission.]

Experimental measurements on the relaxation kinetics of the complex of proflavine with DNA[69] and with poly A · poly U[140] show two well-resolved relaxation times. According to Fig. 7-35 the faster one shows a linear variation with the concentration of binding sites, while the slower is initially linear at low concentration, but reaches a plateau at high concentration. This is the behavior predicted by mechanism (7-116) under the condition that the first-order conversion of $(P-D)_{out}$ to $(P-D)_{in}$ is slow compared with equilibrium between P + D and $(P-D)_{out}$. The lines in Fig. 7-35 were calculated from Eq. (7-117), with the kinetic constants adjusted for best fit. In this manner, all four rate constants can be determined from the relaxation experiments.

The general observation is that the kinetic data for proflavine binding can be interpreted in terms of an initial outside-bound complex, followed by a slower, first-order insertion reaction. On the basis of such experiments it is not possible to eliminate reaction mechanisms in which the minor form is on a side path, not on the direct reaction route from free to intercalated dye, or

$$P + D \begin{array}{l} \rightleftharpoons (P-D)_{out} \\ \rightleftharpoons (P-D)_{in} \end{array} \qquad (7\text{-}118)$$

Li and Crothers[69] discuss reasons for preferring the mechanism (7-116), as well as the bases for identifying the species $(P-D)_{out}$ with an outside-bound, nonintercalated complex.

The kinetic results, which are presently rather limited, show a considerable variation in the rate of intercalation reactions depending on the nature of the dye and nucleic acid. The rate constant for outside binding is usually a few orders of magnitude slower than expected for a diffusion-limited reaction, while the insertion reaction requires times of the order of milliseconds or longer.

A further point from the kinetic experiments is the near balance in the free energy of outside-bound and intercalated forms. When r is small, anywhere from 3% to 30% of the bound dye is attached to the outside of the double helix, depending on environmental conditions and the nature of the nucleic acid. This supports the X-ray diffraction observations of Neville and Davies,[149] which indicated preponderant outside binding at lower fiber hydrations.

VIII BINDING OF SPECIFIC SUBSTANCES

A. Acridines

In this section we consider briefly the evidence on the nature of the complexes of several small molecules with nucleic acids. These considerations are summarized in Table 7-2. In some instances the physical evidence is scanty, and it is not possible even to make a guess at the nature of the complex; a few such cases are listed in Table 7-2 but not discussed further here.

The most extensive physicochemical work has been done on the acridine dyes, including proflavine, acridine orange, quinacrine, and acriflavine. Some kind of intercalated complex is highly probable, although a direct proof, in the form of a crystal structure of the complex, is lacking. A number of techniques show that the length of a DNA molecule increases when an acridine derivative is bound; among these are autoradiography,[150] low-angle X-ray[141] and light[139,142] scattering, and hydrodynamic measurements on rodlike pieces of DNA.[138] X-ray diffraction studies on fibers[148,149] are consistent with lengthening and/or unwinding of the double helix on complex formation. Work with closed circular DNA shows that the helix is definitely unwound,[153] although the literature is more extensive on ethidium[151,152] in this connection. In addition to a length increase and unwinding, any model for the complex must account for the geometry of placement of the chromophore relative to the helix axis. Fluorescence polarization and dichroism under flow[127] indicate that the dye molecule is roughly perpendicular to the helix axis, and that the base pairs also remain approximately perpendicular in the complex.

All of these observations are predicted by the intercalation model, and on that basis it must be considered successful. It is, however, possible to invent other models that also account for the properties of the complex. For example, one could suppose that the dye binds to the outside of the double helix, perpendicular to the axis, and causes the helix to unwind and adjacent base pairs to separate along the helix axis, producing a length increase. This would be structurally analogous to intercalation, but the intercalation space would be occupied by solvent, not the dye. Such models are far-fetched, have little predictive value, and can therefore generally be discounted.

A few problems remain in deciding on a final detailed model for the intercalated complex. Some structural basis must be found for the observed restriction on the number of binding sites, to a saturation level of about one per two base pairs. The models built so far do not predict such a phenomenon in any obvious way. We discussed earlier the problem of the apparent latitude in model building, leading to different estimates of the degree of unwinding on intercalation. Further experiments with closed circular DNA will make the unwinding number a fixed, experimental quantity, which should be taken into account in refinement of the model. Finally, some of the assumptions that went into earlier models should be relaxed. For example, it is conceivable that the Watson-Crick base pairing is altered in the complex, possibly forming a smaller number of hydrogen bonds, and involving a kind of "wobble" of one or both bases from their usual positions. Experimental resolution of this problem will probably require a crystalline complex between the dye and a double helical nucleic acid or oligonucleotide.

TABLE 7-2 SUBSTANCES KNOWN TO BIND TO NUCLEIC ACIDS

Substance	*Binding mechanism*
1. Acridines 168	110, 118, 167
proflavine cation	intercalation 69, 75, 112, 114, 120, 122, 128, 129, 130, 132, 133, 137, 138, 139, 140, 141, 153
acridine orange cation	intercalation 112, 120, 122, 124, 128, 131, 149
quinacrine cation (atebrin, atabrine)	intercalation 120, 127, 171
acriflàvine	intercalation 120, 126
2. Phenanthridines	110, 118, 167
ethidium	intercalation 113, 123, 151, 152, 153
3. Phenazines	
5-methylphenazinium cation and 5-methylphenazinium cation radical	intercalation probable 135
4. Phenothiazines	
chlorpromazine	intercalation probable for radical ion 134
5. Quinolines	
chloroquine	intercalation probable 169–171, 153
quinine	171
6. Aflatoxins	
Aflaxotin B, G	unknown 173–176
7. Polycyclic hydrocarbons	
for example, 3,4-benzpyrene	possible intercalation 177–184 also, covalent binding 185, 186

TABLE 7-2 (*Cont.*)

Substance	*Binding mechanism*
8. Actinomycins 187	
for example, Actinomycin D (= C_1)	intercalation probable, prefers G among common bases 110, 111, 167, 73, 78–81, 188–197, 153, 156
9. Chromomycin (and related compounds mithramycin, oliomycin)	simple intercalation unlikely, does not unwind DNA, prefers G among common bases, requires Mg^{2+} or equivalent for spectral effect 111, 167, 194, 153, 198–204
10. Anthracyclines 205	
β-Rhodomycin A, Daunomycin	intercalation probable 111, 167, 156, 153
11. Thiaxanthenones Miracil D	intercalation probable 206–208, 153
12. Rubiflavine ($C_{23}H_{29-31}NO_5$)	
Hedamycin ($C_{41}H_{52}N_2O_{11}$)	209
13. Anthramycin 210, 211	probably covalent 212
14. Mitomycin	covalent, crosslinks 213–216
15. Steroid hormones	
for example, estradiol, testosterone, progesterone	unknown 217–219
16. Alkyl amines and diamines	binding largely electrostatic, weak preference for A · T pairs in some cases 162, 163, 220–222
17. Steroid diamines	
irehdiamine A, malouetine	electrostatic binding, alters DNA structure 164, 165, 223, 153

Other modes of binding acridines to nucleic acids unquestionably exist. Stacking of the dyes on the outside of the double helix is well documented,[75,76,115] a reaction that occurs with a variety of polyanions. In addition, kinetic studies[69] show that isolated dyes can be bound to the outside of the double helix, accounting for several percent of the bound dye in the strong binding region. Little is yet known about the properties of this outside complex, except that it has a difference spectrum on binding that is very similar to that seen on intercalation.[69]

The question of the biological significance of intercalation remains open. Intercalating dyes cause frameshift mutations, in which a base is either inserted or deleted, resulting in loss of register for the triplet code. However, intercalating dyes differ greatly in their mutagenic activity, and the physical basis for this variation is still being sought.

B. Other Planar Dyes: Phenanthridines, Phenazines, Phenothiazines, and Quinolines

Less extensive work has been done on the physical chemistry of binding these dyes to DNA, although because of the close structural analogy with acridines and the similar characteristics of the complex, intercalation seems the likely model of binding in all instances. One general class of compounds whose binding to DNA is clearly important are those with the structure

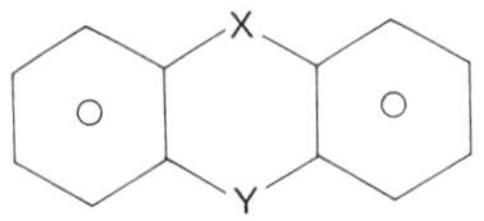

where X or Y, or both, can be hetero-atoms. In the acridines, X is N and Y is C, while the phenazines have N for both X and Y, and the phenothiazines have N and S. In other examples of biological importance, the combination is N and O in the actinomycin chromophore, and C and S in the chromophore of Miracil D (all in Table 7-2). It seems likely that all these chromophores can intercalate between the bases in the double helix, with important biochemical consequences. The structures of the phenanthridines and the quinolines are also reasonably similar to the acridines, and a conclusion of intercalation based on limited experimental evidence seems therefore justified.

C. Polycyclic Hydrocarbons

A number of carcinogenic agents, of which 3,4-benzpyrene is an example, are to be found among the polycyclic hydrocarbons. Physical studies on these materials are hampered by their limited solubility in water, which is, however, increased by the presence of DNA. Polarized fluorescence and dichroism under flow[181,182] show that the absorption moment of the chromophore is perpendicular to the helix axis. This is consistent with, but of course does not prove, the intercalation mode of binding proposed for these substances.[177] The spectral changes on binding are also consistent with intercalation.[184] Because of the limited solubility, only small amounts of the hydrocarbon can be bound to DNA, and it is therefore very difficult to detect a length increase on complex formation.

In addition to the proposed intercalation mode of binding, which is reversible, there is also evidence for covalent attachment of polycyclic hydrocarbons to DNA.[185,186] The nature of the attachment is unclear; it can be photoactivated, or brought about under mild oxidation-reduction conditions. A correlation is found between carcinogenic activity and the reactivity for covalent attachment.

D. Actinomycins

There are several groups of antibiotics that bind tightly to DNA; the most studied of these are the actinomycins. The naturally occurring members of the group have the same chromophore, but differ in amino acids in the peptide rings.[187] Several synthetic derivatives have also been studied, including substitution on the chromophore and replacement of the peptide rings by simpler groups.[73] These peptide groups are essential for the biological activity, which is to block the RNA polymerase, but not for the binding reaction with DNA.

This class of compounds shows several important contrasts with the acridines and phenanthridines, to which the actinomycin chromophore is structurally related. One aspect of particular interest is the selectivity of the binding process. Strong binding is found only with double helical DNA, and specifically does not occur with double helical RNA.[193] Furthermore, in polynucleotides containing only the common bases A, T, G, and C, no binding occurs unless G is present.[189] A similar selectivity is found for complex formation with monomers: deoxynucleosides or deoxynucleotides of guanosine form a complex[197] that is one to two orders of magnitude more stable than that of the ribo analog or nucleotides containing other bases. However, the selection rules are not quite so simple when other bases are considered. A polymer containing 2–NH_2 adenosine is found to bind actinomycin,[194] as does poly dI.[196] Furthermore, a deoxy polymer containing G, namely poly d(A · T · C) poly d(G · A · T), does not bind actinomycin appreciably.[196] In contrast to all this, the strength of binding of proflavine or ethidium is nearly independent of base composition or whether the double helical polymer is RNA or DNA.

The origin of this specificity is at present unknown; understanding its source is an important physicochemical problem. The peptide rings cannot be entirely responsible, since their removal does not cause complete loss of the preference for G.[73] In most cases "stacking" type interactions seem to be relatively nonspecific; what influences could give rise to selectivity is a largely unexplored problem.

In addition to base selectivity, there is another experimental fact concerning actinomycin binding on which there is general agreement, but little consensus concerning physical origin. Scatchard plots of binding isotherms show that the number of binding sites is even more restricted than is the case of intercalating dyes. Extrapolation of the initial linear region of the isotherm through the horizontal axis leads to a value of B_{ap} [see Eq. (7-70)] of about 0.1, which means that the apparent number of binding sites is 1 for each 10 base pairs. This is much less than the fraction of G · C pairs, indicating that for some reason binding cannot proceed to the extent of one actinomycin for each G · C pair.

There seem to be no simple rules for exclusion of binding based on local base sequence, so the most acceptable present model is that the degree of binding is limited by neighbor exclusion.[73] According to this concept, binding of an actinomycin at one site greatly reduces the affinity of neighboring sites. A theoretical binding isotherm was calculated by Müller and Crothers[73] on the assumption that bound actinomycins intercalate between the base pairs (when one pair is G · C), but there must be at least five empty intercalation sites between bound actinomycins. Agreement is very good except at high levels of binding. Again, there are different ways to interpret this disagreement. It is possible that there is another class of "weak" binding sites that are quite different, or one can suppose that the neighbor exclusion is not absolute, and therefore binding can be forced to higher levels. In this latter case, the binding sites are the same, but their affinity is greatly reduced by the presence of the adjacent bound molecules. We stress once again that curved Scatchard plots can arise either from heterogeneity of binding sites or from interaction between bound molecules; there is no way to distinguish these two possibilities simply on the basis of equilibrium binding isotherms.

A primary issue in the study of the actinomycin-DNA complex is whether or not the chromophore is intercalated between the base pairs. The similarity of its structure to that of acridines and other intercalating dyes makes logical the hypothesis that intercalation occurs in this case also. However, some of the early evidence spoke against this possibility, particularly the observation[81,156] that the viscosity of the DNA sample decreases when actinomycin is added, and the sedimentation coefficient rises. Subsequent work with rodlike pieces of DNA[73] showed that there is a length increase when the antibiotic is bound, of the magnitude expected for intercalation, and that the effects observed with high molecular weight DNA are due mainly to a weak "cross-linking" of the polymer, making its hydrodynamic volume smaller. Actinomycin analogs lacking the peptide rings produce hydrodynamic changes exactly like those of the intercalating dyes, implicating the peptide rings in the (noncovalent) "cross-linking."

Added to the observation from optical studies of DNA under flow[81,181] that the chromophore and base pairs probably remain roughly perpendicular to the helix axis in the complex, the length increase makes an intercalated structure the most attractive model. This hypothesis is further supported by the finding that actinomycin unwinds closed circular DNA to about the same extent that ethidium does.[153,224] Furthermore, the rate of binding of an actinomycin derivative having a bulky group on the chromophore is slowed down by nearly three orders of magnitude relative to the parent compound. This observation would be predicted by an intercalation model, since the chromophore must be slipped between the base pairs, but can only be added as an afterthought to an external complex model. For these reasons the intercalation model of Müller and Crothers[73] seems presently preferable to the earlier hypothesis of Hamilton, Fuller, and Reich[116] that the chromophore is hydrogen bonded to the outside of the double helix.

Sobell et al.[155] recently proposed a modified intercalation model for the actinomycin–DNA complex based on their solution of the crystal structure of the actinomycin–deoxyguanosine complex, in which they found actinomycin intercalated between two guanine rings, with a hydrogen bond between the guanine 2-amino group and the carbonyl oxygen of the actinomycin L-threonine residue. A weaker hydrogen bond connects the guanine N(3) ring nitrogen with the NH group on this same threonine residue. This stereochemistry is the basis for the structure of the DNA complex proposed by Sobell et al.[155] Deoxycytidine 5′-monophosphate is placed opposite each deoxyguanosine to form a hydrogen-bonded pair. Hence the binding involves a double helical region with sequence GpC, giving the whole complex twofold symmetry. The structure is highly plausible; however, there is ample evidence that DNA binding requires only a single guanosine residue and not a GpC sequence, so the twofold symmetry cannot be of crucial importance. A further problem is the nature of the kinetically detected conformational change, probably involving the peptide rings, when actinomycin binds to DNA. This property would seem to require that the peptide conformation be different when bound to DNA or when free in solution.

One interesting physical characteristic of the actinomycin-DNA complex is its very slow rate of dissociation. The least labile form of the complex has a time constant for dissociation of roughly 1000 seconds, compared with about 0.1 second for an actinomycin derivative lacking the peptide rings[73] and 0.01 second for proflavine.[69] Müller and Crothers argue that this slow dissociation is essential for the biological activity, since a rapidly dissociating molecule would not be so well suited to block the progress of the RNA polymerase along the DNA template. The time that the polymerase has to wait for dissociation of the antibiotic must be added to the average uninhibited polymerization time to get the total time required to copy a given number of base pairs in the presence of actinomycin. Only slowly dissociating molecules can greatly increase this time at low levels of binding. Other inhibitors of the RNA polymerase, such as the chromomycin antibiotics, also show a slow dissociation rate.[198]

E. Chromomycin, and the Related Compounds Mithramycin and Olivomycin

Chromomycin contains a chromophore to which are attached several sugar residues; the structure of chromomycin A_3 is shown in Table 7-2. This antibiotic is very similar in action to actinomycin, since the two are equally efficient in blocking the DNA-dependent RNA polymerase,[200] but both have much less effectiveness against DNA polymerase. Furthermore, chromomycin does not block synthesis from an RNA template. Its base selectivity is very similar to that of actinomycin, requiring G among the common bases, with a more general requirement for the purine 2-NH_2 group.[194,198,201] An unusual feature is the requirement for Mg^{2+} for strong complex formation.[198,199] Mg^{2+} ions lead to aggregation of the free antibiotic, and Hayasaka and Inoue[199] suggest that it binds to DNA as aggregates. However, Behr et al.[198] report no cooperative character for equilibrium binding isotherms, as would be anticipated if there were strong interaction of the bound molecules.

It seems unlikely that the chromomycin chromophore is intercalated between the base pairs. Kersten et al.[156] found no effect of chromomycin, mithramycin, or olivomycin on the viscosity of DNA. However, these experiments were done in the absence of Mg^{2+}, and may not indicate the character of the complex in the presence of divalent metal ions. More definitive is the observation of Waring[153] that binding does not unwind the double helix.

The kinetic properties of the chromomycin-DNA complex were studied by Behr et al.[198] who found that the antibiotic and several derivatives lacking one or more of the sugar residues reacted with DNA to form complex at essentially the same rate. The substances were, however, quite different in dissociation rate, with half-times of many hours in some cases. The derivatives with the greatest ability to block the RNA polymerase were also slowest to dissociate, reinforcing the conclusion, discussed earlier in connection with actinomycin binding, that a long-lived complex is essential for an effective inhibitor of the RNA polymerase.

F. Steroid Hormones

A number of steroid hormones bind to nucleic acids.[217-219] In all cases, only the denatured or single-strand form serves as a binding site, and attachment is observed only to polymers that contain either G or I. Of the steroids examined,[219] all will attach to poly G, but only estradiol and estrone show binding to poly I. An unusual feature of the binding is a Scatchard plot that indicates a very small apparent number of binding sites, approximately 1 per 10^4 nucleotides. The reason for this behavior is unknown at present, as is the structure of the complex.

G. Amines and Diamines

The main technique that has been used for examining the interaction of alkyl amines and diamines with nucleic acids is the influence on melting transitions. The general observation is that at low salt concentrations these materials cause a considerable stabilization of the double helix structure.[162,163,220] Since the amines are protonated at the pH values used, this behavior is expected by analogy with the effect of mono- and divalent metal ions like Na^+ and Mg^{2+}. Furthermore, since the binding disappears at high salt concentrations, electrostatic terms must be mainly responsible for the interaction energy. The fact that the transition temperature is raised indicates that the amines bind more strongly to helix than to coil forms, which, again, is to be expected on the basis of the higher local charge density in the double helix.

There is, however, evidence for some additional specific features of the binding process. In the case of diamines there is some structural specificity in the reaction, since the series $H_2N(CH_2)_nNH_2$ shows variation with n of the ability to stabilize DNA.[162] A maximum in the interaction is indicated at $n = 5$ (the diamine cadaverine). This may represent an optimum in the bridging distance between two phosphates. However, with poly A · poly U and poly I · poly C, this maximum is not observed;[220] in this case Mg^{2+} and diaminoethane are about equally effective in stabilization, and both are better than diaminopentane (cadaverine), which is better than diaminooctane.

A general observation that seems to have wide validity is that many alkyl amines have greater binding affinity for A · T-rich than for G · C-rich DNA. The most direct demonstration of this is provided by the experiments of Shapiro et al.,[221] who showed that whereas ions such as Li^+, Na^+, K^+, Cs^+, lysine, arginine, and tetralysine bind equally tightly to DNA of any base composition, the quaternary alkyl amines $(CH_3)_4N^+$ and $(CH_3CH_2)_4N^+$ show a definite preference for binding to DNA rich in A · T. The thermodynamic magnitude of the selectivity is not large, however; they calculated a free energy change of just 200 cal for transferring tetramethyl ammonium from a G · C pair to an A · T pair.

Results that bear on this question are experiments on stabilization of the double helix by alkyl amines. For example, Mahler and Mehrotra[162] found, for the series $NH_2(CH_2)_nNH_2$, a greater increase in T_m for DNAs rich in A · T when the diamine was added than was the case for G · C-rich samples. The experiments of Venner et al.[163] can be interpreted in an analogous manner. They measured the thermal stability of DNA when the counterion was changed from one amine to another, at a given counterion concentration. The finding was that the DNA was less stable in the presence of such alkyl amines as $CH_3(CH_2)_nNH_2$, where $n = 4$, 6, or 7, than in the presence of NH_4^+. The simple interpretation is that the alkyl amines do not bind as tightly to DNA as does ammonium. (Why this should be is not entirely clear; it may be that the ion activity is reduced by self-association of the alkyl amines.) Furthermore, the T_m of DNAs rich in G · C was reduced more by exchanging for alkyl amines than was the case for A · T-rich DNA. Another example is spermine $[NH_2(CH_2)_4NH(CH_2)_3NH_2]$, which, as Mandel[222] showed, stabilizes A · T-rich DNA more than G · C rich, and hence binds more tightly to A · T pairs. In this case Hirschman et al.[225] showed that the effect is due to the decreased ionization of spermine as temperature increases. At the T_m for G · C-rich DNA the degree of spermine protonation is smaller than at the T_m for A · T-rich DNA. Therefore the A · T-rich sample experiences stronger binding and greater influence of spermine concentration on T_m than does the G · C-rich sample at its T_m. They further found that at 4°C there is no composition dependence of the binding affinity for spermine, using equilibrium dialysis to measure the binding constant.

H. Steroid Diamines

A class of compounds of particular interest are the steroid diamines, represented in Table 7-2 by two examples, irehdiamine A and malouetine, studied in considerable detail by Mahler and his collaborators.[164,165,223] Like most amines, these stabilize the double helix when added in small amounts at low salt concentrations, but the complex has a number of additional properties not shared by diamines that are connected by a flexible alkyl chain. For example, irehdiamine A alters the absorbance and optical rotation properties of DNA,[165] and also unwinds the double helix,[153] although intercalation can surely be ruled out for these nonplanar molecules. In addition, when the diamino groups are either primary or secondary (but not tertiary or quaternary), addition of larger amounts of diamine results in a decrease of the double helix stability. The data for irehdiamine A indicate a change in the nature of the complex when the ratio of DNA phosphate to steroid is about 2.

The postulated mechanism of binding of these materials involves first a phosphate-bridging, charge-neutralizing attachment of the diamine, producing a stabilization of the double helix.[164] There is presumably some structural alteration of the DNA because of the rigid steroid framework on which the two positively charged amino groups are placed; the DNA adapts its structure to the diamine to some extent. At higher degrees of binding a nonbridging, charge-reversing interaction is postulated, in which only one amino group from each steroid interacts directly with a phosphate. Since this binding reverses the charge on the polyelectrolyte, it should be stronger with the coil than with the helix because of the lesser local charge density in the former case. Tighter binding to the coil produces a reduction in stability of the double helix. As the binding proceeds further, it is thought that stacking or aggregation of the steroids occurs on the surface of the DNA, followed ultimately by aggregation and precipitation of the polymer.

I. Polyamino Acids and Basic Proteins

There is a very substantial literature on the interaction of basic proteins with DNA. (See, for example, the book on nucleohistones by Bonner and Ts'o.[226]) We restrict ourselves here to consideration of physical experiments on well-defined systems. Much of the work in this category has been done on model systems, in which homogeneous polyamino acids are used in place of the natural histones or other basic proteins. The materials most commonly used are polyarginine, polylysine, polyornithine, and polyhomoarginine.

At neutral pH these polyamino acids are extensively protonated, and hence carry a charge opposite to that of the nucleic acid. One expects, therefore, that there will be a considerable electrostatic contribution to the binding free energy. Complexes are readily observed at salt concentrations of 1 M, but frequently begin to dissociate above this level. At much lower salt concentrations, binding is so tight as to be irreversible,[159] although freely reversible in 1 M salt. As a consequence, complexes are often annealed by formation at high salt concentration, followed by gradual dialysis into lower salt concentrations.

A basic physical feature of complex formation with polyamino acids is a phase separation that produces micellelike particles. Suppose that polyamino acid is added to a DNA solution, with less than one amino acid per phosphate residue. Brief high-speed centrifugation serves to separate a pellet containing a 1:1 complex between polyamino acid and DNA phosphate, leaving supernatant containing mostly uncomplexed DNA.[227] Physical studies[228] show that the aggregated phase consists of spherical particles that are highly solvated and of remarkably uniform size, with average radius 1700 Å. These particles can be readily resuspended, and hence studied as a "solution." They do, of course, scatter light appreciably, accounting for the turbidity of polyamino acid–DNA complexes. The optical rotation of the aggregate suspension is greatly increased over DNA in the wavelength region 250–290 nm, possibly due to perturbation of the DNA structure, or more likely because of the formation of long-range order in the micelle particles.[228] Some alteration of DNA structure is indicated by an increase in the number of rapidly exchangeable DNA protons in the complex, compared to DNA alone.[229]

Several of the properties of the complex are more readily understood if the phase separation property is kept in mind. For one, it is found that there is a sharp selectivity by polylysine for DNA rich in A · T pairs,[227] although this specificity can be reversed by using tetramethylammonium as the counterion. Other polyamino acids, polyarginine for example, show a slight preference for G · C pairs. Since the aggregate formation involves whole molecules at a time, it has a highly cooperative character, and a slight association preference for one or another base pair, when summed over a whole molecule, can produce an enormous base selectivity. It is probable that the binding specificity of these polyamino acids arises in just this way, with only a very small discriminatory ability on a residue basis.

Melting profiles of polyamino acid–DNA complexes also reflect the separation into two phases.[159–161] As shown earlier, in Fig. 7-33, the usual behavior with homogeneous polyamino acid–nucleic acid complexes is two separate melting transitions, each at its characteristic temperature, with only the amplitude of each influenced by the input ratio of amino acid to phosphate. As expected, the transition temperature of the complex is higher than that of DNA alone, although the melting behavior of naked DNA can be seen until sufficient polyamino acid is added to complex all the phosphates. In agreement with the selectivity for micelle formation with A · T-rich DNA, polylysine preferentially stabilizes A · T-rich DNA, as does polyornithine, while polyarginine and polyhomoarginine are less discriminating.[160] The extent of stabilization, on the other hand, decreases in the order ornithine, lysine, arginine, homoarginine.[160]

It is interesting that natural histone fractions seem to have less tendency to form an aggregated phase. Olins[230] examined the lysine-rich histone fraction F1 in complex with calf thymus DNA. He found that at high dilution the sedimentation properties were those of individually dispersed molecules, although there was a rather strong tendency to aggregate. Furthermore, in contrast to the complex of polylysine with DNA, the optical rotation properties were not greatly changed from those of DNA alone. Biphasic transition curves were found for the complex, but again with differences from the polylysine complex. In the case of the histone fraction, the temperature of the lower transition did not remain constant as more histone was added, indicating that uncomplexed DNA molecules were not present in the solution. This is probably largely a consequence of the lesser tendency to phase separation, producing a broader distribution of bound histone molecules over the population of DNA molecules. The melting transition of protamine–DNA complexes behaves similarly,[161] with a shifting lower transition temperature as more protein is added.

Olins[230] also examined the ability of the histone fraction to inhibit actinomycin binding and to block glucosylation of a phage DNA. Actinomycin, thought to bind in the small groove of DNA, binds readily to the histone complex, whereas glucosylation, occurring in the large groove, is blocked by the histone. The conclusion drawn was that the histone binds in the large groove of the DNA.

J. Repressor Proteins

Nucleic acid interactions of central importance to biology are those in which special proteins bind to a particular base sequence and thereby control the expression or repression of genetic information. Protein molecules that bind to a particular DNA locus and block the expression of an adjacent genetic message are called repressors; two that have been widely studied are those for the *lac* operon in *E. coli*,[231-236] and for λ-bacteriophage.[237-240]

In an elegant series of experiments, Riggs, Bourgeois, and several collaborators[234-236] have examined the physical characteristics of the interaction of the *lac* repressor (R) with the operator (O) region of the DNA, estimated to consist of at least 12 base pairs. The stoichiometry is one repressor per operator, so that the dissociation of the repressor-operator complex (RO) may be written

$$\mathrm{RO} \underset{k_a}{\overset{k_b}{\rightleftharpoons}} \mathrm{R} + \mathrm{O} \tag{7-119}$$

The dissociation constant $K = k_b/k_a$ is found to be about 10^{-13} M at 0.05 M ionic strength, decreasing strongly as salt concentration is raised. The association rate constant under the same conditions is $k_a = 7 \times 10^9\ \mathrm{M}^{-1}\ \mathrm{sec}^{-1}$, a remarkably high value exceeded only in such simple processes as proton transfer reactions. The dissociation rate constant is $k_b = 6 \times 10^{-4}\ \mathrm{sec}^{-1}$, giving the complex a half-life of about 20 minutes. The association reaction has a small activation energy of about 8.5 kcal, and the activation energy for the dissociation process is nearly zero. The enthalpy change on binding is therefore +8.5 kcal/mole, and since the standard free energy of binding is −18 kcal/mole at 25°C, the binding reaction is entropically driven, with a positive standard entropy change of 90 entropy units.

Perhaps the most remarkable of these physical properties is the very large association rate constant, much greater than found for binding simple dyes or antibiotics to DNA. In cases where two substances react at essentially every collision, the rate is said to be diffusion limited, and the diffusion-limited rate constant can be estimated from an equation first given by Smoluchowski[241] (applying to uncharged particles):

$$k_a = \frac{4\pi N}{1000} r_{12} D_{12} \tag{7-120}$$

where r_{12} is the reaction radius, D_{12} is the sum of diffusion coefficients for the reacting particles, and N is Avogadro's number. Riggs et al. estimate from expected diffusion constants, with $r_{12} = 5$ Å, that $k_a = 10^8$ M^{-1} sec^{-1}, nearly two orders of magnitude smaller than observed. They propose that the difference is made up by electrostatic attraction between repressor and operator. The net charge observed for repressors is small and slightly negative at neutral pH. However, because of the dependence of binding strength on salt concentration, the repressor probably contains a region of net positive charge which interacts with the (negatively charged) operator. It is physically plausible that these electrostatic effects would lead to acceleration of the reaction beyond that predicted by Eq. (7-120) but the theory remains to be worked out for treating diffusion-limited reaction between a polyion and a strongly dipolar macromolecule.

REFERENCES

1 R. Franklin and R. G. Gosling, *Acta Cryst.*, **6**, 673 (1953).
2 M. Falk, K. A. Hartman, Jr., and R. C. Lord, *J. Amer. Chem. Soc.*, **85**, 391 (1963).
3 M. J. B. Tunis and J. E. Hearst, *Biopolymers*, **6**, 1325 (1968).
4 M. Falk, K. A. Hartman, Jr., and R. C. Lord, *J. Amer. Chem. Soc.*, **84**, 3843 (1962).
5 M. Falk, K. A. Hartman, Jr., and R. C. Lord, *J. Amer. Chem. Soc.*, **85**, 387 (1963).
6 S. Brunauer, P. H. Emmett, and E. Teller, *J. Amer. Chem. Soc.*, **60**, 309 (1938).
7 R. Langridge, H. R. Wilson, C. W. Hooper, M. H. F. Wilkins, and L. D. Hamilton, *J. Mol. Biol.*, **2**, 19 (1960).
8 A. C. T. North and A. Rich, *Nature*, **191**, 1242 (1961).
9 M. Falk, *Canad. J. Chem.*, **43**, 314 (1965).
10 J. H. Wang, *J. Amer. Chem. Soc.*, **76**, 4755 (1954).
11 J. H. Wang, *J. Amer. Chem. Soc.*, **77**, 258 (1955).
12 H. B. Gray, Jr., V. A. Bloomfield, and J. E. Hearst, *J. Chem. Phys.*, **46**, 1493 (1967).
13 P. L. Privalov and G. M. Mrevlishvili, *Biofizika*, **12**, 22 (1967).
14 H. Spring, D. Dollstadt, and G. Hubner, *Biopolymers*, **7**, 447 (1969).

15 I. D. Kuntz, Jr., T. S. Brassfield, G. D. Law, and G. V. Purcell, *Science,* **163**, 1329 (1969).
16 J. E. Hearst and J. Vinograd, *Proc. Nat. Acad. Sci., U.S.,* **47**, 825, 1306 (1961).
17 J. E. Hearst and J. Vinograd, *Proc. Nat. Acad. Sci., U.S.,* **47**, 999 (1961).
18 J. E. Hearst and J. Vinograd, *Proc. Nat. Acad. Sci., U.S.,* **47**, 1005 (1961).
19 J. E. Hearst, J. B. Ifft, and J. Vinograd, *Proc. Nat. Acad. Sci., U.S.,* **47**, 1015 (1961).
20 J. Vinograd and J. E. Hearst, *Fortsch. Chem. Org. Naturstoffe,* **20**, 372 (1962).
21 M. J. B. Tunis and J. E. Hearst, *Biopolymers,* **6**, 1345 (1968).
22 J. E. Hearst, *Biopolymers,* **3**, 57 (1965).
23 G. Cohen and H. Eisenberg, *Biopolymers,* **6**, 1077 (1968).
24 R. Bruner and J. Vinograd, *Biochim. Biophys. Acta,* **108**, 18 (1965).
25 V. Luzzati, A. Nicolaieff, and F. Masson, *J. Mol. Biol.,* **3**, 185 (1961).
26 R. A. Robinson and R. H. Stokes, *Electrolyte Solutions*, Academic Press, New York (1955).
27 R. B. Inman and D. O. Jordan, *Biochim. Biophys. Acta,* **42**, 421 (1960).
28 G. B. B. M. Sutherland and M. Tsuboi, *Proc. Roy. Soc. (London)*, **A239**, 446 (1957).
29 E. M. Bradbury, W. C. Price, and G. R. Wilkinson, *J. Mol. Biol.,* **3**, 301 (1961).
30 M. Tsuboi, *J. Amer. Chem. Soc.,* **79**, 1351 (1957).
31 M. Tsuboi, *Prog. Theor. Phys. (Kyoto) Supp.,* **17**, 99 (1961).
32 Y. Kyogoku, M. Tsuboi, T. Shimanouchi, and I. Watanabe, *J. Mol. Biol.,* **3**, 741 (1961).
33 S. Lewin, *J. Theoret. Biol.,* **17**, 181 (1967).
34 W. Kauzmann, *Advan. Protein Chem.,* **14**, 1 (1959).
35 O. Sinanoğlu and S. Abdulnur, *Photochem. Photobiol.,* **3**, 333 (1964).
36 O. Sinanoğlu and S. Abdulnur, *Fed. Proc.,* **24**, 5–12 (1965).
37 S. A. Rice and M. Nagasawa, *Polyelectrolyte Solutions*, Academic Press, London and New York (1961); see especially Chaps. 5, 8–11.
38 H. Morawetz, *Macromolecules in Solution*, Interscience, New York (1965), Chap. 7.
39 F. E. Harris and S. A. Rice, *J. Chem. Phys.,* **25**, 955 (1956).
40 G. S. Manning and B. H. Zimm, *J. Chem. Phys.,* **43**, 4250 (1965).
41 G. S. Manning, *J. Chem. Phys.,* **51**, 924 (1969).
42 F. G. Donnan, *Z. Electrochem.,* **17**, 572 (1911).
43 C. Tanford, *Physical Chemistry of Macromolecules*, Wiley, New York (1961), p. 225.
44 G. Cohen and H. Eisenberg, *Biopolymers,* **6**, 1077 (1968).
45 U. P. Strauss, C. Helfgott, and H. Pin, *J. Phys. Chem.,* **71**, 2550 (1967).
46 J. A. Harpst, A. I. Krasna, and B. H. Zimm, *Biopolymers,* **6**, 595 (1968).
47 J. W. Lyons and L. Kotin, *J. Amer. Chem. Soc.,* **87**, 1670 (1965).
48 N. Bjerrum, *Kgl. Danske Vidensk. Selskab.,* **7**, No. 9 (1926).
49 J. J. Hermans, *J. Polymer Sci.,* **18**, 527 (1955).
50 P. Debye and A. M. Bueche, *J. Chem. Phys.,* **16**, 573 (1948).
51 H. C. Brinkman, *Proc. Acad. Amsterdam,* **50**, 618 (1947).
52 B. M. Olivera, P. Baine, and N. Davidson, *Biopolymers,* **2**, 245 (1964).
53 J. J. Hermans and H. Fujita, *Proc. Roy. Netherland Acad. Sci.,* **B58**, 182 (1955).
54 P. D. Ross and R. L. Scruggs, *Biopolymers,* **2**, 231 (1964).
55 U. P. Strauss and P. D. Ross, *J. Amer. Chem. Soc.,* **81**, 5295 (1959).
56 L. Costantino, A. M. Liquori, and V. Vitagliano, *Biopolymers,* **2**, 1 (1964).
57 Z. Alexandrowicz, *J. Chem. Phys.,* **47**, 4377 (1967).

58 A. Katchalsky and S. Lifson, *J. Polymer Sci.*, **11**, 409 (1956).
59 P. J. Flory, *J. Chem. Phys.*, **21**, 162 (1953).
60 H. Triebel and K. E. Reinert, *Studia Biophys.*, **10**, 57 (1968).
61 P. D. Ross and R. L. Scruggs, *Biopolymers*, **6**, 1005 (1968).
62 R. A. Cox, *J. Polymer Sci.*, **47**, 441 (1960).
63 R. J. Douthart and V. A. Bloomfield, *Biopolymers*, **6**, 1297 (1968).
64 A. H. Rosenberg and F. W. Studier, *Biopolymers*, **7**, 765 (1969).
65 Z. Alexandrowicz and E. Daniel, *Biopolymers*, **1**, 447, 473 (1963); **6**, 1500 (1968).
66 C. S. Lee and N. Davidson, *Biopolymers*, **6**, 531 (1968).
67 G. Scatchard, *Ann. New York Acad. Sci.*, **51**, 660 (1949).
68 H. A. Benesi and J. H. Hildebrand, *J. Amer. Chem. Soc.*, **71**, 2703 (1949).
69 H. J. Li and D. M. Crothers, *J. Mol. Biol.*, **39**, 461 (1969).
70 D. F. Bradley and S. Lifson, in *Molecular Associations in Biology*, B. Pullman, ed., Academic Press, New York (1968).
71 S. A. Latt and H. A. Sober, *Biochemistry*, **6**, 3293 (1967).
72 D. M. Crothers, *Biopolymers*, **6**, 575 (1968).
73 W. Müller and D. M. Crothers, *J. Mol. Biol.*, **35**, 251 (1968).
74 W. Bauer and J. Vinograd, *J. Mol. Biol.*, **47**, 419 (1970).
75 A. R. Peacocke and J. N. H. Skerrett, *Trans. Farad. Soc.*, **52**, 261 (1956).
76 D. F. Bradley and M. K. Wolf, *Proc. Nat. Acad. Sci., U.S.*, **45**, 944 (1959).
77 U. S. Nandi, J. C. Wang, and N. Davidson, *Biochemistry*, **4**, 1687 (1965).
78 W. Kersten, *Biochim. Biophys. Acta*, **47**, 610 (1961).
79 E. Reich, J. H. Goldberg, and M. Rabinowitz, *Nature*, **196**, 743 (1962).
80 E. Kahan, F. M. Kahan, and J. Hurwitz, *J. Biol. Chem.*, **238**, 2491 (1963).
81 M. Gellert, C. E. Smith, D. Neville, and G. Felsenfeld, *J. Mol. Biol.*, **11**, 445 (1965).
82 D. M. Crothers, *Biopolymers*, **4**, 1025 (1966).
83 R. B. Simpson, *J. Amer. Chem. Soc.*, **86**, 2059 (1964).
84 P. Y. Cheng, D. S. Hondo, and J. Rozsnyai, *Biochem.*, **8**, 4470 (1969).
85 K. Gillen, R. Jensen, and N. Davidson, *J. Amer. Chem. Soc.*, **86**, 2792 (1964).
86 G. L. Eichhorn, J. J. Butzow, P. Clark, and E. Tarien, *Biopolymers*, **5**, 283 (1967).
87 A. T. Tu and J. A. Reinosa, *Biochemistry*, **5**, 3375 (1966).
88 G. L. Eichhorn, P. Clark, and E. A. Becker, *Biochemistry*, **5**, 245 (1966).
89 S. Katz, *J. Amer. Chem. Soc.*, **74**, 2258 (1952).
90 C. A. Thomas, *J. Amer. Chem. Soc.*, **76**, 2052 (1954).
91 T. Yamane and N. Davidson, *J. Amer. Chem. Soc.*, **83**, 2599 (1961).
92 T. Yamane and N. Davidson, *Biochim. Biophys. Acta*, **55**, 700 (1962).
93 S. Katz, *Biochim. Biophys. Acta*, **68**, 240 (1963).
94 U. S. Nandi, J. C. Wang, and N. Davidson, *Biochemistry*, **4**, 1687 (1965).
95 D. W. Gruenwedel and N. Davidson, *J. Mol. Biol.*, **21**, 129 (1966).
96 M. N. Williams, Thesis, Yale University (1968).
97 T. Yamane and N. Davidson, *Biochim. Biophys. Acta*, **55**, 609 (1962).
98 R. H. Jensen and N. Davidson, *Biopolymers*, **4**, 17 (1966).
99 M. Daune, C. A. Dekker, and H. K. Schachman, *Biopolymers*, **4**, 51 (1966).
100 L. E. Minchenkova and V. I. Ivanov, *Biopolymers*, **5**, 615 (1967).
101 S. Hiai, *J. Mol. Biol.*, **11**, 672 (1965).
102 H. Venner and Ch. Zimmer, *Biopolymers*, **4**, 321 (1966).
103 G. L. Eichhorn and P. Clark, *Proc. Nat. Acad. Sci., U.S.*, **53**, 586 (1965).
104 G. L. Eichhorn, *Nature*, **194**, 474 (1962).
105 G. L. Eichhorn and E. Tarien, *Biopolymers*, **5**, 273 (1967).
106 D. Bach and I. Miller, *Biopolymers*, **5**, 161 (1967).
107 A. M. Fiskin and M. Beer, *Biochemistry*, **4**, 1289 (1965).

108 J. P. Schreiber and M. Daune, *Biopolymers,* 8, 139 (1969).
109 Y. A. Shin and G. L. Eichhorn, *Biochemistry,* 7, 1026 (1968).
110 M. J. Waring, *Nature,* **219**, 1320 (1968).
111 G. Hartmann, W. Behr, K.-A. Beissner, K. Honikel, and A. Sippel, *Angew. Chem.*, English ed., 7, 693 (1968).
112 L. S. Lerman, *J. Mol. Biol.*, **3**, 18 (1961).
113 W. Fuller and M. J. Waring, *Ber. Bunsenges. Physik. Chem.*, **68**, 805 (1964).
114 L. S. Lerman, *J. Cell. Comp. Physiol.*, **64** (Suppl. 1), 1 (1964).
115 A. L. Stone and D. F. Bradley, *J. Amer. Chem. Soc.*, **83**, 3627 (1961).
116 L. Hamilton, W. Fuller, and E. Reich, *Nature,* **198**, 538 (1963).
117 V. N. Iyer and W. Szybalski, *Proc. Nat. Acad. Sci., U.S.*, **50**, 355 (1963).
118 A. Blake and A. R. Peacocke, *Biopolymers,* **6**, 1225 (1968).
119 R. Bersohn and I. Isenberg, *J. Chem. Phys.*, **40**, 3175 (1964).
120 I. Isenberg, R. B. Leslie, S. L. Baird, Jr., R. Rosenbluth, and R. Bersohn, *Proc. Nat. Acad. Sci., U.S.*, **52**, 379 (1964).
121 A. A. Lamola, M. Gueran, T. Tamane, J. Eisinger, and R. G. Shulman, *J. Chem. Phys.*, **47**, 2210 (1967).
122 G. Weill and M. Calvin, *Biopolymers,* **1**, 401 (1963).
123 J.-B. LePecq and C. Paoletti, *J. Mol. Biol.*, **27**, 87 (1967).
124 W. C. Galley, *Biopolymers,* **6**, 1279 (1968).
125 B. J. Gardner and S. F. Mason, *Biopolymers,* **5**, 79 (1967).
126 R. K. Tubbs, W. E. Ditman, Jr., and Q. Van Winkle, *J. Mol. Biol.*, **9**, 545 (1964).
127 L. S. Lerman, *Proc. Nat. Acad. Sci., U.S.*, **49**, 94 (1963).
128 A. Blake and A. R. Peacocke, *Biopolymers,* **4**, 1091 (1966).
129 A. Blake and A. R. Peacocke, *Biopolymers,* **5**, 383 (1967).
130 A. Blake and A. R. Peacocke, *Biopolymers,* **5**, 871 (1967).
131 S. F. Mason and A. J. McCaffery, *Nature,* **204**, 468 (1964).
132 K. Yamaoka and R. Resnik, *Nature,* **213**, 1031 (1967).
133 H. J. Li and D. M. Crothers, *Biopolymers,* 8, 217 (1969).
134 S. Ohnishi and H. M. McConnell, *J. Amer. Chem. Soc.*, 87, 2293 (1964).
135 K. Ishizu, H. H. Dearman, M. T. Huang, and J. R. White, *Biochemistry,* **8**, 1238 (1969).
136 D. J. Blears and S. S. Danyluk, *Biopolymers,* **5**, 535 (1967).
137 D. S. Drummond, N. J. Pritchard, V. F. W. Simpson-Gildmeister, and A. R. Peacocke, *Biopolymers,* **4**, 971 (1966).
138 G. Cohen and H. Eisenberg, *Biopolymers,* **8**, 45 (1969).
139 Y. Mauss, J. Chambron, M. Daune, and H. Benoit, *J. Mol. Biol.*, **27**, 579 (1967).
140 D. E. Schmechel and D. M. Crothers, *Biopolymers,* **10**, 465 (1971).
141 V. Luzzati, F. Masson, and L. S. Lerman, *J. Mol. Biol.*, **3**, 634 (1961).
142 H. Eisenberg and G. Cohen, *J. Mol. Biol.*, **37**, 355 (1968).
143 E. F. Casassa, *J. Chem. Phys.*, **23**, 596 (1955).
144 A. Holtzer, *J. Polymer Sci.*, **17**, 432 (1955).
145 V. Luzzati and H. Benoit, *Acta Cryst.*, **14**, 297 (1961).
146 V. Luzzati, A. Nicolaieff, and F. Masson, *J. Mol. Biol.*, **3**, 185 (1961).
147 V. Luzzati, F. Masson, A. Mathis, and P. Saludjian, *Biopolymers,* **5**, 491 (1967).
148 L. S. Lerman, *Proc. Nat. Acad. Sci., U.S.*, **49**, 94 (1963).
149 D. M. Neville, Jr., and D. R. Davies, *J. Mol. Biol.*, **17**, 57 (1966).
150 J. Cairns, *Cold Spring Harbor Symp. Quant. Biol.*, **27**, 311 (1962).
151 L. V. Crawford and M. J. Waring, *J. Mol. Biol.*, **25**, 23 (1967).
152 W. Bauer and J. Vinograd, *J. Mol. Biol.*, **33**, 141 (1968).
153 M. J. Waring, *J. Mol. Biol.*, **54**, 247 (1970).

154 J. Vinograd, J. Lebowitz, and R. Watson, *J. Mol. Biol.*, **33**, 173 (1968).
155 H. M. Sobell, S. C. Jain, T. D. Sakore, and C. E. Nordman, *Nature,* **231**, 200 (1971).
156 W. Kersten, H. Kersten, and W. Szybalski, *Biochemistry,* **5**, 236 (1966).
157 D. Freifelder, P. F. Davison, and E. P. Geiduschek, *Biophys. J.,* **1**, 389 (1961).
158 C. R. Stewart, *Biopolymers,* **6**, 1737 (1968).
159 M. Tsuboi, K. Matsuo, and P. O. P. Ts'o, *J. Mol. Biol.,* **15**, 256 (1966).
160 D. E. Olins, A. L. Olins, and P. H. von Hippel, *J. Mol. Biol.,* **24**, 157 (1967).
161 D. E. Olins, A. L. Olins, and P. H. von Hippel, *J. Mol. Biol.,* **33**, 265 (1968).
162 H. R. Mahler and B. D. Mehrotra, *Biochim. Biophys. Acta,* **68**, 211 (1963).
163 H. Venner, Ch. Zimmer, and S. Schroder, *Biochim. Biophys. Acta,* **76**, 312 (1963).
164 H. R. Mahler, R. Goutarel, Q. Khuong-Huu, and M. T. Ho, *Biochemistry,* **5**, 2177 (1966).
165 H. R. Mahler, G. Green, R. Goutarel, and Q. Khuong-Huu, *Biochemistry,* **7**, 1568 (1968).
166 M. Eigen and L. DeMaeyer, in *Technique of Organic Chemistry*, S. L. Freiss, E. S. Lewis, and A. Weissberger, eds. Vol. 8, Part II, Interscience, New York (1963).
167 G. Lober, *Z. Chemie,* **9**, 252 (1969).
168 A. Albert, *The Acridines*, Arnold, London (1951, 1966).
169 S. N. Cohen and K. L. Yielding, *J. Biol. Chem.,* **240**, 3123 (1965).
170 R. L. O'Brien, J. L. Allison, and F. E. Hahn, *Biochim. Biophys. Acta,* **129**, 622 (1966).
171 R. L. O'Brien, J. G. Olenick, and F. E. Hahn, *Proc. Nat. Acad. Sci., U.S.,* **55**,1511 (1966).
172 T. Asao, G. Buchi, M. M. Abdel-Kader, S. B. Chang, E. L. Wick, and G. N. Wogan, *J. Amer. Chem. Soc.,* **85**, 1706 (1963).
173 J. I. Clifford and K. Rees, *Nature,* **209**, 319 (1966).
174 J. I. Clifford and K. Rees, *Biochem. J.,* **102**, 65 (1967).
175 J. I. Clifford and K. Rees, *Biochem. J.,* **103**, 467 (1967).
176 W. C. Neely, J. A. Lansden, and J. R. McDuffie, *Biochemistry* (in press).
177 E. Boyland and B. Green, *Brit. J. Cancer,* **16**, 507 (1962).
178 A. M. Liquori, B. DeLerma, F. Ascoli, C. Botre, and M. Trasciatti, *J. Mol. Biol.,* **5**, 521 (1962).
179 L. S. Lerman, *Proc. 5th Nat. Cancer Conf.*, Philadelphia, **39** (1964).
180 J. K. Ball, J. A. McCarter, and M. F. Smith, *Biochim. Biophys. Acta,* **103**, 275 (1965).
181 C. Nagata, M. Kodama, Y. Tagashira, and A. Imamura, *Biopolymers,* **4**, 409 (1966).
182 B. Green and J. A. McCarter, *J. Mol. Biol.,* **29**, 447 (1967).
183 I. Isenberg, S. L. Baird, and R. Bersohn, *Biopolymers,* **5**, 477 (1967).
184 S. A. Lesko, A. Smith, P. O. P. Ts'o, and R. S. Umans, *Biochemistry,* **7**, 434 (1968).
185 P. O. P. Ts'o and P. Lu, *Proc. Nat. Acad. Sci., U.S.,* **51**, 272 (1964).
186 S. A. Lesko, Jr., P. O. P. Ts'o, and R. S. Umans, *Biochemistry,* **8**, 2291 (1969).
187 H. Brockmann, *Fortschr. Chem. Org. Naturstoffe,* **18**, 1 (1960).
188 W. Kersten, H. Kersten, and H. M. Rauen, *Nature,* **187**, 60 (1960).
189 I. H. Goldberg, M. Rabinowitz, and E. Reich, *Proc. Nat. Acad. Sci., U.S.,* **48**, 2094 (1962).
190 J. Hurwitz, J. J. Furth, M. Malamy, and M. Alexander, *Proc. Nat. Acad. Sci., U.S.,* **48**, 122 (1962).
191 G. Hartman and U. Coy, *Angew. Chem.,* **74**, 501 (1962).
192 E. Harbers and W. Müller, *Biochem. Biophys. Res. Commun.,* **7**, 107 (1962).

193 R. Haselkorn, *Science,* **143**, 682 (1964).
194 A. Cerami, E. Reich, D. C. Ward, and I. H. Goldberg, *Proc. Nat. Acad. Sci., U.S.,* **57**, 1036 (1967).
195 R. D. Wells, *Science,* **165**, 75 (1969).
196 R. D. Wells and J. E. Larson, *J. Mol. Biol.* (in press).
197 W. Kersten, *Biochim. Biophys. Acta,* **47**, 610 (1961).
198 W. Behr, K. Honikel, and G. Hartmann, *European J. Biochem.,* 9, 82 (1969).
199 T. Hayasaka and Y. Inoue, *Biochemistry,* 8, 2342 (1969).
200 K. Koschel, G. Hartmann, W. Kersten, and H. Kersten, *Biochemistry,* **344**, 76 (1966).
201 D. Ward, E. Reich, and I. Goldberg, *Science,* **149**, 1259 (1965).
202 W. Behr and G. Hartmann, *Biochem. Z.,* **343**, 519 (1965).
203 W. Kersten and H. Kersten, *Biochem. Z.,* **341**, 174 (1965).
204 M. Kamiyama, *J. Biochem.* (Tokyo), **63**, 566 (1968).
205 N. Brockmann, Fortschr. *Chem. Org. Naturstoffe,* **21**, 121 (1963).
206 I. B. Weinstein, R. Carchman, E. Marner, and E. Hirschberg, *Biochim. Biophys. Acta,* **142**, 440 (1967).
207 I. B. Weinstein, R. Chernoff, I. Finkelstein, and E. Hirschberg, *Mol. Pharmacol.,* **1**, 297 (1965).
208 E. Hirschberg, I. B. Weinstein, N. Gersten, E. Marner, T. Finkelstein, and R. Carchmann, *Cancer Res.,* **28**, 601 (1968).
209 H. L. White and J. R. White, *Biochemistry,* 8, 1030 (1969).
210 W. Leimgruber, V. Stafanovic, F. Schenker, A. Karr, and J. Berger, *J. Amer. Chem. Soc.,* **87**, 5791 (1965).
211 W. Leimgruber, A. D. Batcho, and F. Schenker, *J. Amer. Chem. Soc.,* **87**, 5793 (1965).
212 K. W. Kohn, V. H. Bono, Jr., and H. E. Kann, Jr., *Biochim. Biophys. Acta,* **155**, 121 (1968).
213 V. N. Iyer and W. Szybalski, *Proc. Nat. Acad. Sci., U.S.,* **50**, 355 (1963).
214 W. Szybalski and V. N. Iyer, *Fed. Proc.,* **23**, 946 (1964).
215 W. Szybalski and V. N. Iyer, *Microbiol. Genetics Bull.,* No. 21, 16 (1964).
216 M. N. Lipsett and A. Weissbach, *Biochemistry,* **4**, 206 (1965).
217 P. Cohen and C. Kidson, *Fed. Proc.,* **28**, 531 (1969).
218 P. Cohen and C. Kidson, *Proc. Nat. Acad. Sci., U.S.,* **63**, 458 (1969).
219 P. Cohen, R. C. Chin, and C. Kidson, *Biochemistry,* **8**, 3603 (1969).
220 B. D. Mehrotra and H. R. Mahler, *Biochim. Biophys. Acta,* **91**, 78 (1964).
221 J. T. Shapiro, B. S. Stannard, and G. Felsenfeld, *Biochemistry*, **8**, 3233 (1969).
222 M. Mandel, *J. Mol. Biol.,* **5**, 435 (1962).
223 H. R. Mahler and G. Dutton, *J. Mol. Biol.,* **10**, 157 (1964).
224 J. C. Wang (personal commun.).
225 S. Z. Hirschman, M. Leng, and G. Felsenfeld, *Biopolymers,* **5**, 227 (1967).
226 J. Bonner and P. O. P. Ts'o, *The Nucleohistones*, Holden Day, San Francisco (1964).
227 M. Leng and G. Felsenfeld, *Proc. Nat. Acad. Sci., U.S.,* **56**, 1325 (1966).
228 J. T. Shapiro, M. Leng, and G. Felsenfeld, *Biochemistry,* **8**, 3219 (1969).
229 C. N. Lees and P. H. von Hippel, *Biochemistry,* **7**, 2480 (1968).
230 D. E. Olins, *J. Mol. Biol.,* **43**, 439 (1969).
231 W. Gilbert and B. Müller-Hill, *Proc. Nat. Acad. Sci., Wash.,* **58**, 2415 (1967).
232 A. D. Riggs and S. Bourgeois, *J. Mol. Biol.,* **34**, 361 (1968).
233 A. D. Riggs, S. Bourgeois, R. F. Newby, and M. Cohn, *J. Mol. Biol.,* **34**, 365 (1968).

234 A. D. Riggs, H. Suzuki, and S. Bourgeois, *J. Mol. Biol.*, **48**, 67 (1970).
235 A. D. Riggs, R. F. Newby, and S. Bourgeois, *J. Mol. Biol.*, **51**, 303 (1970).
236 A. D. Riggs, S. Bourgeois, and M. Cohn, *J. Mol. Biol.*, **53**, 401 (1970).
237 M. Ptashne, *Nature,* **214**, 232 (1967).
238 M. Ptashne and N. Hopkins, *Proc. Nat. Acad. Sci., Wash.*, **60**, 1282 (1968).
239 P. Chadwick, V. Pirotta, R. Steinberg, N. Hopkins, and M. Ptashne, *Cold Spring Harbor Symp. Quant. Biol.,* **35**, 283 (1970).
240 V. Pirotta, P. Chadwick, and M. Ptashne, *Nature,* **227**, 41 (1970).
241 M. V. Smoluchowski, *Z. Phys. Chem.*, **92**, 129 (1917).

chapter 8
transfer rna

I PRIMARY STRUCTURE

Transfer RNAs are the adaptors which read the three-letter message on messenger RNA and translate it into the correct amino acid. Although the genetic dictionary (Chapter 1, Tables 1-1 and 1-2) seems to be universal, the transfer RNAs are not identical for all species. They are similar, however. For a given species one might expect 61 transfer RNAs, because the three stop signals are recognized by proteins, not tRNAs. In fact there may be closer to 30 tRNAs, because each tRNA can recognize more than one triplet of bases on the messenger RNA.

The attachment of an amino acid such as alanine to a tRNA can be represented by the structure shown on the bottom of p. 479.

A specific synthetase activates an amino acid by catalyzing its reaction with adenosine triphosphate (ATP) to form the AMP-amino acid anhydride shown in the equation. The same enzyme catalyzes the transfer of the amino acid to the 3′-hydroxyl of the terminal adenosine found in all tRNAs. This "loaded" tRNA then finds its way to the appropriate location on a ribosome where the amino acid is connected to the growing polypeptide chain. If the transfer RNA is to initiate the polypeptide chain, it must be a formyl methionine transfer RNA in bacteria. After the tRNA has been loaded with methionine, it is formylated to give:

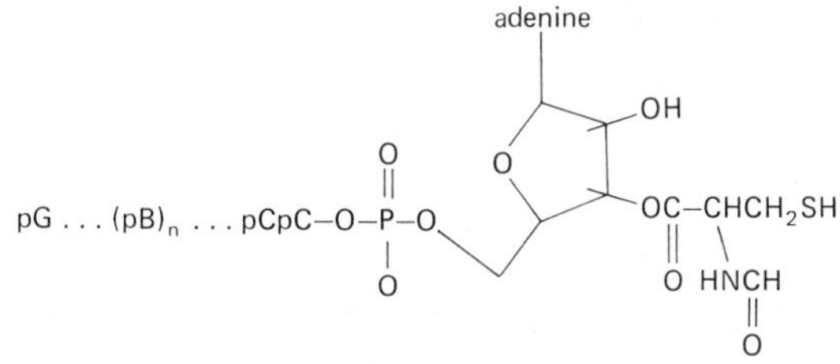

adenine

pG . . . $(pB)_n$ . . . pCpC–O–P(=O)(O)–O–[ribose]–OH, –OC(=O)–CHCH$_2$SH (HNCH=O)

formylmethionyl tRNA$_F^{met}$

It is not known how each specific synthetase and corresponding tRNA recognize each other. However, once the tRNA is loaded, it recognizes the base triplet on the messenger RNA (the codon) through a complementary base triplet (the anticodon or nodoc). Figure 8-1 shows the complete sequences of some of the transfer RNAs which have been studied. They are very similar in many respects. They contain about 75 nucleotides including many bases other than the four coding bases. Some of the bases which have been identified are shown in Fig. 8-2. There are various substitutions on the bases; in pseudouridine the change is the site of attachment to the ribose (at C5 instead of N1); there can also be methylation of the 2′ oxygen of the sugar. All these changes are made after the polynucleotide has been

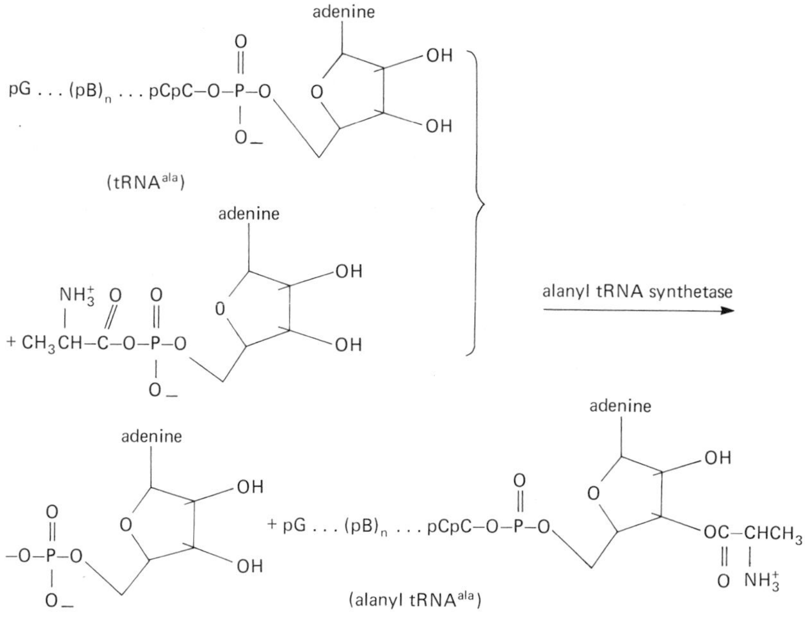

adenine

pG . . . $(pB)_n$ . . . pCpC–O–P(=O)(O$^-$)–O–[ribose](OH)(OH)

$(tRNA^{ala})$

adenine

NH_3^+ O O
+ CH_3CH–C–O–P(O$^-$)–O–[ribose](OH)(OH)

alanyl tRNA synthetase →

adenine

–O–P(=O)(O$^-$)–O–[ribose](OH)(OH)

adenine

+ pG . . . $(pB)_n$ . . . pCpC–O–P(=O)–O–[ribose]–OH, –OC(=O)–CHCH$_3$ (NH_3^+)

$(alanyl\ tRNA^{ala})$

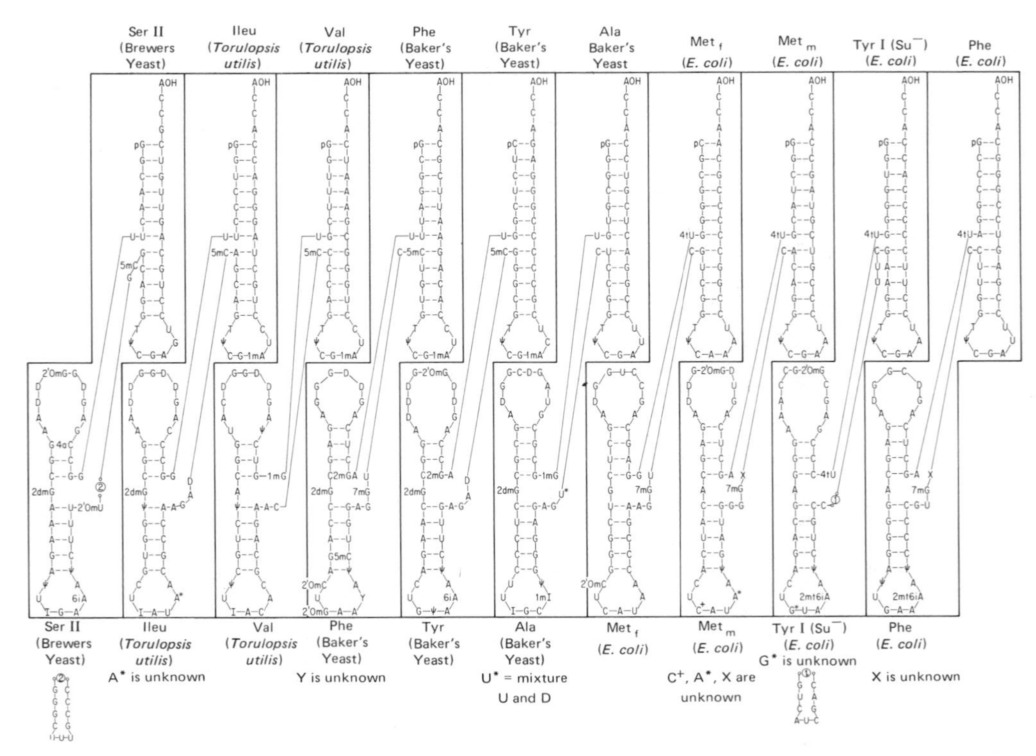

FIGURE 8-1 Base sequences of ten transfer RNA molecules. The abbreviations for the minor bases are given in Fig. 8-2. The double strand regions are those of the standard clover-leaf model. The anticodons are the three bases at the bottom of each molecule. The amino acids are attached at the hydroxyl shown at the top of each molecule. The sequences of these and other tRNA molecules, plus evidence for the secondary structure, can be found in the review article of H. Zachau, *Angewandte Chemie, Intern. Ed.*, 8, 711 (1969).

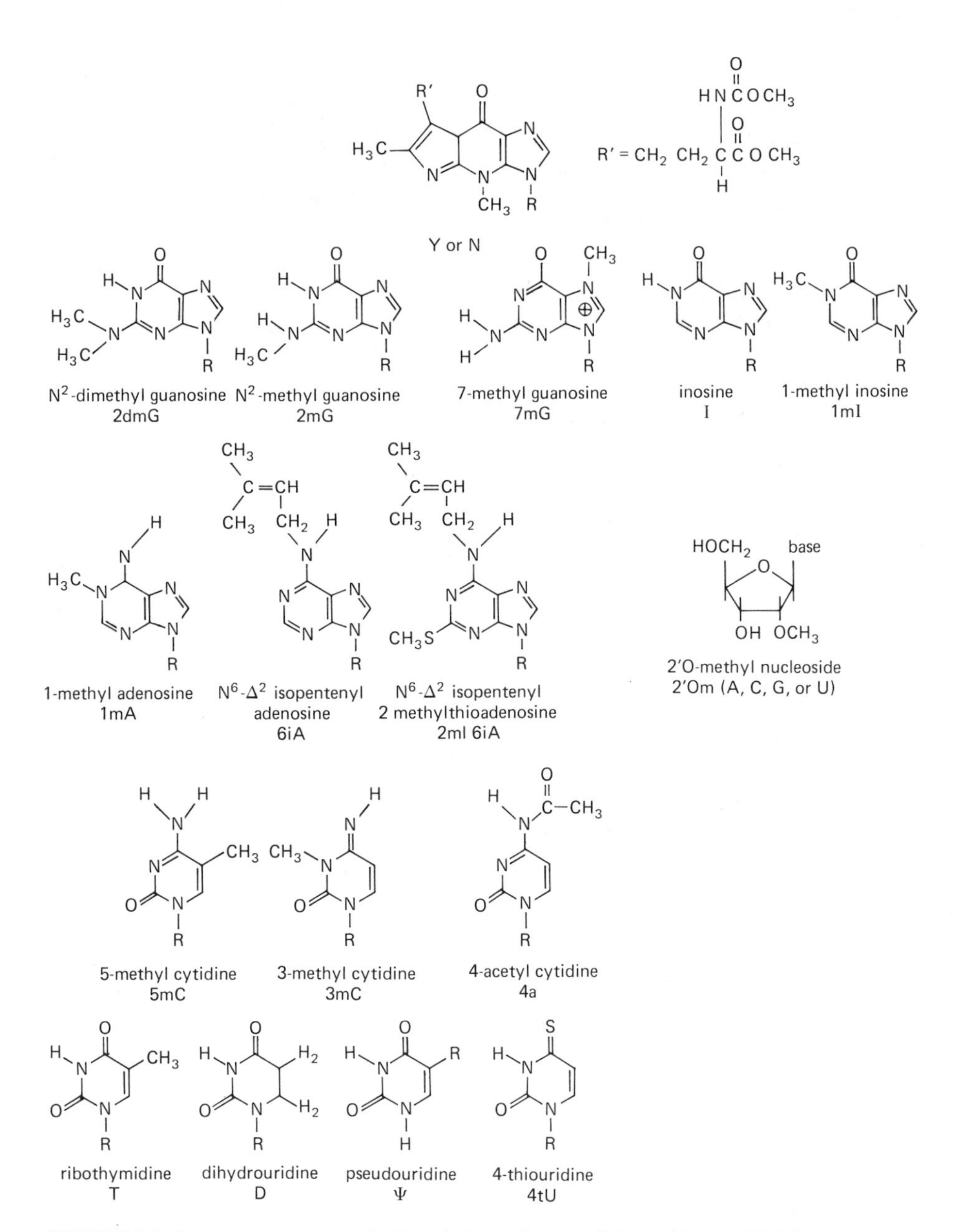

FIGURE 8-2 Structures, names, and abbreviations of many of the odd bases which have been found in tRNA molecules.

synthesized. The function of these odd bases is not known; however we note which are not capable of forming standard Watson-Crick base pairs. These are N^2-dimethylguanine, 7-methylguanine, 1-methylhypoxanthine (from 1-methylinosine), 1-methyladenine, and 3-methylcytosine. Some of the others are capable of forming standard base pairs, but could not fit in a continuous double-stranded region.

The anticodon triplet in each transfer RNA is indicated in Fig. 8-1. There is much indirect evidence to indicate which bases are the anticodon, but the best direct evidence is the sequences of *E. coli* tyrosine tRNA and suppressor (SU^+_{III}) tRNA.[1] Genetic experiments[2] showed that in certain *E. coli* a transfer RNA exists (SU^+_{III}) which can add tyrosine instead of allowing normal polypeptide chain termination as coded for by UAG on the messenger RNA. The normal tyrosine tRNA has GUA as its anticodon complementary to UAC on the messenger RNA; the suppressor (SU^+_{III}) tRNA is exactly the same as tyrosine tRNA except that it has CUA as its anticodon. It thus records the UAG stop signal as tyrosine.

The codon-anticodon relation was at first thought to involve only Watson-Crick standard base pairs. Exceptions were eventually noticed, so Crick[3] systematized and generalized the data with his "wobble" hypothesis. He stated that the first two positions of the codon would use standard base pairs: A · U and G · C, but that the third position would allow wobble, that is, other base pairs. For the third position of the codon reading from the 5′ end to the 3′ end, the dictionary would be:

Anticodon (first base)	Codon (third base)
U	A or G
ψ	A or G
C	G
A	U
G	U or C
I	A or U or C

This wobble hypothesis is consistent with the known information on the genetic code and the sequences of transfer RNAs.

II SECONDARY STRUCTURE

As soon as the first sequence of a tRNA was determined,[4] possible double-stranded structures were proposed. As more sequences were determined it was found that a cloverleaf model was consistent with all the sequences. This model, first proposed by Holley et al.[4] is shown in Fig. 8-3. It

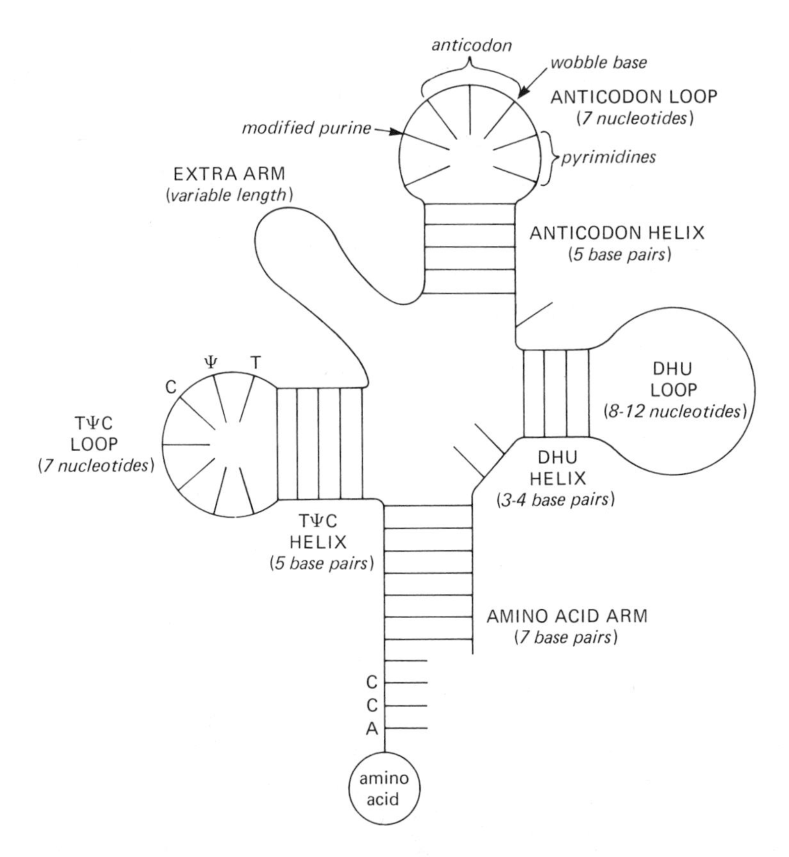

FIGURE 8-3 Cloverleaf diagram of the secondary structure for all transfer RNA molecules with the nomenclature for the different regions. [Reprinted with permission from S. Arnott, *Progress in Biophysics and Molecular Biophysics,* **22**, 181 (1971).]

should be very clear that this figure only represents the base pairing (secondary structure) and says nothing about the tertiary structure. The model has 20 to 21 base pairs; the fact that it is consistent with so many sequences makes it very attractive.

Concurrent with the sequence work, physical chemical studies were being done to try to measure the amount of base pairing in transfer RNAs in aqueous solution, including spectroscopic measurements and hydrogen-tritium exchange studies.[5,6] The questions asked were: How many and what type of base pairs? What is their dependence on temperature and salt conditions?

The optical methods will be discussed first; they are ultraviolet absorption, ultraviolet CD or ORD, and infrared absorption. Characteristic differences in spectra occur between the single-stranded and double-stranded conformations (see Chapters 3 and 4). Therefore quantitative analysis of the spectrum at any condition of temperature and salt can lead to the number and type of base pairs. In the published work the authors assumed that the spectrum of the double-stranded region was independent of base sequence. This serious approximation should be eliminated before quantitative results can be confidently accepted. With this approximation, however, only two double-stranded polynucleotides are needed to calibrate a spectrum – one gives the A · U contribution the other the G · C.

From ultraviolet absorption measurements, Fresco and co-workers[7,8] deduced that yeast transfer RNA was roughly 50% double-stranded at pH 7, 0.20 M Na^+. There was a broad melting range with a midpoint around 60°C. Some tRNAs showed biphasic melting curves.

ORD measurements[9,10] on yeast alananine tRNA gave similar results. The ORD curves of poly A · poly U and poly G · poly C were used for the double-stranded regions and the ORD curves of the dinucleoside phosphates were used for the single-stranded regions. A base-paired model with about as many base pairs as the cloverleaf gave results consistent with experiment at room temperature, pH 6.8, 0.15 M KCl, 0.1 M phosphate.

A quantitative infrared absorption study in the 1450–1750 cm^{-1} range has been made on *E. coli* formylmethionine transfer RNA in D_2O.[11] Seven peaks were seen corresponding to free and base-paired residues. At 33°C in 0.2 M Na^+ and 0.005 M Mg^{2+}, the cloverleaf model gave a calculated spectrum consistent with the observed spectrum. That is 17 ± 1G · C + 2 ± 1 A · U + 13A + 6U + 7G + 8C were found. As the temperature was raised there was a broad melting beginning at about 60°C. The two A · U base pairs disappeared first; by 85°C only half the G · C base pairs were left. The percent double-stranded bases at 33°C was about 53%.

We can conclude that the various optical studies agree well with the cloverleaf secondary structure and that they can be used to follow changes in conformation with changes in environment. To obtain more detailed information from spectra it will be necessary to study more model compounds.

Nuclear magnetic resonance of tRNA in H_2O has also been used to study secondary structure[12]; the observed low-field spectrum for yeast phenylalanine tRNA is shown in Fig. 8-4. The peaks are assigned to exchangeable protons because they disappear in D_2O. By analogy with the nucleosides in dimethyl sulfoxide the peaks between −12 ppm and −16 ppm are identified as hydrogen-bonded N–H ring protons. This means that each Watson-Crick base pair contributes one proton to this spectral region. At 24°C the integrated area in this region gives 21 ± 3 base pairs in good agreement with the cloverleaf model. As the temperature is raised the peaks broaden and decrease in magnitude. The position of each peak should correspond to the

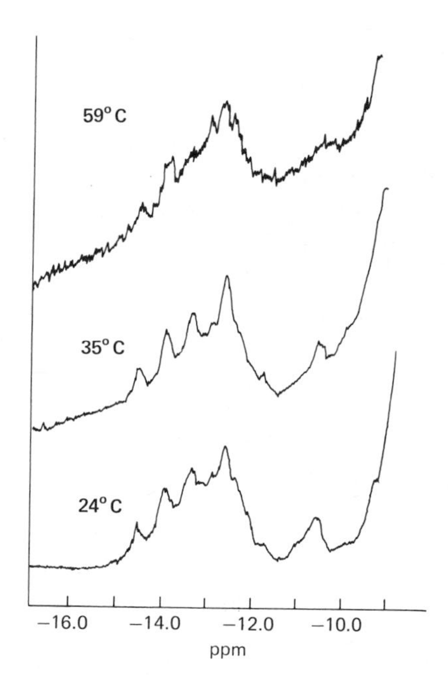

FIGURE 8-4 The low field 220 MHz nuclear magnetic resonance spectrum of yeast phenylalanine transfer RNA as a function of temperature. [From D. R. Kearns, D. J. Patel, and R. G. Shulman, *Nature,* **229**, 338 (1971). Reprinted by permission.]

hydrogen bond strength. When the individual peaks can be assigned to particular base pairs, it should be possible to determine their kinetic and thermodynamic stability.

A powerful thermodynamic method for studying secondary structure is measurement of binding of complementary oligonucleotides to the RNA. The sites of strong binding should characterize single-stranded RNA regions, provided they are not involved in some tertiary structure. Any method of measuring the binding of oligonucleotides can be employed, but simple equilibrium dialysis of radioactive oligonucleotides has been most used.[13,14] Transfer RNA is put on one side of a dialysis membrane and a radioactive tri- or tetranucleotide is allowed to come to equilibrium with it. If only one oligonucleotide is bound per tRNA, the equilibrium association constant is:

$$\text{oligonucleotide} + \text{tRNA} \underset{}{\overset{K}{\rightleftharpoons}} \text{complex}$$

For a large excess of tRNA the equilibrium constant is simply related to the ratio (R) of oligonucleotide on the same side of the membrane as the tRNA to oligonucleotide on the other side.

$$K = \frac{R - 1}{\text{tRNA conc}}$$

A very thorough study using this method has been made of *E. coli* tyrosine tRNA.[14] Measurements were mainly done at 0°C, 1 M NaCl, 0.01 M $MgCl_2$, 0.01 M Na_2HPO_4, pH 7. All 64 possible trinucleotides (except GGG) were tested. Twenty-three gave association constants greater than 300 liters/mole, the error in these experiments. The lowest significant constant was 500 for UGU; the highest was 105,000 for GGC. If possible, Scatchard plots were done to show that only one trimer was bound per tRNA, at least at the low concentrations used in the binding studies.

The criterion for positive binding of a tetramer was chosen to be that the tetramer had at least twice the value of K of its constituent trimers. On this basis 13 out of 70 tetramers bound to the tRNA. The results of this study are summarized in Fig. 8-5, where the boxed regions indicate binding regions. The fact that none of the bases in the four standard arms of the cloverleaf showed binding is very good evidence for the correctness of the cloverleaf model. As expected from the cloverleaf model, the ACCA end on the amino acid arm, the dihydrouracil loop, and the anticodon loop were found to be available for complementary binding. However, the TψC loop, shown free in the cloverleaf, was not available and the extra arm, shown base paired, was available. The good agreement in general with the model (and with other organic chemical experiments) allows us to interpret the slight disagreements with confidence. Presumably the extra arm is not base paired. The TψC loop is either base paired and/or it is involved in some tertiary structure in the RNA.

III TERTIARY STRUCTURE

The biological functions of transfer RNA very probably require a definite three-dimensional structure for the molecule. All transfer RNAs must interact identically with messenger RNA and ribosomes. They must provide a constant distance between the site on the ribosome of the messenger RNA and the site of the growing polypeptide chain. Furthermore, the thermodynamics and kinetics of interaction should be similar. However, the recognition of the specific aminoacyl synthetase must be unique. An understanding of the structure of tRNA molecules should therefore include knowledge of the three-dimensional arrangement of the atoms, plus knowledge of the various sites involved in each biological function.

Early evidence for a compact (tertiary), native structure came from comparison of UV absorption and hydrodynamic measurements.[8] Between 20° and 40°C there was an increase in intrinsic viscosity, a decrease in

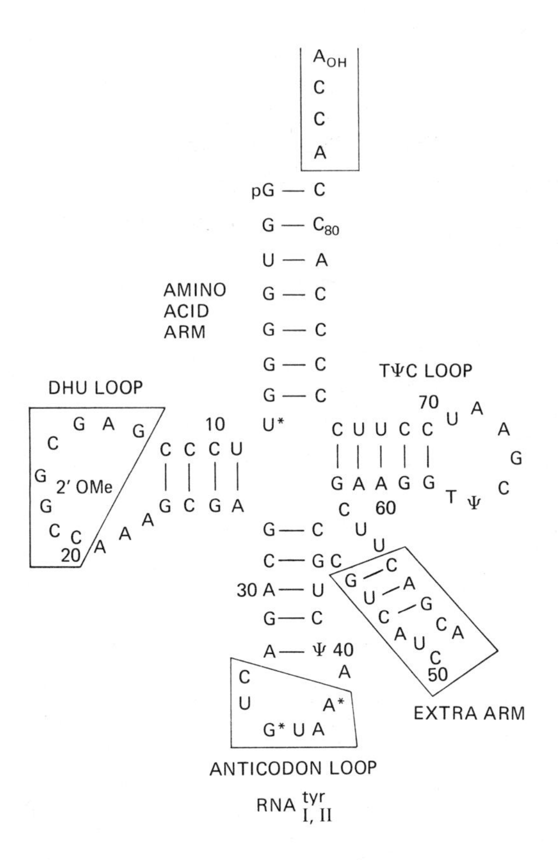

FIGURE 8-5 The regions of *E. coli* tyrosine transfer RNA which bind to complementary oligonucleotides are shown boxed. [From O. Uhlenbeck, *J. Mol. Biol.*, **65**, 25 (1972).]

sedimentation coefficient, but only a slight increase in absorbance (shown in Fig. 8-6). This is interpreted as a loss of tertiary structure in this temperature range which does not involve breaking more than two or three base pairs. Tertiary structure can also be destroyed by lowering the salt concentration or removing Mg^{2+} with EDTA. More direct evidence for a compact structure comes from small-angle X-ray scattering[15] studies of phenylalanine yeast tRNA. The radius of gyration is 24 Å at 20°; it can be fit by a structure in which the arms of the cloverleaf are folded together somehow. By 70°C the radius of gyration has more than doubled to 55 Å; this corresponds to a random coil. The complete scattering curve at 20°C can be fit by an ellipsoid with dimensions of 22 x 36 x 92 Å. Small-angle X-ray scattering of other transfer RNAs at room temperature leads to similar dimensions of 25 x 35 x 85 Å.[16]

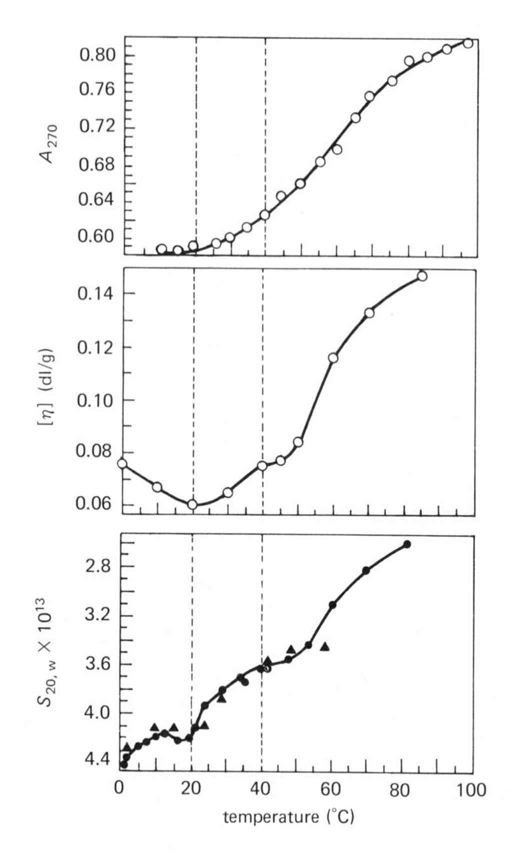

FIGURE 8-6 Temperature dependence of some physical properties of yeast tRNA in 0.2 M NaCl + 0.01 M phosphate (Na^+) + 0.0005 M EDTA, pH 6.85. Top: absorbance-temperature profiles. Middle: intrinsic viscosity. Bottom: sedimentation coefficients. [From J. R. Fresco et al., *Symp. Quant. Biol.*, **31**, 527 (1966). Reprinted with permission.]

To obtain more detailed information, methods which probe a particular region of the tRNA molecule can be used. Electron spin resonance has been used to study the valine group on a loaded valyl transfer RNA.[17] A nitroxide probe was bonded to the α-amino group of the attached valine. The shape and width of the three-line spectrum from the nitroxide can be analyzed to give a relaxation time (τ) for the rate of tumbling of the probe. This τ measures how rigidly the probe is attached to the rest of the molecule, therefore it is a measure of the freedom of motion of the amino acid acceptor end of the tRNA. Figure 8-7 shows some results. The significant observation is that there is an abrupt change in the value of τ at a certain temperature, dependent on salt concentration. Furthermore the slope of the log τ vs. $1/T$ line (a measure

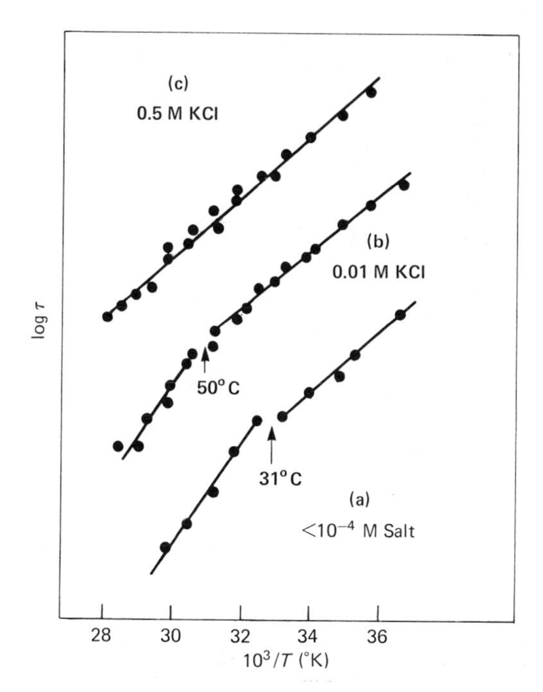

FIGURE 8-7 The rotational relaxation time τ for a nitroxide spin-labeled valine on *E. coli* valyl transfer RNA. Log τ is plotted versus inverse absolute temperature at three different salt concentrations: (a) less than 10^{-4} M salt; (b) 0.01 M Na acetate, pH 5, 0.01 M KCl; (c) 0.01 M Tris, pH 7.4, 0.5 M KCl. [From B. M. Hoffman, P. Schofield, and A. Rich, *Proc. Nat. Acad. Sci.*, **166**, 1528 (1969).]

of the activation energy for tumbling) also changes abruptly. In the presence of urea or dimethylsulfoxide an abrupt break still occurs. Parallel observations of the absorbance as a function of temperature show the gradual loss of secondary structure as expected, but with no definite relation between absorbance and spin signal. The interpretation of these experiments is that the amino acid acceptor end of the tRNA molecule is in a definite structure at low temperature which melts abruptly as the temperature is raised. The amino acid acceptor end is probably on the periphery of the tRNA because it can become free before the rest of the molecule is disrupted.

Fluorescence energy transfer can be used to measure distances between groups on a tRNA. The efficiency of energy transfer can be measured either as a quenching of the donor or a sensitized emission of the acceptor. The efficiency (E) of energy transfer depends on the quantum yield (Q) of the donor, the overlap (J) between donor emission and acceptor absorbance, the average relative orientation (K) and distance (R) between donor and acceptor, and finally the refractive index (n) of the medium. The equations used are[18]:

$$E = \frac{\text{no. of photons transferred}}{\text{no. of photons absorbed by donor}}$$

$$E = \frac{R_0{}^6}{R_0{}^6 + R^6}$$

$$R_0(\text{cm}) = 9.79 \times 10^3 \, (Jn^{-4} K^2 Q)^{1/6}$$

K is the angular part of the dipole interaction energy
The distance between the "Y" base (a fluorescent modified guanine) next to the anticodon on yeast phenylalanine tRNA and its amino acid acceptor end has been measured this way.[19] The Y base acted as the donor and the acceptor was each of three different dyes covalently attached to the periodate-oxidized 3′ end of the tRNA. Three different dyes were used to try to average the effects of orientation and to measure primarily the effect of distance. The efficiency of energy transfer was small and corresponds to a distance of between 40 and 60 Å. Thus the amino acid acceptor end is at least 40 Å away from the anticodon loop.

Finally, the only method we know which can eventually give a complete three-dimensional structure for transfer RNA is single crystal X-ray diffraction. This method of course gives the structure of tRNA in a crystal, but most tRNA crystals have so much water in them that the structure should be very similar to that in solution. Furthermore, it is encouraging to learn that mixed tRNA molecules can crystallize,[20] indicating that all have very similar shapes. Many groups have obtained crystals and are doing the tedious X-ray work.[20–23] Determination of a structure by X-ray diffraction is unique in that little preliminary information is obtained. Either a structure is present or it is not. The MIT group lead by Rich have published a structure for yeast phenylalanine transfer RNA with 4 Å resolution. This is shown in Figure 8-8. The molecule is in the form of an L with each arm of the L made up of approximately one turn of a double strand helix. One arm contains the 12 base pairs of the amino acid arm and the TψC arm; the other arm contains the 9 base pairs of the anticodon arm and the dihydrouracil arm.[24–25] The distance from the anticodon loop to the TψC loop is approximately 77 Å. The distance from the 3′ end to the TψC loop is approximately 55 Å. The entire structure is 20 Å thick, just the thickness of an RNA double helix. It is encouraging to note that this structure is consistent with the earlier physical chemical studies in solution. In addition to the studies mentioned here other tRNA work has been discussed and compared with earlier three dimensional models.[26] It is instructive to compare those extensive data with the structure shown in Figure 8-8.

Some progress has been made in the study of the various biologically active sites of transfer RNA. Fuller and Hodgson[27] proposed a very definite

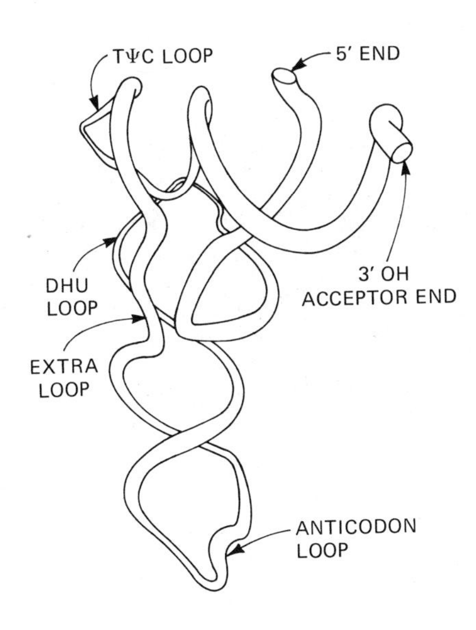

FIGURE 8-8 A perspective diagram of the three-dimensional structure obtained for yeast phenylalanine t-RNA from single crystal X-ray diffraction. [From S. H. Kim, et. al., *Science*, **179**, 285–288 (1973). Copyright 1973 by the American Association for the Advancement of Science. Reprinted by permission.]

structure (Fig. 8-9) for the anticodon site based on model building. Whether this specific model is correct or not, it is probable that a rigid structure exists at the anticodon site. Complementary binding of a trinucleotide to the anticodon of a tRNA is much more stable than double strand formation between two complementary trinucleotides.[14,28] The main difference in the free energies of binding is the smaller loss of entropy on tRNA-trinucleotide binding. If two transfer RNAs with complementary anticodons are mixed, a very stable complex is formed which is partly stabilized by a *positive* entropy of complex formation.[29] It is obvious that the anticodon loop and its surrounding solvent must be more ordered in the natural, unbound state than in the complex.

The region (or regions) of transfer RNA which is specifically recognized by the aminoacyl synthetase is not known. Dudock et al.[30] have compared the sequences of four different tRNAs (yeast, wheat, and *E. coli* phenylalanine; and *E. coli* valine) and have proposed the dihydrouracil arm as the common recognition site. Chambers[31] has used photochemical modification to argue that the first three base pairs of the amio acid arm form a recognition site for yeast alanine tRNA. Much further work is still needed before either or both of these suggestions are accepted.

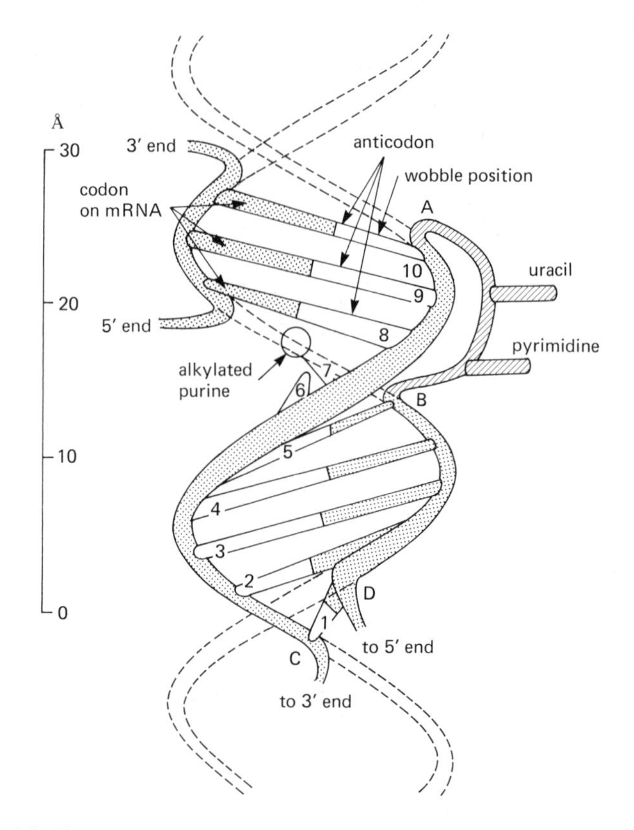

FIGURE 8-9 The Fuller-Hodgson anticodon loop model showing the proposed interaction with the messenger RNA. [From W. Fuller and A. Hodgson, *Nature,* **215**, 817 (1967). Reprinted by permission.]

No other sites have been suggested very convincingly. The TψC loop has been identified as a ribosome binding site, only because it is a constant feature of all transfer RNAs.

Many other physical and chemical studies have been made on transfer RNA molecules.[26,32] Their binding of ions and dye molecules has been studied. Their interactions with enzymes, ribosomes, messenger RNA, and so on have been investigated. However, much of this latter work is very preliminary. We will not describe it but instead will leave the reader with the hope that he will be able to do the definitive (physical chemical) experiments to solve the problem of tRNA structure and function.

REFERENCES

1 H. M. Goodman, J. Abelson, A. Landy, S. Brenner, and J. D. Smith, *Nature,* **217**, 1019 (1968).
2 J. D. Smith, J. N. Abelson, B. F. C. Clark, H. M. Goodman, and S. Brenner, *Symp. Quant. Biol.,* **31**, 479 (1966).
3 F. H. C. Crick, *J. Mol. Biol.,* **19**, 548 (1966).
4 R. W. Holley, J. Apgar, G. A. Everett, J. T. Madison, M. Marquisee, S. H. Merrill, J. R. Penswick, and A. Zamer, *Science,* **147**, 1462 (1965).
5 S. W. Englander and J. J. Englander, *Proc. Nat. Acad. Sci., U.S.,* **53**, 370 (1965).
6 R. R. Gantt, S. W. Englander, and M. V. Simpson, *Biochemistry,* **8**, 475 (1969).
7 J. R. Fresco, in *Informational Macromolecules*, H. J. Vogel, V. Bryson, and J. O. Lampen, eds., Academic Press, New York (1963), p. 121.
8 J. R. Fresco, A. Adams, R. Ascione, D. Henley, and T. Lindahl, *Symp. Quant. Biol.,* **31**, 527 (1966).
9 J. N. Vournakis and H. A. Scheraga, *Biochemistry,* **5**, 2997 (1966).
10 C. R. Cantor, S. R. Jaskunas, and I. Tinoco, Jr., *J. Mol. Biol.,* **20**, 39 (1966).
11 M. Tsuboi, S. Higuchi, Y. Kyogoku, and S. Nishimura, *Biochim. Biophys. Acta,* **195**, 23 (1969).
12 D. R. Kearns, D. J. Patel, and R. G. Shulman, *Nature*, **229**, 338 (1971); D. R. Kearns, D. Patel, R. G. Shulman, and T. Yamane, *J. Mol. Biol.*, **61**, 265 (1971).
13 O. C. Uhlenbeck, J. Baller, and P. Doty, *Nature,* **225**, 508 (1970).
14 O. C. Uhlenbeck, *J. Mol. Biol.,* **65**, 25 (1972).
15 O. Kratky, I. Pilz, F. Cramer, F. von der Haar, and E. Schlimme, *Monatsh. Chem.,* **100**, 748 (1969); I. Pilz, O. Kratky, F. Cramer, F. von der Haar, and E. Schlimme, *Europ. J. Biochem.,* **15**, 401 (1970).
16 P. G. Connors, M. Labanauskas, and W. W. Beeman, *Science,* **166**, 1528 (1969).
17 B. M. Hoffman, P. Schofield, and A. Rich, *Proc. Nat. Acad. Sci., U.S.,* **62**, 1195 (1969).
18 T. Förster, *Ann. Physik,* **2**, 55 (1948); S. A. Latt, H. T. Cheung, and E. R. Blout, *J. Amer. Chem. Soc.,* **87**, 995 (1965).
19 K. Beardsley and C. R. Cantor, *Proc. Nat. Acad. Sci., U.S.,* **65**, 39 (1970).
20 R. D. Blake, J. R. Fresco, and R. Langridge, *Nature,* **225**, 32 (1970).
21 B. F. C. Clark, B. P. Doctor, K. C. Holmes, A. Klug, K. A. Marcher, S. J. Morris, and H. H. Paradies, *Nature,* **219**, 1222 (1968).
22 J. D. Young, R. M. Bock, S. Nishimura, H. Ishikura, Y. Yamada, U. L. Raj Bhandary, M. Labanauskas, and P. G. Connors, *Science,* **166**, 1527 (1969).
23 F. Cramer, F. von der Haar, K. C. Holmes, W. Saenger, E. Schlimme, and G. E. Schultz, *J. Mol. Biol.,* **51**, 523 (1970).
24 S. H. Kim, G. Quigley, F. L. Suddath, A. McPherson, D. Sneden, J. J. Kim, J. Weinzierl, P. Blattmann, and A. Rich, *Proc. Nat. Acad. Sci. U.S.,* **69**, 3746 (1972).
25 S. H. Kim, G. J. Quigley, F. L. Suddath, A. McPherson, D. Sneden, J. J. Kim, J. Weinzierl, and A. Rich, *Science*, **179**, 288 (1973).
26 F. Cramer, *Progress in Nucleic Acid Research and Molecular Biology*, Vol. 11, Academic Press, New York (1971), p. 391.
27 W. Fuller and A. Hodgson, *Nature,* **215**, 817 (1967).
28 J. Eisinger, B. Feuer, and T. Yamane, *Nature New Biology,* **231**, 127 (1971).
29 J. Eisinger, *Biochem. Biophys. Res. Commun.,* **43**, 854 (1971).

30 B. Dudock, C. DiPeri, K. Scillepi, and R. Reszelbach, *Proc. Nat. Acad. Sci., U.S.,* **68**, 681 (1971).
31 R. W. Chambers, *Progress in Nucleic Acid Research and Molecular Biology,* Vol. 11, Academic Press, New York (1971), p. 489.
32 S. Arnott, *Progress in Biophysics and Molecular Biology*, Vol. 22, Pergamon Press, New York (1971), p. 181.

author index

subject index

75 76 76 5 4 3 2